"This volume is timely in a world significantly unsettled by converging crises including the Coronavirus pandemic, the climate emergency, and deep-seated racial injustices. This radical publication recognizes that in this context, existing ways of knowing, being in, and experiencing tourism are outdated and must be disrupted and transformed. The text's 16 chapters rise expertly to this challenge with a degree of honesty and clarity, rarely witnessed in our field. I congratulate the editors for successfully combining such novel contributions into an exciting and coherent publication. Undoubtedly, this ambitious volume will become a classic in the field of critical tourism studies."

– **Professor Donna Chambers**, *University of Sunderland*

"While ostensibly about qualitative research in tourism studies, this book goes much wider and deeper. It addresses innovative, creative and disruptive methodologies in ways that will be transferable to many fields and disciplines in the social sciences and beyond. This highly ethical work challenges the domination of knowledges from the global North and amplifies voices from the margins. I would recommend it for any researcher, from academia or elsewhere, who wants to deepen their understanding of research methods."

– **Dr. Helen Kara**, *Researcher, and Author of 'Creative Research Methods in the Social Sciences: A Practical Guide'*

Qualitative Methodologies in Tourism Studies

Disruptive and creative research methodologies proposed in this book are designed to dismantle neoliberal narratives deployed in tourism studies and wider social sciences. Progressing criticality in tourism studies, this volume showcases cutting-edge contributions ranging from reflexivity, subjectivities, and dreams; to messy emotions in auto-ethnographic accounts of fieldwork; 'motherhood capital' accessing Inuit communities; collective memory work; ethnodrama and creative non-fiction, amongst others.

Disruption and creativity are the two ideas around which tourism geographers challenge and begin dismantling hegemonic ideologies in tourism studies. The chapters in this book provide a vantage point from where to disrupt first, before tourism geographers can engender progress and transformation within and outside of the field. In tourism studies in general, and tourism geography in particular, the years of the 2000s have witnessed an emphasis on qualitative methodological research, both in terms of the topics addressed and the types of methodological tools. In many ways, this legitimisation of qualitative work mirrors developments in other areas such as human geography, sociology and anthropology, in which this book is anchored. The authors debate in more depth how tourism studies offer multidimensional, multilogical, and multi-emotional approaches to research design.

The chapters were originally published as a special issue of the journal, *Tourism Geographies*.

Milka Ivanova is a qualitative researcher who focuses on the ways '*non-dominant*' narratives are re/created through tourism in the cases of dissonant and communist heritage. As such, Milka published her research in *The Routledge Handbook of Cultural Tourism*, *Tourism Culture & Communication*, and *Sustainability of Tourism: Cultural and Environmental Perspectives*.

Dorina-Maria Buda conducts interdisciplinary research focusing on the interconnections between tourist spaces, people and emotions in times and places of socio-political conflict. She conducts ethnographic work in such places of on-going conflicts and turmoil like Jordan, Israel, and Palestine. She is the author of *Affective Tourism: Dark Routes in Conflict*.

Elisa Burrai offers robust and thought-provoking critiques of concepts such as volunteer tourism and responsible tourism developed through ethnographic, critical, and qualitative methodological approaches that explore power and research methodologies, reflexivity, and positionality. Her work is published in *Tourism Geographies*, *International Journal of Tourism Research*, and *Journal of Sustainable Tourism*.

Qualitative Methodologies in Tourism Studies

Disrupting and Co-creating Critical Research

Edited by
Milka Ivanova, Dorina-Maria Buda and Elisa Burrai

Routledge
Taylor & Francis Group
LONDON AND NEW YORK

First published 2023
by Routledge
4 Park Square, Milton Park, Abingdon, Oxon OX14 4RN

and by Routledge
605 Third Avenue, New York, NY 10158

Routledge is an imprint of the Taylor & Francis Group, an informa business

British Library Cataloguing in Publication Data
A catalogue record for this book is available from the British Library

ISBN13: 978-1-032-22774-0 (hbk)
ISBN13: 978-1-032-22775-7 (pbk)
ISBN13: 978-1-003-27411-7 (ebk)

DOI: 10.4324/9781003274117

Typeset in Myriad Pro
by Newgen Publishing UK

Publisher's Note
The publisher accepts responsibility for any inconsistencies that may have arisen during the conversion of this book from journal articles to book chapters, namely the inclusion of journal terminology.

Contents

Citation Information ix
Notes on Contributors xii
Foreword xvii
Professor Jennie Germann Molz

Introduction: Creative and disruptive methodologies in tourism studies 1
Milka Ivanova, Dorina-Maria Buda and Elisa Burrai

Part I
Dis-rupting Methodologies 13

1 Collective memory work as an unsettling methodology in tourism 15
Bryan S. R. Grimwood and Corey W. Johnson

2 'Motherhood capital' in tourism fieldwork: experiences from Arctic Canada 37
Roslyn Kerr and Emma J. Stewart

3 Social constructionism as a tool to maintain an advantage in tourism research 57
Leszek Butowski, Jacek Kaczmarek, Joanna Kowalczyk-Anioł and Ewa Szafrańska

4 Disruptive and Adaptive Methods in Activist Tourism Studies: Socio-Spatial Imaginaries of Dissent 79
Ian R. Lamond

5 The disruptive 'other'? Exploring human-animal relations in tourism through videography 101
Minni Haanpää, Tarja Salmela, José-Carlos García-Rosell and Mikko Äijälä

6 Emplacing non-human voices in tourism research: the role of dissensus as a qualitative method 122
Abhik Chakraborty

7 Hanging out on Snapchat: disrupting passive covert netnography in tourism research 148
Heather L. Jeffrey, Hamna Ashraf and Cody Morris Paris

Part II
Re/Creating Methodologies 167

8 A critical consideration of LEGO® SERIOUS PLAY® methodology for tourism studies 169
Yana Wengel, Alison McIntosh and Cheryl Cockburn-Wootten

9 Stakeholder engagement in sustainable tourism planning through serious gaming 192
Lidija Lalicic and Jessika Weber-Sabil

10 Deep reflexivity in tourism research 213
Émilie Crossley

11 Challenges in outdoor tourism explorations: an embodied approach 235
Jelena Farkic

12 Leveraging digital and physical spaces to 'de-risk' and access Rio's favela communities 256
Nicola Cade, Sally Everett and Michael Duignan

13 Walking methodologies, digital platforms and the interrogation of Olympic spaces: the '#RioZones-Approach' 282
Michael B. Duignan and David McGillivray

14 'Que será, será!': creative analytical practice within the critical sports event tourism discourse 303
Richard Keith Wright

15 Why is research–practice collaboration so challenging to achieve? A creative tourism experiment 325
Nancy Duxbury, Fiona Eva Bakas and Cláudia Pato de Carvalho

16 The case for linguistic narrative analysis, illustrated studying small firms in tourism 351
Lucia Tomassini, Xavier Font and Rhodri Thomas

Afterword 367
Professor Ajay Bailey

Index 368

Citation Information

The following chapters were originally published in the journal, *Tourism Geographies*, volume 23, issue 1–2 (2021). When citing this material, please use the original page numbering for each article, as follows:

Chapter 1
Collective memory work as an unsettling methodology in tourism
Bryan S. R. Grimwood and Corey W. Johnson
Tourism Geographies, volume 23, issue 1–2 (2021), pp. 11–32

Chapter 2
'Motherhood capital' in tourism fieldwork: experiences from Arctic Canada
Roslyn Kerr and Emma J. Stewart
Tourism Geographies, volume 23, issue 1–2 (2021), pp. 33–52

Chapter 3
Social constructionism as a tool to maintain an advantage in tourism research
Leszek Butowski, Jacek Kaczmarek, Joanna Kowalczyk-Anioł and Ewa Szafrańska
Tourism Geographies, volume 23, issue 1–2 (2021), pp. 53–74

Chapter 4
Disruptive and Adaptive Methods in Activist Tourism Studies: Socio-Spatial Imaginaries of Dissent
Ian R. Lamond
Tourism Geographies, volume 23, issue 1–2 (2021), pp. 75–96

Chapter 5
The disruptive 'other'? Exploring human-animal relations in tourism through videography
Minni Haanpää, Tarja Salmela, José-Carlos García-Rosell and Mikko Äijälä
Tourism Geographies, volume 23, issue 1–2 (2021), pp. 97–117

Chapter 6
Emplacing non-human voices in tourism research: the role of dissensus as a qualitative method
Abhik Chakraborty
Tourism Geographies, volume 23, issue 1–2 (2021), pp. 118–143

Chapter 7

Hanging out on Snapchat: disrupting passive covert netnography in tourism research
Heather L. Jeffrey, Hamna Ashraf and Cody Morris Paris
Tourism Geographies, volume 23, issue 1–2 (2021), pp. 144–161

Chapter 8

A critical consideration of LEGO® SERIOUS PLAY® methodology for tourism studies
Yana Wengel, Alison McIntosh and Cheryl Cockburn-Wootten
Tourism Geographies, volume 23, issue 1–2 (2021), pp. 162–184

Chapter 9

Stakeholder engagement in sustainable tourism planning through serious gaming
Lidija Lalicic and Jessika Weber-Sabil
Tourism Geographies, volume 23, issue 1–2 (2021), pp. 185–205

Chapter 10

Deep reflexivity in tourism research
Émilie Crossley
Tourism Geographies, volume 23, issue 1–2 (2021), pp. 206–227

Chapter 11

Challenges in outdoor tourism explorations: an embodied approach
Jelena Farkic
Tourism Geographies, volume 23, issue 1–2 (2021), pp. 228–248

Chapter 12

Leveraging digital and physical spaces to 'de-risk' and access Rio's favela communities
Nicola Cade, Sally Everett and Michael Duignan
Tourism Geographies, volume 23, issue 1–2 (2021), pp. 249–274

Chapter 13

Walking methodologies, digital platforms and the interrogation of Olympic spaces: the '#RioZones-Approach'
Michael B. Duignan and David McGillivray
Tourism Geographies, volume 23, issue 1–2 (2021), pp. 275–295

Chapter 14

'Que será, será!': creative analytical practice within the critical sports event tourism discourse
Richard Keith Wright
Tourism Geographies, volume 23, issue 1–2 (2021), pp. 296–317

Chapter 15

Why is research–practice collaboration so challenging to achieve? A creative tourism experiment
Nancy Duxbury, Fiona Eva Bakas and Cláudia Pato de Carvalho
Tourism Geographies, volume 23, issue 1–2 (2021), pp. 318–343

Chapter 16

The case for linguistic narrative analysis, illustrated studying small firms in tourism
L. Tomassini, X. Font and R. Thomas
Tourism Geographies, volume 23, issue 1–2 (2021), pp. 344–359

Notes on Contributors

Mikko Äijälä is a junior researcher and PhD Candidate at the University of Lapland, Faculty of Social Sciences, Multidimensional Tourism Institute (MTI). In his PhD project he studies the agency of sled dogs by exploring human-sled dog encounters in tourism.

Hamna Ashraf completed her BA Honours in International Tourism Management with Distinction at Middlesex University Dubai. She is an accomplished freelance photographer and a digital marketing specialist based in the United Arab Emirates.

Fiona Eva Bakas is a critical tourism researcher with international teaching experience. She is contracted postdoctoral researcher in a nation-wide project on creative tourism in rural areas and small cities (CREATOUR), at the Centre for Social Studies of the University of Coimbra, Portugal. Her research interests are: creative and cultural tourism, gender in tourism labour, qualitative methodologies, cultural mapping, handicrafts, entrepreneurship, rural tourism, and ecotourism.

Dorina-Maria Buda conducts interdisciplinary research focusing on the interconnections between tourist spaces, people and emotions in times and places of socio-political conflict. She conducts ethnographic work in such places of on-going conflicts and turmoil like Jordan, Israel and Palestine. She is the author of *Affective Tourism: Dark Routes in Conflict*.

Elisa Burrai offers robust and thought-provoking critiques of concepts such as volunteer tourism and responsible tourism developed through ethnographic, critical and qualitative methodological approaches that explore power and research methodologies, reflexivity and positionality. Her work is published in *Tourism Geographies*, *International Journal of Tourism Research*, and *Journal of Sustainable Tourism*.

Leszek Butowski is Associate Professor at the University of Lodz (Poland), Institute of Urban Geography and Tourism Studies. Research interests: philosophy of science, methodology of social sciences, human geography and tourism research, sustainable development and tourism, maritime & sailing tourism.

Nicola Cade is Doctoral Student and Associate Lecturer in the Faculty of Business and Law, Anglia Ruskin University. Nikki's research focuses on Rio's recent mega events and the socio-economic/-cultural impacts on favela communities during the planning, delivery and legacy stages. Nikki's interests also include cultural and heritage tourism namely dark and favela tourism; and local tourism development within the slow tourism context as well as local festivals and sporting events.

Abhik Chakraborty is Lecturer at Center for Tourism Research, Wakayama University in Japan. He has worked as the lead editor and author for 'Natural Heritage of Japan', the first edited monograph in English on the geology, geomorphology and ecological aspects of the Japanese Islands. He has also edited the Routledge volume 'Rivers & Society'.

Cheryl Cockburn-Wootten is Senior Lecturer, University of Waikato, New Zealand. Her work adopts an interdisciplinary approach but mainly draws on organisational communication theories to apply to tourism, hospitality and contexts involving social issues.

Émilie Crossley is a tourism researcher whose work explores tourist subjectivity from a psychosocial perspective. Her research draws on psychoanalytic theory to approach the psychological in tourism studies in a non-reductive and culturally engaged way. She holds a PhD from Cardiff University and is based at Otago Polytechnic in Aotearoa New Zealand.

Michael Duignan's research spans across the field of tourism management and marketing, applied economics, major/mega-event and festival impact and legacy studies, spatial governance and political economy, with a case study emphasis on food festivals and the Olympic Games.

Michael B. Duignan looks at the organizational, management and strategic aspects of tourism industries, visitor economies, large- and small-scale sporting events and cultural festivals. Specifically, he draws on the study of field configuring events to examine how and why events serve as a key driver and mechanism of change that enable and constrain urban, place, tourism and regional development, management and strategy.

Nancy Duxbury is Senior Researcher and Co-coordinator of the Cities, Cultures and Architecture Research Group at the Centre for Social Studies, University of Coimbra. She is Principal Investigator of CREATOUR, and a member of the European Expert Network on Culture. Her research has examined culture in local sustainable development; culture-based development models in smaller communities; and cultural mapping, which bridges academic inquiry, community practice, and artistic approaches to understand and articulate place.

Sally Everett is Professor of Business Education at King's Business School, King's College London. She was formally the Deputy Dean for the Faculty of Business and Law, Anglia Ruskin University. Sally is a Principal Fellow of the Higher Education Academy and a National Teaching Fellow. Her research focuses on tourism development, food and drink tourism, destination management and marketing, learning and teaching, and inclusive education.

Jelena Farkic holds PhD in tourism studies. Working at the intersections of adventure tourism and hospitality studies, she explores sociological and philosophical dimensions of being in and guiding through the outdoors.

Xavier Font is Professor of Sustainability Marketing at the University of Surrey, and Head of Impact at Travindy. His research focuses on understanding reasons for pro-sustainability behaviour and market-based mechanisms to encourage sustainable production and consumption. Xavier has run over 130 training courses on sustainability marketing with over 2500 delegates, and has consulted for UNWTO, UNEP, EC, IFC, Rainforest Alliance and numerous governments and businesses.

José-Carlos García-Rosell is Senior Lecturer at the University of Lapland, Faculty of Social Sciences, Multidimensional Tourism Institute (MTI). His research interests are in the areas of sustainable business development, corporate social responsibility, multi-stakeholder processes, tourism service development, tourism and management education, action research and ethnographic research.

Bryan S. R. Grimwood is Settler Canadian of English descent and Associate Professor in the Department of Recreation and Leisure Studies at the University of Waterloo. His research and teaching examines human-nature relationships and advocates social justice and sustainability in contexts of tourism, leisure, and livelihoods.

Minni Haanpää is Senior Lecturer at the University of Lapland, Faculty of Social Sciences, the University of Lapland, Multidimensional Tourism Institute (MTI). Her research interests are in the areas of tourism and event experiences, affect, co-creation, ethnographic methodologies, especially videography and autoethnography, and tourism as a means of social change.

Milka Ivanova is a qualitative researcher who focuses on the ways *'non-dominant'* narratives are re/created through tourism in the cases of dissonant and communist heritage. As such, Milka published her research in *The Routledge Handbook of Cultural Tourism*, *Tourism Culture & Communication*, and *Sustainability of Tourism: Cultural and Environmental Perspectives.*

Heather L. Jeffrey is a social researcher with interests in equality and ethics. She completed her doctoral studies in 2017, with a thesis centred on women and tourism in Tunisia. Heather is Senior Lecturer in social sciences at Middlesex University Dubai and an associate at Equality in Tourism.

Corey W. Johnson is Professor in the Department of Recreation and Leisure Studies at the University of Waterloo. He teaches courses on inclusive recreation, social justice, gender and sexuality, qualitative research methods, and the philosophy of science.

Jacek Kaczmarek is Associate Professor at the University of Lodz (Poland), Institute of Urban Geography and Tourism Studies. Research interests: tourism marketing, cultural tourism, urban management. Currently works on measure and management of tourism flows in urban areas.

Roslyn Kerr is Lecturer in Sociology at the University of Canterbury, Christchurch, NZ. Research interests include the integration of technology into coaching programmes, the use of mundane technologies in everyday life, talent identification and the aesthetics of sport.

Joanna Kowalczyk-Anioł is Assistant Professor at the University of Lodz (Poland), Institute of Urban Geography and Tourism Studies. Research interests: tourism geography, urban tourism, heritage city, social and spatial aspects of tourism development, dysfunctions of tourism development and tourism-community relations.

Lidija Lalicic is Assistant Professor in the Department of Tourism and Service Management at MODUL University Vienna, Austria. Her research interest are related to technology-enhanced tourist experiences, innovative marketing and entrepreneurial practices and stakeholder management in the field of tourism. She is a board member of the International Federation for Information Technology in Travel and Tourism (IFITT).

Ian R. Lamond is a critical scholar within the field of event, leisure and tourism studies at Leeds Beckett University, at which he is a part of International Centre for Research in Tourism, Events and Hospitality. His work ranges from inquiries into protest as event within a critical leisure and tourism framework, the mediated representation of political events and the philosophical foundations of events and leisure studies.

David McGillivray holds a Chair in Event and Digital Cultures. His research focuses on two main areas of activity. He has published extensively on the topic of accelerated leisure cultures and digital technology, focusing on the wider social and cultural implications for producers, consumers and regulators across the lifespan. He is currently Deputy Editor of the *Annals of Leisure Research* and sits on the Editorial Board of *Leisure Studies*.

Alison McIntosh is Professor of Tourism and Hospitality at Auckland University of Technology, New Zealand. As a critical scholar, her main research interests are in social justice, inclusion and advocacy through tourism and hospitality; humanitarian travel, community hospitality and volunteering; reflexivity, creative and participative research approaches, and community engagement methods.

Cody Morris Paris is Deputy Director for Academic Planning and Research at Middlesex University Dubai and an Associate Professor in the School of Law and the Business School. He is also a Senior Research Fellow in the School of Hospitality and Tourism at the University of Johannesburg.

Cláudia Pato de Carvalho is a permanent researcher at the Centre for Social Studies of the University of Coimbra under the post-doctoral fellowship (2013–2018) and under the ARTERIA project. She completed her PhD in Sociology, with a specialization in Sociology of Culture, Knowledge and Communication, at the Faculty of Economics of the University of Coimbra in October 2010, in collaboration with the Center for Reflective Community Practice (DUSP, MIT).

Tarja Salmela is a researcher and critical organization scholar in the University of Lapland. Her work centers around topics of disruptive and more-than-human methodologies, the body, ethics and affects, fueled by her multidisciplinary background at the intersection of organization studies, feminist studies, critical animal studies, tourism studies and academic activism.

Emma J. Stewart is Associate Professor in Department of Tourism at Lincoln University in Christchurch, New Zealand. Her research focuses on the human dimensions of environmental change and implications for the tourism sector, neighbouring communities, underlying resources and governance structures in climate-sensitive nature-based settings such as polar, coastal and high mountain environments.

Ewa Szafrańska is Associate Professor at the University of Lodz (Poland), Institute of Urban Geography and Tourism Studies. Research interests: urban geography, urban sociology, spatial aspects of human behavior, methodology of social sciences.

Rhodri Thomas is Dean of the School of Events, Tourism and Hospitality Management and Professor of Tourism and Events Policy at Leeds Beckett University, UK.

Lucia Tomassini is a PhD graduate from the School of Events, Tourism and Hospitality Management at Leeds Beckett University. An Urban Planner, with an MSc in Urban Development in developing countries, Lucia is the co-founder and Destination Manager at Amavido. Her academic research focuses on small tourism firms committed to an ethical vision of tourism, and the implications on the narrativisation of their identity, sense of self, and role.

Jessika Weber-Sabil is a senior researcher at the Faculty of Digital Entertainment at the University of Applied Sciences Breda under the professorship of Applied Games, Innovation and Society, where she conducts research on games with the focus on travel and tourism organizations. Her research interests range from the application of games and new technologies in travel and tourism as a tool to explain complex realities to the evaluation of game experience in (location-based) games.

Yana Wengel is Lecturer at the Work, Employment & Organisation department at the University of Strathclyde Business School. Yana takes a critical approach to tourism studies and her interests include volunteer tourism, non-profit tourism, tourism in developing economies, creative methodologies and mountain tourism.

Richard Keith Wright is a member of the Sports Performance Research Institute New Zealand, the New Zealand Tourism Research Institute and the AUT Centre for Active Aging. Dr Wright has presented at over a dozen international conferences, had his research published in sport, event and tourism management journals and textbooks and sits on the board of the Australian and New Zealand Association of Leisure Studies (ANZALS). He is a strong advocate for the use of problem-based learning and creative analytical practice within sport event tourism education.

Foreword

How is knowledge produced at the intersection of creativity and disruption? We tend to celebrate creativity, as if methodological innovations are inherently positive, but disruption cuts both ways. It can signal devastating consequences, especially for an already suffering planet or a precarious population, but it can also clear a space where our creative imaginings for a better world might flourish. The COVID-19 pandemic brought this relationship between disruption and creativity into sharp focus, first by interrupting business-as-usual tourism and fieldwork and then by catalyzing creative responses from critical tourism theory. For many tourism scholars, themselves stuck at home, images of empty airports, beaches, and city streets opened up a moment to think about how we might do tourism *otherwise* going forward. Alternative visions of an anti-racist, decolonized, non-extractive, just and sustainable tourism filled the air. The disruption visited on tourism by the global pandemic did not necessarily generate these visions, but it did give them breathing room. The lockdown made tourism's uneven social, environmental, and economic impact undeniably visible and reinvigorated tourism scholars' calls to action on overtourism, racism, the climate crisis, and entrenched colonial structures.

As I listened to those powerful calls and read the contributions in this volume, I was reminded of the concept of outlaw emotions. Feminist sociologist Alison Jaggar defines outlaw emotions as those feelings that don't quite fit the situation or live up to the hype, like a hot flush of anger at a racist joke or uneasiness in the face of sexist banter. Or the inability to take pleasure in tourism's colonial projects and climate-killing activities. Outlaw emotions can feel painful and isolating, but Jaggar reminds us that they are also epistemologically subversive. By exposing the reality that the slick comforts of the *status quo* are not comfortable for everyone, outlaw emotions clue us in to new avenues for investigation and new ways of knowing. The epistemological project behind disruptive and creative methodologies thus goes beyond reworking the technical aspects of collecting and analyzing social data about tourism and toward an understanding of how our methods also make the worlds we study. It is in the messy and embodied, multimedia, multisensual and multimodal, active and playful engagement with human and nonhuman participants that these methods make and make knowable the emerging world of tourism.

The reflexive insight that ways of knowing are intertwined with ways of being underscores the urgent need for knowledge practices that decolonize tourism, both as a way of inhabiting an unequal world and as a field of academic inquiry. By disrupting the flat inevitability of the single narrative that we have told and retold and been told about tourism, these researchers allow for the creation of knowledge in multiple forms and formats. The problem is not that our traditional methods have left some stories untold but that they have made some stories untellable. This is not to say that disruption and creativity necessarily mean discarding tried-and-true methodological techniques, but rather to recognize our methods for the world-making tools they are. And to then use them in the vital projects of destabilizing hegemonic narratives and writing alternative narratives from multiple perspectives. From there, new ways of doing tourism research can go hand in hand with new ways of doing tourism.

Professor Jennie Germann Molz
College of the Holy Cross

Introduction

Creative and disruptive methodologies in tourism studies

Milka Ivanova, Dorina-Maria Buda and Elisa Burrai

Prelude

At the turn of the millennium most tourism researchers still regarded qualitative methods as soft and non-scientific compared to the perceived rigour of quantitative tools (Phillimore & Goodson, 2004). Debates ensued regarding the use of qualitative methodologies in tourism research especially within, what is identified as, the critical turn (Ateljevic, et al., 2005). Progress has been made since, and tourism researchers have moved from times when our work, from doctoral theses to peer reviewed articles, needed to be prefaced with explanations of why qualitative methodology was chosen over quantitative. All three of us editing this book conducted our first research project, namely our PhD, in the early/mid-2000s, and we remember all too well having to legitimise our choice of qualitative over quantitative methods, or being advised to write our ethnographies in the third person singular. Thankfully those days appear to be long gone. Indeed, new debates have now developed within qualitative methods calling for robust considerations of co-creative and experiential methods (Nunkoo et al., 2020; Wilson et al., 2020).

We are three migrant women, two of us originating in Eastern Europe, Bulgaria and Romania respectively, and one in south-central Europe, Italy to be precise. We operate from an anglophone centre, being based at English universities, yet in our works, we critique and aim to disrupt white, "western", post/colonial underpinnings of tourism knowledge production. Two of us have lived through the communist revolution and the fall of the Iron Curtain in 1989, and as such the instinct to disrupt oppressive and stale establishments has shaped our worldviews. The toppling of the communist regimes disrupted the singular, binary narrative of our lives and opened possibilities for transformation towards plurality, agency, and participation. We remember the potential of that time with joy, happiness, hope and seek to bring those emotions into our work. We embrace disruption not as a destructive but as a creative force.

We start with our lived histories of communist revolutions and disruptions, because the ethos of those times and places have opened up a new world of knowledge for us, and it is that ethos that drives this book. It is in this spirit of disruption, innovation and creation that we invited contributors to explore in this edited collection.

Intermezzo

Disrupt the status-quo. Subvert the establishment. Dare to go against the stream. Confront the mainstream. Be non-linear, non-binary thinking (Figure 1). This is the spirit that energised us in 2018 in the English city of Leeds on Headingley Campus surrounded by Georgian and classic renaissance architectural buildings which served as military hospitals during World War I. The past and visions of the future spiralled into one in our effervesce to disrupt the stale status-quo and charge ahead creating new knowledges. We enthusiastically discussed ways to disrupt what we perceived as run of the mill things, techniques, tools and methods in tourism studies. We wanted to see who was with us, who were our comrades in mischief and disruption, as it were. So, we launched a call to challenge and disrupt methodological conventions and make way for new co-created knowledges. This is not to say that disruptive and creative methods and knowledges have been totally absent from tourism studies, but we wanted to welcome the movers and shakers, the subverters of the establishment, the non-linear thinkers.

The overwhelming response to our call affirmed that tourism scholars are eager to join us in our mission to shake up the qualitative status quo, are eager to go against the grain, challenge, and interrogate new ways to create knowledges. In the chapters included here, readers will not find settled definitions for, say, interviewing, focus groups, fieldwork, or surveys. The sixteen chapters included here offer multidimensional, multi-logical and multi-emotional, ontological, epistemological, and methodological approaches to research, many inspired by critical theories. Our contributors tackle the ways in which research methodologies can be creative and disruptive to the seemingly prevalent narratives through deep reflexivity, subjectivities, and dreams; messy emotions in auto-ethnographic accounts of fieldwork; "motherhood capital" accessing Inuit communities; collective memory work; ethnodrama and creative non-fiction, amongst others.

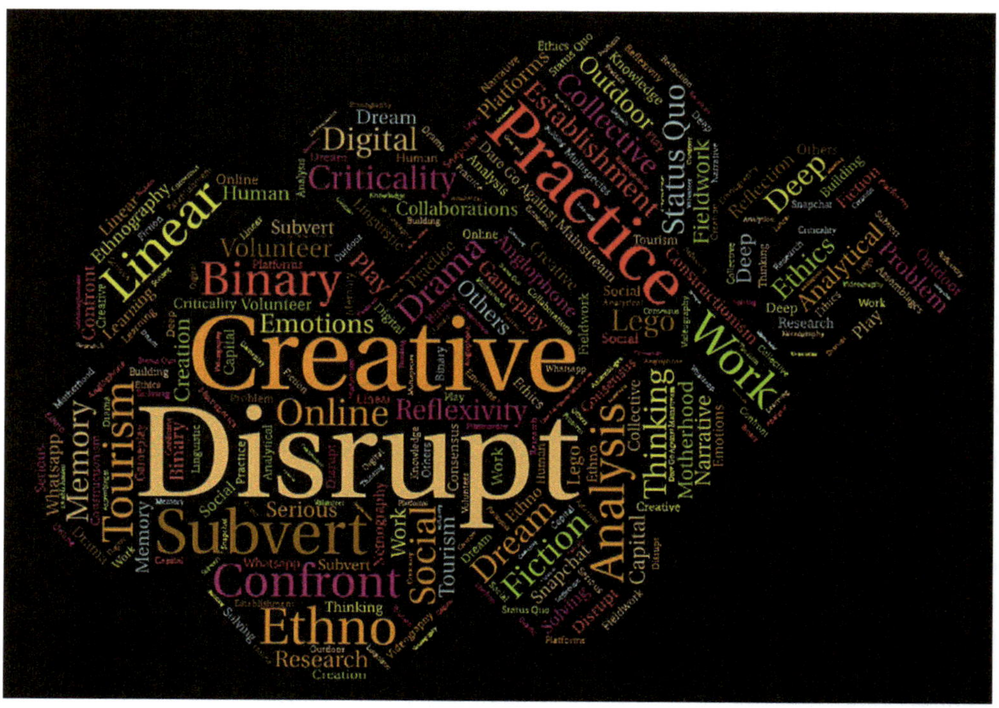

Figure 1.

We have organised the chapters in four central themes: criticality; fieldwork and ethnography; digital platforms; and research–practice collaborations. As readers will quickly realize, however, the chapters included here overlap and inform one another in multiple ways.

I. Criticality

The turn to criticality, which refers to meaningful engagements with the body of critical theories, such as poststructuralism, gender and feminist theories, decolonial theories, socio-spatial theories of emotion and affect, was meant to stimulate radical change in tourism research. Indeed, as early as the 1980s, tourism scholars have drawn on critical theories developed by the Frankfurt School to examine how tourism intersects with capitalism in contemporary societies (Mura and Wijesinghe, 2021). The Frankfurt School of critical theory established in the 1930s proposed three key concerns: (1) conducting research to examine fundamental philosophical questions through sophisticated research methods; (2) re-introducing and re-thinking Marxism so as to enable it to more fully achieve its objectives; and (3) developing social theory so as to analyse the mediation of economic, psychological, and cultural facets of social life (Phillips, 2009, p.385).

Critical theory and a critical approach refer ways to "understand the world in such a manner that enables people to become more emancipated" (Phillips, 2009, p.385). Criticality, therefore, recognises the importance of centring indigenous knowledges within the field of tourism. This opens up possibilities to move away from the subjugating and colonising nature of tourism giving space to a "radical shift away from the dominant western paradigm of knowledge production and binary anthropocentric thinking" (Everingham et al. 2021, p.3).

Indigenous scholarship becomes increasingly prominent and proves to be a needed tool in challenging "western" academic discourses which have, so far, perpetuated and legitimised taken-for-granted colonial structures and narratives. However, indigenous scholars have increasingly focused on the nexus between "western" academic research and colonial power. Disrupting this nexus, Bryan Grimwood and Corey Johnson (Chapter 1) examine how their disruptive methodological approach of *collective memory work* contributes to decolonising tourism research and education. A participatory and participant-focused methodology, collective memory work, is deployed to disrupt Settler colonialism narratives amongst Canadian graduate students in a process of critiquing individual memories and collective experiences of tourism and Indigenous–Settler relationships.

Grimwood and Johnson's methodological approach resonates with Ian Lamond's critical work (Chapter 4) on activist mobilities, events of dissent, and activist tourism. Lamond showcases the use of four novel techniques in the project titled *disrUPt*: the creation of a bicycle based mobile film projector; the presentation of augmented film screenings; activist film making with a group of female asylum seekers; and a series of conversations that brought together participants more commonly in opposition during events of dissent. These techniques facilitated the articulation of imaginaries of dissent attached to protest in ways that traditional methods would have not been adequate. Yet, much more remains to be done to unsettle conventional ways of capturing material, and to give enough space to disruptive and adaptive perspectives on participatory data capture.

In a similar vein, Richard Keith Wright (Chapter 14) invites readers to critique the production and consumption of sports event discourse. Wright proposes *socially-constructed creative analytical practice* as a methodological tool to unpack global, complex and deeply political narratives entrenched in sporting events. He showcases the potential of creative analytical practice, creative non-fiction and ethno-drama to present an alternative approach that potentially disrupts how we perceive and analyse the future of critical sports event tourism research. Wright's work embeds post-disciplinary methodological approaches in tourism research through "strengthening the dialogue with business approaches, developing research agendas that contemplate multiple worldviews and engaging in reflexivity as the main challenges faced by critical tourism scholars in the future" (Mura and Wijesinghe, 2021, p.7).

Criticality ought to be created in spaces where multilingual voices at the margins become co-producers of knowledge, and where anglophone domination in tourism research, and the wider social sciences, is meaningfully challenged. Leszek Butowski, Jacek Kaczmarek, Joanna Kowalczyk-Anioł, and Ewa Szafrańska (Chapter 3) make a poignant case for the cost of lost epistemological opportunities. The authors point to the contradictory nature of critical approaches such as social constructionism based on pluralism of perspectives, which is paradoxically mainly published in English. Progress has indeed been made in crossing the ontological, epistemological, and methodological boundaries, yet we recognise that disruptive, diverse, and innovative ways of presenting and producing critical tourism knowledges still have a long way to go.

II. Fieldwork and ethnography

The beginnings of ethnography in the wider social sciences, was a colonial project of exploration of other people's cultures and ways of life. Historically, the ethnographer was considered a heroic figure, a detached observer of "foreign" cultures who collected material culture, and stories, to advance scientific research on behalf of the colonial project. Much seems to have changed since in relation to the role of the researcher in the field, their position and the challenges related to relationships between people and place, and amongst various individuals, including researchers and participants.

Ethnography is a research strategy rather than methodology per se, as it entails a variety of research designs, methods, analytical tools, theoretical perspectives, and representational forms (Denzin & Lincoln, 2017). Ethnography includes observing, collecting material during fieldwork, listening, reflecting, experiencing, writing, interviewing (either one to one or in groups), commissioning diaries from participants, and learning from peoples, places, situations, institutions, landscapes, and things (Till, 2009).

The transforming and shifting positions of tourism ethnographers affect the fieldwork experience and, consequently, the production of knowledge. Indeed, centring the researcher and the participants with their feelings, emotions, and relationships at the core of the research process enables a meaningful and more inclusive reconsideration of the ways knowledge is crafted. Tourism ethnographers become, therefore, co-creators of knowledge, and, as such, include in their collected material their deep reflections.

Roslyn Kerr and Emma Stewart (Chapter 2) bring together fieldwork considerations with *motherhood capital* to present their reflections on their own positionality while conducting fieldwork in a remote community in the Canadian Arctic. In this context, gathering material with the nine-month-old son of one of the two researchers helped to get access to the community. The researcher's motherhood status was instrumental in building trust with members of the Innuit community. To capture the complexities related to the context where the fieldwork took place, a reflexive field journal was kept throughout the journey. The reflexive notes focused on "her social position in relation to the research; her axiological position; her reflections on building trust with community members; her daily experiences of the highs and lows of accompanied research, including the potential bias of being a highly visible mother in the field; as well as the adaptations she made in the field as a result of her personal circumstances" (Chapter 2). Being a mother in the field had a positive effect on the way the research was perceived by her informants, hence her perceived status shifted from being "a high-status researcher" to being an "equal-status mother".

Reflexivity is also central in Emily Crossley's methodological approach (Chapter 10). More specifically the author's "deep reflexivity" which intersects with psycho-social research enabling the reflexive practice to encompass the researcher's emotions, dreams, and unconscious drives which may impact upon research. In her research on the experiences and identities of volunteer tourists in Kenya, Crossley (2019) reflects on her subjectivity, emotions, feelings, and unconscious desires to interpret and make sense of the topic being studied. Similarly, to Kerr and Steward's methodological approach, Crossley's epistemological and ontological positions become central in the ways the research material is gathered and reflected upon. The unsettling dream she had whilst in the field, made her consider the importance of dream analysis, fantasies, and emotions which underpin many qualitative studies.

Ethnography is often a messy business, and Jelena Farkic (Chapter 11) explains the tensions she encountered in conducting ethnographic research in outdoor adventure tourism. Her background, knowledge of concepts, lack of experience in doing research, and her researcher-as-tourist position affected the way she engaged with the human/non-human field. In doing so, the senses and the body became part of the process enabling her "to corporeally enter various social and physical spaces, more easily take part in activities and become more personal with others" (Chapter 11). Thus, the body, in Farkic's reflections, becomes a meaningful tool to produce knowledge. Her own feelings, emotions, and bodily reactions underpin new and more holistic ways of developing tourism knowledges. Embracing the complexities of embodied ethnography which involves the inseparable relationships between senses, place and research, has deepened the ways in which we view and analyse tourist experiences in providing nuanced knowledges of the consumption and production of tourist spaces in the context of outdoor adventure. The ways researchers engage with the field through the body, emotions, senses, dreams, and pre-acquired subjective knowledges will advance the ethnographic co-creation of knowledge in a less rigid and linear manner.

III. Digital platforms

The study of online communities as "cultures" has been framed by an array of terms some conflicting, some unclear. These include "media ethnography", "cyberethnography", "network ethnography", and "virtual ethnography". Sometimes these appear as ground-breaking new approaches or, as others have argued, can be seen as generally continuous with anthropological ethnography. In the digital era, interactions with others happen both in the physical and the digital environment (Laurell & Björner, 2018). Digital technology is being used to support and enhance all stages of the research process, but most commonly for data gathering, transcription and analysis (Kara et al., 2020). Recently, in the digital ethnographic turn, ethnographers have embraced the possibilities offered by technology to both enhance the conventional ethnographic ethos and to shift the boundaries of ethnography itself (Kara et al., 2020; Lupton, 2017; Pink et al., 2015; van Doorn, 2013). A key attraction to using digital technologies and the internet is tourist generated content which has facilitated the ways researchers access tourist activities. Embracing such digital fieldwork will only continue to grow in tourism research (Chambers & Rakić, 2015).

Web 2.0 digital ethnography is one of the most utilised digital research approaches in tourism studies (Tavakoli & Wijesinghe, 2019). Platforms such as websites, blogs, social media, YouTube, and TripAdvisor remain the most popular sites for digital fieldwork (Tavakoli & Mura, 2018). Epistemologically most digital researchers in tourism studies have a passive role in conducting netnographic studies as they are limited to analysing conversations after they have occurred (Kozinets et al., 2014). In that respect, qualitative researchers believe that a deep understanding and rich insights can be achieved by prolonged presence and engagement with communities which can include within the digital world (Kara et al., 2020; Kozinets et al., 2014). A further concern is that few digital researchers engage in reflexivity and provide a reflexive account to explain their position and experience in the study context (Mkono et al., 2015; Tavakoli & Wijesinghe, 2019).

Consequently, digital research in tourism is ripe for new and creative approaches that disrupt and problematise the static and passive production of knowledge. In this book we present works that engage in creative and innovative ways to embed the qualitative ethical ethos of knowledge co-production, openness, transformation, and empowerment in the realm of the digital.

Engagement and empowerment of participants and their active participation in the knowledge creation process is a central part of the discussions about digital methods and research. Building relationships of trust and mutual understanding between researchers, local stakeholders, and gatekeepers is widely regarded as a critical factor in successful research (Cade, Everett & Duignan, Chapter 12). Participatory digital methods, such as the ones examined in these chapters, bring together academic, industry, and policy stakeholders to debate and influence change (Duignan & McGillivray, Chapter 13). Specifically, researchers' adoption and application of these methods enable them to access often marginalised or silenced voices. Consequently, social networks including social messaging platforms such as WhatsApp and Snapchat are leveraged in these chapters to bridge the gap between researchers and community. Such platforms help to negotiate and facilitate access and establish trustworthiness with often marginalised groups in high-risk community settings in ways that minimise risk to the researchers (Chapter 13).

Co-creating new and multivocal insights enables participatory digital approaches to give greater value to the knowledge, capabilities, and experiences of all those involved in, or affected by a situation. As a result, these approaches represent powerful means of exposing power relations, and offering pathways to enact positive change for marginalised groups. An important element of disruptive digital research that facilitates collaborative approaches is the blurring of the binary distinction between the digital and the physical. In Chapters 12 and 13 the authors emphasise the porous nature of the digital and the physical. They offer a hybrid way of knowing that is bodily immersive, sensory, and affectively interlinked with digital methods for capturing, documenting, and communicating. Hybrid approaches such as the "Digi-cal" in Chapter 12, and the "#RioZones-Approach" in Chapter 13 help build and maintain trust with highly respected community leaders, local stakeholders, and industry before, during, and after the research to maintain longer-term engagement consistent with the longitudinal ethos of qualitative research.

The digital elements of such hybrid approaches facilitate researcher reflexivity and can serve as a platform for reflection of the researcher's journey, positionality, and personal safety. Indeed, researchers' reflexivity and their ability to write themselves into the research are vital elements in establishing access, trustworthiness, and transparency, as Chapters 12 and 13 demonstrate.

Qualitative research is often preoccupied with ethics. In the case of digital qualitative research, the discussion is around privacy and the ethics of online data collection. Privacy concerns in social media research are problematised via the use of Snapchat as a method of data generation and collection by Heather Jeffrey, Hamna Ashraf and Cody Paris (Chapter 7). The authors challenge the notion of social media as public space and demonstrate that the ethics of online research is intimately linked with arguments of public and private space alongside issues of informed consent.

The theme of ethics in online research is also tackled by Nicola Cade, Sally Everett and Michael Duignan (Chapter 12) where the "digi-cal" model is deployed to facilitate access and establish trustworthiness with often marginalised groups in high-risk community settings in ways that minimise risk to the researchers. The authors further establish that online research underpinned by a qualitative ethical ethos can be a more open and collaborative process that invites and empowers participants in the research process. This can help to negotiate access with difficult to reach communities and minimise risk. It also transforms the researcher from passive observer into active participant in a way that empowers the participants and their communities, building trust and mutual understanding between researchers, local stakeholders, and gatekeepers.

Matters of ethics of online research remains a primary concern for future digital ethnographers. Legislation such as general data protection regulation (GDPR) demands all organisations obtain consent from European social media participants in the retrieval and analysis of their data (EUGDPR.ORG., 2018). The contributors to this book view the future of netnography as dominated not by big data and finished downloaded conversations, but by deep contextual understandings afforded by consensual synchronistic co-presence. Our contributors further see the future of digital research as a hybrid, porous approach between the physical and digital and not one dominated by the purely technological, virtual, and disembodied realm. Such "digi-cal" approaches would help researchers with embedding themselves in diverse community contexts for more significant timescales to access deeper insights.

IV. Research–practice collaborations

The research–practice nexus is central to creative and disruptive approaches to research. Collaborative approaches between researchers and practitioners are being driven by the desire to confront issues of praxis in research and engage in forms of inquiry that are participatory, activist-oriented, and grounded in local realities (Wilson et al., 2020). The connection, or lack thereof, between research and practice have been debated and explored in the educational, business, health, and social work fields for some time (Kara et al., 2020). Practice-based and practice-led research methodologies have also been extensively discussed in artistic fields (Freeman, 2010; Leavy, 2015). Within tourism research, there has been little attention to research–practice knowledge exchange during the research process nor to practice-based research (Chapter 15). Developing closer links between research and practice in tourism has been discussed and, indeed, there is a call for academics to simplify and share the findings of research projects that may inform strategic decision-making by developing an understanding of the stylistic requirements for translating their knowledge for industry (Walters, Burns, & Stettler, 2015). Although some research does produce material useful for practitioners, more often than not the knowledge transfer occurs in the post-research period, rather than during the research period (Anderson & Sanga, 2019).

Enabling practitioners to take on the role of co-researchers has been identified as a key strategic arena of enabling practice-based knowledge creation (Nancy Duxbury, Fiona Eva Bakas & Cláudia Pato de Carvalho, Chapter 15). Collaborative research centres the perspectives, experiences, and needs of the practitioners throughout the research process (Duxbury et al., 2021). Such processes of co-production of knowledge require appropriate methodologies that weave together both academic research and practice-based knowledge (Jeannotte & Duxbury, 2015). Yana Wengel, Alison McIntosh and Cheryl Cockburn-Wootten (Chapter 8) propose that capturing practice-based knowledge requires consideration of three strategic areas: developing spaces for ongoing knowledge exchange, enabling practitioners to take on the role of co-researcher, and fostering researchers' close attention to the application side of the project.

Chapters 8 and 9 propose the use of games as a research approach that embraces research–practice collaboration and creates a space for knowledge exchange. The authors demonstrate how participants can creatively identify, represent, reflect upon, and communicate complex and often sensitive issues by centring them around their relationships "to objects, landscapes, people and identities – aspects that may otherwise be silenced by traditional research methods" (Chapter 8). Similarly, Lidija Lalicic and Jessika Weber-Sabil show in Chapter 9 how gameplay as a collaborative tool is a powerful method to engage stakeholders in reflection, learning, problem-solving, consensus-building, and knowledge creation. This approach provides opportunities for researchers to co-create spaces for knowledge exchange and develop an in-depth understanding of socially constructed relationships and realities.

Learning is an essential mechanism of the creation of the research–practice nexus. Learning occurs during the co-creation process, through gameplay or other collaborative research–practice approaches. In this process participants do not only learn from each other but also learn as a community (Chapter 9). During the collaborative process of knowledge creation various game mechanisms trigger individual, social, and collective learning outcomes of the players, which in the end can lead to a change in the individual belief system towards the subject that can act as consensus building processes (Lalicic, 2018).

Gaming, whether in the form of serious gaming (Chapter 9) or the Lego Serious Play (Chapter 8), allows the players to explore issues outside of their usual parameters. The game helps participants to identify and articulate their perspectives and pinpoint areas of tensions in playful activities, collective problem-solving, and collaboration during the gameplay. These gameplays provide great insight into the perspectives of others and thus allow actors to jointly identify suitable solutions. This also improves social inclusiveness for those who would normally not engage in tourism matters, such as local communities, NGOs, and the mobility industry.

Nancy Duxbury, Fiona Eva Bakas & Cláudia Pato de Carvalho (Chapter 15) emphasize that in the research–practice nexus the process of knowledge creation, storytelling, sharing, learning, and consensus building needs to not only be analytical and reflexive, but also performative. This needs to be actualised through testing things out and through doing, prompting actors to take actions that implement the research findings. Practice is thus positioned in a historical and social context that gives it structure and meaning.

The process of empowering participants to become co-researchers in the research–practice nexus is unpacked via storytelling, reflection, and performativity. Lucia Tomassini, Xavier Font and Rhodri Thomas (Chapter 16) propose *linguistic narrative analysis* as an innovative approach that moves beyond what the participants' report through stories and narratives to interpreting the subtle understandings expressed via linguistic choices, non-verbal utterances, and the temporal and causal sequences of their storytelling. Such an approach foregrounds the meanings participants attach to their own actions and positions themselves and others through the stories. Participants' narratives become an alternative way of knowing in which participants understand themselves "as subjects critically reflecting on tourism, contemporary society and development" (Chapter 16).

The researcher is integral in the examination of research–practice collaboration as they do the "unpacking" and interpreting of narrative data. Similarly, the researcher needs to embed the use of language, sense-making, role interpretations, and decision-making processes in the interpretation and knowledge creation. In collaborative research, the role of the researcher shifts to that of a facilitator.

Intensive and ongoing reflexive position by all parties involved is fundamental in many qualitative approaches to research. Research–practice collaboration projects, however, can also entail a series of actions beyond ongoing reflective practices and include more in-depth mentorship and training to enable practitioners to take on the role of co-researcher. It is proposed that training on research–practice collaboration, knowledge transfer, and mentorship techniques for both researchers and practitioners are needed to facilitate the transition to co-researchers (Chapter 15).

Future studies that adopt such hybrid research–practice approaches can co-create a new way of "knowing" that has been largely disregarded within tourism studies thus far. Such approaches bring together different perspectives to consider an issue with the aim of reciprocal learning, communication and developing solutions. Our contributing authors provide us with robust, nuanced research approaches to help answer fundamental questions about research and practice. As interest in this nexus continues to grow, it is important that the critical knowledge and practice developed through such approaches create better governance of destinations that re-imagine tourism as diverse, empowering, and responsive.

Future research avenues

Qualitative approaches have been preoccupied with and have embraced inquiry-based struggles to represent and give voice to the so-called "Other" since the advent of the "crisis of representation" (Denzin & Lincoln, 2017; Wilson et al., 2020). Most of the works dismantling the notion of the "Other" centre the human, guided by constructivist, feminist, ethnic, Marxist, queer, decolonising, and cultural studies paradigms. Such privileging of human experiences is disrupted by Abhik Chakraborty (Chapter 6) and Minni Haanpää, Tarja Salmela, José-Carlos García-Rosell, Mikko Äijälä (Chapter 7) as they turn their attention to the non-human and posthuman other. Their work proposes an ontological, epistemological, methodological and ethical rationale for the unmooring of qualitative research from its anthropocentric preoccupation, something that is scant in tourism studies. They invite researchers to embrace "alternative, fluid and speculative inquiries into the inherent plurality and complexity of non-human others" (Chapter 6) as "multispecies assemblages" (Chapter 5). The multiplicity of non-human "actants", such as ice, snow, water, animals, landslide, snowmelt, climate shift, is discussed to highlight that engagement with non-human voices is a challenging, and sometimes messy process without a common verbal language. As such research of multispecies assemblages is an ethno-political act that questions the agency and the power imbalance between the researcher and the non-human subjects as well as the ability to break completely from the anthropocentric research tradition in tourism studies.

Abhik Chakraborty (Chapter 6) and Minni Haanpää, Tarja Salmela, José-Carlos García-Rosell and Mikko Äijälä (Chapter 7), deploy methods that are both traditional (interviews, participant observation, reflection) and emerging (videography, in relation to the non-human stakeholders). Chakraborty, in particular, is careful to avoid being prescriptive and ascertains that "the means of engaging with non-human others should remain fluid and perennially evolving by necessity" (Chapter 6). Here we agree with the two authors that such post-humanist perspectives can disrupt the linearity of qualitative research that centres the human experience. Non-human approaches can provide a "creative-rupture" and offer more fluid ways of engaging with tourism-related phenomena to further the increasingly reflexive and transformative tourism field already shaped by the critical and hopeful turns in tourism.

Disruption and creativity are the two ideas around which we wanted to challenge and begin dismantling white, "western", neoliberal hegemonic narratives and ideologies in qualitative tourism methodologies. In tourism studies in general, and tourism geography in particular, we have witnessed an emphasis on qualitative methodological research, both in terms of the topics addressed and the types of methodological tools employed. In many ways, this legitimisation of qualitative work mirrors developments in other areas such as human geography (Crang, 2003; Davies & Dwyer, 2007), sociology and anthropology (Hennink, Hutter, & Bailey, 2020; Molz, 2018). Such explorations contribute to the critical understanding of the responsibility of tourism research to be firstly disruptive, before it can engender progress and transformation within and outside of our field.

It is clear that in order to progress and diversify tourism qualitative enquiry we need to create spaces where diverse voices, practices and experiences can be undertaken. The world, and us as researchers, face unique challenges such as the global Covid-19 pandemic with its need for physical distancing that renders a lot of our traditional methods unhelpful and even risky. Societies are grappling with our colonial pasts and confronting racism. Thus, decolonising the ways we produce knowledge is no longer optional. Examining and adopting creative, innovative, and disruptive methodologies will help us face these challenges.

Research avenues outlined in this book are designed to disrupt and co-create tourism knowledges. To this end, we pose an invitation to qualitative scholars in tourism to embrace the messy, fluid, and complex new horizons emerging in tourism.

References

Anderson, W., & Sanga, J. J. (2019). Academia–industry partnerships for hospitality and tourism education in Tanzania. *Journal of Hospitality & Tourism Education*, *31*(1), 34–48.

Ateljevic, I., Harris, C., Wilson, E., & Collins, F. L. (2005). Getting "entangled": Reflexivity and the 'critical turn'in tourism studies. *Tourism Recreation Research*, *30*(2), 9–21.

Denzin, N. K., & Lincoln, Y. S. (2017). *The sage handbook of qualitative research* (5rd ed.). Sage.

Chambers, D., & Rakic, T. (Eds.). (2015). *Tourism research frontiers: Beyond the boundaries of knowledge*. Emerald Group Publishing.

Crang, M. (2003). Qualitative methods: touchy, feely, look-see? *Progress in Human Geography*, *27*(4), 494–504.

Crossley, É., (2019). Deep reflexivity in tourism research. *Tourism Geographies*, *23*(1–2), 206–227.

Davies, G., & Dwyer, C. (2007). Qualitative methods: are you enchanted or are you alienated? *Progress in Human Geography*, *31*(2), 257–266.

Everingham, P., Peters, A., and Higgins-Desbiolles, F. (2021). The (im) possibilities of doing tourism otherwise: The case of settler colonial Australia and the closure of the climb at Uluru. *Annals of Tourism Research*, *88*, 103178.

Freeman, R. E. (2010). *Strategic management: A stakeholder approach*. Cambridge University Press.

Hennink, M., Hutter, I., & Bailey, A. (2020). *Qualitative research methods*. Sage.

Kara, Gergen, K. J., & Gergen, M. M. (2020). *Creative research methods in the social sciences: a practical guide*. Bristol: Policy Press (2nd ed., Issue May). www.gbv.de/dms/bowker/toc/9781447316275.pdf

Kozinets, R. V, Dolbec, P.-Y., & Earley, A. (2014). Netnographic analysis: Understanding culture through social media data. The SAGE Handbook of Qualitative Data Analysis, 262–276.

Lalicic, L. (2018). Open innovation platforms in tourism: how do stakeholders engage and reach consensus? *International Journal of Contemporary Hospitality Management*, *30*(6), 2517–2536. https://doi.org/10.1108/IJCHM-04-2016-0233

Laurell, C., & Björner, E. (2018). Digital festival engagement: On the interplay between festivals, place brands, and social media. *Event Management*, *22*(4), 527–540.

Leavy, P. (2015). *Method Meets Art*. New York: The Guilford Press.

Lupton, D. (2017). Feeling your data: Touch and making sense of personal digital data. *New Media & Society*, *19*(10), 1599–1614. https://doi.org/10.1177/1461444817717515

Mkono, M., Ruhanen, L., & Markwell, K. (2015). From netnography to autonetnography in tourism studies. *Annals of Tourism Research*, *52*, 167–169.

Molz, J. G. (2018). *Mobile Virtual Ethnography: Studying Interactive Travelers Online and On-the-Move*. Sage.

Mura, P., and Wijesinghe, S. N. (2021). Critical theories in tourism – S systematic literature review. *Tourism Geographies*, 1–21.

Nunkoo, R., Thelwall, M., Ladsawut, J., & Goolaup, S. (2020). Three decades of tourism scholarship: Gender, collaboration and research methods. *Tourism Management*, *78*, 104056.

Phillips, M. (2009). Critical theory (after Habermas). International Encyclopaedia of Human Geography, pp. 626–631.

Phillimore, J., & Goodson, L. (2004). Progress in qualitative research in tourism: Epistemology, ontology and methodology. In *Qualitative research in tourism* (pp. 21–23). Routledge.

Pink, S., Horst, H., Postill, J., Hjorth, L., Lewis, T., & Tacchi, J. (2015). *Digital ethnography: Principles and practice*. Sage.

Simoni, V., and McCabe, S. (2008). From ethnographers to tourists and back again. On positioning issues in the anthropology of tourism (No. 57, pp. 173–189). Université libre de Bruxelles.

Tavakoli, R., & Mura, P. (2018). Netnography in tourism – Beyond Web 2.0. *Annals of Tourism Research*, *73*, 190–192. https://doi.org/10.1016/J.ANNALS.2018.06.002

Tavakoli, R., & Wijesinghe, S. N. R. (2019). The evolution of the web and netnography in tourism: A systematic review. *Tourism Management Perspectives*, *29*, 48–55. https://doi.org/10.1016/J.TMP.2018.10.008

Till, K.E (2009). Ethnography. *International Encyclopaedia of Human Geography*, 385–397.

van Doorn, N. (2013). Assembling the Affective Field: How Smartphone Technology Impacts Ethnographic Research Practice. *Qualitative Inquiry*, *19*(5), 385–396. https://doi.org/10.1177/1077800413479566

Walters, G., Burns, P., & Stettler, J. (2015). Fostering collaboration between academia and the tourism sector. *Tourism Planning & Development*, *12*(4), 489–494.

Whyte, K. P. (2010). An environmental justice framework for indigenous tourism. *Environmental Philosophy*, *7*(2), 75–92.

Wilson, E., Mura, P., Sharif, S. P., & Wijesinghe, S. N. (2020). Beyond the third moment? Mapping the state of qualitative tourism research. *Current Issues in Tourism*, *23*(7), 795–810.

Wright, R.K., (2021) 'Que será, será!': creative analytical practice within the critical sports event tourism discourse. *Tourism Geographies*, *23*(1–2), 296–317.

Part I

Dis-rupting Methodologies

Collective memory work as an unsettling methodology in tourism

Bryan S. R. Grimwood and Corey W. Johnson

ABSTRACT

Research has exposed how colonial power relations operate in and through various domains of tourism. As byproducts of Western academia, tourism research and education are significant sites where the structures, systems, and narratives of Settler colonialism can become further entrenched and legitimized. What research methodologies can challenge the colonial complexion of tourism research and enable tourism students and scholars to confront how their identities and responsibilities are tethered to (Settler) colonization? We argue that collective memory work (CMW), a participatory and participant-focused methodology, can contribute to these disruptive aims by examining individual experience as embedded and imbued with social meaning. Our ultimate objective is to situate, articulate, and reflect on the use of CMW as an *unsettling methodology* in tourism research and education contexts. Since 2016, we have used CMW to engage Settler Canadian graduate students in a process of critically analyzing individual memories and collective experiences of tourism and Indigenous–Settler relationships. After establishing theoretical and political contexts of Settler colonialism, we present an overview of CMW's feminist and transformative underpinnings and explain how these are being adapted into the methods of our ongoing research with students. Preliminary insights from this research illuminate CMW as a consciousness-raising pedagogical methodology that, in focusing in on Settler memory narratives, helps make space for decolonization in tourism and tourism research.

摘要

已有研究揭示了殖民权力关系是如何在旅游的各个领域以及通过旅游的各个领域运作的。作为西方学术界的副产品，旅游研究和教育是定居者(Settler)殖民主义的结构、体系和叙事得以进一步巩固和合法化的重要场所。什么样的研究方法可以挑战旅游研究的殖民性，并使旅游专业的学生和学者能够面对他们的身份和责任是如何与定居者殖民联系在一起的? 我们认为，集体记忆作品是一种参与性的、以参与者为中心的方法，它可以通过考察嵌入或渗透着社会意义的个人经验来实现这些颠覆性的目标。我们的最终目标是定位、阐明和反思集体记忆作品作为一种令人不安的方法在旅游研究和教育背景下的应用。自2016年以来，我们利用集体记忆作品让加拿大移民研究生参与到一个批判性分析个人记忆、集体旅游体验以及土著与移民关系的过程之中。在建立定居者殖民主义的理论和政治背景后，我们概述了集体记忆作品的女权主义和变革的基础，并解释了这些理论和政治背景是如何适应我们与学生共同进行的研究方法。从这项研究的初步认识，说明集体记忆作品作为一种提高意识的教学方法，在关注定居者的记忆叙事方面，有助于在旅游和旅游研究中去殖民化的空间。

Introduction

> Jack's sister brought sweet grass. She knew it was sacred. Jack didn't bring it. Nor did he know its scent, its softness when braided, its dusty green shade – let alone its significance to peoples and places in the land he inhabits. Jack, his sis, and others were nine or ten days into the river trip when they came to the "Meeting Place", a point on Indian House Lake in northeastern Quebec where Cree, Innu, and Settler allies had come together to engage in collaborative dialogue and to live: a different sort of camping from what Jack and his companions were doing. The Elder at the camp was the camp organizer's mother; a gentle and wise soul whose years of living (hard at times no doubt) were revealed in the smiles that illuminated her face and echoed in her glowing and care-full eyes. Jack's sister knew. The gift of sweet grass – to show gratitude to the Elder for sharing land, sharing knowledge, sharing those kind smiles – could not possibly be matched.
>
> (*Author's memory about an indigenous travel experience*)

As two white, male Settlers socialized in Western academic discourse, we realize our knowledge production practices are invariably hitched to historical and contemporary workings of colonialism. While there are multiple meanings and models of colonialism, our entanglements within particular, North American geographies shape our experiences and understandings of colonialism as an enduring relational process of domination and dispossession often associated with the oppression of Indigenous lands and life and the exercise of Eurocentric imperialist power. Colonialism becomes manifest in many ways: through the expropriation, fragmentation, and external control of Indigenous worlds, bodies, plants, animals, and resources that feed the wealth and privilege of colonizers (Tuck & Yang, 2012); through social and cultural processes that naturalize meanings and narratives of people and place (Said, 1978); through particular, internalized modes of surveillance and management of people, resources, and borders designed to ensure nation-state authority and white privilege (Fanon, 2004; Tuck & Yang, 2012); and as tourism scholars have observed, through cycles of tourism production and consumption that perpetuate ideologies and imaginaries (Chambers & Buzinde, 2015; Hollinshead, 2004; Grimwood, Yudina, Muldoon, & Qiu, 2015).

For some time now, Indigenous scholars, communities, activists, and allies have called attention to the links between Western academic research and colonial power. In the first volume of *Decolonizing methodologies*, Maori scholar Linda Tuhiwai Smith (1999) famously refers to research as 'one of the dirtiest words in the indigenous world's vocabulary' (p. 1). Smith explains how the pursuit of knowledge is so intricately embedded in imperial and colonial domination. Researchers have extracted knowledge, artifacts, narratives, and other 'data' from Indigenous communities—often without meaningful consultation or locally appropriate permissions—and leveraged these for career advancement with little benefit flowing to communities studied (Smith, 2012). What is more, considerable weight is afforded to researcher interpretations of Indigenous knowledge and experience as they circulate within academic and public discourses as the reality of Indigenous thinking and being, contributing to the pathologization of Indigenous peoples and communities on the one hand (Strega & Brown, 2015), or grossly over-romanticized representations of primitive, anti-modern cultures on the other (Braun, 2002). Such research contributes to the violence of fixing Indigenous identity as uncivilized 'Other', a move of essentializing difference that marks 'Us' (i.e., Western/European/white) and our objectifying perspectives as both 'normal' and 'superior' (Tucker & Akama, 2009; Hollinshead, 1992; Said, 1978). Arrogance abounds in such performances of expertise, as does a sense of entitlement to 'know the Other' and access marginalized communities (Strega & Brown, 2015). Indeed, as Tuck and Yang (2014) assert, 'the relationship of research to other human ways of knowing resembles a colonizing formation, acquiring, claiming, absorbing, consuming' (p. 237).

According to *trawlwulwuy* scholar, Emma Lee (2017) of *tebrakunna* country, the majority of such work has been done, and continues to be done, by 'establishment men'. Lee names this subject position as one that privileges Western epistemologies and seeks to advance an authoritative universal voice in research. Establishment men represent—and provocatively so—systems of value often embodied by, though hardly exclusive to, white Eurocentric masculinity. Behind veils of objectivity, neutrality, order, and linear worldviews, those performing as establishment men conceal their reification of structures that exclude, exploit, and harm Indigenous bodies and diminish difference. Researchers, in this vein, have played significant roles in the destruction and devaluation of Indigenous ways of knowing and being (Strega & Brown, 2015), all the while attempting to preserve their dominantly located selves from scrutiny (Fine, 1998). And despite being more acutely attuned to issues of exploitation of Indigenous peoples and devastation of Indigenous knowledge in recent decades, scholarly infrastructures are still built to service these colonial dispositions and inscriptions.[1] Those coded as establishment men still tend to stand not only as the gatekeepers of academic knowledge—for example, arbitrating quality and controlling circulation in their disproportionate numbers as journal editors, and more often than not from Western and English-speaking metropoles—but also as the engineers who define, construct, and manage the rules of scholarly disciplines and paradigms (Lee, 2017; see also Munar, 2017; Smith, 2012; Wijesignhe, Mura, & Bouchon, 2017).

This article represents our initial attempts at collaborating in tourism research that is attentive to and disruptive of the colonial underbelly of much academic research, including the production of tourism knowledge (Tribe & Liburd, 2016). Our aim is to situate, articulate, and reflect on our use of collective memory work (CMW) as an *unsettling methodology* in tourism research and education contexts. CMW, which has received modest uptake in tourism studies, is a participant-focused methodology that examines individual experience as it is embedded and imbued with social meaning (Johnson, 2018). In this article, we delve into the epistemological, methodological, and transformative elements of CMW, arguing and illustrating how it can be useful to disrupting—perhaps decolonizing—tourism research. Beforehand, however, we situate this article, the pilot research we have carried out, and ourselves within Settler colonialism, the particular mode of colonial domination that structures our lives, work, and stories here in Canada.

Settler colonialism, identity, and stories

Colonial domination operates not in any singular, monolithic sense, but rather in distinct and contextually specific ways (Braun, 2002; Veracini, 2015). Tourism scholars have made important contributions to exposing the multiple incarnations of colonialism, including how colonial power relations playing out in and through tourism differ across temporal, spatial, experiential, and affective domains (see e.g., Chambers & Buzinde, 2015; Cooke, 2017; d'Hautesarre, 2004; Grimwood, Muldoon, & Stevens, 2019; Kim & Prideaux, 2012; Lee, 2017; Tucker, 2009). Postcolonial tourism literature has been especially alert to interrupting the various configurations and recurrent effects of colonialism with critical and hybrid interpretations of touristic understandings, practices, and representations (Amoamo & Thompson, 2010; Grimwood et al., 2015; Hall & Tucker, 2004; Tucker & Akama, 2009).

Our application of CMW is informed by Settler colonial theory and the overt and elusive ways Settler colonialism has been advanced in Canada. Settler colonialism is distinguishable from exogenous modes of colonialism 'in that colonizers arrive at a place ("discovering" it) and make it a permanent home (claiming it)' (Tuck & Yang, 2014, p. 224). Veracini (2015) refers metaphorically to Settler colonialism as a bacterial form—one that attaches to a surface and forms aggregations without need of the host. Exogenous colonialism, in contrast, assumes the viral properties of penetrating host cells and maintaining dependence on them for metabolism and replication (Veracini, 2015). Occupying land – in the sense of earth/water/air/plants/animals (Tuck & Yang, 2012) and territories saturated with social meanings and essential to political economies and cultural livelihoods (Lowman & Barker, 2015) – is therefore the central and most valuable concern within Settler colonialism. To ensure the permanence of Settler society, Settler colonialism must ultimately erase Indigenous inhabitation from occupied lands.

Notions of permanence and erasure weave through what Lowman and Barker (2015) identify as Settler colonialism's three main pillars. First, settler colonial 'invasion is a structure not an event' (Wolfe, 2006, p. 388). Invasion, in other words, is not a historical moment of imperial conquest of Indigenous lands, but rather an ongoing form of occupation exerted through the enduring social, political, and economic structures built by the invading society. Structures of invasion also include cultural norms and practices that become institutionalized through laws (e.g., private property) and social discourse (e.g., racist or sexist ideas of 'development') (Lowman & Barker, 2015). Second, Settlers come to stay. As Settlers, we are not simply concerned with the selective expropriation of resources and profit-potential from colonized lands, but rather with 'making a new home on the land, a homemaking that insists on Settler sovereignty over all things in their new domain' (Tuck & Yang, 2012, p. 5). This *taking of space* is intended to be permanent and our claims to land beyond question, which requires Settlers to deny Indigenous presence, insist that history begins with our national inception, and assert narratives of progress and cultural superiority (Lowman & Barker, 2015). Third, settler colonialism aims to transcend colonialism. This transcendence is not one of decolonization, but rather of securing Settler sovereignty by eliminating Indigenous peoples and their connections to land and establishing Settler society so completely that Settler claims to place go unquestioned. Lowman and Barker explain how this violence gets obscured by national myths of benevolence, peacemaking, treaties, and gift-giving. In effect, Settler colonialism renders itself invisible (Veracini, 2010). For Settler colonialism to succeed then, Indigenous peoples 'must be made into ghosts' (Tuck & Yang, 2012, p. 6) and the logics and structures of Settler domination must be exercised to such an extent that they become naturalized, seemingly self-evident, and woven through virtually every aspect of Settler society (Rifkin, 2013).

Power relations enacted through Settler colonialism are productive in a generative rather than positive sense. In addition to enabling the transfer of land from Indigenous peoples to Settler colonial nation-state control, and legitimizing such occupation through law and narrative, Settler colonialism produces an entirely new people – a Settler society (Veracini, 2010). Over time, Settler societies foster distinct identities that are situated in location-specific relationships to land and characterized by dynamic practices and processes (Lowman & Barker, 2015). Settler identity is, therefore, expressed in different and changing ways as Settlers come to identify by doing things that service their assumptions and end goals. Here, on the land that is now often called Canada, Settler identity tends to be disavowed; part of the process of rendering Settler colonialism, and our positionality as beneficiaries of settler colonization, invisible. According to Lowman and Barker (2015), 'disavowal is a key part of the Settler identity and marks Settler people as benefitting from the dispossession and destruction of Indigenous peoples while at the same time vehemently denying complicity in the events and processes that make that happen' (pp. 15–16). We have conditioned ourselves to refuse (consciously or not), at almost every possible turn, the histories and contemporary practices of Settler colonial invasion so integral to the making of Canada. In contrast to Indigenous identities,[2] which for many years in Canada (and elsewhere to be sure) have struggled to articulate an empowering identity to resist elimination, our Settler identity is denied, even as people pervasively attach themselves to Settler processes of being and becoming (Lowman & Barker, 2015). Denial is what makes it sensible for our former Prime Minister Stephen Harper to offer up a formal apology to Indigenous peoples and families in Canada for the government sponsored atrocities of the Indian Residential School system[3] and then, shortly thereafter, state that 'Canada has no history of colonialism' (Hui, 2009).

Theorists stress the central role that stories play in Settler colonialism and the construction of Settler identity that is denied. As Lowman and Barker (2015) explain, narratives 'are the means through which violent colonization is transformed into the story of heroic struggle and the inevitable establishment of an exceptionally successful, just, and distinct society' (p. 33). Settler Canadian stories serve the dual purpose of obscuring Indigenous presence, agency, and claims to land, and memorializing events and achievements in nation-making. They facilitate 'historical amnesia' about our colonial complicity (Regan, 2010) and make us feel good about being on the land, virtuous and legitimate. They allow us to articulate to ourselves and to others who we are as a society. They define and normalize our sense of belonging. They attempt to justify our histories and the authority of our country. They comfort us (Lowman & Barker, 2015).

Perhaps the most exemplary story of Settler Canadians is the one that starts with the land being perceived by European explorers as a unoccupied canvas of nature – that is, *terra nullius*. Through hard work and struggle, as the story goes, this 'new', 'blank' space was mapped and cultivated by Settlers and ultimately civilized through the establishment of what is deemed to be a coherent and progressive society

(Lowman & Barker, 2015). The associated frontier logics here code Indigeneity as a tragic obstacle that had to be overcome or assimilated to ensure development on the expropriated land base. Of course, some of that original 'wild' land we try to set aside for conservation, science, and touristic jaunts that afford not only a sense of rejuvenation but also a sense of what once was. To realize and secure this dimension of the *terra nullius* narrative, Settler structures and institutions have used various strategies to remove Indigenous peoples from their homelands (see e.g., Sandlos, 2007's historical account of how establishing protected areas in northern Canada entailed prohibiting Indigenous hunting and trapping) or cast Indigenous societies as 'anachronistic others', what Erickson (2018) explains as the representation of Indigenous peoples and their influence as existing outside the progress of modernity and therefore incapable of managing lands or owning property now or in the future.

Regan (2010) pinpoints several other stories in the Canadian Settler repertoire. The peacemaker myth, she argues, is one that lies especially close to the heart of our Settler identity. In this narrative, Settler Canadians pridefully describe their historical relationship with Indigenous peoples as non-violent and their settlement process as benign, both of which are used to distinguish us from the horrors of settlement wars in the United States. To quote Regan at length,

> we cast ourselves in the role of benevolent peacemakers – neutral arbitrators of British law and justice, Christian messengers of the peaceable kingdom – who collaborated together in various ways to negotiate treaties and implement Indian policy intended to bestow upon Indigenous people the generous benefits or gifts of peace, order, good government, and Western education that were the hallmarks of the colonial project of civilizing 'savages'. (p. 83)

This story is maintained in the long list of contemporary Canadian bureaucratic solutions to so-called 'Indigenous problems' of poverty, poor health, political and social dysfunction, and low education levels. These nation-state efforts perpetuate Canadian society's self-characterization as a citizenry of 'heroes on a mythical quest to save Indians' (Regan, 2010, p. 34). They also extend into the way we cast ourselves internationally as well-meaning peacekeepers (Lowman & Barker, 2015).

The clear problem with such Settler stories is that they rest upon certain refusals and omissions. In particular, they refuse to acknowledge the violence of warfare, disease, assimilation, upheaval, dispossession, and fragmentation that were requisite to the making of Canada as a Settler colonial nation; and they obscure the generosity, knowledge, skills, and labour of Indigenous communities so pivotal to Settler survival (Lowman & Barker, 2015). Settler stories, therefore, misguide us into claiming ignorance of our history and ongoing participation in colonization, believing Indigenous communities are beneficiaries of our altruism, and leveraging our empathetic and caring impulses to resolve—always from a neutral and safe distance—the ostensible plight of Indigenous peoples (Regan, 2010). The stories effectively deflect attention away from viewing ourselves, and our complicity in Settler colonialism, as the problem. To do so, as Regan (2010) observes, would 'engender our own collective identify crisis and expose us to the trauma of admitting uncomfortable truths' (p. 34).

Given this discussion, a guiding premise of this article—that Settler colonial invasion, identity, and stories are enmeshed with the contours of our academic research—should be apparent. But to explicate this clearly, we borrow from Macoun and Strakosch (2013, p. 433) who suggest various symptoms of Settler colonial moves in and through academic research. These include assertions that:

- Academics legitimately control the entire space of knowledge production;
- Academics have clear access to the knowledge of Indigenous people;
- Academic research is represented as the only site of political activity and social change;
- Indigenous peoples are receptive and passive;
- Indigenous peoples are violent and uncivilized;
- Indigenous peoples are a homogeneous minority that can be incorporated into tolerant multicultural frameworks;
- Academics act in Indigenous interests when seeking to change their lives;
- Academics are generally benevolent and tolerant;
- Academics are neutral observers of people and place;
- Academics are responsible for mediating difference and translating objective knowledge claims about Indigenous peoples for use in Settler colonial systems (e.g., legal, policy, or indeed, tourism).

How can we, as Settler scholars who embody characteristics of establishment men, not just avoid these traps in our research but also work to disrupt them? What options do we have in research to transform our Settler selves? What methodologies might direct us towards exposing Settler stories, confronting our identities and responsibilities as beneficiaries of Settler colonialism, and subverting the colonial complexion of tourism research? The balance of this article is dedicated to describing and reflecting on our application of CMW as an unsettling methodology.

What is collective memory work?

CMW is a participatory action-oriented methodology that encourages and assists a group of research participants (i.e., 'co-researchers', as per below) to make sense of how, unconsciously and through the internalization of taken-for-granted beliefs, they help create social and ideological dimensions of identity, including gender, race, sexual orientation, and other socially relevant categories. Grounded in social constructionism, the theoretical foundations of CMW rest on generating memory narratives that allow co-researchers to illuminate the effects of ideology and discourses (the metaphorical point where culture and language converge). They also let co-researchers see how they are positioned in relation to a variety of social forces; that is, how these forces *subject* us. In other words, memory narratives are used to enable us to see the ideological positions in cultural institutions and language. Rather than merely describing or mirroring reality, these memory narratives are understood to constitute and shape our concepts of identity (Campbell & Kean, 2016).

In CMW, co-researchers engage in a process of discourse analysis whereby they examine and deconstruct their own use of language in written narratives as 'language is not simply a tool' but rather a means to 'convey … the construction of meaning' (Haug, 1997, p. 9). The goal of this deconstructive analysis is to recognize the complexities obscured by the implied and tacit knowledge found in conversation, and therefore politicize speech. In the end, CMW seeks to unravel the ways in which individual's storytelling collaborate with discourse and ideology by implicating themselves into the social structures that act to oppress them (Haug et al., 1987). This unraveling allows us to see hegemonic identities at work. The result is the collective's articulation of a theory that explains how everyday life is the site where society, along with its narratives, reproduces itself. This theory is grounded in the experiences of individuals constructed through cultural ideologies by processes of hegemony; in other words, CMW allows for the personal sphere of experience to be articulated in political terms (Haug et al., 1987).

Much of the research around 'experience' in tourism and related fields has been focused at the individual level, with extended analysis from the researcher to societal or ideological contexts (see e.g., Sharpley & Stone, 2011). While this approach illuminates the life stories of the individual, it does little to give wider context to that experience in a real way for the participant of the research. Sociologist Dorothy Smith (1987) argued for a contextualization of experience that is based on an examination of social relations and institutional structures. She asserted that, 'rather than explaining behavior, we begin from where people are in the world, explaining the social relations of the society of which we are part, explaining an organization that is not fully present in any one individual's everyday experience' (p. 89). She argued that while we may not 'see' these institutional structures, they operate at various levels and in ways that influence our everyday experiences. Consequently, experience is never simply a reflection of what someone has done, feels, or thinks—experience is always constructed through discourses of *a priori* knowledge and power (Smith, 1987). Similarly, individuals 'become' in and through various ideologies and discourses of power that revolve around a variety of identity markers including gender, race, sexuality, and colonizer/colonized. Thus, how can scholars examine individuals and their experiences apart from the memories (ideologies and discourses) that shape everyday lives? Indeed, 'experiences' need to be contextualized and theorized in relation to these important social factors.

It is here where CMW can help us fill the void left by more traditional research methodologies. The collaborative nature of the methodology, along with a natural inclination toward social justice in its practice, allows CMW to not only contextualize individual experience within a collective social experience, but to have participants share, question, and transform from their participation as co-researchers. In this way, CMW complements emerging tourism research utilizing co-creational (Ren, Jóhannesson, & van der Duim, 2017) and transformative learning (Pritchard, Morgan, & Ateljevic, 2011; Stone & Duffy, 2015) perspectives.

The development of collective memory work

CMW was originally developed by Frigga Haug who was working as part of a socialist collective in the 1970s and 1980s. First published in English and in complex detail in *Female Sexualization: A Collective Work of Memory* (1987), the methodology was a way of 'reconstructing scientific work along feminist lines, and that of remodeling Marxism to open up a place within it for issues concerning women' (Haug et al., 1987, p. 23). Haug's collective were heavily involved in women's movement groups in Germany, including the Women's Liberation Movement, and founders of the Socialist Women's Association. The development of this research methodology was first published with the first volume of *Frauenformen*, a reflection on the socialization of women. The collective continued their development of CMW with the publication in 1983 of *Sexualisierung: Frauenformen 2*, the volume which was translated into the 1987 English title noted above with Haug presented as the principle author at that time. Haug continued to develop the methodology in her extensive, subsequent work in released in 1992, *Beyond Female Masochism: Memory-Work and Politics*. Given the heavy influence of Marx on the socialist and women's movements of the 1970s and 1980s in Germany, it is understandable that Haug and other members of the collective would approach research methods and data with the fundamental concepts of social constructionism in mind. The construction of self at any given time plays an important part in how events are constructed and how individuals might remember those memories. Because of this, Haug (1987) used memories as their initial data; hence the name of the methodology.

With its feminist orientation towards the exploration, emancipation, and development of women in society, the advancement of CMW was not apolitical, but intended as a research methodology that investigated the depths of women's experience, and provided a voice to their lived experiences contextualized by the social realities of their lives. This focus on the lived experiences of women, the legitimation of women's experiences as valid historical data, and gender-conscious representations of reality, created a feminist space which was designed to be used by women for the development of women's rights and possibilities. In working towards these goals, Haug et al. (1987) and Haug (1992) reflected on both the role of the participant within research, as well as the role of the researcher throughout the course of a study. They determined the importance of viewing participants in CMW as equal partners in the research process, creating a co-researcher relationship between participants and with the research team of the project. Seen as a way to both name and level the hierarchies which are inherent in much academic research, CMW thus conceives of participants as collaborators in the conduct of research. Additionally, to accomplish the emancipatory goal of less power-laden research practices, the co-researchers must of course volunteer their participation and consent to having their experiences be part of the research. The researcher must also presume that all participants in the research, whether lay or academic, are capable of generating, analyzing, interpreting, and discussing one-another's experiences (Johnson, 2018).

These feminist intentions of CMW have been carried forward by some researchers who see this methodology as a highly gendered exploration, and designed to be that way (e.g., Fraser & Michell, 2015). Haug and her colleagues would not have shied away from that characterization; indeed, it was enshrined in how they contextualized their own approach to this research: 'The question we want to raise is thus an empirical one; it is the "how" of lived feminine practice' (Haug et al., 1987, p. 33). But, in the process of answering that question, it is possible to reassess and reconstitute the original feminist ethos within current social practices. Feminism is now recognized as a complex value-orientation of knowing, listening, resisting, and actioning against relations that oppress; it is not strictly for and about the category 'woman' (Munar, 2017). So, while contemporary applications of CMW remain grounded in feminism, like the work of Haug and her colleagues, many of these are alert to the failure of the founding white women to recognize, for instance, their racialized identities (Johnson, 2018). The acknowledgement of these oversights has allowed the methodology to retain its feminist ethos and justice orientation while being used with various groups. As Johnson (2018) notes:

> Although Haug et al. (1987) conceived the method as an emancipatory exercise for women and women's voices, the epistemological roots of this method allow for its application across populations with the use of alternate trigger concepts... Collective memory work and its offshoots have been used in a targeted way with gay and straight men, transgendered individuals, marketing research groups, lesbian women, foodies, among others (p. 15)

Ultimately, we see within CMW the potential to support co-researchers to peel back layers of hegemonic identities such that they could begin to discuss how they have come to see themselves, their relationships with others, and the world. The opportunity to engage with questions like 'Who am I?', 'Why do I believe what I do?', and 'How do I fit into society?', makes this research methodology an incredibly powerful tool for social justice work. Haug's work was, in part, about consciousness-raising and that aspect of the work – raising awareness and consciousness so that the work of social justice and social change begins with the individual – continues to be critical to today. Further, Haug's ability to bridge the gap between agency and structure by creating a methodology in which the individual links their experience to ideologies and hegemonies, as well as social structures such as sexism, classism, racism, and specific to our aims herein, Settler colonialism, further helps to bridge the gap between researcher and researched, and theory and practice in more empowering ways that acknowledge different ways of knowing and being, embracing multiple perspectives and contradictions to lay the foundation for a 'justice to come' (Derrida, 1994).

Of course, like any methodology, CMW does have its limitations and potential for pitfalls. Laird and Johnson (2018), for example, outline several cautions to scholars considering the approach, including: participants' geographic accessibility; participants' knowledge, skills and abilities (i.e., literacy, use of technology); participants time investment; participants' willingness to discuss social issues; researcher's positionality as an 'in group' or 'out group' member; and ethical considerations for representation. And though not addressed in their list, we also would suggest scholars consider the significant level of trust and existing relationships among co-researchers that may be difficult to develop or navigate, especially in Settler colonial contexts.

Towards an unsettling methodology in tourism

As mentioned earlier, CMW has received some notable attention in tourism research. McCormack (1998) and Small (1999) were the earliest to use it to interrogate how leisure and tourism experiences manifest differently among women. Small's other significant contributions include outlining memory-work as method (Onyx & Small, 2001) and exploring differences for older women as tourists (Small, 2003), women and girls' holidays (Small, 2005), the emergence of the body for women and girls on holiday (Small, 2007), and work/life balance of women tourism scholars (Small, Harris, Wilson, & Ateljevic, 2011). More recently, Mooney (2017) used memory-work to explore gender and power dynamics in the hospitality industry, while Rouzrokh, Muldoon, Torabian, and Mair (2017) engaged it to explore women's travel memories and as transformational tourism pedagogy.

Our consideration of CMW as an unsettling methodology builds on these previous tourism applications and oriented our collaboration in a pilot project completed between 2016 and 2018. The project's central objective was to engage non-Indigenous graduate students in a participatory process of critically reflecting on individual and collective experiences of tourism and Indigenous-Settler relationships in Canada. In other words, we aimed to invite our students to join us in analyzing personal memories and narratives of tourism and Indigeneity in a Settler colonial context. As it will become clear, these intentions deliberately shift us away from research practices that reinforce desires to know and represent 'others' or that place undue burden on Indigenous peoples for helping Settler's understand and resist their colonizing (Lowman & Barker, 2015). The critical self-reflection necessary in our CMW involved unpacking how one is being constructed and how one participates in constructing our world (Potts & Brown, 2015). We anticipated this work would be uncomfortable, both for us and our students. Test-driving the process, therefore, seemed appropriate so that we could learn through experience how participants would respond to the exercise and then refine the style and procedures of our approach for future use.

The pilot project involved two groups of graduate students that we supervised and/or instructed at a large comprehensive Canadian university. The first group of six—which actually included one upper year undergraduate and one international graduate student[4]—joined Bryan on a weekend-long field trip in October 2016 to nature-based and Indigenous tourism sites in the province of Ontario. Details of this trip are peripheral to the aims of this article. Suffice to say: the excursion was voluntary; involved visiting a casino, provincial park, and community festival; and included two nights' accommodation at an outdoor education facility. The group included two students who identify as female and four who identify as male. The CMW exercise with this group commenced during the field trip. The second group of seven students was enrolled in Bryan's advanced qualitative methodology course between January and April 2018 and included five students who identify as female and two who identify as male. These students were invited to participate in the CMW exercise for the additional purposes of (a) generating qualitative data for later experimentation with different analysis methods, and (b) living and reflecting on the experience of being simultaneously a research participant and co-researcher. Activities transpired inside and outside of class time during the months of February and March 2018.

CMW with both student groups modified procedures outlined by Haug et al. (1987) and Johnson (2018). The first step was to conceptualize the trigger (or cue) word or phrase that would be the focus for the memory work. For our pilot project, we identified two triggers: first, *travel experience with a genuinely Canadian quality*, and second, *travel experience with a genuinely indigenous quality*. Next, the co-researchers were tasked with independently writing about a memory elicited by the triggers. The writing of these memories came with guidelines in order to create robust and useful memory texts. These included: (1) write for 30 minutes using each trigger as a focal point; (2) write in third person and use a pseudonym – the purpose being to anonymize the experience for the reader and shed pre-conceived understandings about the writer; (3) write as descriptively as possible, including details that might seem mundane or trivial; and (4) the memory text should be a description or factual account of an event, as void as possible of personal interpretation or analysis. The first group wrote their memory texts on the first night of their field trip while sitting individually at tables in the outdoor education facility's dining hall. Participants in the second group wrote their memory texts during class time. 'Jack's'[5] narrative, which opens this article, is an example of an elicited memory.

The second phase in the process focuses simultaneously on additional data generation and analysis. After copying and sharing the written memories with participants, each memory is read by other participants, and each in turn, the memories are discussed and analyzed by the group as a whole. This group analysis forms an important participatory crux of the methodology; it's the point where the participation of the group in the collective analysis of each participant's memories create a collective memory space, as well as a collective analysis of the meaning of the memories and memory trigger for the group. In the case of the first group, participants convened around a campfire on the second evening of the field trip. Each memory was read out loud and followed by a free-form discussion about their manifest content and latent meanings. Several weeks later, the group reconvened on campus and was guided in a more structured collective analysis of the memory texts. This stage engaged the co-researchers in a process of collectively constructing social understandings of the memory narratives drawn from their own earlier experiences. This collective theorizing occurred through processes of debate about individual memories, and acknowledged that meanings could be contested, that contradiction and dispute might be common, and that multiple meanings are possible. The focus group session was facilitated by Corey and lasted approximately 3 hours. Corey guided the discussion by asking participants to (a) express opinions and ideas about each story, (b) look for similarities and differences in each story, and (c) identify generalizations and overarching themes regarding Canadian and indigenous tourism experiences. The entire facilitation guide is shown in Table 1. The second group of co-researchers used class time to read, share, and discuss their memory in the free-form style noted above. Time constraints and other course requirements made it impractical to carry out the second, more structured group analysis.

Disrupting Settler colonialism in academic research

The aim of this article is not to leverage truth claims verified with empirical findings, but instead to chart out how we are advancing CMW as a methodological and pedagogical process to disrupt Settler colonialism. Nevertheless, showcasing some of our empirical outcomes is instructive to contextualizing our process and arguments. Accordingly, in the subsections that follows, we utilize author memories and some of the associated group analysis, integrating these into a discussion of three inter-related

Table 1. Collective Memory Work Facilitation Guide.

Analysis of each story begins with these questions:

1. What is this story about – what does it mean?
2. What is the author's theory on what it means to be Canadian in the story? (Theory should be explained as a way of explaining the phenomenon)
3. What does Indigenous mean in the story? (Theory should be explained as a way of explaining the phenomenon)

After gaining consensus from the group on these two questions, the facilitator moves on to the literal interpretation of the text of each memory narrative. Participants are asked to underline verbs (actions) and circle adjectives (emotions), and to consider each of the following:

Activity of the author (Verbs)	Emotions of the author (Adjectives)	Interests/ Wishes of the author	Activity of others	Emotions of others	Interests/ Wishes of others	Contradictions	Empty spaces/ silences	Observations about the use of language

Once the group completes this list, the facilitator then asks:

1. How has the author/narrator been created in the story?
2. How have others in the text been created in the story?
3. What does the story mean?
4. How can the author's understanding of what it means to be Canadian or indigenous inform our perspectives of Settler colonialism? What's the larger message of the story beyond the author's individual experience?

Once all the stories have been analyzed, the facilitator should ask:

1. How do this group of stories of tourism inform our understandings of Canadian identity? Indigenous identity? What are the implications of those messages (individually, structurally, etc.)?

ways that CMW can work as an unsettling methodology. These are preliminary and suggestive insights that warrant corroboration and critical debate in future research.

A consciousness-raising pedagogical methodology

Similar to previous work with marginalized populations (Johnson, 2018) and even more directly in tourism studies (Rouzrokh et al., 2017), we see how CMW can be a powerful pedagogical maneuver when used appropriately. In fact, Rouzrokh et al. (2017) describe some key components about using CMW with students, including building trust, setting aside enough time, letting things unfold naturally, and getting student co-researcher buy-in. They also suggest the relevance of asking who might and who might not have similar tourism experiences (Rouzrokh et al., 2017). These important considerations informed the context of our pilot project. The triggers we used to prompt tourism-related memories (i.e., genuinely Canadian and genuinely indigenous travel experiences) were deliberately broad so as to be inclusive yet implicitly tethered to the Settler colonial relationships we wanted to explore. Student trust and buy-in was enhanced either through shared experience of the weekend field trip or the weeks of shared learning in an intensive graduate seminar. We facilitated the process to give students time to reflect on their memories prior to sharing and analyzing them: the field trip group, for example, wrote memories the first night, shared them around the storytelling atmosphere of the second night's campfire, and proceeded with collective analysis weeks later back on campus.

Gulley, Patterson, and Johnson (2018) suggest that 'CMW seeks to understand an overarching meaning from multiple individuals with shared lived experiences and avoids singling out one individual as a representative for a group of people' (p. 81).

It's not about shaming or venerating an individual, but rather about revealing and critically unravelling meanings we have in common (for better or worse) and how these circulate. Importantly, we positioned ourselves alongside students in this process, situating ourselves as facilitators yet participating in our own critical consciousness-raising with other co-researchers. This was an uncomfortable positionality to adopt since, for instance, author memories were included within data generated and collectively analyzed. We do, however, see it as consistent with the participatory character of CMW and utterly crucial if we are serious about destabilizing conventional power relations associated with academic expertise and which we know sit all too comfortably within colonial logics (Smith, 2012). To illustrate how we incorporated these features as part of our pedagogical/methodological strategy, we note the following points derived from co-researcher discussions about the qualities of Canadian identity and experience as conveyed in relation to 'Charlie's' memory.

> Friday night, the second week of August. It seemed routine for Charlie. He and his family – all five of them – would buckle up in the Volvo station wagon, bags and bikes packed on the roof or off the trunk. Dad would drive. Sis was in the middle of the back-bench seat between Charlie and his brother. Sis would doze off sooner than later, head using a brother's shoulder as a pillow. Mom was upfront in the passenger seat, probably glad it was late and that her babes were fed, tired, and stoked about the next two weeks at the cottage.
>
> Charlie would eventually doze off, missing much of the four-hour drive north to a service centre. When they arrive, Charlie and his family would hop out of the car and into the building to use the washroom. No snacks though. It was time for bed. Back at the car, the bench seat was folded down and blankets laid, converting the entire wagon area into a bed. Mom and kids snuggled in. Dad slept across the front driver and passenger seat. Somehow.
>
> Charlie always felt refreshed on the Saturday morning. Waking up to mist, the sweet smells of mid-summer, and anticipation for the lake, for play, for family time, for freedom. They'd be at the cottage an hour after breakfast, arriving early in their first of 14 days in order to soak up as much holiday time as they could.

Co-researchers were quick to identify that the relevance of family and nature in Charlie's memory resonated with their broader understanding of Canadian travel experience. So too did Charlie's anticipation for adventure and his sense of nostalgia and freedom tied to holiday time spent outdoors. The car, cottage, and lake were familiar spaces among co-researchers, even with the vagueness in Charlie's description – it was 'the lake' and 'the cottage', generic and unspecified yet known. Analysis shifted towards unearthing other layers within the memory: the relative class-based privilege of owning cars and bicycles and visiting a cottage each year for 2 weeks; the clear gendering of roles that Charlie recalls being played out by his mom (the care-giving passenger) and dad (the obstacle-confronting driver); the nuclear family. These insights from Charlie's memory helped co-researchers trace similar features in other memory texts and articulate that while nature experiences are naturalized within meanings of Canadian identity and travel experience, they remain contingent upon social circumstances.

Some of the specific ways co-researchers highlighted their 'learning moments' included statements such as: 'I had never considered that perspective before'; 'now my memories make me really uncomfortable'; and, related especially well to our aims, 'Canadian stories reinforce a lot of stereotypes that leave Indigenous people out of being Canadian'. One co-researcher highlighted the pedagogical process well, stating:

> There's something about the process. I guess it's about our investment of energy and time into thinking about the memory and putting it on paper in a very descriptive way. It is not just the telling of the story, you invest in that. And it's also yours to own and not own. So there's the investment in the actual memory and the experience that happened, but also the investment in the storytelling and processing of it... That investment in the process creates some opportunities for learning what it means to be Canadian, and what it means to be Indigenous or not Indigenous alike. Working through this process seems like the right thing to do, and there's power in that in terms of thinking about how we might rewrite those narratives or think them in a different way.

As for exploring shared aspects of communities of privilege, CMW can foster enhanced understanding of systemic and structured forms of privilege across a participant's singular identity. By collectively analyzing aggregated experiences of individuals who share identity categories, we can create meaning that works to avoid singularizing, while not shying away from discomfort. It allows us to explore and disrupt experiences across shared identities without shaming them, which is suggestive of CMW's potential when focusing in on Settler stories.

Focusing on Settler stories

In addition to living the recurrent traumas associated with Settler colonization, Indigenous peoples in Canada and elsewhere are often looked to for identifying colonial counter-narratives. We see this occurring in qualitative tourism research—including some of our own—designed to elevate Indigenous knowledges, experiences, and relationships as part of the project to decolonize our field (Chambers & Buzinde, 2015). One effect of this is that Indigenous peoples are faced with additional burden of shouldering responsibility for Settler colonial resistance. The transformation of Settler colonial structures, systems, and stories is mistakenly cast as an Indigenous problem, while Settler scholars doing such work risk re-inscribing stories of benevolence and avoiding uncomfortable questions that trouble our complicity in Settler colonialization or reveal our responsibilities in undoing it (Regan, 2010).

We suggest that CMW as an approach to research relating to colonization, Indigeneity, and Settler identity troubles conventions of what tourism research *does*. In particular, it can shift us away from collecting and representing information about 'Others' and/or cultural landscapes and towards questioning and transforming ourselves. In this light, tourism research is less about compiling facts or understanding truths external to us and more about interrogating how the stories we tell of ourselves are situated, partial, and contested. Recalling an event during previous field research within an Inuit community in the Canadian Arctic, Charlie wrote the following memory to represent an indigenous travel experience:

> Charlie picked up the phone. "Hello".
>
> Lucy's raspy words of broken English identified her as the caller. "You want to go to the cabin, Charlie? I'm leaving now".
>
> It was a no-brainer. Travelling on the land with Lucy was a precious opportunity; something Charlie had done several times previously and, each time, returned back to his

room in town having learned profoundly about Inuit life and relations and land. This time though was a bit different. Lucy called him. Lucy invited him. Other journeys with Lucy were prompted by Charlie and, in part, by the pressure he felt to get on-the-land with Lucy and other hunters to satisfy research plans, expectations, and desires.

Charlie collected what he felt he needed for the day on the trail and out at the river's terminus where Lucy's cabin was perched. It was March. Cold. White. World of ice and snow. He raced to Lucy's house to help load up and lash down the sled, fuel the snowmobile, and pull on extra layers, including Lucy's husband's old caribou pants. Lucy wandered off to the neighbour's house.

"Might as well invite Hilary and her kids along", she thought out loud.

Engaging in group analysis around this story, co-researchers noticed Charlie's role as a Settler researcher eager to gain privileged access to a particular site for research purposes, but also the positive and negative implications of his Settler identity as he did so. Charlie's memory was deemed to convey Settler desires *to know about* Indigenous lives and the value we place on authentic cultural encounters in remote places. Co-researchers detected some uncertainty about how Charlie recognized his role within recurrent invasions of Settler colonialism, and linked this to their own limited consideration about things like ownership and occupation of land in relation to tourism. One co-researcher indicated these discussions highlighted:

> the complexity and dynamic nature of what it means to be Inuit. And there's uncertainty when Charlie's telling this story, but also more than that, better than that ... a sensitivity to Inuit power and authority and expertise, right? And so, now I'm likely able to be more sensitive to being invited to go on Inuit or other Indigenous lands.

It would seem that the collective dialogue around Charlie's memory encouraged this co-researcher to articulate an alternative storyline to those that essentialize and erase Indigenous presence.

Given statements such as these, we cautiously argue that CMW can be deployed in ways that enable Settlers to begin taking responsibility for Settler colonialism—in Regan's (2010) words, 'to name and then transform the Settler—the colonizer who lurks within—not just in words but by our actions' (p. 11). As we have suggested above, Settler colonialism does not only obscure Indigeneity; it also affects the colonizer. If decolonization is our aim, Settlers must deal with the limits we place on ourselves and our ways of knowing (Regan, 2010). The emphasis in CMW on participants analyzing their experience in relation to broader ideological, discursive, and narrative fields means that Settlers who engage in the methodology might learn to unravel their personal stories as expressions of Settler colonial power. In this way, CMW provides us with a methodological and pedagogical maneuver for critically analyzing how we are implicated in systems and stories that sustain inequality and oppression (Potts & Brown, 2015).

Making space for decolonization

'Decolonization' has taken on different meanings in relation to different places and subject positions (see e.g., Chambers & Buzinde, 2015; Fanon, 2004; Kovach, 2009; Smith, 2012). The most politically and materially charged conceptualization that we have come across is Tuck and Yang's (2012), who argue decolonization entails the return of autonomous control of Indigenous lands to Indigenous peoples, and consequently, the revitalization of Indigenous lifeways. To realize such Indigenous futurity here in Canada not only demands destabilizing Settler colonial structures and relations, but also necessitates fostering practices, knowledges, ambitions, and life stories grounded elsewhere (Cameron, 2015). Indeed, decolonization requires an 'impossible but necessary turn both toward and away from colonial relations' (Cameron, 2015, p. 19).

Through our initial collaboration with CMW, we have observed its potential as a methodology that can contribute to these aims. In particular, using CMW to focus in on Settler memory narratives can destabilize the imprints of Settler colonialism and help make space for Indigenous stories, knowledges, meanings, practices, and relations to gain traction within tourism. It should be clear that we do not mean our use of CMW empowers or amplifies Indigenous presence, or reproduces the struggle to create 'better' representations of Indigenous peoples and communities. Rather, as a process that allows us to critically examine ourselves and connect our stories to those of other Settlers, as well as the social structures and systems enabling our privilege, CMW alerts us to how we are socialized (e.g., as students of tourism) and the limits of what and how we know. Take as an example the following excerpt from a co-researcher:

> One of the things I observed around the campfire was the memories of an indigenous travel experience were all linked to an Indigenous "other". It wasn't about the self or the places that we're indigenous to. Or thinking about our own indigeneity. So it's the separation ... that we're visitors on the land or we're not indigenous to this place. Or maybe it's around the recognition that the term "Indigenous" has a political connotation, and it's an important term that's used by certain communities to advocate and to claim power and identity. Something that we're cautious of appropriating, because we need to take that seriously.

Here, the co-researcher gestures to how Settlers are conditioned to represent Indigeneity as 'other' within their stories, but also suggests an attention to debates about why it may be unfitting to associate the 'indigenous' concept with himself. There is a clear contrast in this passage to Settler colonial desires for knowledge based on objective certainty and neutral observation. The knowledge conveyed is cautious and tentative – suspicious of dichotomies like Indigenous/non-Indigenous that scaffold Settler frameworks yet uncertain about the appropriateness of other stories that can be told.

In this and other related moments, we interpret co-researchers working through the discomfort that comes with disrupting, or unsettling, the underlying assumptions and socially configured meanings embedded within their memory narratives. More often than not, we tell our Settler stories to justify our own histories, actions, and knowledges; to feel good about ourselves and our claims to the land we occupy (Lowman & Barker, 2015). CMW can enable Settlers to recognize, articulate, and push back against the narratives that comfort us. According to Carlson (2017), this is exactly the sort of decolonizing work that Settlers can and should be doing. Our task is not to understand, recover, or represent Indigenous ways of knowing and being—Indigenous peoples and communities are in control of these efforts—but rather position ourselves more proximate to Settler colonial structures, systems, and stories where we can strive to subvert Settler colonial power and make space for Indigenous resurgence (Carlson, 2017).

Conclusion

Colonial violence is interwoven with the fabrics of tourism research and stories of tourism. In Canada and other Settler colonial states, this violence is characterized as a structure of invasion that erases Indigenous presence and establishes a Settler society that denies its legacy of transgressions. For Indigenous peoples, these dynamics of erasure and settlement are 'reasserted each day of occupation' to suppress epistemic, ontological, and cosmological systems (Tuck & Yang, 2012, p. 5). In this article, we argue for using CMW as an unsettling methodology, one that helps Settler co-researchers unravel how comfortable assumptions embedded in our individual tourism experiences are often welded to hegemonic social narratives, ideologies, and discourses. As we have laid out, CMW's collaborative nature, along with its inclination toward social justice in its practice, allows the methodology to not only contextualize individual memory narratives within broader social experience, but to encourage participants to share, question, and transform from their participation as co-researchers. In addition to raising Settler consciousness and focusing on Settler stories—both of which jar the disavowal of Settler identity and may prompt Settler responsibility (Lowman & Barker, 2015)—we see much promise in CMW for making decolonizing space in tourism for Indigenous knowledges told by Indigenous peoples themselves (Regan, 2010).

Challenges should of course be expected in using CMW as an unsettling methodology in tourism. We noted above the importance of trust among co-researchers, which in our case was facilitated by generally positive and existing student–student and student–instructor relationships. Establishing and maintaining trust among co-researchers in other contexts where colonial violence is much more palpable, or directly felt and experienced, would demand significant relationship building on the part of the researcher. Additionally, even in contexts of trust and participatory research, power relations are never eschewed. Tending to power relations in all their manifestations, from historical to intra-group for instance, must be a consistent feature in practicing CMW. Doing so requires that researchers are knowledgeable of the specific colonial geographies they are working within, and reflexive facilitators adept at crafting trigger questions that effectively exhume memories relevant to study intentions (unsettling or otherwise), identifying Settler stories in co-researcher text and talk, and prompting groups to respond critically and constructively. In our case, being reflexive has also meant remaining consistently alert to how, when, and for what ends we enact our relative privilege as beneficiaries of Settler colonialism.

Moving forward, we are carrying with us several lessons learned. Ongoing and future research includes implementing CMW with students participating in an ecotourism field course and tourism students and researchers involved in a series of land camp programs. The latter is evolving through a collaboration with our university's Indigenous Student Centre, and includes both Indigenous and Settler students and researchers curious about tourism and relations to land. As part of the experience, land camp participants will engage in a CMW process similar to the one outlined herein, albeit with the facilitator role and facilitation process being determined in collaboration with our Indigenous colleagues. Using CMW as an unsettling methodology with tourists and tourism operator poses some clear challenges in terms of its alignment, or apparent lack thereof, with holiday or business desires. The promise for such research seems genuine though, especially among consumers and providers of reconciliation tourism (Higgins-Desbiolles, 2003), ethical tourism

(Lovelock & Lovelock, 2013), or culturally sensitive tourism (see e.g., http://sensitivetourism.interreg-npa.eu/). As we develop opportunities for such work, theoretical curiosities are moving us to consider the material conditions of CMW given that memories and materialities are often fused in the construction of tourist identities and performances (Morgan & Pritchard, 2005). How might attention to materiality within CMW prompt transformative—perhaps unsettling—ways of relating to land or attending to material inequalities encountered while on tour?

Some might interpret that what drives our use of CMW may be a sense of Settler guilt or shame, the latter of which Tucker (2009) shows to have positive disruptive potential in colonial tourism encounters. For us, though, engaging ourselves and other Settlers in CMW has more to do with understanding and pulling back the structures of our privilege, learning to take responsibility for our complicity in Settler colonialism and its particular imprints on Canada, and collaborating with each other and our students in social justice inquiry. These motives echo Higgins-Desbiolles and Whyte's (2013) assertion that hopeful tourism researchers must consistently, critically, and reflexively interrogate their situated positionalities and partial knowledges, and Denzin's (2005) contention that all inquiry is moral and political. They are also suggestive of the work still yet to come. Our next paper—and ensuing papers by other scholars (assuming this article inspires further applications of CMW as an unsettling methodology in tourism)—should delve more deeply into empirical matters to more fully illuminate and contextualize Settlers coming to grips with the stories they tell.

Notes

1. Noteworthy in Canada are the ethics guidelines established by our principle federal research funding agencies (see http://www.pre.ethics.gc.ca/eng/policy-politique/initiatives/tcps2-eptc2/Default/). Chapter 9 speaks to the colonizing legacy of research and generates awareness of the dynamic ethical relationships and accountabilities associated with contemporary research involving Indigenous peoples and Indigenous knowledges.
2. Indigenous peoples in Canada include diverse First Nations, Inuit, and Métis peoples.
3. Indian Residential Schools existed in Canada for over 100 years—the last one closing in 1996—and were designed to indoctrinate Indigenous children into the Euro-Christian Canadian mainstream by separating them from families and cultures. Approximately 150,000 children were taken from their families and placed in residential schools, many of them being subjected to physical, emotional, verbal, and sexual abuse (Truth and Reconciliation Commission of Canada (TRC), 2015). Much has been written on this chapter of Canadian nation-making. See Regan (2010) for an influential source.
4. The undergraduate student was doing related independent research and therefore included in the pilot project. The international graduate student's inclusion was premised on the fact that, despite not being Settler Canadian, he is a beneficiary of Settler colonialism in Canada (e.g., his host university exists on colonized lands, his visa is stamped by settler-state governmental officials).
5. Consistent with CMW, we use pseudonyms for the authors/co-researchers of memories reported.

References

Amoamo, M., & Thompson, A. (2010). (re)Imagingin Māori tourism: Representation and cultural hybridity in postcolonial New Zealand. *Tourist Studies*, *10*(1), 35–55.

Braun, B. (2002). *The intemperate rainforest: Nature, culture, and power on Canada's West Coast*. Minneapolis: University of Minnesota Press. doi:10.1086/ahr/57.2.471

Cameron, E. (2015). *Far off Metal River: Inuit lands, settler stories, and the making of the contemporary Arctic*. Vancouver, BC: UBC Press.

Campbell, N., & Kean, A. (2016). *American cultural studies: An introduction to American culture*. New York: Routledge.

Carlson, E. (2017). Anti-colonial methodologies and practices for settler colonial studies. *Settler Colonial Studies*, *7*(4), 496–517. doi:10.1080/2201473X.2016.1241213

Chambers, D., & Buzinde, C. (2015). Tourism and decolonisation: Locating research and self. *Annals of Tourism Research*, *51*, 1–16.

Cooke, L. (2017). Carving "turns" and unsettling the ground under our feet (and skis): A reading of Sun Peaks resort as a settler colonial moral terrain. *Tourist Studies*, *17*(1), 36–53. doi: 10.1177/1468797616685643

Denzin, N. K. (2005). Emancipatory discourses and the ethics and politics of interpretation. In N. K. Denzin & Y. S. Lincoln (Eds.), *Handbook of qualitative research* (3rd ed., pp. 933–954). Thousand Oaks, CA: Sage.

Derrida, J. (1994). *Spectres of Marx*. New York: Routledge.

d'Hautesarre, A. M. (2004). Postcolonialism, colonialism, and tourism. In A. A. Lew, M. C. Hall, & A. M. Williams (Eds.), *A companion to tourism* (pp. 235–245). Malden, MA: Blackwell.

Erickson, B. (2018). Anachronistic others and embedded dangers: Race and the logic of whiteness in nature tourism. In B. S. R. Grimwood, K. Caton, & L. Cooke (Eds.), *New moral natures in tourism* (pp. 43–56). New York: Routledge.

Fanon, F. (2004). *The wretched of the earth*. New York: Grove Press.

Fine, M. (1998). Working the hyphens: Reinventing self and other in qualitative research. In N. K. Denzin, & Y. S. Lincoln (Eds.), *Handbook of qualitative research* (pp. 130–155). Thousand Oaks, CA: Sage.

Fraser, H., & Michell, D. (2015). Feminist memory work in action: Method and practicalities. *Qualitative Social Work*, *14*(3), 321–337. doi:10.1177/1473325014539374

Grimwood, B. S. R., Yudina, O., Muldoon, M., & Qiu, J. (2015). Responsibility in tourism: A discursive analysis. *Annals of Tourism Research*, *50*, 22–38. doi:10.1016/j.annals.2014.10.006

Grimwood, B. S. R., Muldoon, M., & Stevens, Z. M. (2019). Indigenous cultures and settler stories within the tourism promotional landscape of Ontario, Canada. *Journal of Heritage Tourism*, *14*(3), 233–248.

Gulley, N. Y., Patterson, T., & Johnson, C. W. (2018). What are the experiences of White faculty at historically Black colleges and universities? In C. W. Johnson (Ed.), *Collective Memory Work: A methodology for learning with and from lived experience*. New York: Routledge.

Hall, M. C., & Tucker, H. (2004). *Tourism and postcolonialism: Contested discourses, identities and representations*. New York: Routledge.

Haug, F. et al. (1987). *Female sexualization: A collective work of memory*. London: Verso.

Haug, F. (1992). *Beyond female masochism: Memory-work and politics*. London: Verso.

Haug, F. (1997). Memory-work as a method of social science research: A detailed rendering of memory-work method. www.friggahaug.inkrit.de/documents/memorywork-researchguidei7.pdf

Higgins-Desbiolles, F. (2003). Reconciliation tourism: Tourism healing divided societies!. *Tourism Recreation Research*, *28*(3), 35–44.

Higgins-Desbiolles, F., & Whyte, K. P. (2013). No high hopes for hopeful tourism: A critical comment. *Annals of Tourism Research*, *40*, 428–433.

Hollinshead, K. (1992). 'White' gaze, 'red' people – shadow visions: The disidentification of 'Indians' in cultural tourism. *Leisure Studies*, *11*(1), 43–64. doi:10.1080/02614369100390301

Hollinshead, K. (2004). Tourism and the new sense: Worldmaking and the enunciative value of tourism. In C. M. Hall & H. Tucker (Eds.), *Tourism and postcolonialism: Contested discourses, identities and representations* (pp. 25–42). New York: Routledge.

Hui, S. (2009). Shawn Atleo criticizes Harper over "no history of colonialism" remark. *The Georgia Straight*, October 2, 2009. Accessed 30 July 2018: https://www.straight.com/blogra/shawn-atleo-criticizes-stephen-harper-over-no-history-colonialism-remark

Johnson, C. W. (2018). *Collective memory work: A methodology for learning with and from lived experience*. Los Angeles: Routledge.

Kim, S., & Prideaux, B. (2012). A post-colonial analysis of bilateral tourism flows: The case of Korea and Japan. *International Journal of Tourism Research*, *14*(6), 586–600.

Kovach, M. (2009). *Indigenous methodologies: Characteristics, conversations, and contexts*. Toronto: University of Toronto Press.

Laird, N., & Johnson, C. W. (2018). Why Shouldn't I Do Collective Memory Work? Potential Challenges and Pitfalls. In C. W. Johnson (Ed.), *Collective Memory Work A Methodology for Learning With and From Lived Experience* (pp. 119–128). New York, NY: Routledge.

Lee, E. (2017). Performing colonisation: The manufacture of Black female bodies in tourism research. *Annals of Tourism Research*, *66*, 95–104.

Lowman, E. B., & Barker, A. (2015). *Settler: Identity and colonialism in 21st century Canada*. Halifax: Fernwood Publishing.

Lovelock, B., & Lovelock, K. M. (2013). *The ethics of tourism*. New York: Routledge.

Macoun, A., & Strakosch, E. (2013). The ethical demands of settler colonial theory. *Settler Colonial Studies*, *3*(3-04), 426–443.

McCormack, C. (1998). Memories bridge the gap between theory and practice in women's leisure research. *Annals of Leisure Research*, *1*(1), 37–50.

Mooney, S. (2017). The hidden power of memory-work for hospitality and tourism researchers. *Journal of Hospitality and Tourism Management*, *33*, 142–148. doi:10.1016/j.jhtm.2017.09.010

Morgan, N., & Pritchard, A. (2005). On souvenirs and metonymy: Narratives of memory, metaphor and materiality. *Tourist Studies*, *5*(1), 29–53. doi:10.1177/1468797605062714

Munar, A. (2017). To be a feminist in (tourism) academia. *Anatolia*, *28*(4), 514–529. doi:10.1080/13032917.2017.1370777

Onyx, J., & Small, J. (2001). Memory-work: The method. *Qualitative Inquiry*, *7*(6), 773–786. doi:10.1177/107780040100700608

Potts, K. L., & Brown, L. (2015). Becoming an anti-oppressive researcher. In S. Strega, & L. Brown (Eds.), *Research as resistance: Revisiting critical, Indigenous, and anti-oppressive approaches* (2nd ed., pp. 17–41). Toronto: Canadian Scholars' Press.

Pritchard, A., Morgan, N., & Ateljevic, I. (2011). Hopeful tourism: A new transformative perspective. *Annals of Tourism Research*, *38*(3), 941–963. doi:10.1016/j.annals.2011.01.004

Regan, P. (2010). *Unsettling the settler within*. Vancouver: UBC Press.

Rifkin, M. (2013). Settler common sense. *Settler Colonial Studies*, *3*(3-04), 322–340. doi:10.1080/2201473X.2013.810702

Ren, C., Jóhannesson, G. T., & van der Duim, R. (2017). *Co-creating tourism research: Towards collaborative ways of knowing*. New York: Routledge.

Rouzrokh, M., Muldoon, M., Torabian, P., & Mair, H. (2017). The memory work sessions: Exploring critical pedagogy in tourism. *Journal of Hospitality, Sport and Tourism Education*, *21*, 163–173. doi:10.1016/j.jhlste.2017.08.006

Said, E. (1978). *Orientalism*. New York: Vintage Books. doi:10.1086/ahr/84.5.1334

Sandlos, J. (2007). *Hunters at the margin: Native people and wildlife conservation in the Northwest Territories*. Vancouver: UBC Press.

Sharpley, R., & Stone, P. R. (Eds.). (2011). *Tourist experience: Contemporary perspectives*. New York: Routledge.

Small, J. (1999). Memory-work: A method for researching women's tourist experiences. *Tourism Management*, *20*(1), 25–35.

Small, J. (2003). The voices of older women tourists. *Tourism Recreation Research*, *28*(2), 31–39. doi:10.1080/02508281.2003.11081402

Small, J. (2005). Women's holidays: Disruption of the motherhood myth. *Tourism Review International*, *9*(2), 139–154.

Small, J. (2007). The emergence of the body in the holiday accounts of women and girls. In A. Pritchard, N. Morgan, I. Ateljevic, & C. Harris (Eds.), *Tourism and gender: Embodiment, sensuality and experience* (pp. 73–91). Oxford: CABI.

Small, J., Harris, C., Wilson, E., & Ateljevic, I. (2011). Voices of women: A memory-work reflection on work-life dis/harmony in tourism academia. *The Journal of Hospitality Leisure Sport and Tourism*, *10*(1), 23–36. doi:10.3794/johlste.101.265

Smith, D. E. (1987). *The everyday world as problematic: A feminist sociology*. Toronto: University of Toronto Press.

Smith, L. T. (1999). *Decolonizing methodologies: Research and Indigenous peoples*. New York: Zed Books.

Smith, L. T. (2012). *Decolonizing methodologies: Research and Indigenous peoples* (2nd ed.). New York: Zed Books.

Stone, G. A., & Duffy, L. N. (2015). Transformative learning theory: A systematic review of travel and tourism scholarship. *Journal of Teaching in Travel & Tourism*, *15*(3), 204–224. doi:10.1080/15313220.2015.1059305

Strega, S., & Brown, L. (2015). From resistance to resurgence. In S. Strega & L. Brown (Eds.), *Research as resistance: Revisiting critical, Indigenous, and anti-oppressive approaches* (2nd ed., pp. 1–16). Toronto: Canadian Scholars' Press.

Tribe, J., & Liburd, J. J. (2016). The tourism knowledge system. *Annals of Tourism Research*, *57*, 44–61.

Truth and Reconciliation Commission of Canada (TRC). (2015). *Honouring the truth, reconciling for the future: Summary of the final report of the Truth and Reconciliation Commission of Canada*. Winnipeg: Library and Archives Canada.

Tuck, E., & Yang, W. K. (2012). Decolonization is not a metaphor. *Decolonization: Indigeneity, Education & Society*, *1*(1), 1–40.

Tuck, E., & Yang, K. W. (2014). R-words: Refusing research. In D. Paris & M. T. Winn (Eds.), *Humanizing research: Decolonizing qualitative inquiry with youth and communities* (pp. 223–248). Thousand Oaks, CA: SAGE.

Tucker, H. (2009). Recognizing emotion and its postcolonial potentialities: Discomfort and shame in a tourism encounter in Turkey. *Tourism Geographies*, *11*(4), 444–461.

Tucker, H., & Akama, J. (2009). Tourism as postcolonialism. In T. Jamal & M. Robinson (Eds.), *The Sage handbook of tourism studies* (pp. 504–520). Thousand Oaks, CA: Sage.

Veracini, L. (2010). *Settler colonialism: A theoretical overview*. London: Palgrave Macmillan.

Veracini, L. (2015). *The settler colonial present*. New York: Palgrave Macmillian.

Wijesignhe, S. N. R., Mura, P., & Bouchon, F. (2017). Tourism knowledge and neocolonialism –a systematic critical review of the literature. *Current Issues in Tourism*, *22*(11), 1263–1279. doi: 10.1080/13683500.2017.1402871

Wolfe, P. (2006). Settler colonialism and the elimination of the Native. *Journal of Genocide Research*, *8*(4), 387–409. doi:10.1080/14623520601056240

'Motherhood capital' in tourism fieldwork: experiences from Arctic Canada

Roslyn Kerr and Emma J. Stewart

ABSTRACT

Drawing on a study of resident attitudes toward tourism development in the Canadian Arctic, this article examines the experiences of one of the authors who was accompanied by her infant son during her fieldwork. Utilising Lo's (Sociological Perspectives 59(3):694–713, 2016) concept of 'motherhood capital', derived from the work of Pierre Bourdieu, we argue that the presence of her son, who joined her on five occasions between the age of 9 and 22 months, disrupted normal research roles and relationships. We describe how her son helped rather than hindered her acceptance into the Inuit communities where she was working. By using Bourdieusian theory to examine researcher experience, we move beyond the descriptive accounts that often dominate methodological discussions. Instead we aim to demonstrate the value of applying theory that is normally only used to examine encounters observed during fieldwork, to include analysing the researcher's access and interactions. Due to the child-centred nature of the communities and the shared common ground of parenthood, we propose that Stewart's demonstration of motherhood had the effect of dismantling several of the barriers some researchers face when entering communities for the first time. Motherhood capital facilitated her privileged access to the field and changed her perceived status from an outside researcher to an equal-status mother. This was particularly important in an indigenous context where she was always going to be seen as 'other'. Drawing on extensive journal entries, this article adopts a reflexive approach to explore the influences her infant had on her experiences in the field.

摘要

本文以一项加拿大北极地区居民对旅游发展的态度研究为例，考察了本文一位作者在实地考察期间与幼子同行的研究经历。利用罗明诚（Ming-Cheng Miriam Lo）(2016)从皮埃尔·布迪厄（Pierre Bourdieu）的著作中衍生出来的"母性资本"的概念，我们认为她儿子的出现扰乱了正常的研究角色和关系。她的儿子在22个月至9岁期间曾5次跟随她做田野研究。我们描述了她的儿子如何帮助而不是阻碍她进入她调查的因纽特人社区。我们通过使用布迪厄斯理论考察了研究人员的经验，超越了通常主导方法论讨论的描述性叙述。相反，我们的目的是展示应用理论的价值，该理论通常只用于检查实地工作中观察到的遭遇，包括分析研究人员接触社区和与社区的互动。由于社区以儿童为中心的性质和为人父母的共同基础，我们建议，斯图尔特对母亲身份的展示消除了一些研究人员在首次进入社区时面临的一些障碍。母性资本为她进入这一领域提供了便利，并使她从一个外部研究人员转变为一个具有平等身份的母亲。这在土著文化中尤为重要，因为在没有孩子陪伴的情况下她经常被视为"外人"。 本文在大量的日记记录的基础上，采用反思性的方法探讨了她的孩子对她在该地的研究经历所产生的影响。

Introduction

When entering the field, researchers must inevitably make decisions around when to enter the field, what to bring with them, and who might accompany them. However, in communities where access may be difficult to obtain, a researcher may find herself in a position where access to the community takes place at a time that is not ideal for her own circumstances. Such was the situation in the case examined in this article, where one of the author's community-based doctoral research (Stewart, 2009) occurred when she had a 9 month old baby boy. Given that she was breastfeeding at the time, she made the decision to take her baby with her into the field. The settings for the fieldwork were three remote communities in the Canadian Arctic, where the researcher was interested in exploring resident attitudes toward tourism development utilising community-based research approaches. This article examines her experiences of accompanied fieldwork with her baby, particularly focusing on the community reactions to her and her child. Drawing on Lo's (2016) concept of 'motherhood capital', we argue that the presence of her baby helped rather than hindered her acceptance into Inuit communities, and that this can be understood through analysing the power dynamics that the presence of a child interrupted.

The power dynamics of this setting formed a particular challenge. In recent decades northern Canadian communities have become the subjects of an increasing number of research projects, yet in many instances the communities have failed to benefit from their involvement. At the time the research was conducted some community leaders and researchers claimed that communities were suffering from 'research fatigue' (Oakes & Riewe, 1996; Parlee, 2006) or even worse, that social science research had perpetuated deeply embedded inequalities of power through a history of Western colonising processes that privileged Western knowledge over indigenous knowledge (Tuhiwai Smith, 1999). These inequalities were manifested in research which is 'done' to indigenous communities, whereby researchers visit communities, obtain data and subsequently fail to report back their findings (Freeman, 1993), or if they do, the indigenous groups do not recognise themselves in the data (Tuhiwai Smith, 1999). As Wachowich (1999) notes, some local people have become passive and obliging recipients in a process they know little about, rather than active and engaging participants, in a process of which they should be fully informed. Not surprisingly, there is lingering mistrust and suspicion of researchers in some of these communities (Simpson, 2001; Tuhiwai Smith, 1999). The history of the power inequalities in the region is well illustrated by this quotation:

> The researchers, most of the time they just did whatever they wanted when they were up here [referring to the community of Pond Inlet, Nunavut]. A lot of the time they didn't

> really bother to explain themselves very well. A lot of the times we didn't really understand what was going on. We just did whatever they told us to do. People in the community, Inuit, would complain to each other during those years. They would say stuff like, 'Oh, here they come again to study us.' I think that maybe even today there are some of the same attitudes in town when a researcher goes into a house and starts asking questions. We might think, 'you again' that kind of thing, but we would still be, what is the word? 'Polite' – I guess that's the word ... Sometimes I wonder why people agreed all the time even when they didn't want to. I guess what it comes down to is that the Qallunaat [referring to non-Inuit people] have always been the ones with the authority. I learned that in school. Even my parents always treated them that way ... We just did whatever they told us to do. They were the ones that ran the town. They were the ones who said what was appropriate and proper and told us what needed to be done. If the Qallunaat say so, then it must be so. We always agreed. Extract from Saqiyuq: Stories from the lives of three Inuit women (Wachowich, 1999, p. 176).

As Tuhiwai Smith (1999) noted, the very act of research provides the researcher with privileged information, and the researcher has the power to determine how that information is interpreted and presented. Consequently, the power relationship between the researcher and the researched always places the researcher in a position of privilege, which in the case of indigenous research by a Westerner, is exacerbated by a history of colonisation (Denzin & Lincoln, 2008). In the tourism context, Porter (2018) described a feeling of embarrassment in relation to her higher socio-economic status in comparison with the indigenous community she was examining. In this article, we aim to explore how these power dynamics are affected, if at all, through the researcher demonstrating motherhood capital in the form of her child accompanying her into the field.

The aim of this article is to use the lens of 'motherhood capital' to reflect on and examine Stewart's experiences of fieldwork while accompanied by an infant. In doing so, we hope to illustrate how theoretical ideas, such as motherhood capital, can be valuable for examining researcher experiences in the field and disrupt the purely descriptive accounts which have become the norm when discussing researcher experience. As Caplan (1993) notes, the ethnographic tradition includes significant reflexivity on the part of the researcher and an acknowledgement that factors, such as gender, ethnicity, sexuality and many others influence researcher encounters. Here, we argue that a way to analyse these factors is to draw on the same theory that we use to analyse the participants being studied. In essence, we suggest that the use of theory provides a way to recognise the role of the researcher as subject as well as object, through providing symmetry in examining both the participants and the researcher through a theoretical lens. In adopting the argument that the researcher and the researched should be examined using the same *theoretical* lens, our epistemological position is to view both the researcher and the participants as teachers and learners. We adopt what Mertens (2007, p. 218) refers to as a transformative epistemological position, defined as leading to: 'a cyclical model of research that includes the establishment of partnerships between researchers and community members, including the recognition of power differences and building trust through the use of culturally competent practices'. This article outlines the role of motherhood capital in establishing trusting relationships that were further cemented through a dedicated phase of the research designed to report back findings to the communities in meaningful and engaging ways.

In what follows, we first explore the theoretical idea of motherhood capital as an expansion on Bourdieu's concept of capital. We then review the few cases where researchers reflect on their experience of accompanied fieldwork. Next we introduce the context and background of Stewart's experience of being accompanied by her infant son while in the field, before analysing her experiences in terms of motherhood capital.

Feminine and motherhood capitals

Bourdieu (1984) conceptualised capital as a resource which can be accumulated and exchanged in order to better one's position in life. Originally, Bourdieu identified three forms of capital: social, cultural and economic, though later added a fourth, symbolic capital. Richards (1996) noted the particular importance of Bourdieu's notion of cultural capital in tourism, with middle-class tourists often seeking unique tourism experiences as a way to demonstrate high cultural capital. Bourdieu (1984) argued that one's possession of these various forms of capital, together with one's habitus or (subjective) disposition, determined one's position in the (objective) field. By the term 'field', he meant an arena in which individuals aim to fight for resources. In emphasising the significance of the field in tourism, Ahmad (2013) and Çakmak, Lie, and McCabe (2018) argue that distinctions such as those between types of tourists or tourist providers are unable to be made without recognising the role of field, since it is only through understand the workings of the field that the value of particular capitals and the positions of individuals can be identified. In terms of the positioning of individuals, Bourdieu coined the term 'bodily hexis' to refer to the way the physical appearance of the body (also known as physical capital, see Shilling 1991, 2017) is read by others as indicating a particular level of social positioning (Noble & Watkins, 2003). Overall, Bourdieu was interested in the way that individuals within any given field use the resources at their disposal (i.e. capitals) to position themselves in particular ways (Çakmak et al., 2018).

In discussing gender, Bourdieu (1984) regarded gender as secondary to these original capitals, both because he saw the original capitals as the primary organising structures, and because he described the impact of gender as hidden (Skeggs, 2004). At the same time, he also argued that gender and class are unable to be separated, thereby acknowledging the importance of gender as an organising principle.

By contrast, other researchers have noted how women in several fields use gendered capital in order to accumulate capital in their own right. Such authors have conceptualised women's gendered capital in a variety of ways. Huppatz (2009) and Ross-Smith and Huppatz (2010) differentiate between feminine capital and female capital, in the same way that gender and sex are similarly differentiated. They use the term 'feminine capital' to refer to the recognition of the appearance (through behaviour or physical appearance) of femininity but differentiate this from 'female capital', which refers to the capital one may acquire simply from being female. Huppatz (2009) found women were able to utilise both feminine and female capital in the fields of nursing and social work, while Ross-Smith and Huppatz (2010) found the same in their study of high level female managers.

However, both studies found that feminine and female capital were limited in their use, echoing Skeggs (1997) who argued that femininity has only has limited use-value. Skeggs (1997, p. 10) argued 'it provides only restricted access to potential forms of power' such as those at a local or tactical level. Demonstrating Skeggs' (1997) argument, Huppatz (2009) similarly found that feminine and female capital may have very limited value outside of the fields under consideration (nursing and social work), noting that the highest management positions in these fields were dominated by men where feminine or female capital did not provide any advantages.

Lo (2016, p. 695) coined a different form of gendered capital, which she deemed 'motherhood capital', describing it as 'the mothers' interactional styles and informal knowledge that signal to institutional gatekeepers their deep caring for and intimate understandings about their children'. This caring mentality is reminiscent of Arlie Hochschild's notion of emotional labour, later picked up by Hakim (2010) as a component of erotic capital encapsulated by the caring demeanour adopted by females in occupations such as flight attendants. Lo (2016) described how domestic care work could be converted into a form of capital outside the home. Lo (2016) found that motherhood capital was successfully deployed in the school context, but in the health-care context where mothers were primarily seen as lower-class patients, they had difficulty deploying motherhood capital. Confirming the argument that the effects of gendered capital are limited to 'tactical rather than strategic ways' (Skeggs, 1997, p. 10) Lo (2016) argued that motherhood capital contributes to small day-to-day life improvements but is not able to create large-scale change.

Continuing the discussion of the devaluing of women's gendered capital, Huppatz (2009) argued that female participants' beliefs about the value of feminine capital were limited because they were still a part of the dominant doxic order that devalues feminine traits in comparison with masculine ones. This is clearly apparent through studies of employment patterns which consistently demonstrate the devaluing of femininity through lower pay rates being applied to professions commonly viewed as feminine (England, 2010), such as nursing and midwifery. Here the term 'motherhood penalty' has been used to describe the gender pay gap which is even stronger for women *with* children (Lips & Lawson, 2009: emphasis added). Although not utilising the terminology of capital, Bays (2016) similarly found that in Western society, mothers were perceived as more caring but also less competent than non-mothers. Confirming the devaluing of feminine traits, Huppatz (2009) also noted her participants viewed feminine traits as innate and natural rather than acquired, and therefore of less value. Similarly, Skeggs (2004) cites Adkins and Lury (1999) as noting that men can acquire advantages from feminine dispositions in a way that women cannot since women are perceived to have such dispositions naturally.

This devaluing of femininity is summed up by Rich's definition of motherhood as a patriarchal institution. Rich (1995) differentiated between the institution of motherhood which she saw as primarily a form of oppressive patriarchal control, and mothering, the experience of being a mother which can be empowering for women. Such a divide mirrors the classic sociological structure/agency binary, where macro level structures such as patriarchy are considered separate from micro level experience such as mothering. In this article, although we adopt the term 'motherhood' in recognition of Lo's (2016) usage, we do not use it in Rich's (1995) sense since one of the dominant goals of Bourdieu's work was to dissolve the structure/agency binary which he arguably achieved through his notions of capital, habitus and field, and a focus on practical action. In our study, we are particularly interested in a form of motherhood capital that has yet to be explored and reflects Bourdieu's concept of bodily hexis. Just as Huppatz (2009) differentiated between perceptions and behaviours that signified a feminine disposition and the female body itself, we similarly differentiate between Lo's (2016) discussion of the behaviours that indicated motherhood and the act of possessing a child that demonstrates motherhood. Combining these two concepts incorporates the notion of an individual's status being judged by both their appearance and behaviour, as encapsulated by the notion of bodily hexis. In acknowledging a combination of factors at work, our usage bears some similarity to Hakim's (2010) notion of erotic capital, which she argued to incorporate a range of factors including social presentation, and the demonstrating of fertility. We are interested in motherhood capital in both the same behavioural sense as Lo (2016) and in the visible appearance of a woman being accompanied by a child.

Accompanied fieldwork

There are a number of references to fieldwork where researchers have been accompanied by family members and they follow a similar theme. All researchers described the initial stress created by going into the field either pregnant, or accompanied by a child. The amount of preparation time required was a particular focus (Canosa, 2018; Porter, 2018) along with the awareness that taking a child into a potentially unsafe and unfamiliar environment could well involve significant risk (Fluehr-Lobban & Lobban, 1986; Porter, 2018; Schrijvers, 1993). Cooke (2018) reflected on the way becoming pregnant in the field disrupted her researcher identity, which encompassed the ideal of researching alone in the anthropological tradition.

However, several authors noted that unexpectedly, their children were found to be assets in the field. They noted that a major contribution that children played was through creating greater access to the community under examination (Canosa, 2018; Cooke, 2018; Fluehr-Lobban & Lobban, 1986; Porter, 2018). All these authors described participants as reacting positively to the presence of the child. Participants remarked positively on the baby (Canosa, 2018), or picked him up for a cuddle (Cooke, 2018) or the children caused playful interruptions that led participants to feel more at ease (Canosa, 2018). They argued deeper connections were formed that might not otherwise have been the case.

Similarly, Oboler (1986) researching in western Kenya, and Porter (2018) researching in the Philippines, reported a tangible difference in their relationship with the community after they became pregnant. Although the pregnancy had a draining influence on them both physically it had a positive effect on their field relationships with the women of the community becoming far more accepting and welcoming. For example, Oboler (1986, p. 45) described:

> As the news of my pregnancy made the rounds, a remarkable thing happened: my rapport with the women, which I already thought was quite good, took a sudden dramatic turn for the better. In people's minds I had moved from the category of 'probably barren woman' to 'childbearing woman' [...] [they] were suddenly much more at ease with me and began to treat me fully as one of them. Many more confidences and jokes were directed my way.

Similarly, Warren and Vincent (2001) described how as a result of pregnancy, the women in her study became far more interested in her as a person, and engaged with her far more readily. In contrast to Oboler (1986), Warren and Vincent (2001) interpret their change in behaviour to the pregnancy resulting in a changed power dynamic between the researcher and the researched. As a woman pregnant for the first time amongst women with several children, the researcher was now the one needing to be taught, thus equalising the 'teacher-taught' relationship. It is these power dynamics, which Warren and Vincent (2001) found pregnancy disrupted, that we are particularly interested in exploring in relation to the presence of a child in Stewart's doctoral work.

The study

The original research upon which this article is based was premised on the idea that complex phenomena such as tourism are best understood through the lived experiences of individuals. The research set out to explore resident attitudes toward tourism, and the role tourism has played and continues to play in their communities, and their aspirations for future tourism development. At the time of the research, the global focus on the Arctic was unprecedented; issues such as climate change, sovereignty, contaminants, species depletion, natural resource development, and indigenous well-being, were receiving broad media coverage. However, within this wider milieu of change, tourism rarely was mentioned resulting in very little being known about the tourism phenomenon in the Canadian Arctic (Maher & Stewart, 2007).

The three communities involved in the research were Churchill, Cambridge Bay and Pond Inlet (see Figure 1). When the fieldwork was conducted, Churchill was internationally renowned as one of the best northern Canadian destinations for nature-based tourism activities, such as polar bear viewing, beluga whale watching and birding. The largely Inuit community of Cambridge Bay was regarded as a hub for sport hunters, as well as being a key stopping-off point for Northwest Passage cruise travellers. Pond Inlet, another mainly Inuit community, was known principally as a destination for cruise ships as well as an important service centre for all season adventure tourism activities in nearby Sirmilik National Park. The research ultimately aimed to compare and contrast how do resident attitudes toward tourism vary across, and within, communities that are at different stages of tourism development.[1] Fieldwork started in June 2005 and concluded in September 2007. Churchill was visited on four occasions and Cambridge Bay and Pond Inlet were visited three times (the visits ranged in duration from 1 to 3 weeks).

At the time of the research, new frameworks for conducting research across many regions of the Canadian Arctic were being established as part of land-claim agreements. The message was clear: research 'can no longer take place in the Canadian

[1]See Stewart (2009) for a full account of the findings.

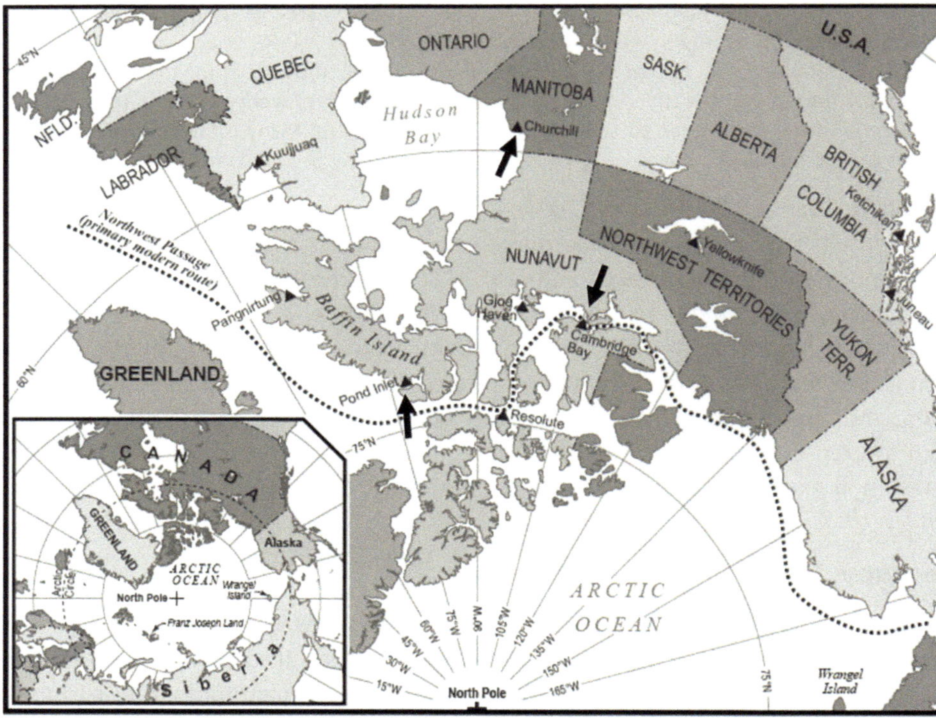

Figure 1. Research locations in northern Canada (Stewart, 2009).

Arctic without some communication and consultation with a local community' (Gearheard & Shirley, 2007, p. 63). With this message at the fore, and the focus on the effects of tourism on communities, Stewart adopted a transformative axiological position through living the principles of 'respect, beneficence and justice' (Mertens, 2007, p. 216). This meant that she aimed to show respect for the people and places where she was working, and also to ensure that there were outcomes that benefitted the communities in meaningful ways. Stewart aligned her work with community-based approaches to research whereby the underlying principles give focus to: the importance of the active participation of community members in all phases of research (DeLemos, 2006; Israel, Schulz, Parker, & Becker, 1998; Taylor et al., 2004); the development of trust between researcher and community (Castleden, Garvin, & Huu-ay-aht First Nation, 2008; Harvey, 2003); the recognition of community as a unit of identity; the strengths, weaknesses and resources existing within the community; integrating knowledge and action for mutual benefit of all partners; a co-learning and empowering course of action; a cyclical and iterative research process; and disseminating findings of research in a respectful manner, where community ownership of knowledge is accredited (Israel et al., 1998).

In designing a research strategy that might reflect such principles, it was paramount that local people were involved with as many different aspects of the research as possible. Stewart knew realistically that as an outsider with no previous contact with the community it would be hard to develop 'community-initiated/directed research', as much as this would be the ideal. Her intent was to develop the project

Table 1. The four stages of the research reported in this article (after Lofland & Lofland,1984).

Stage of research	Descriptor of stage	Purpose	Research activities
Stage 0:	*'Getting there'*	Initial process of finding potential case study communities	• Making contacts • Preliminary site visit • Reflexive journaling
Stage 1:	*'Getting in'*	Preliminary research visits to gain acceptance of community being studied and to gather background information	• Stakeholder interviews • Pilot resident interviews • Participant observation • Reflexive journaling
Stage 2:	*'Getting along'*	Main data collection period with residents in the community	• Resident interviews • Participant observation • Reflexive journaling
Stage 3:	*'Getting out'*	Period of reporting back research findings to the community and departing the field	• Poster display • Weblog • Stakeholder reporting • Community research talks

with input from the local community in defining the project objectives, and in the field her intention was to involve as far as possible local residents in method development, data collection and analysis. These intentions were in stark contrast to previous tourism research in the Canadian Arctic which tended to be characterised by top-down and expert driven approaches to data collection. In part these intentions were influenced by the critical turn in tourism studies whereby researchers are encouraged to challenge the field's dominant discourses (Ateljevic, Pritchard & Morgan, 2007).

Since the research was comparative it was important to develop a research process that could be replicated, as far as possible, in each of the three case study communities. Table 1 illustrates the iterative, multi-staged and multi-method research process that evolved. Qualitative interviews based on a series of open-ended questions with a cross-section of the hamlets' residents and stakeholders formed the basis to the data collection. The topics of interest in the interviews included, but were not restricted to: residents' involvement in the tourism industry; how cruise tourism had evolved over time; residents' attitudes toward the consequences, both negative and positive, of cruise development and; the challenges facing the tourism industry in the future. Between 2006 and 2007, approximately 70 residents were interviewed in each of the case study communities. In addition, participant observation transpired in two ways: (a) being an active member of the community (e.g., Stewart took part in daily routines such as taking her son to community day care, going out on the land (an important aspect of Inuit life) to collect water, and to camp and to fish with her hosts, and to volunteer her time at local events), and (b) observing and participating in activities directly or indirectly related to tourism in each of the three communities. Interview transcripts and observation notes were analysed through a process of open and axial coding (Strauss & Corbin, 1998).

Journaling

A reflexive field journal was kept throughout the research period, and it served three purposes: (a) to record appointments, daily activities and other practicalities; (b) to enter hand written field notes from participant observation activities and; (c) to reflect on emerging patterns from data and sketch out evolving ideas and concepts. Reflexivity, which has a long history in social inquiry (May, 1998; May & Perry, 2017) was clearly critical as Stewart immersed herself in the field. The field journal became a key tool to capture Stewart's reflexive practice enabling both an examination of 'the grounds upon which claims to know the social world are based and an exploration of the strengths and limitations of forms of knowledge' (May & Perry, 2014, p. 109). While not a method in itself, reflexivity, is a way of thinking to 'aid interpretation, translation and representation … It is an iterative and continuous characteristic of good research practice' (May & Perry, 2014, p. 111). Given the context of the Canadian Arctic where modes of knowledge production are (rightly) changing with communities now demanding researchers work to co-produce knowledge, it is critical that reflexive practice is central to the research process. As such Stewart's journal is a rich descriptive account of her experiences in the field, replete with reflexive commentary on wide ranging matters such as her ontological and epistemological basis; her social position in relation to the research (Dowling, 2000); her axiological position (Mertens, 2007); her reflections on building trust with community members; her daily experiences of the highs and lows of accompanied research (Frohlick, 2002) including the potential bias of being a highly visible mother in the field; as well as the adaptations she made in the field as a result of her personal circumstances.

Data analysis

In this article, we draw on the text in Stewart's reflexive journal to provide some direct perspective on the experience of 'accompanied' fieldwork (Frohlick, 2002). Stewart reread her journals, noting any references to her son. She then generated a descriptive account that combined these references into a cohesive 'story' which was published as a book chapter (see Stewart, 2018). Kerr read this account and made notes about how aspects of the account could be interpreted as evidence of gendered or motherhood capital at work. These were then shared with Stewart, and several discussions were held to agree on how the account could be interpreted, before a draft was generated by Kerr, and refined by Stewart.

Results

Getting-there

The insight that community-based fieldwork is never fixed, but rather a partial and ongoing process (Caine, Davison, & Stewart, 2009), required Stewart to be reflexive even before she arrived in the three case study communities. In the preliminary stages of Stewart's doctoral studies, establishing a case study site for her emerging research turned out to be a long and frustrating one. As a non-Canadian and first time visitor to Canada, she was acutely aware of her 'otherness' as a researcher. She found the key was to stay active and reflexive and eventually important openings presented themselves. An opening came as a result of her attendance at a conference. By chance, one lunch time Stewart sat next to a sister-duo from Cambridge Bay who were eager for her to visit their community. By that time she was 6 months pregnant with her second child, Benjamin, and the recognition of her pregnancy, and the conversations shared between them as mothers, in part facilitated this all important invitation to the community. This had a snow-balling effect, and as a result of direct field experience in Cambridge Bay she was able to develop contacts in Pond Inlet, which became another case study site.

As Longhurst (1999) noted, pregnant bodies frequently inspire the attention of strangers, who might ask when the baby is due, provide advice or even touch the mother's body. While for some women, such attention is seen as intrusive and unwanted, for Stewart, the attention, in the form of a conversation and an invitation to share experiences of pregnancy and motherhood, resulted in a desired outcome: access to a potential study site.

The pregnant body therefore acted as a form of physical capital. Bourdieu (1984, 1989) argued that in any interaction, an individual's physical body (also referred to as bodily hexis) is read and valued in terms of its physical capital. For Bourdieu, physical capital incorporates the idea that the body has symbolic value that can be converted into other forms of capital (Bourdieu, 1989; Shilling, 1991, 2017). In the context of gaining access to the field, the desirable form of capital is social capital, since it is social capital that ensures invitations to further interactions. In Stewart's case, it was her pregnant body that acted as a form of physical capital that allowed the conversion to social capital to take place.

Getting in

Once in the field itself, the presence of the baby had the same effect as the pregnant body:

> Within moments Ben and I were surrounded by new faces; smiling. Benji (as he is already called here) is a real hit, a magnet and an *icebreaker*. I am introduced to folk [...] and before I know it I have an interview with the Assistant Mayor and a dinner date with someone else... I am convinced that access to key people such as these in the community would have been longer in coming if Benjamin was not with me. (July 21st 2005).

Participants were as interested in the baby as they were in the pregnant body, with both eliciting the same positive, interested reaction, and the same conversion to social capital. Indeed, the conversion to social capital is almost more prominent in this context given that the attention Ben received included that of prominent key informants such as the Assistant Mayor.

Goffman (1963, p. 126) argued that in any social interaction, children hold a unique role in being what he referred to as 'open persons'. By this, Goffman meant that they are not bound by the usual rules of behaviour. For example, where it is not acceptable to stare at an adult, it is acceptable to stare at a child (Collett, 2005). Using this example, it is understandable that members of the community curious about a new researcher would feel more comfortable approaching a mother with a baby as opposed to a lone researcher as they would not feel bound by any rules when interacting with the baby. In a cross-cultural interaction, the presence of a baby that is outside the rules therefore makes interactions easier and less daunting for both sides.

This phenomenon was also noted by Porter (2018) who found women were keen to interact with her and talk about her baby which eased the data collection process considerably. For Stewart, she described how when she met the elders, she was very nervous, but that it 'felt right' to have her son with her in her arms as she introduced herself to them and this alleviated some of her tension. In this example, enacting the role of a mother and demonstrating motherhood capital was effective for both Stewart and the participants in smoothing the interaction. Her sense of rightness may also have come from a sense of the importance of kinship and family bonds within Inuit culture which has been emphasised in wellbeing studies (see Kral et al., 2011; Kral & Idlout, 2012). While the notion of motherhood capital is of course a Western concept essentially stemming from capitalist logic, it is likely that in the Inuit environment, any expressions or demonstrations of a female motherhood would be met with approval given the significance of family within Inuit communities.

Getting along

Once in the field for some time, Stewart reflected on her role as a mother and how she thought this was perceived in the community:

> I think folk saw me as just a normal person getting on with life – juggling responsibilities and work. The common ground between me and many people I met was obvious. I think it made folk more relaxed knowing that I was just a mum, not just another white southern researcher asking questions. We had something in common. It also gave me a sense of normality (April 4th, 2006).

In this context, Stewart's comments suggest a power dynamic where the researcher was perceived as an equal in the community, in being a 'mum' rather than a 'white southern researcher'. In understanding how this power dynamic developed, Stewart refers not to the immediate presence of her son, but to the community's perceptions of herself as a mother. In essence, her 'juggling' acted as a form of motherhood capital. She also demonstrated motherhood capital in asking for advice about caring for her son while within the community. She recalls:

> When quizzed how Inuit deal with a teething child, they said a rubbery orange ducks foot was the answer [a real one with bone-in and sealed at the ankle. Nothing wasted in this part of the world]. One was promptly produced and Ben gnawed happily away (July 24th, 2005).

In this example, Stewart's concern for her child acted as a form of motherhood capital since she was expressing an appropriate level of concern for her child's wellbeing. As in the health context in Lo's (2016) study, the demonstration of motherhood capital received a positive response, with the community feeling more comfortable talking with someone who exhibited behaviour consistent with that of a *good* mother. Further, and perhaps more powerfully, Stewart's question of her hosts immediately inverted the usual researcher/participant relationship, with Stewart needing to ask the hosts for advice. She placed herself in a subservient position, requiring the help of her hosts in order to solve a problem for her child. Such a position was consistent with the transformative axiological and epistemological position (see Mertens, 2007) adopted by the researcher in acknowledging the researcher as learner. Indeed, during

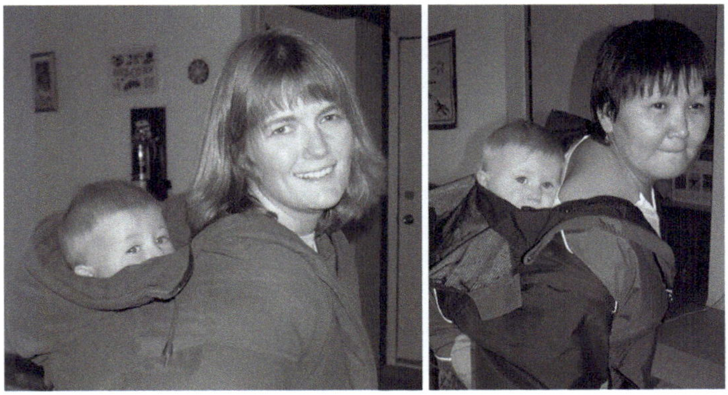

Figure 2. Carrying Ben 'Inuit-style' – Note the look of fear in Benjamin's eyes in his mother's parka compared to the expression of contentment when carried by the Inuit host (May 2006). Source: Stewart.

her time with the Inuit community, Stewart attempted to learn a number of the Inuit parenting practices. One was her attempt to carry her son 'Inuit-style' which involved packing him into her parka on her back (see Figure 2). The way she never entirely mastered this practice, choosing to compromise through the use of a Western style backpack baby carrier, furthered her position as a mother who was trying her best to fit in, but falling short.

In relation to her motherhood, she also required childcare for Ben. She was able to find a mix of daycare and babysitting arrangements that allowed her to spend time completing fieldwork without Ben, which was particularly important in her later trips once Ben was older and more active. In demonstrating a form of motherhood that relied on local services and assistance, the motherhood capital Stewart demonstrated was a local kind, thereby again reducing her status to a level equal with the community. The most graphic example of Stewart demonstrating local motherhood capital as opposed to Western motherhood came early on. In her descriptive account of her experiences, Stewart recalls:

> When my Inuit host unexpectedly regurgitated a mouthful of caribou stew into the obliging mouth of my nine month old son I was momentarily horrified. It was our first evening in an Inuit community and the first of many experiences of doing things 'Inuit-style' during my two field seasons of doctoral research in Arctic Canada. At the dinner table that evening, health concerns were at the forefront of my mind, but I tried to remain calm. I took the lead from Benjamin who was not in the least bit phased. To the contrary, he was relishing this new feeding experience. The half chewed food was the right temperature, the perfect consistency and the salty meat taste was clearly something he was enjoying. It was like watching a mother bird feed her chick. I was mildly irritated that he no longer had any interest in the meal I had lovingly prepared for him. I had spent the morning creating his supposedly favourite meal: a sloppy concoction of sweet potato, chicken and onion (ingredients that are not cheap in the Arctic) and freezing it in to small portions to see him through our stay (Stewart, 2018, p. 155).

In this account, Stewart directly compares the Western and Inuit methods of feeding a baby. While in Western terms, the preparation of her child's food would likely be seen as appropriate and laudable behaviour, and thus a good demonstration of motherhood capital, this behaviour did not translate into motherhood capital within the Inuit community. Instead, Stewart was beginning to learn what constituted motherhood capital within this environment, and the power relations between her and her hosts were reversed as she became the learner. This example is also indicative of the communal living practices that are argued to have developed in many Arctic communities due to their effectiveness during the freezing winter months (Dawson, 2006). By 'remaining calm' and accepting this cultural practice, Stewart demonstrated her respectful axiological position towards the community.

Getting out

The respect with which Stewart was held within the community, in contrast to earlier researchers who were felt to have exploited their participants, was demonstrated on the day that she left the field. Her journal describes:

> As I left the house [for the airport] my host handed me a gift bag containing one of her mother's ulu's (a traditional Inuit all-purpose knife with a curved blade) as a memento of my stay. She said it was 'highly honourable' to receive such a gift since the ulu is used by women on a daily basis and connects people to their food, and its source, the land. It was from Qikiqtarjuaq (Broughton Island), where her mother was from, so it was a double honour (June 2nd, 2007).

At the time, Stewart felt awed by receiving a gift of such personal and significant value to her host. But we suggest that this gift was indicative of the importance of Stewart's role as a mother throughout her stay. The gift was something of her own mother's, but also created for the purpose of enacting Inuit motherhood, in preparing food for her family. In gifting it to Stewart, we suggest that it could be considered as a symbol of Inuit motherhood, which Stewart had attempted to learn throughout her stay.

Later, Stewart further extended her care for these communities by returning to them and discussing the findings of the research with them. As outlined in Stewart and Draper (2009), Stewart went to considerable lengths to engage with the communities about the research findings. She physically returned to each of the three communities and gave presentations and displayed posters outlining the key results, she set up a blog to engage the community (which received a large number of responses) and she met with particular groups such as high school students and elders. She also translated the findings into local dialects to ensure comprehension. She explained:

> In each of the three communities, there was a genuine sense of appreciation that a researcher had returned findings. As one gatekeeper in Cambridge Bay said 'people are really pleased and surprised you came back, as it doesn't happen very often'. In Pond Inlet, the interpreter said that for many of the residents she had spoken to this was the first time they had seen research results 'come back' and that this was a good thing because 'this material is part of *us* and we gave it to you' (Stewart & Draper, 2009, pp. 139–140).

In this quotation, there is a clear recognition that there had been a gifting of cultural capital, with their knowledge being given to the researcher. However, by Stewart returning and reporting the findings, the relationship became more reciprocal, as the communities felt that the knowledge had been returned in a different form. By presenting it in this form, some members of the community recognised that it was now in a form that could be exchanged for economic capital that would benefit for the community. For example, a number of years after completing her doctoral research, Stewart along with colleagues, returned to the community of Pond Inlet where in collaboration with community members, a 'visitor code of conduct' was written and later published and disseminated to all cruise ship passengers arriving into the community (see Dawson, Stewart, Johnston, & Lemieux, 2016).

The reporting back process was also significant in allowing Stewart to validate the research findings (Stewart & Draper, 2009). During this process, the Inuit communities recognised their own voices in the findings, which was significant for Stewart in indicating that she had succeeded in her goal of avoiding her research being 'done to' the communities. Indeed, consistent with the transformative axiological position she adopted for the research, the recognition of the findings by the community confirmed that she had built the all-important 'trust', which motherhood capital facilitated, and illustrated ways and means to conduct research 'with' rather than 'on' communities (see Mertens, 2007) where community members were informed, active and engaged participants in the research relationship rather than passive recipients.

Discussion

In order to analyse the power relations at work in the case setting, we draw attention to Harrington's (2003) notion that the same theories we use to analyse our data can similarly be used for examining the research process. Harrington (2003) explored the idea of using theory to understand ethnography through her use of symbolic interactionism. She argued that theory can be invaluable for understanding ethnographic access given that qualitative research involves the analysis of interpersonal interactions. She notes the value of an approach that focuses on the use of symbolic resources, given that it is these resources that produce the power dynamics between individuals.

The use of theory for analysing researcher positioning and experience also contributes to what Lincoln (2010, p. 5) refers to as 'working the hyphen', which she describes as 'studying the Self-Other conjunction, that fragile and sometimes fractious splice between ourselves as subject and object and those for whom we work, as subject and object.' In a traditional social science analysis, the participants are treated as the object and analysed through social theory. Yet despite the acknowledgement of researchers as both subjects and objects, social theory is rarely used to analyse the researchers as objects. Instead, descriptive terminology is generally used to provide identifying markers, such as gender, ethnicity or class. We suggest that applying social theory to both the participants and to the researcher is a fruitful way of ensuring that both the researcher and researched are treated symmetrically as both subject and object. As Mertens (2007) points out, the symmetrical treatment of researcher and participant is necessary for research aiming for or advocating social justice, since it acknowledges and potentially mitigates the uneven power relations that can exist between researchers and participants.

To illustrate the potential value of theory for analysing researcher interactions and experience, we followed Harrington (2003) through similarly utilising a theoretically driven resource-orientated approach to examine how Stewart gained access and interacted with a research community. Our interest in analysing her experience as a mother in the field led us to the concepts of gendered (Huppatz, 2009, 2010) and motherhood capital (Lo, 2016). These concepts were developed in order to understand how being female, acting feminine or possessing a child, can act as a resource, or capital, within interactions with others. Further, as an extension of Harrington's (2003) work using symbolic interactionism, we noted that the work of Bourdieu can be more relevant for drawing attention to power relationships, which were not a specific focus of symbolic interactionists (Parker & Aggleton, 2003).

The experiences of accompanied fieldwork illustrate the way Stewart demonstrated both visible evidence of being a mother, through having her child with her, and also behaviours consistent with the community's understanding of motherhood. These behaviours were not known to Stewart prior to her research, and it was therefore necessary for her to become a learner while collecting learning herself. We argue that her behaviour *reduced* Stewart to being perceived as a mother, rather than a white southern researcher. We emphasise the word reduce because in terms of the power dynamics, it was the reduction in status that allowed Stewart to disrupt the usual researcher/researched relationship and become a trusted member of the research community, which was heightened when she returned to report back her findings. Though we use the word 'reduce' with care, this interpretation stems from the authors' axiological and epistemological positions, which we acknowledge may not be shared by other researchers.

This 'reduction' is also important because of the evidence that feminine capital usually holds a lesser or reduced status (Lo, 2016; Huppatz, 2009; Ross-Smith & Huppatz, 2010; Skeggs, 1997). In other Western contexts, the lower status has been framed as problematic, with feminine or motherhood capital acting to disempower women (Skeggs, 1997). However, in the indigenous research context, where a problem is the high status of the researcher, an ironic situation develops. The lesser status can work to equalise the power relations between researcher and participants, yet in doing so, the status of researcher is therefore heightened within the academic field through the completion of a successful research project.

Motherhood and gendered capital are argued to not be able to be transferred to other arenas (Skeggs, 1997; Huppatz, 2009), but in this case, their lesser status was the very reason why they could be transformed into other forms of capital. If we apply a classic Bourdieusian analysis, we can note how motherhood capital was transferred to social capital, which allowed access to the field and the completion of the research project. As a consequence, the researcher was also able to gain cultural and economic capital, both through the publication of her PhD, and through later gaining grant funding which was applied for together with colleagues and the community. Essentially, through a number of transactions, motherhood and gendered capital were responsible for a rise in her status in the Western academic field of academia.

At the same time, the transaction appeared to also work for the communities, with Stewart's return to the communities to disseminate and discuss the findings working to provide the communities with cultural capital which they perceived to be valuable owing to its ability to be exchanged for economic capital. Stewart's decision to return to the communities could also be read as an aspect of feminine capital in the sense that it was done, in the light of the social justice position adopted by the researchers, to care for the communities. It was both expensive and time consuming for her to return to all three of the Arctic communities, and in that sense incorporated the same sense of self-sacrifice and care that is usually associated with mothering and femininity (Bays, 2016; Huppatz, 2009).

Conclusion

This study has examined the experiences of a female researcher in bringing her child with her into the field. We found that far from this being an impediment, her demonstration of motherhood capital facilitated her access to the field which unintentionally disrupted the established way of conducting field research within tourism studies. We suggest that this was effective due to motherhood capital changing her perceived status from a high status researcher to an equal-status mother. This caused her to be perceived as an equal to those in the community, which was particularly important in an indigenous context where she was always going to be seen as 'other'.

This article has also demonstrated the used of theory in analysing researcher-participant encounters. Where such encounters are most often outlined in a descriptive manner, the use of theory can assist with providing explanations for understanding why interactions proceeded as they did. Just as we may apply theory to examine the encounters observed during fieldwork, we argue that, following Harrington (2003), the theoretical explanation should extend to include the researcher as part of the interactions. In this article, we have focused on the theoretical concept of motherhood capital, but equally a researcher could also be understood as possessing a variety of forms of other capital that influence their encounters. Just as researchers describe themselves in terms of gender, class and ethnicity (Schrijvers, 1993), we argue that researchers could also describe themselves as possessing particular levels of capital. While this study has focused on engagement with an indigenous community, such an approach may be even more fruitful in ethnography at-home given that there are a variety of Bourdieusian studies mapping the types of capitals valued in different Western fields that could be drawn upon for analysis.

This study specifically focused on the account of a mother; however an interesting question, which further study could examine, would be the impact of fathers being accompanied by their children. Studies have shown how women and men are perceived very differently in relation to their roles as parents and traditionally fathers are not expected to demonstrate the same intense level of care as mothers (Bays, 2016). Consequently, we suggest that men would be unlikely to experience the same acceptance as a result of being accompanied by children.

Acknowledgements

Both authors would like to thank the Editor and the anonymous reviewers for their constructive feedback. Emma warmly acknowledges the support she received as both a mother and a researcher in the Canadian Arctic. She would also like to thank her Doctoral Supervisor Professor Dianne Draper for encouraging her to continue with her studies while pregnant. Her thanks also go to the Commonwealth Association of Universities and the Pierre Elliot Trudeau Foundation for research funding. Finally she would like to thank her son, Benjamin, for breaking the ice.

References

Adkins, L., & Lury, C. (1999). The labour of identity: performing identities, performing economies. *Economy and Society, 28*, 598–614.

Ahmad, R. (2013). Working with Pierre Bourdieu in the tourism field making a case for 'third world' tourism. *Cultural Studies, 27*, 519–539. doi:10.1080/09502386.2012.707222

Ateljevic, I., Pritchard, A., & Morgan, N. (Eds.). (2007). *The critical turn in tourism studies*. London: Routledge.

Bays, A. (2016). Perceptions, emotions, and behaviors toward women based on parental status. *Sex Roles, 76*, 138–155.

Bourdieu, P. (1984). *Distinction: A social critique of the judgement of taste*. Cambridge, MA: Harvard University Press.

Bourdieu, P. (1989). Social space and symbolic power. *Sociological Theory, 7*, 14–25. doi:10.2307/202060

Caine, K. J., Davison, C. M., & Stewart, E. J. (2009). Preliminary field-work: Methodological reflections from northern Canadian research. *Qualitative Research, 9*, 489–513. doi:10.1177/1468794109337880

Çakmak, E., Lie, R., & McCabe, S. (2018). Reframing informal tourism entrepreneurial practices: Capital and field relations structuring the informal tourism economy of Chiang Mai. *Annals of Tourism Research, 72*, 37–47. doi:10.1016/j.annals.2018.06.003

Canosa, A. (2018). 'Mummy, When are We Getting to the Fields?' Doing fieldwork with three children. In B. A. Porter & H. A. Schänzel (Eds.), *Femininities in the field: Tourism and transdisciplinary research* (pp. 84–95). Bristol: Channel View publications.

Caplan, P. (1993). Introduction 2: The volume. In D. Bell, P. Caplan, & W. J. Karim (Eds.) *Gendered fields: Women, men and ethnography* (pp. 19–27). London: Routledge.

Castleden, H., Garvin, T., & Huu-ay-aht First Nation. (2008). Modifying photovoice for community-based participatory indigenous research. *Social Science and Medicine, 66*, 1393–1405. doi:10.1016/j.socscimed.2007.11.030

Collett, J. L. (2005). What kind of mother am I? Impression management and the social construction of motherhood. *Symbolic Interaction, 28*, 327–347. doi:10.1525/si.2005.28.3.327

Cooke, L. (2018). Subjectivities implode: When 'The Lone Male' ethnographer is actually nursing mother… In B. A. Porter & H. A. Schänzel (Eds.), *Femininities in the field: Tourism and transdisciplinary research* (pp. 140–153). Bristol: Channel View publications.

Dawson, J., Stewart, E. J., Johnston, M. E., & Lemieux, C. J. (2016). Identifying and evaluating adaptation strategies for cruise tourism in Arctic Canada. *Journal of Sustainable Tourism, 24*, 1425–1441. doi:10.1080/09669582.2015.1125358

Dawson, P. (2006). Seeing like an Inuit family: The relationship between house form and culture in northern Canada. *Études/Inuit/Studies, 30*, 113–135. doi:10.7202/017568ar

DeLemos, J. L. (2006). Community-based participatory research: Changing scientific practice from research on communities to research with and for communities. *Local Environment, 11*, 329–338. doi:10.1080/13549830600558838

Denzin, N. K., & Lincoln, Y. S. (2008). Introduction: Critical methodologies and indigenous inquiry. In N. K. Denzin, Y. S. Lincoln, & L. Tuhiwai Smith (Eds.), *Handbook of critical and indigenous methodologies* (pp. 1–20). Thousand Oaks: Sage.

Dowling, R. (2000). Power, subjectivity and ethics in qualitative research. In I. Hay (Ed), *Qualitative research methods in human geography* (pp. 23–36). Oxford: Oxford University Press.

England, P. (2010). The gender revolution: Uneven and stalled. *Gender & Society, 24*, 149–166. doi:10.1177/0891243210361475

Fluehr-Lobban, C., & Lobban, R. A. (1986). Families, gender and methodology in the Sudan. In T. L. Whitehead & M. E. Conaway (Eds.) *Self, sex and gender in cross-cultural fieldwork* (pp. 152–195). Urbana: University of Illinois Press.

Freeman, W. L. (1993). Research in rural native communities. In M. J. Bass, E. V. Dunn, P. G. Norton, M. Stewart, & F. Tudiver (Eds), *Conducting research in the practice setting* (pp. 179–196). London: Sage Publications.

Frohlick, S. E. (2002). "You brought your baby to base camp?" Families and Field Sites. *The Great Lakes Geographer*, *9*, 49–58.

Gearheard, S., & Shirley, J. (2007). Challenges in community-research relationships: Learning from natural science in Nunavut. *Arctic*, *60*, 62–74.

Goffman, E. (1963). *Behavior in public place*. New York: Free Press of Glencoe.

Hakim, C. (2010). Erotic capital. *European Sociological Review*, *26*, 499–518. doi:10.1093/esr/jcq014

Harrington, B. (2003). The social psychology of access in ethnographic research. *Journal of Contemporary Ethnography*, *32*, 592–625. doi:10.1177/0891241603255677

Harvey, F. (2003). Developing geographic information infrastructures for local government: The role of trust. *The Canadian Geographer*, *47*, 28–36. doi:10.1111/1541-0064.02e10

Huppatz, K. (2009). Reworking Bourdieu's 'Capital': Feminine and female capitals in the field of paid caring work. *Sociology*, *41*, 45–66. doi:10.1177/0038038508099097

Israel, B. A., Schulz, A. J., Parker, E. A., & Becker, A. B. (1998). Review of community-based research: Assessing partnership approaches to improve public health. *Annual Review of Public Health*, *19*, 173–202. doi:10.1146/annurev.publhealth.19.1.173

Kral, M. J., & Idlout, L. (2012). It's all in the family: Wellbeing among Inuit in Arctic Canada. In *Happiness across cultures* (pp. 387–398). Dordrecht: Springer.

Kral, M. J., Idlout, L., Minore, J. B., Dyck, R. J., & Kirmayer, L. J. (2011). Unikkaartuit: Meanings of well-being, unhappiness, health, and community change among Inuit in Nunavut, Canada. *American Journal of Community Psychology*, *48*(3-4), 426–438. doi:10.1007/s10464-011-9431-4

Lincoln, Y. S. (2010). "What a long, strange trip it's been ... ": Twenty-five years of qualitative and new paradigm research. *Qualitative Inquiry*, *16*, 3–9. doi:10.1177/1077800409349754

Lips, H., & Lawson, K. (2009). Work values, gender, and expectations about work commitment and pay: Laying the groundwork for the "motherhood penalty"? *Sex Roles*, *61*(9-10), 667–676. doi:10.1007/s11199-009-9670-0

Lo, M.-C. M. (2016). Cultural capital, motherhood capital, and low-income immigrant mothers' institutional negotiations. *Sociological Perspectives*, *59*, 694–713. doi:10.1177/0731121416629988

Lofland, J., & Lofland, L. H. (1984). *Analyzing social settings: A guide to qualitative observation and analysis*. Belmont: Wadsworth Publishing Company.

Longhurst, R. (1999). Pregnant bodies, public scrutiny: 'Giving' advice to pregnant women. In E. K. Teather (Ed.), *Embodied geographies* (pp. 77–90). London: Routledge.

Noble, G., & Watkins, M. (2003). So, how did Bourdieu learn to play tennis? Habitus, consciousness and habituation. *Cultural Studies*, *17*(3-4), 520–539. doi:10.1080/0950238032000083926

Maher, P. T., & Stewart, E. J. (2007). Polar tourism: Research directions for current realities and future possibilities. *Tourism in Marine Environments*, *4*, 65–68. doi:10.3727/154427307784772048

May, T. (1998). Reflexivity in the age of reconstructive social science. *International Journal of Social Research Methodology*, *1*, 7–24. doi:10.1080/13645579.1998.10846860

May, T., & Perry, B. (2014). Reflexivity and the practice of qualitative research. In *The SAGE handbook of qualitative data analysis* (vol. 109, p. 122). London: Sage.

May, T., & Perry, B. (2017). *Reflexivity: The essential guide*. London: Sage.

Mertens, D. M. (2007). Transformative paradigm: Mixed methods and social justice. *Journal of Mixed Methods Research*, *1*, 212–225. doi:10.1177/1558689807302811

Oakes, J., & Riewe, R. (1996). Communicating Inuit perspectives on research. In J. Oakes and R. Riewe (Eds), *Issues in the North: Volume 1* (Occasional publication series 406, pp. 987–981). Edmonton: Canadian Circumpolar Institute.

Oboler, R. S. (1986). For better or worse: Anthropologists and husbands in the field. In T. L. Whitehead & M. E. Conaway (Eds.), *Self, sex and gender in cross-cultural fieldwork* (pp. 28–51). Urbana: University of Illinois Press.

Parker, R., & Aggleton, P. (2003). HIV and AIDS-related stigma and discrimination: A conceptual framework and implications for action. *Social Science & Medicine, 57*, 13–24. doi:10.1016/S0277-9536(02)00304-0

Parlee, B. L. (2006). Dealing with ecological variability and change: Perspectives from the Denesoline and Gwich'in of Northern Canada. *Unpublished Doctor of Philosophy*, University of Manitoba, Winnipeg.

Porter, B. A. (2018). Early motherhood and research in the Philippines: From bump to baby in the field. In B. A. Porter & H. A. Schänzel (Eds.), *Femininities in the field: Tourism and transdisciplinary research* (pp. 68–83). Bristol: Channel View Publications.

Rich, A. (1995). *Of woman born: Motherhood as experience and institution*. New York: WW Norton & Company.

Richards, G. (1996). Production and consumption of European cultural tourism. *Annals of Tourism Research, 23*, 261–283. doi:10.1016/0160-7383(95)00063-1

Ross-Smith, A., & Huppatz, K. (2010). Management, women and gender capital. *Gender, Work & Organization, 17*, 547–566. doi:10.1111/j.1468-0432.2010.00523.x

Schrijvers, J. (1993). Motherhood experienced and conceptualised: Changing images in Sri Lanka and the Netherlands. In D. Bell, P. Caplan, & W. J. Karim (Eds.). *Gendered fields: Women, men and ethnography* (pp. 143–158). London: Routledge.

Shilling, C. (1991). Educating the body: Physical capital and the production of social inequalities. *Sociology, 25*, 653–672. doi:10.1177/0038038591025004006

Shilling, C. (2017). Physical capital and situated action: a new direction for corporeal sociology. In M. D. Giardina & M. K. Donnelly (Eds.), *Physical culture, ethnography and the body* (pp. 47–62). London: Routledge.

Simpson, L. (2001). Aboriginal peoples and knowledge: Decolonizing our processes. *The Canadian Journal of Native Studies, 21*, 137–148.

Skeggs, B. (1997). *Formations of class & gender: Becoming respectable*. London: Sage.

Skeggs, B. (2004). Context and background: Pierre Bourdieu's analysis of class, gender and sexuality. *The Sociological Review, 52*(2_suppl), 19–33. doi:10.1111/j.1467-954X.2005.00522.x

Stewart, E. J. (2009). Resident Attitudes toward Tourism: Community-based cases from Arctic Canada. Doctor of Philosophy (Geography) University of Calgary, Alberta, Canada.

Stewart, E. J. (2018). Icebreaker: Experiences of conducting fieldwork in Arctic Canada with my infant son. In B. A. Porter & H. A. Schänzel (Eds.), *Femininities in the field: Tourism and transdisciplinary research* (pp. 154–169). Bristol: Channel View Publications.

Stewart, E. J., & Draper, D. (2009). Reporting back research findings: A case study of community-based tourism research in northern Canada. *Journal of Ecotourism, 8*, 128–143. doi:10.1080/14724040802696023

Strauss, A. L., & Corbin, J. (1998) *Basics of qualitative research: Grounded theory procedures and techniques* (2nd ed.). Thousand Oaks, CA: Sage.

Taylor, R. R., Jason, L. A., Keys, C. B., Suarez-Balcazar, Y., Davis, M. I., Durlak, J. A., & Holtz Isenberg, D. (2004). Capturing theory and methodology in participatory research. In L. A. Jason, C. B. Keys, Y. Suarez-Balcazar, R. R. Taylor & M. I. Davis (Eds), *Participatory community research: Theories and methods in action* (pp. 3–14). Washington, DC: American Psychological Association.

Tuhiwai Smith, L. (1999). *Decolonizing methodologies: Research and indigenous peoples*: Zed books.

Wachowich, N. (1999). *Saqiyuq: Stories from the lives of three Inuit women*. Montreal: McGill-Queen's University Press.

Warren, S., Vincent, C. (2001). "This won't take long ... ": Interviewing, ethics and diversity. *International Journal of Qualitative Studies in Education, 14*, 39–53. doi:10.1080/09518390010007674

Social constructionism as a tool to maintain an advantage in tourism research

Leszek Butowski, Jacek Kaczmarek, Joanna Kowalczyk-Anioł and Ewa Szafrańska

ABSTRACT

Social constructionism is commonly understood as one of qualitative approaches in the social sciences. It takes the ontological viewpoint that the meaning of the social world cannot be discovered, but is rather constructed by humans along with their histories, societies, ideas and languages. Historically, social constructionism is anchored in Kuhn's idea of the paradigm, but its foundations can be traced back to Kant's philosophy of knowledge and his rejection of the naturalistic 'tabula rasa'. Social constructionism as the main idea of 'non-classical sociology of science' has assumed relativity of knowledge, where knowledge is seen primarily as a social product. Thereby, it acknowledges the role of various perspectives created in different social circumstances. In tourism research, social constructionism started to develop in the 1990s and has become one of the dominant approaches among qualitative methodologies in the recent years. Paradoxically, social constructionism, which is based on the pluralism of perspectives, can also be used as a tool to maintain an advantage. This can be easily observed with respect to the production of knowledge on tourism. Analyzing the content of leading journals, we note that the dominant position is occupied (all too often) by one language, one tradition, one chosen set of research topics, and one group of scholars, who serve as the 'gate keepers' or 'power-brokers' of tourism knowledge. We find this problem important and believe that the described phenomenon can limit the freedom of research on tourism, which is illustrated through quantitative and qualitative analyses of the leading journals in terms of the represented traditions, languages, authors, and editors.

摘要

社会建构主义是社会科学中公认的一种定性研究方法。它的本体论的观点认为，社会世界的意义是无法被发现的，而是由人类及其历史、社会、思想和语言共同建构的。从历史上看，库恩的范式思想是社会建构主义的基石，但其基础可以追溯到康德的知识哲学及其对自然主义"白板"的排斥。社会建构主义作为"非经典科学社会学"的主要思想，假定了知识的相对性，认为知识主要是一种社会产物。因此，它承认在不同的社会环中创立的各种观点的作用。在旅游研究中，社会建构主义从20世纪90年代开始发展，近年来已成为定性方法论的主流研究方法之一。吊诡的是，以多元视角为基础的社会建构主义也可以作为一种维护利益的工具。这一点在旅游知识的生产方面很容易观察到。通过对主流期刊内容的分析，我们注意到，社会建构主义旅游研究中占据主导地位的往往是一种语言、一种传统、一套选定的研究主题，以及一群充当旅游知识"看门人"或"权力掮客"的学者。我们发现这个问题很重要，并且相信所描述的现象会限制旅游研究的自由。为了说明这一问题，我们将从所代表的传统、语言、作者、编辑等方面对主要期刊进行定量和定性分析。

In lieu of an introduction, or towards a new paradigm in the social sciences

At least a century old now, sociology of knowledge since its very beginning struggles with the problem of relativism. It can be seen in the main thesis put forward by its proponents – that the human cognition depends on the social and cultural environment in which it takes place. This problem, in itself related to ontology, has profound epistemological consequences and the ways in which it was addressed marked the distinctions between various schools of sociology of knowledge, starting with quite moderate views of Mannheim (1960) and Merton (1947, 1973) and ending with postulates of the strong program of sociology of knowledge (Barnes & Bloor, 1982; Bloor, 1976, 1984) or anarchism of science presented in the late works of Feyerband (1974). It is worth stressing that at least some supporters of the aforementioned schools attached key importance to ideas of Kant, who rejected the naturalistic view of the human mind as a 'tabula rasa'. Many of them referred also to the views of Kuhn (1962). This last author, introducing the notion of the 'scientific paradigm', has laid the theoretical foundation – in all probabilities in a way and range which he himself did not intend – for the development of contemporary post-modernist conceptions in social science. These conceptions seem to be in opposition to post-positivism and, in a broader context, modernism, understood here as a continuation of these currents which treat empirical facts and relations between them expressed in the formal language of mathematics and logic as the foundation of scientific cognition. On the other hand, many post-modernist conceptions address relations between human knowledge and power and have been considered by many thinkers from different (sometimes even opposing) perspectives, among others by Husserl's phenomenological method (Husserl, 1913, 1999), the theory of discourse by Foucault (1961, 1971, 1980), the theory of reproduction by Bourdieu (1973), the theory of commodification by Habermans (1978) or the theory of representation by Said (1994), as well as works of the Polish sociologist Zybertowicz (1995). Among them, a separate group of authors investigated relations between knowledge and power from the postcolonial and decolonial perspectives, including Smith (1999), Mignolo (2007), Maldonado-Torres (2007) and Grosfoguel (2007).

Still, all of them were inclined to believe that science is (or can be) a product and a tool of power and that violence, embodied in different ways, can play a significant role in the process of cognition and impact transformations of knowledge and communication thereof in public circulation.

At the same time, it appears that the connecting factor of otherwise internally diverse post-modern schools of the contemporary social sciences is the constructionist model of cognition (Machnikowski, 2011). It assumes that scientific knowledge is not a reflection of an independent reality or a product of application of historically fixed rules of scientific conduct, as the objectivist model of cognition would have it. On the contrary, constructionism sees development of science as a product of social activity. According to the advocates of this view, there is no way of knowing the world (not only social, but also natural) which would not be dependent on our culture and somehow mediated by it. Consequently, attainment of an objective truth, independent of local social conditions and conventions, is impossible. In this sense, the dualism of cognition becomes an empty term, because there are only cultural sciences through which the world of nature is also observed. Instead, attention is focused on the question: 'What have the historical, social, ideological and linguistic influences upon perceptions of the social world been?' (Moses & Knutsen, 2007).

On the other hand, the opponents of constructionism – its extreme post-modern version in particular – argue that it assumes that instead of talking about the truth it directs the discussion to a plethora of individual narrations or subjectivist interpretations, all of which are equally justified. Thus, the truth itself is taken to be just an arbitrarily established product of power – also power held by mass media (Szklarska, 2017). This line of criticism has been expressed sarcastically by the Polish methodologist Grobler (2006, pp. 274–275), who said that according to the supporters of constructionism 'the true is not that which is true, but rather that which is seen as true according to the norms of the given culture' and that 'science does not discover facts, but constructs them; nor does it investigate the reality, but creates it'.

In the methodological context, social constructionism is undoubtedly rooted in the qualitative model of scientific cognition, which uses hermeneutical methods of reasoning. This fact has prompted us to address the issue of the role of the constructionist paradigm in the contemporary tourism research and to submit this text to the Special Issue in Tourism Geographies Journal on Creative and Disruptive Methodologies in Tourism Studies. Special attention has been paid to the danger this approach poses in the context of the domination of one (Western – mostly Anglo-American) tradition, understood not only in the linguistic and geographical, but primarily in the economic, political and cultural-historical sense. This domination is expressed mainly through the neo-liberal ideology or, at least, the neo-liberal political rationality, which 'casts the political and social spheres both as appropriately dominated by market concerns and as themselves organized by market rationality' (Brown, 2006, p. 694). It is worthy of note that critical analysis of [the impact of?] the neoliberal ideology on the contemporary society has been carried out both from the perspective of Catholic social teaching and using the neo-Marxist approach. The former 'school' is represented, primarily, in such source documents as encyclicals by the last popes, in particular: *Laborem exercens* by John Paul II (1981), *Caritas in veritate* by Benedict XVI (2009) together with the document published by the Pontifical Council for Justice and Peace entitled "Towards reforming the International Financial" (2011), or the apostolic exhortation *Evangelii gaudium* (Francis, 2013) and the encyclical *Laudato si* (Francis, 2015) by Pope Francis.

The opposing pole of the critique of neoliberalism features publications in the neo-Marxist 'school' of critical theory. They are often based on the classical (for the proponents of this approach) views of Foucault (1975, 2001, 2004a, 2004b). These were used, among others, by Lazzarato (2009, p. 109), who offered a critique of the neoliberal ideology, reconstructing the process of transformation of the society into an 'enterprise society' based on 'the market, competition, inequality, the privilege of the individual, and a decisive role of financialization, as a key apparatus in achieving this transformation'.

Taking into consideration the characteristics of neoliberalism and its specific expansionism, it is clear that this ideology had to affect the way of functioning of the Western academic systems and institutions (and those from outside the 'Western World' which model themselves after them); in consequence, it was also bound to dominate the process of knowledge production and dissemination (which will be discussed below).

Now, returning to social constructionism – according to its assumptions, truth is but a specific cultural convention, in particular in social research, including the study of tourism. Thus, it can be easily imposed by those in power and claimed valid, regardless of its objectivity or lack thereof. In this context, the questions posed by Said (1994, p. 88): '[H]ow does one speak the truth? What truth? For whom and where?' Remain as valid as ever. This issue is particularly important when re-thinking the role of the neoliberal ideology as the foundation of social, economic and political systems, but also as the basis for (tourism) knowledge production.

This brings obvious threats, which can be boiled down to the problem of lost epistemological opportunities; we devote one section of this article solely to this problem. This loss is related to the ousting or non-admittance of researchers from different schools, including from countries such as Germany, France, Poland as well as Spain together with the entire Latin American legacy, which all have long histories of tourism research, to the broad (global) scientific circulation, unless they accept the dominant convention – or, to use Kuhn's name, the paradigm in force. In order to present works of researchers from outside the Anglo-American tradition, we have sought to include many such publications in the references, also in languages other than English.

Obviously, the aforementioned phenomenon does not concern tourism research only. It has been identified and well documented for other social sciences and humanities, notably, in the decolonial and postcolonial academic discourses, also with respect to socio-geographical studies – to remain within the main field of interest of the 'Tourism Geographies' journal. Given the profile of the journal to which this work is submitted, these two areas of study i.e. human geography and tourism have been analyzed in detail. The focal point throughout has been the Anglo-American (which belongs to the Western tradition, but does not exhaust it) domination in the fields of geography and tourism, which is manifested in the significant advantage of authors representing this cultural tradition, dominated characteristically by the neo-liberal ideology and the English language. Even though in this case language is just a tool, as Grosfoguel (2007, p. 217) rightly notes, 'a linguistic hierarchy between European languages and non-European languages privileges communication and knowledge/theoretical production'; thus, it contributes significantly to the establishment of a linguistic and institutional monopoly over the international publishing spaces.

Our study covered both selected bibliometric aspects of the leading journals and more qualitative analyses, including their theoretical backgrounds and contents, the functioning of the editorial boards and the role of scholars from the 'peripheries'. We have also consulted results of quite numerous studies in this field, in particular with respect to socio-economic (human) geography.

Anglo-American domination in geography and tourism research

At the very beginning, it should be noted that the Anglo-American tradition (and domination), as it is understood in this article, belongs to a wider Western tradition, but does not exhaust it completely. Conversely, one can hypothesize that the strength of the Anglo-American tradition contributes to the weakening or even disappearance of other Western achievements in human geography and, mostly, in tourism research.

The issue of the Anglo-American domination in geography and its consequences for the development of this discipline has – for at least several decades – been studied by researchers mainly from outside this historical-cultural and linguistic environment (Bański & Ferenc, 2012; Guttierez & Lopez-Nieva, 2001; Rodríguez-Pose, 2006). The studies conducted have often been set in the school of critical geography and referred to globalization processes and academic capitalism. The topic has been addressed, among others, by Bajerski (2010), Berg (2004), Desbiens (2002), Garcia-Ramon (2004), Gregson, Simonsen, and Vaiou (2003), Kitchin (2005), Kitchin and Fuller (2003), Olds (2001), Paasi (2005, 2015), Potter (2001), Short, Boniche, Kim, and Li (2001), Slater (1992). The research, which started in the 1980s (Foster, Muellerleile, Olds, & Peck, 2007) and intensified in the following decades (Garcia-Ramon, 2003; Harris, 2001), suggests unambiguously that such domination has persisted for at least several decades. Evidence can be found in the results of bibliometric studies of articles published in the leading geographic journals. For example, Bański and Ferenc (2012) have shown that in the period 2000–2010 in six selected geographic journals included in the database of the Institute for Scientific Information (ISI), British scholars published 40% of the articles and Americans – another 32%. Altogether, authors from English-speaking countries wrote in the analyzed period as much as 88% of the published works. Similar results, pointing unequivocally to the persisting domination of the Anglo-American tradition, are reported also by other scholars. Short et al. (2001) studied the percentage of authors from outside the Anglo-American environment in the 16 most frequently cited journals in the field of socio-economic geography in the period 1981–1996. It amounted to only 6.3%. Guttiérez and López-Nieva (2001) calculated the percentage of authors from individual countries in the 19 leading geographic journals indexed by the ISI in the period 1991–1997. It turned out that 74% of the articles were written by scholars from the USA and UK, while another 13% were produced by geographers from Canada, Australia and New Zealand. Authors from non-English speaking countries accounted only for 13% of the works, with French and German scholars publishing only 0.5% each. Even bigger domination of Anglo-American geographers was observed in the editorial boards of the analyzed journals (as of 1999, it exceeded 90%). Similar conclusions are reached in tourism geographies by Gibson

(2008), who believes that over 80% of articles published in so-called international journals come from English-speaking countries (after: Viken, 2013).

The quantitative data concerning the several decades-long persistence of the Anglo-American domination over geography, in particular socio-economic (human) geography and tourism geography, leave no room for doubt. The discussion can now turn to its causes and consequences. According to Minca (2000), the domination of the Anglo-American tradition is reinforced by the way of functioning of editorial boards of international journals and appointed reviewers. Minca points to the common practice of rejecting articles whose theoretical backgrounds and bibliographies do not fit in the accepted pattern. This forces geographer from outside the Anglo-American environment to refer to the debate held currently therein, often at the expense of their own traditions. Aalbers (2004), on the other hand, stresses the fact that in order to publish in prestigious journals, one has to cite primarily works by British and American writers, which increases the number of their citations. Another, particularly discriminative practice is described by Bajerski (2010), Kitchin (2005), Meriläinen, Tienari, Thomas, and Davies (2008) and Simonsen (2004), who state that authors from outside the Anglo-American world are expected to submit descriptive-exploratory articles and not theoretical-methodological ones. In practice, this often means that a scholar from the 'peripheries' is included as a coauthor and assigned the task of presenting at most a case study. On the other hand, one should also consider the opinion of Rodriguez-Pose (2006), who claims that the reason for the Anglo-American domination lies also in the organization of science outside this circle, where it is characterized by rigidity, nepotism, hierarchical character, pauperization of scholars and low outlays on science.

The hegemony of the Western, mostly Anglo-American, tradition in tourism research is also conspicuous – recently it has been addressed, among others, by Chambers and Buzinde (2015) and Wijesinghe, Mura, and Bouchon (2017) – although perhaps not as clearly documented as in the case of geography. Nevertheless, many authors have pointed to the persistence of such imbalance in specific disciplines dealing with tourism or tourism research as a whole; apart from the aforementioned works of Gibson (2008) and Viken (2013), one should also mention a publication by Dann and Liebman Parrinello (2009), who found out that 97% of the citations in the 'Annals of Tourism Research' were from English-language sources. Hall, Williams, and Lew (2004) demonstrated that 77% of the editors of journals dealing with tourism were affiliated with American, British, Australian, Canadian and New Zealand universities. Another text by Dann (2011) contains great bibliometric evidence showing the decades-long domination of the Anglo-American tradition in tourism-related studies published in international journals. At the same time, the author argues that the contemporary academic investigation of tourism owes much to researchers from outside the Anglo-American circle, in particular from Germany and France.

The thesis about the persisting hegemony of the Anglo-American world in tourism research has also been documented in empirical research conducted specifically for this article. It was based on the analysis of international journals and articles published therein. The journals have been selected using two criteria; the first one was related to the journal and concerned its presence on the ISI list and having an Impact Factor above 1 in 2016; the second criterion was related to topics addressed by the journal; thus, the analysis covered all journals whose titles included words such as: Tourism and/or Hospitality and/or Travel. The study was carried out for the year 2017 and covered all articles published in these journals, 1075 texts in total. Due to the fact that the majority of the articles had more than one author, the analysis took into consideration not the articles, but their 2933 authors (the number is bigger than the actual number of authors because each instance of the same author appearing was counted anew), considering the first affiliation declared by the author to the editorial office of the journal; please note that for the purposes of this study, we have adopted the authors' institutional affiliation as the main indicator of participation in a given social, cultural and, in consequence, also ideological tradition of knowledge production[1]. The combined results of the study are presented in Tables 1 and 2. The juxtaposition of the countries where the authors of the articles are affiliated reveals unanimous domination of English-speaking countries, in particular the USA and either UK or Australia, depending on the journal. In six out of seven best journals with the impact factor above 2.5, the total percentage of authors from English-speaking countries (USA, UK, Canada, New Zealand, Australia, Ireland) exceeded 50%, reaching even more – over 65% – in the 'Annals of Tourism Research'. It is worthy of note that Editors-in-Chief of these journals represent almost exclusively English-speaking countries, mostly USA, UK, NZ, Australia. One exception is a journal with the geographic indication in its very title, namely 'Asia Pacific Journal of Tourism Research' with its office in Hong Kong.

Elaborating on this analysis, one can note that all top tourism journals are published by major Western publishing houses, including above all: Taylor and Francis (4), Elsevier (3), Sage (3), Emerald (1), and Wiley (1); access (also on-line) to their contents and, consequently, the currently established canon of knowledge, is limited and paid (which definitely constitutes a significant barrier for many potential readers). This

Table 1. Percentage of authors from English-speaking countries among authors of articles in journals from the field of tourism studies with IF $\geq$1 in 2017 – in total.

Journal name	IF 2016	% of authors from English-speaking countries*	Including:		
			USA [%]	UK [%]	Australia [%]
Tourism Management	4.707	41.2	18.4	10.7	10.3
Journal of Travel Research	4.564	54.4	23.5	12.4	16.4
Annals of Tourism Research	3.194	65.6	10.0	37.5	10.6
Journal of Sustainable Tourism	2.987	59.0	22.7	15.7	14.7
International Journal of Hospitality Management	2.787	52.3	42.2	4.5	3.8
Cornell Hospitality Quarterly	2.657	63.6	62.5	0.0	1.1
Journal of Hospitality & Tourism Research	2.646	58.0	45.0	0.0	12.0
Current Issues in Tourism	2.451	24.2	4.3	6.6	13.3
International Journal of Contemporary Hospitality Management	2.176	49.4	32.5	8.3	5.8
International Journal of Tourism Research	1.857	29.0	10.8	10.8	7.4
Tourism Geographies	1.663	22.8	7.6	3.3	6.5
Asia Pacific Journal of Tourism Research	1.051	13.7	6.6	1.1	5.5

*USA, UK, Canada, New Zealand, Australia, Ireland.
Source: Authors.

Table 2. Percentage of authors from Asia and Europe among authors of articles in journals from the field of tourism studies with IF ≥1 in 2017.

Journal name	IF 2016	Asia (excluding China and Taiwan) [%]	China [%]	Taiwan [%]	Europe* (excluding UK and Ireland) [%]
Tourism Management	4.707	15.0	14.7	3.0	22.2
Journal of Travel Research	4.564	9.3	7.1	0.9	11.9
Annals of Tourism Research	3.194	4.4	3.1	0.6	25.0
Journal of Sustainable Tourism	2.987	9.3	3.7	0.3	21.3
International Journal of Hospitality Management	2.787	9.8	9.8	4.5	17.4
Cornell Hospitality Quarterly	2.657	9,1	12.5	0.0	14.8
Journal of Hospitality & Tourism Research	2.646	24.0	6.0	4,0	8.0
Current Issues in Tourism	2.451	23.7	8.1	9,5	27.5
International Journal of Contemporary Hospitality Management	2.176	16.1	10.3	6.1	11.9
International Journal of Tourism Research	1.857	18.2	4.0	1.7	29.0
Tourism Geographies	1.663	16.3	2.2	0.0	44.6
Asia Pacific Journal of Tourism Research	1.051	35.8	27.7	14.0	5.9

* Considering cultural criteria, Turkey and Israel were counted among European countries.
Source: Authors.

evidently is one of the manifestations of the absolute hegemony of giant Western-based and directed publishing companies (houses), which exploit their monopolistic position to ruthlessly impose their, primarily Anglo-American, Western-centric, and mostly neo-liberal vision of knowledge production (Cohen, Cohen, & King, 2018; Hall, 2013).

Furthermore, obtaining 'academic legitimacy' by potential authors who wish to publish in the most frequently cited journals requires them to undergo some sort of acculturation, i.e. to submit themselves to the rules applicable there, often at the expense of giving up their own concepts stemming from other traditions. This particularly dangerous phenomenon (present also in journals from the field of human geography) is also noticed, i.e, by Tribe (2010).

Another author who draws attention to another consequence of the monopolization of the publishing market of academic tourism journals is McKercher (2015). He noticed that in the course of the last three decades, the publishing landscape has changed in the direction of its domination by a relatively small group of journals with the biggest IF. This conclusion follows, among others, from the analysis of the influence ratio, which reveals that the 'top tier journals generate a disproportionately higher share of citations per paper published' (McKercher, 2015, p. 307). Hall and Page (2015, p. 310), on the other hand, stress that '(t)he growth in the significance attached to the IF cannot be isolated from the rapid expansion and commercialization of academic publishing'. They go on to say that '(t)he present audit culture and systems of academic surveillance (…) combine the neoliberal economic imperatives of competition and academic entrepreneurship with the technologies of public management and publishing'.

These opinions are supported by Ayikoru, Tribe, and Airey (2009), who notice that with the extension of neoliberal policies and managerialism to higher education institutions, the pressure of the 'metrification' of scientific production has reinforced journal metrics (in particular, the IF) as a principal indicator of scientific achievements (see also: Perdue, 2015).

Moreover, one can also mention the conclusions reached by Cohen et al. (2018, p. 2035), who argue that, 'as the Western publication regime is becoming increasingly globalized, local intellectual voices tend to be silenced by the regime's growing hegemony, and, call for the integration of non-Western perspectives into the framework of the publication regime'.

Yet another aspect of the debate concerning the framing of tourism knowledge within the neoliberal publication regime (or, to use Tribe's (2018) term, 'creating and curating' of tourism knowledge within neoliberal systems) is brought up by Dredge and Jamal (2015), Airey, Tribe, Benckendorff, and Xiao (2015) and Tribe (2006); these authors mention the problem of '(t)he influence of neoliberal economic management on what type of knowledge is generated', indicating threats connected with this situation. On the other hand, Berg, Huijbens, and Larsen (2016), commenting on the functioning of many Western universities, point out audit systems employed there, which, in their view, are key mechanisms of neoliberalization of the academy.

In this context, one may be tempted to cite also the call by the proponents of the 'critical' theory in tourism studies, like Ateljevic, Harris, Wilson, and Collins (2005, p.13), which, admittedly, was made from the perspective of social sciences and 'cultural turn': to 'unmask the cultural politics and knowledge-making processes of academia'. This call fits in the debate over power and economic relations in tourism as well as power and tourism knowledge production and dissemination.

Importantly, a relatively high percentage of authors in the analyzed journals were also writers from Asian countries, primarily from China (close to 15% in the journal with the biggest IF), and, to a slightly lesser extent, from Taiwan, South Korea and Hong Kong. Other listed countries included Malaysia and Singapore. Arguably, this phenomenon could, to a large extent, be explained by the growing advancement of the processes of globalization, Americanization and neo-liberal ideologization of culture (including science), which, since the middle of the 20th century, have been noted in Asian countries (Haney, 2008). In the field of tourism studies, attention to this phenomenon is drawn, among others, by Tucker and Zhang (2016) and Wijesinghe, Mura, and Culala (2019). These authors offer a discussion of Eurocentric and capitalist (neoliberal) academic structures of power that have been influencing and molding the next generation of Asian tourism scholars. Their work aims to highlight the ways in which Eurocentric ideologies, consciously or un/sub-consciously still shape the minds of non-Western academics.

Equally worthy of notice is the fact that European countries, excluding the UK, were represented in the analyzed journals rather poorly (among those with the highest IF ≥ 2.5, the percentage was no more than 25%). One exception is 'Tourism Geographies', which in 2017, as the only journal included in the analysis, gathered nearly 45% of European non-English countries. This observation is particularly striking given the historical contribution of European countries (Germany, France, Poland, Czech Republic, Italy) to tourism research. The only non-English country whose authors clearly marked their presence (although they never dominated) was Spain. Authors from Spain were especially noticeable in journals such as 'Tourism Management', 'International Journal of Hospitality Management', 'Current Issues in Tourism' and 'International Journal of Tourism Research'. This fact does not appear to be related to the language, as the same cannot be said about other Spanish-speaking countries, which were hardly represented in the analyzed journals in 2017. Even less frequent were contributions from African countries (with the specific exception of the South African Republic).

Therefore, to quote Hall (2013, p. 608), the English language is a 'part of the 'ideological complex' that produces and maintains the increasing hegemony of the English speaking academy'. On the basis of a systematic critical review of the literature, Wijesinghe et al. (2017) conclude that despite the several epistemological and methodological turns in tourism, the observable institutional and linguistic monopolization of the international publishing spaces (with Western/Anglo-centric standards) is still an acceptable norm worldwide. In the *status quo*, they discern a new form of colonialism evinced, among others, in neo-colonial academic structures. For Chambers and Buzinde (2015), the so-called top journals are based on the Western ideologies and standardizations – in relation to both the paradigms, and the methods applied.

Constructionist approach in tourism research

According to many scholars, constructionism is one of the dominant methodological schools in the current study of tourism (Botteril & Platenkamp, 2012; Hollinshead, 2004, 2006; Jamal & Everett, 2004; Tribe, 2006, 2008), in particular in the context of the qualitative approach (Hollinshead & Jamal, 2007; Westwood, Morgan, & Pritchard, 2006). Its detailed methodological characteristic is presented by Pernecky (2012). This author has offered a careful discussion of the opportunities connected with the use of the constructionist approach in tourism research, pointing to both its advantages and disadvantages. He also ventured a controversial claim that this school should be both realist and relativist at the same time, which provoked strong protest of Botteril (2014), who declared himself a critical realist. Referring to the works in the field of contemporary sociology of knowledge, Pernecky (2012) in his analysis distinguishes between strong constructionism and weak constructionism. The former, being close to the strong program of sociology of knowledge, is considered by this author to be rather impractical in tourism research and akin to philosophical solipsism. Weak constructionism, however, is pronounced very important and underused in the contemporary tourism research. Pernecky stresses the opportunities connected with this school, stating that tourism as an activity of contemporary societies is nothing other than constructionism in action, as evidenced by socially constructed meanings of things (e.g. airline tickets, souvenirs), places (e.g. tourism destinations, resorts) or social roles (e.g. host, guest, tour guide). The view is shared, among others, by Ren, Pritchard, and Morgan (2010, p. 885), who claim that constructionism enables 'conceptualisation of tourisms that encompass multiple worldviews and cultural differences as well as research praxis that recognizes and reflects the plurality of multiple positions, practices and insights'. Pritchard, Morgan, and Ateljevic (2011), the inventors of the term 'hopeful tourism', go one step further to state that the task of tourism researchers is to 'disturb and critique hegemonic socio-political practices', which is indeed very close to the critical theory understood as the need for radical emancipatory change (Kołakowski, 1978).

Another interesting perspective is offered by Tribe (2006) in a seminal text written from the constructionist viewpoint entitled 'The Truth about Tourism'. The author attempts to apply the force-field conception of Lewin (1935) to understand the factors which resist and promote truth creation in tourism. Here, the notion of the 'force-field' is understood as a tool describing those factors which mediate in the process where the phenomenal world of tourism is translated into its known world, in accordance with the diagram (Tribe, 2006, p. 362):

Tourism phenomenon ⟺ knowledge force-field ⟺ tourism knowledge

The author analyses five factors constituting the knowledge force-field: person, position, rules, ends, ideology. Among these, position and rules deserve particular attention, given the issue addressed in this article. The former factor determines the position of the researchers both in the geographic sense (country) and in relation to the academic community or even the wider linguistic and cultural community. It appears that in this context the notion of ethnocentrism is of key importance; it denotes an attitude in which the interests of one's own group are put in the first place and those of others are discriminated against. Interestingly, in the Anglo-American literature, this notion is usually referred to in the neocolonial perspective (Berno & Butler, 1996; Dyer, Aberdeen, & Schuler, 2003; Hollinshead, 1999; Whittaker, 1999). Another, equally important issue in the context of the position factor is mentioned by Becher (1989). This author, pointing to academic communities, touches upon to problem of 'academic tribes' and roles they play in the creation of behavioral patterns, rules, norms and hierarchies. In practice, this translates directly to who holds chairs, editorships of key journals, seats on panels, keynote invitations, memberships in assessment bodies, etc. Undoubtedly, on the international level, as far as tourism research is concerned, there is a huge dominance of scholars from the Anglo-American tradition, whom McKarcher (2002) dubs the 'power-brokers'. A similar opinion is formulated by Hall (2004, p. 143), who claims that 'in tourism there is very little discussion of the role that … gate-keepers have in determining the scope and direction of tourism knowledge'. It is noteworthy that this author is one of the most prominent scholars who keep those chairs, editorships of key journals, seats on panels, keynote invitations, memberships in assessment bodies.

As regards the other factor of the force-field (listed by Tribe), namely the rules, special attention should be paid to the role of discourse in Foucaludian meaning (Foucault, 1971, 1980), which has been determined by Hall (1997, p. 44) as 'a group of statements which provide a language for talking about … a particular topic at a particular historical moment'. It is discourse which determines what counts as knowledge and what does not – obviously, in accordance with the interests of the strongest group involved.

Finally, one should also underline the role of ideology, in particular in the context of the power of neoliberal ideology often backed by liberal democracy, both of which have dominated the majority of contemporary academic systems and institutions and promote certain ideas, values and methodological orientations at the expense of

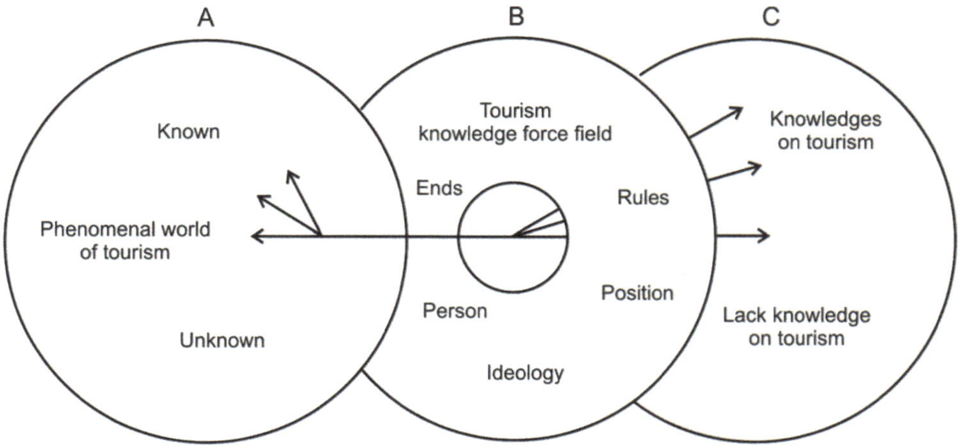

Figure 1. Dominant position of the Anglo-American tradition in tourism research according to the knowledge force-field conception.
Source: Authors based on Tribe (2006, p. 363).

others. This can lead to the limitation of the research field and interpretation of results in the light of the adopted ideology.

In the conclusion, Tribe notes that 'tourism knowledge cannot provide a faithful representation of reality' and adds in the justification that the canon of knowledge on tourism is 'heavily contingent on the power of those who speak for tourism, their spatial and temporal situatedness and the social practices that sustain their position and authority' (Tribe, 2006, p. 376). We hold this view dear.

Using the concept of the force-field and Tribe's (2006, p. 363) diagram, which shows the role of the tourism knowledge force-field in the process of tourism knowledge production (Figure 1), one can outline the current situation in tourism research, pointing to the dominant position of the Anglo-American tradition. It can be perceived as a specific force-field (Figure 1(B)), which determines and transforms (using various factors, with a particular role of the position and rules – discourse) the phenomenal world of tourism (Figure 1(A)) into tourism knowledges (Figure 1(C)); these are, of course, social constructs characterized by certain traits (Figure 1).

The constructionist approach, as an analytical tool and a method of reasoning, was also applied by Butowski in his studies devoted to disciplinarity and non-disciplinarity of tourism research (studies), the results of which were published in a series of articles (Butowski, 2014a, 2014b, 2016a, 2016b).

Cost of lost epistemological opportunities

The conducted analysis has allowed us to make a diagnosis of the contemporary tourism research. Undoubtedly, in tourism research we are dealing with an institutional, geographic, linguistic and ideological advantage of one group over the others. Let us go beyond the mere diagnosis – the next step is necessary. We have confirmed the suggestions about one viewpoint maintaining domination in terms of the language, institutions and methods. First, we look at the world through 'epistemological glasses' – shaped in a particular way in sociological and cultural terms. We obtain a mediated image of the world in the form of accepted and negotiated social constructs. The structure of thinking has already been established. Then, we try to understand the described reality, but we adjust our activity to the already established schemata. It should be added that these methodical structures have already grown old. Like a spell – we keep citing similar publications and often the same authors. Let's refer to Wittgenstein's (2001) opus – 'Tractatus Logico-Philosophicus', marked by the author with the inscription 'Vienna 1918':

> 6.54 My propositions serve as elucidations in the following way: anyone who understands me eventually recognizes them as nonsensical, when he has used them-as steps-to climb up beyond them. (He must, so to speak, throw away the ladder after he has climbed up it.). He must transcend these propositions, and then he will see the world aright.

Referring to the statement of the famous philosopher (made in a different context, to be sure), let us, for the sake of argument, take our knowledge of works in the field of tourism research for granted. Now throw the ladder which has helped us climb the fortifications of knowledge of tourism phenomena away. What about authenticity, if we assume that authenticity is just a fact following from the existence of something or someone. It is enough for it to be authentic. Such examples are numerous. We just need to go beyond the safe institutional (and, therefore, epistemological) walls and enter those geographic, epistemological, grammatical areas from which the academic tourist establishment has separated itself. This separation has brought, most likely, a safe academic life, but it has led to loss in terms of our understanding of the world of tourism phenomena. In our discussion, we call this phenomenon the *cost of lost epistemological opportunities*. Let us once again quote Wittgenstein (2001, Preface):

> I do not wish to judge how far my efforts coincide with those of other philosophers. Indeed, what I have written here makes no claim to novelty in detail, and the reason why I give no sources is that it is a matter of indifference to me whether the thoughts that I have had have been anticipated by someone else.

Let us listen to the sound of the grammar of our thoughts and let us not be bothered by what has been cited over and over again. It surely is important, but it brings no novel, original thoughts to our view and understanding of the world. Repeatable citations of literature bring about geometric increase in the number of citations, but usually they do not have an impact on the qualitative development of knowledge in tourism research. Commonly acknowledged works contain discussions of prominent theoretical concepts or present results of empirical studies. But we are dealing with separate, isolated research areas. They are significant on the map of achievements of scholars of tourism-related phenomena, but they do not construct a system of knowledge. Researchers pursue trains of thought which are isolated in the space of knowledge. Familiarity with works of others does not lead to the integration of knowledge. These are the contexts of the concept of lost epistemological opportunities. We would have achieved much more in our investigation of tourism phenomena, if we accepted wanderers from other fields of cognition without fear. The ubiquitous English language is indispensable in universal scientific communication. It undoubtedly is an effective and available tool of communication. But the undeniable advantages of the communicativeness of the English language deprive us of important linguistic sensitivity, which is responsiveness to ontological differentiation. There is no way to experience the uniqueness of the streets of Lisbon other than to immerse yourself in the nostalgic sound of Portuguese. Forgetting about everything while dancing salsa on the streets of Habana Vieja would be nothing like Cuba without shouts in Spanish. For these reasons, the English form of the scientific message leaves out other worlds with their unique sounds on the margin, both in terms of their natural existence and cognition – and, ultimately, explanation and understanding in analyses of tourism phenomena. One effect of such thinking and conduct is the relatively small number of genuine works written in the field of humanities. The grammar of English expresses well the meaning of economic, political, marketing and social phenomena. But capturing the individuality of the subject cannot be taken to mean only producing a homogenous scientific message.

Other costs of lost epistemological opportunities are related to limited opportunities of comparative studies. A comparison of diverse processes usually can cover only areas which attract the attention of well-situated researchers. Lost opportunities include also omission of other fields in the procedure of falsification of various hypotheses. What lesson are we to learn? The true meaning of science is to discover ontological diversity and to explain it and, in the context of the humanities – to understand it. Limiting oneself to the circle of acknowledged works, correct institutions and one grammar leads the academic world to build an epistemological tower – with decreasing genetic diversity and increasing cognitive entropy.

In our discussion, we should return to the valuable French term of *terroir*. It allows us to see what is important and where it comes from. In this context, the place can be the source of all ontology and all cognition. It determines the language of the researcher and the logic of his or her thinking. *Terroir* should be counted among lost opportunities. Global thinking and uniform grammatical structure flatten the world and reject diversities of the studied realities. *Terroir* determines the meaning of the place, it exposes the diversities.

In this article, we suggest considering the meaning of social constructionism and emphasizing the role of thinking originating in the place understood as a specific geographic location. We believe that the methodological foundation for tourism research can be found in topoconstructionism (geoconstructionism), which in our opinion is close to the phenomenological approach, both in the naturalist view of Husserl and in Heidegger's hermeneutics (Buczyńska-Garewicz, 2006). It is here that we see the costs of lost epistemological opportunities, i.e. how much we have lost by leaving out the diversity of places (with all their cultural-linguistic attributes and bodies of academic work) on the margin of the globalist Anglo-American attitudes. Mercier (2004), when traveling to foreign countries (cities), first visited bookstores. He glimpsed through books written in languages he did not know and could not read. This was his first impression of a foreign country, this was how he came to know the place. He cherished the *terroir*. Let us, then, return to the place by practicing topoconstructionism.

In his the latest paper titled 'The two vistas' Hägerstrand (2004, p. 315) places a human being in between diverse horizons of our cognition:

> We human have two vistas towards which we can turn our eyes. One consists of the heavens and the other the neighbouring landscape all the way to the horizon. In the sky above, the sun, the moon, the planets regularly pass by against a backcloth of seemingly fixed stars. The view towards the heavens exhibits stability and regularity. By contrast, the most characteristic trait of the view of the earth is the immense wealth of forms and

> complexities through which matter spontaneously manifest itself. The paraphernalia of human culture displays similar variation. The surrounding landscape meets us as a remarkable mixture of natural and manmade things, unpredictable life form, conflicts and compromises. The part of the world enclosed with the skin of the observer also belongs to the landscape. It can partly be reached through a perspective eye, but it remains more concealed than conspicuous. Difficulties comprehending that nearby us seem to stand in contrast to the clarity of what is distant. The special features of these two vistas have ... had a huge impact on scientific thinking.

Geography seeks to explain and understand spatial relations in nature and in the social environment. In general qualitative terms, (tourism) geography can be described as 'the art of reading the space'. To a geographer but also, to a certain extent, to a tourist, geographic environment is a book which he/she is trying to read, its content, the meanings, the symbols and relations between the events. In order to read the space one needs intuition, going beyond existing schemes. Such a thinking, understood in a qualitative topoconstructionistic context, can be a key to explain the world in its whole diversity and, consequently, to understand the place occupied by humans. Therefore, we need topocostuctionism. This should be the window into the world of human cognition.

Summary and closing remarks

Summing our discussion up, it appears to us that the reasons for the decades-long persistence of the Anglo-American domination in the social sciences, and in particular in geography and tourism research, reported by many different authors do not get to the heart of the matter. These analyses are rather superficial and point to certain mechanisms which at most contribute to the imbalance (or result from it) but not to the fundamental factors which determine this state of affairs. We believe that the true causes are deeper and related to mutual relations of science and power (understood as political, economic and cultural domination), addressed in the literature most frequently from Neo-Marxist positions (Foucault, 1961, 1980; Gramscii, 1971). The way in which Zybertowicz (1995), a Polish sociologist, addressed this problem from a perspective different than the Neo-Marxist one deserves particular attention; he adopted the constructionist view to highlight the relations between knowledge and power without resorting to Marx's theory (which he, admittedly, had used in his early works, but later considered, together with the positivist thought which it was supposed to criticize, a product the Enlightenment (Domarecki, 2017)). Zybertowicz's ultimate position, which is close to ours, assumes that science as a cultural product can be successfully used to gain and maintain power: one only has to impose a certain cultural pattern (most often together with domination in the economic and military spheres). The consequences of this state of affairs are well expressed by Viken (2013, pp. 38–39):

> [t]he power system in within international academia is such that what matters is not only what is written and by whom it is written but from where the authors and articles originate, and particularly in what journals or books things are published.

A corresponding stance can be found in the notion of 'writing spaces' introduced by Gregson et al. (2003, p. 8), meaning 'hierarchies of power, which position some journals as mattering rather more than others ... capturing, controlling and regulating the international (read theoretical and/or conceptual) high ground, its cutting edges'.

We are well aware of the paradox underlying the reasoning applied. With this article, we are making an attempt to demonstrate that the relativist approach, which rejects 'objectivity' and declaratively assumes a multitude of equally-justified scientific discourses, can lead to their ousting and domination of one convention imposed by the strong actors on the scientific stage. It is noteworthy that similar processes are likely even more noticeable in other spheres of social life. They are manifested in the domination of the neoliberal option, among others, in the global media, university curricula and political life of the 'cultural West'. The hypocrisy of the proponents of this approach lies in the fact that while declaring diversity of values, traditions and discourses, they at the same time try to impose their point of view as the only valid one. All of these seem to be possible once the notion of 'objectivity' as a point of relevance has been rejected.

As the authors of the this article, we are aware of the fact that the adopted assumption can only by treated as a hypothetical and not verifiable or – staying within the convention of the qualitative approach – credible, transferable, dependable and confirmable (Bryman, 2008) statement (if such matters are even relevant in the context of social meta-studies). At the same time, it should be stressed that it fits surprisingly well in the paradigmatic model of science development proposed by Kuhn (1962). It appears that, in line with this perspective (assuming Kuhn's model accurately explains the developmental processes of science) in tourism research we can point to the increasing role of the relativist paradigm, accepted, or rather imposed, by the Anglo-American tradition or, more broadly, by the Western neoliberal system of values. This does not, however, follow from an objective advantage of the internal logic of the studies (and their results) but is related to social external factors (economic, cultural domination, etc.).

In this context, we have decided that the threats inherent in the relativist-constructionist approach (in particular, in its extreme post-modernist version) should be clearly shown, especially as regards the lost epistemological opportunities. At the same time, we encourage searching for (or returning to) ontological and epistemological foundations which can be an alternative to the post-modernist approach in the qualitative tourism research. One such proposition can be a return to the phenomenological approach with one particular variety we call topoconstructionism (geoconstructionism) – which we see as particularly useful in tourism research. This could open broad perspectives to entirely new spaces, not only in the geographical sense, but also in the way of seeing the world and understanding it.

Surely, the postulate to weaken the dominant position of one tradition can lead to a loss of measurable benefits, which would affect some cliques currently enjoying monopoly over the international market of tourism research – the aforementioned 'gate keepers' and 'power-brokers' of tourism knowledge. This is because Hooks's (1990, p. 151–152) words are still valid:

> No need to hear your voice when I can talk about you better than you can speak about yourself. No need to hear your voice. Only tell me about your pain. I want to know your story. And then I will tell it back to you in a new way. Tell it back to you in such a way that it has become mine, my own. Re-writing you, I write myself anew. I am still author, authority.

This does not mean that we should remain silent. On the contrary, our integrity as researchers forces us to put forward the claim we have tried to substantiate in this short text. When working on it, we kept in mind Foucault's (1980, p. 85) assertion that 'where there is power there is resistance'.

Note

1. We are well aware of the limitations connected with this assumption. However, for the purposes of the article, we have assumed that it is sufficient to illustrate the nature of the phenomenon. Moreover, this aspect is usually omitted in bibliometric analyses of the literature. In our view, affiliation offers a good indicator of the interconnected linguistic and institutional hegemony/monopoly over the international publishing space.

Disclosure statement

No potential conflict of interest was reported by the authors.

References

Aalbers, M. (2004). Creative destruction through the Anglo-American hegemony: A non-Anglo-American view on publications, referees and language. *Area*, *36*(3), 319–322. doi:10.1111/j.0004-0894.2004.00229.x

Airey, D., Tribe, J., Benckendorff, P., & Xiao, H. (2015). The managerial gaze: The long tail of tourism education and research. *Journal of Travel Research*, *54*(2), 139–151. doi:10.1177/0047287514522877

Ateljevic, I., Harris, C., Wilson, E., & Collins, F. L. (2005). Getting 'entangled': Reflexivity and the 'critical turn'in tourism studies. *Tourism Recreation Research*, *30*(2), 9–21. doi:10.1080/02508281.2005.11081469

Ayikoru, M., Tribe, J., & Airey, D. (2009). Reading tourism education: Neoliberalism unveiled. *Annals of Tourism Research*, *36*(2), 191–221. doi:10.1016/j.annals.2008.11.001

Bajerski, A. (2010). Anglo-amerykańska dominacja w geografii: Główne wątki dyskusji prowadzonej w ramach geografii krytycznej [Anglo-American domination in geography. The main threads of a discussion conducted within the framework of critical geography]. *Przegląd Geograficzny*, *82*(2), 143–158. doi:10.7163/PrzG.2010.2.1

Bański, J., & Ferenc, M. (2012). Analiza międzynarodowych czasopism geograficznych ze szczególnym uwzględnieniem geografii społeczno-ekonomicznej [Analysis of international geographical journals with particular emphasis on socio-economic geography]. *Przegląd Geograficzny*, *84*(4), 617–638. doi:10.7163/PrzG.2012.4.7

Barnes, B., & Bloor, D. (1982). Relativism, rationalism and the sociology of knowledge. In M. Hollis & S. Lukes (Eds.), *Rationality and relativism* (pp. 21–47). Oxford: Basil Blackwell.

Becher, T. (1989). *Academic tribes and territories*. Buckingham: Open University Press.

Benedict XVI. (2009). Encyclical Letter Caritas in Veritate [On integral human development in charity and truth]. Vatican City, Benedict XVI.

Berg, L. D. (2004). Scaling knowledge: Towards a critical geography of critical geographies. *Geoforum*, *35*(5), 553–558. doi:10.1016/j.geoforum.2004.01.005

Berg, L. D., Huijbens, E. H., & Larsen, H. G. (2016). Producing anxiety in the neoliberal university. *The Canadian Geographer/Le Géographe Canadien*, *60*(2), 168–180. doi:10.1111/cag.12261

Berno, T., & Butler, R. (1996). Cross-cultural research method: Content or context? A cook Island example. In R. Butler & T. Hinch (Eds.), *Tourism and indigenous people* (pp. 376–395). London: International Thomson Business Press.

Bloor, D. (1976). *Knowledge and social imaginery*. London: Routledge & Kegan Paul.

Bloor, D. (1984). The strengths of the strong programme. In J. R. Brown (Ed.), *The Western Ontario series in philosophy of science: 25. Scientific rationality: The sociological turn* (pp.75–94). Dordrecht, The Netherlands: Springer.

Botteril, D. (2014). Constructionism—A critical realist reply. *Annals of Tourism Research*, *48*, 292–298.

Botteril, D., & Platenkamp, V. (2012). *Key concepts in tourism research*. Los Angeles, CA: Sage.

Bourdieu, P. (1973). Cultural reproduction and social reproduction. In R. Brown (Ed.), *Knowledge, education and social change* (pp. 71–112). London: Tavistock.

Brown, W. (2006). American nightmare: Neoliberalism, neoconservatism, and de-democratization. *Political Theory*, *34*(6), 690–714. doi:10.1177/0090591706293016

Bryman, A. (2008). *Social Research Methods*. New York: Oxford University Press.

Buczyńska-Garewicz, H. (2006). *Miejsca, strony, okolice. Przyczynek do fenomenologii przestrzeni [Places, sites, localities. A contribution in the phenomenology of space]*. Kraków: Universitas.

Butowski, L. (2014a). The ontological and epistemological foundations of tourism research. Scientific identity of Polish scholars of tourism. *Economic Problems of Tourism*, *4*(28), 23–38.

Butowski, L. (Ed.). (2014b). *In search of the scientific identity of tourism. Theoretical and methodological aspects of research*. Warszawa: Warsaw School of Tourism and Hospitality Management.

Butowski, L. (2016a). Tourism as a discipline in light of the non-classical sociology of science. *Journal of Tourism and Cultural Change*, *16*, 436–454. doi:10.1080/14766825.2016.1237959

Butowski, L. (2016b). The issue of disciplinarity and non-disciplinarity of tourism studies. *Téoros: Revue de Recherche en Tourisme*, *35*(1). Retrieved from http://teoros.revues.org/2899. doi:10.7202/1040236ar

Chambers, D., & Buzinde, C. (2015). Tourism and decolonisation: Locating research and self. *Annals of Tourism Research*, *51*, 1–16. doi:10.1016/j.annals.2014.12.002

Cohen, E., Cohen, S. A., & King, V. T. (2018). The global permutations of the Western publication regime. *Current Issues in Tourism*, *21*(17), 2035–2051. doi:10.1080/13683500.2016.1272556

Dann, G. M. S. (2011). Anglophone hegemony in tourism studies today. *Enlightening Tourism*: A Pathmaking Journal, *1*(1), 1–30.

Dann, G. M. S., & Liebman Parrinello, G. (2009). *Sociology of tourism: European origins and developments*. New York: Emerald Group.

Desbiens, C. (2002). Speaking in tongues, making geographies. *Environment and Planning D: Society and Space*, *20*, 1–25.

Domarecki, C. (2017, September 4). *Ideologia liberalna w pułapce utopijności [Liberal ideology in the trap of utopia; an interview with A. Zybertowicz]*. Retrieved from http://www.fronda.pl/a/prof-andrzej-zybertowicz-ideologia-liberalna-w-pulapce-utopijnosci,98977.html

Dredge, D., & Jamal, T. (2015). Progress in tourism planning and policy: A post-structural perspective on knowledge production. *Tourism Management, 51*, 285–297. doi:10.1016/j.tourman.2015.06.002

Dyer, P., Aberdeen, L., & Schuler, S. (2003). Tourism impacts on an Australian Indiggenous community: A Djabugay case study. *Tourism Management, 24*(1), 83–95. doi:10.1016/S0261-5177(02)00049-3

Feyerband, P. K. (1974). *Against method*. London: New Left Books.

Foster, J., Muellerleile, C., Olds, K., & Peck, J. (2007). Circulating economic geographies: Citations, pattern and citation behaviour in economic economic geography, 1982–2006. *Transactions of the Institute of British Geographers, 32*(3), 295–312. doi:10.1111/j.1475-5661.2007.00239.x

Foucault, M. (1961). *Histoire de la Folie [History of Madness]*. Paris: Librairie Plon.

Foucault, M. (1971). *L'ordre du discours*. Paris: Gallimard.

Foucault, M. (1975). *Surveiller et punir*. Paris: Gallimard.

Foucault, M. (1980). *Power/Knowledge: Selected Interviews and Other Writings, 1972–1977*. New York: Pantheon Books.

Foucault, M. (2001). *Dits et écrits*, Vol. 2. Paris: Gallimard.

Foucault, M. (2004a). *Naissance de la biopolitique*. Paris: Gallimard-Seuil.

Foucault, M. (2004b). *Securité, territoire, population*. Paris: Gallimard-Seuil.

Francis. (2013). *Apostolic Exhortation Evangelii gaudium [On the Proclamation of the Gospel in Today's World]*. Vatican City, Francis.

Francis. (2015). *Encyclical Letter Laudato si' [On Care on our Common Home]*. Vatican City, Francis.

Garcia-Ramon, M. D. (2003). Globalization and international geography: The questions of languages and scholarly traditions. *Progress in Human Geography, 27*(1), 1–5. doi:10.1191/0309132503ph409xx

Garcia-Ramon, M. D. (2004). The spaces of critical geography: An introduction. *Geoforum, 35*(5), 523–524. doi:10.1016/j.geoforum.2004.01.006

Gibson, C. (2008). Locating geographies of tourism. *Progress in Human Geography, 32*(3), 407–422. doi:10.1177/0309132507086877

Gramscii, A. (1971). *Selections from the prison notebooks*. New York: International Publishers.

Gregson, N., Simonsen, K., & Vaiou, D. (2003). Writing (across) Europe: On writing spaces and writing practices. *European Urban and Regional Studies, 10*(1), 5–22. doi:10.1177/a032521

Grobler, A. (2006). *Metodologia nauk [Methodology of Sciences]*. Kraków: Wydawnictwo Aureus—Wydawnictwo Znak.

Grosfoguel, R. (2007). The epistemic decolonial turn: Beyond political-economy paradigms. *Cultural Studies, 21*(2-3), 211–223. doi:10.1080/09502380601162514

Guttiérez, J., & López-Nieva, P. (2001). An international journals of human geography really international?. *Progress in Human Geography, 25*(1), 53–69. doi:10.1191/030913201666823316

Habermans, J. (1978). *Knowledge and human interests*. London: Heinemann.

Hägerstrand, T. (2004). The two vistas. *Geografiska Annaler. Series B, Human Geography, Special Issue: Path, Prism, Project, Pocket and Population, 86*(4), 315–323. doi:10.1111/j.0435-3684.2004.00170.x

Hall, M. (2004). Reflexivity and tourism research. In J. Phillimore & L. Godson (Eds.), *Qualitative research in tourism* (pp. 137–155). London: Routledge.

Hall, M. (2013). Framing tourism geography: Notes from the underground. *Annals of Tourism Research, 43*, 601–623.

Hall, M., Williams, A., & Lew, A. (2004). Tourism: Conceptualisations, institutions, and issues. In A. Lew, M. Hall, & A. Williams (Eds.), *A companion to tourism* (pp. 3–21). Oxford: Balckwell.

Hall, M., & Page, S. J. (2015). Following the impact factor: Utilitarianism or academic compliance? *Tourism Management, 51*, 309–312. doi:10.1016/j.tourman.2015.05.013

Hall, S. (1997). The work of representation. In S. Hall (Ed.), *Representation: Cultural Representations and Signifying Practices* (pp. 13–74). London: Sage.

Haney, D. (2008). *The Americanization of social science: Intellectuals and public responsibility in the Postwar United States*. Philadelphia, PA: Temple University Press.

Harris, C. D. (2001). English as international language in geography development and limitations. *Geographical Review, 91*(4), 675–689. doi:10.2307/3594725

Hollinshead, K. (1999). Tourism as a public culture: Horne's ideological commentary on The Legerdemain of Tourism. *International Journal of Tourism Research, 1a*, 267–292. doi:10.1002/(SICI)1522-1970(199907/08)1:4<267::AID-JTR171>3.0.CO;2-Z

Hollinshead, K. (2004). A primer in ontological craft. In J. Philimore & L. Goodson (Eds.), *Qualitative research in tourism* (pp. 63–82). London: Routledge.

Hollinshead, K. (2006). The shift to constructivism in social inquiry: Some pointers for tourism studies. *Tourism Recreation Research, 31*(2), 43–58. doi:10.1080/02508281.2006.11081261

Hollinshead, K., & Jamal, T. B. (2007). Tourism and the third ear: Further prospects for qualiatative enquiry. *Tourism Analysis, 12*(1), 85–129. doi:10.3727/108354207780956735

Hooks, B. (1990). *Yearning: Race, gender and cultural politics*. Toronto: Between the Lines Press.

Husserl, E. (1913). *Ideas*. New York: Collier.

Husserl, E. (1999). *Kryzys nauk europejskich i fenomenologia transcendentalna [The Crisis of European Sciences and Transcendental Phenomenology]*. Toruń: Wydawnictwo Rolewski.

Jamal, T. B., & Everett, J. (2004). Resisting rationalisation in the natural and academic life-world: Critical tourism research or hegemonic charity? *Current Issues in Tourism, 7*(1), 1–19. doi:10.1080/13683500408667970

John Paul II. (1981). *Encyclical Letter Laborem exercens [Through Work]*. Vatican City, John Paul II.

Kitchin, R. (2005). Disrupting and destabilizing Anglo-American and English-language hegemony in geography. *Social & Cultural Geography, 6*(1), 1–15. doi:10.1080/1464936052000335937

Kitchin, R., & Fuller, D. (2003). Making the 'black box' transparent: Publishing and presenting geographic knowledge. *Area, 35*(3), 313–315. doi:10.1111/1475-4762.00180

Kołakowski, L. (1978). *Main currents of marxism: Its origins, growth and dissolution*. Oxford: Oxford University Press.

Kuhn, T. (1962). *The structure of scientific revolutions*. Chicago, IL: Chicago University Press.

Lazzarato, M. (2009). Neoliberalism in action: Inequality, insecurity and *The Reconstitution of the Social. Theory, Culture & Society, 26*(6), 109–133. doi:10.1177/0263276409350283

Lewin, K. (1935). *A dynamic: Theory of personality*. New York: McGraw-Hill.

Machnikowski, R. M. (2011). Spór o relatywizm konstruktywistycznej socjologii wiedzy naukowej [Controversy about the relativism of the constructivist sociology of scientific knowledge]. In P. Bytniewski & M. Chałubiński (Eds.), *Teoretyczne podstawy socjologii wiedzy [Theoretical foundations of sociology of knowledge]* (pp. 61–77). Lublin: Wydawnictwo Uniwersytetu Marii Curie-Skłodowskiej.

Maldonado-Torres, N. (2007). On the coloniality of being. *Cultural Studies, 21*(2–3), 240–270. doi:10.1080/09502380601162548

Mannheim, K. (1960). *Ideology and utopia: An introduction to the sociology of knowledge*. London: Routledge and Kegan Paul.

McKarcher, B. (2002). The privileges and responsibilities of being a referee. *Annals of Tourism Research, 29*, 856–859.

McKercher, B. (2015). Why and where to publish. *Tourism Management, 51*, 306–308. doi:10.1016/j.tourman.2015.05.012

Mercier, P. (2004). *Nachtzug nach Lissabon [Night Train to Lisbon]*. Munchen, Wien: Carl Hansen Verlaag.

Meriläinen, S., Tienari, J., Thomas, R., & Davies, A. (2008). Hegemonic academic practices: Experiences of publishing from the periphery. *Organization, 15*(4), 584–597. doi:10.1177/1350508408091008

Merton, R. K. (1947). *Social theory and social structure: Toward the codification of theory and research*. New York: Columbia University Free Press.

Merton, R. K. (1973). *The sociology of science: Theoretical and empirical investigations*. Chicago, IL: The University of Chicago Press.

Mignolo, W. D. (2007). Delinking: The rhetoric of modernity, the logic of coloniality and the grammar of de-coloniality. *Cultural Studies, 21*(2-3), 449–514.

Minca, C. (2000). Guest editorial: Venetian geographical praxis. *Environment and Planning D: Society and Space, 18*(3), 285–289. doi:10.1068/d1803ed

Moses, J. W., & Knutsen, T. L. (2007). *Ways of Knowing: Competing methodologies in social and political research*. Basingstoke: Palgrave MacMillan.

Olds, K. (2001). Guest editorial: Practices for 'process geographies': A view from within and outside the periphery. *Environment and Planning D: Society and Space, 19*(2), 127–136. doi:10.1068/d1902ed

Paasi, A. (2005). Globalisation, academic capitalism, and the uneven geographies of international journal publishing spaces. *Environment and Planning A: Economy and Space, 37*(5), 769–789. doi:10.1068/a3769

Paasi, A. (2015). Academic capitalism and the geopolitics of knowledge. In J. A. Agnew, V. Mamadouh, A. Secor, & J. Sharp (Eds.), *Political geography* (pp. 509–523). Oxford: Wiley Blackwell.

Perdue, R. R. (2015). The unintended consequences of impact factors on tourism research: Thoughts and comments. *Tourism Management, 51*, 303–305. doi:10.1016/j.tourman.2015.05.011

Pernecky, T. (2012). Constructionism. Critical pointers for tourism studies. *Annals of Tourism Research, 39*(2), 1116–1137. doi:10.1016/j.annals.2011.12.010

Potter, R. (2001). Geography and development: 'core and periphery'? *Area, 33*, 422–439. doi:10.1111/1475-4762.00074

Pritchard, A., Morgan, N., & Ateljevic, I. (2011). Hopeful tourism: A new transformative perspective. *Annals of Tourism Research, 38*(3), 941–963. doi:10.1016/j.annals.2011.01.004

Ren, C., Pritchard, A., & Morgan, N. (2010). Constructing tourism research: A critical inquiry. *Annals of Tourism Research, 37*(4), 885–904. doi:10.1016/j.annals.2009.11.006

Rodríguez-Pose, A. (2006). Is there an 'Anglo-American' domination in human geography? And, is it bad? – Commentary. *Environment and Planning A: Economy and Space, 38*(4), 603–610. doi:10.1068/a38280

Said, E. (1994). *Representation of the intellectual*. New York: Pantheon.

Short, R., Boniche, A., Kim, Y., & Li, P. (2001). Cultural globalization, global English, and geography journals. *The Professional Geographer, 53*(1), 1–11. doi:10.1111/0033-0124.00265

Simonsen, K. (2004). Differential spaces of critical geography. *Geoforum, 35*(5), 525–528. doi:10.1016/j.geoforum.2004.01.008

Slater, D. (1992). On the borders of social theory: Learning from other regions. *Environment and Planning D: Society and Space, 10*(3), 307–327. doi:10.1068/d100307

Smith, L. T. (1999). *Decolonizing methodologies: Research and indigenous peoples*. New York: Zed Books.

Szklarska, A. (2017). Czy prawda może być przemocą? [Can Truth Be Equal to Violence?]. *Studia Philosophiae Christianae, 51*(3), 139–157. doi:10.21697/2015.51.3.17

Towards reforming the International Financial and Monetary Systems in the context of World Political Authority. (2011). Vatican City: Pontifical Council for Justice and Peace.

Tribe, J. (2006). The truth about tourism. *Annals of Tourism Research, 33*(2), 360–381. doi:10.1016/j.annals.2005.11.001

Tribe, J. (2008). The art of tourism. *Annals of Tourism Research, 35*(4), 924–944. doi:10.1016/j.annals.2008.07.003

Tribe, J. (2010). Tribes, territories and networks in the tourism academy. *Annals of Tourism Research, 37*(1), 7–33. doi:10.1016/j.annals.2009.05.001

Tribe, J. (2018). Creating and curating tourism knowledge. *Annals of Tourism Research, 73*, 14–25. doi:10.1016/j.annals.2018.08.004

Tucker, H., & Zhang, J. (2016). On western-centrism and Chineseness in tourism studies. *Annals of Tourism Research, 61*, 250–252. doi:10.1016/j.annals.2016.09.007

Viken, A. (2013). What is Arctic tourism, and who should define it? Implication of an Anglo-American hegemony in academic writing. In D. K. Müller, L. Lundmark & R. H. Lamelin (Eds.),

New issues in polar tourism. Communities, environments, politics (pp. 37–50). Dordrecht: Springer.

Westwood, S., Morgan, N., & Pritchard, A. (2006). Situation, participation and reflexivity in tourism research: Furthering interpretive approaches to tourism enquiry. *Tourism Recreation Research, 31*(2), 33–41. doi:10.1080/02508281.2006.11081260

Whittaker, E. (1999). Indigenous tourism: Reclaiming knowledge. Culture and intellectual property in Australia. In M. Robinson & P. Boniface (Eds.), *Tourism and cultural conflicts* (pp. 33–45). Wallingford: CABI Publishing.

Wijesinghe, S. N., Mura, P., & Bouchon, F. (2017). Tourism knowledge and neocolonialism—A systematic critical review of the literature. *Current Issues in Tourism, 22*, 1263–1279. doi: 10.1080/13683500.2017.1402871

Wijesinghe, S. N., Mura, P., & Culala, H. J. (2019). Eurocentrism, capitalism and tourism knowledge. *Tourism Management, 70*, 178–187. doi:10.1016/j.tourman.2018.07.016

Wittgenstein, L. (2001). *Tractatus logico-philosophicus*. London: Routledge.

Zybertowicz, A. (1995). *Przemoc i poznanie: Studium z nie-klasycznej socjologii wiedzy [Violence and cognition: A study in non-classical sociology of knowledge]*. Toruń: Uniwersytet Mikołaja Kopernika.

Disruptive and Adaptive Methods in Activist Tourism Studies: Socio-Spatial Imaginaries of Dissent

Ian R. Lamond

ABSTRACT

Around the world the twenty-first century has seen growth in the number and scale of protest events, mobilising substantial numbers of people to gather in acts of dissent. Central to understanding participation in such large-scale activist mobilities is an examination of those imaginaries of dissent, space and place associated with them. However, attempts to examine such imaginaries are hampered by traditional social science approaches that depoliticise participation and are often treated with suspicion by the protesters and those tasked with mitigating the impact of their activism. The *disrUPt* project confronted those issues by exploring disruptive and adaptive methods that could bring those imaginaries to the fore. The theoretical foundations of the project were rooted in those ideas in contemporary European thought around the philosophy of the event that conceptualise 'events' as sites of multiplicity and contestation (Badiou; Deleuze), articulated through the emerging field of critical event studies (Spracklen and Lamond) and the literature of activist tourism (Pezzullo). Four methodological approaches are explored, these are: the creation of a bicycle based mobile film projector, used to make visible spaces hidden by developments in the contemporary city; the presentation of augmented film screenings, which combine film presentation with non-traditional elements in non-traditional venues; activist film making with a group of female asylum seekers; and a series of conversations that brought together participants more commonly in opposition during events of dissent. The paper concludes that whilst the methods deployed were successful in facilitating the articulation of imaginaries of dissent, space and place, attached to protest and activist tourism, much more needs to be done to both draw such research approaches together and in the development of a deeper understanding of the use of disruptive and adaptive approaches to participatory data analysis.

摘要

在21世纪的世界各地，抗议活动的数量和规模都在增长，动员了大量的人聚集在一起进行抗议活动。理解参与如此大规模的激进分子活动的核心，是对与之相关的异见人士、空间和地点的想象力进行考察。然而，检验这些想象力的尝试受到了传统社会科学方法的阻碍，这些方法将参与抗议活动去政治化。抗议者和那些肩负着减轻其激进主义冲击任务的人，往往对这些方法持怀疑态度。disrUPt项目通过探索颠覆性和适应性的方法来面对这些问题, 这些方法可以将这些想象带到讨论的中心。该项目的理论基础根植于当代欧洲关于抗议事件哲学的思想, 即将"抗议事件"概念化为具有多样性和争议性的场所(Badiou, 2017;Deleuze, 2014), 运用新近的批判事件研究(Spracklen and Lamond, 2016)和激进主义旅游文献对之进行了解释 and the literature of activist tourism (Pezzullo, 2007)。本文探索了四种方法方法, 它们是:创建一个基于自行车的移动电影放映机, 用于使隐藏在当代城市发展中的空间可见化; 将电影展示与非传统元素在非传统场地结合起来的强效的电影展示形式;与一群寻求庇护的女性一起积极拍摄激进主义电影;以及一系列的对话, 这些对话使持不同意见的参与者更经常地聚集在一起。本文得出结论,所用的方法成功地解释了与抗议活动及激进旅游有关的异议、空间和地方的诸多想象, 除此之外, 还需要做更多的工作一起把这样的研究方法聚拢在一起, 深入理解运用颠覆性和适应性方法进行参与性的数据分析。

Introduction

The beginning of the twenty-first century has seen a significant growth in large scale, global, events of dissent. It is estimated that almost 11 million people in around 60 countries and more than 600 cities (Walgrave & Rucht, 2010) participated in demonstrations against going to war with Iraq on February 15th and 16th, 2003. In a period of just over three weeks, following the establishment of the Occupy encampment on Wall Street on the 17th September 2011, *The Washington Post* suggested there were camps in over 950 cities in 82 countries. On April 22nd, 2017 (World Earth Day), protesters in more than 500 cities around the world (Buncombe, 2017) declared that Donald Trump's position on climate change was a mistake that could have catastrophic, global, repercussions. Such large scale, time bound, mobilities, associated with activist events are a significant dimension in the geopolitical turn emerging in tourism geographies, symptomatic of what Mostafanezhad (2018) has referred to as "…the multi-scalar nature of geopolitics" (p345). Whilst ideas associated with activist tourism have been discussed since Dixon's (2001) consideration of activist mobilities in the formation of anti-globalisation movements, emergent from the World Trade Organization (WTO) protest in Seattle in 1999 (See, also, Bevington and Dixon (2005)), *activist tourism* is a little explored area within event or tourism research. The first explicit discussion of the activist tourist appears in Phaedra Pezzullo's (2009) book *Toxic Tourism*, which addresses the layered and complex discourse of pollution and tourism from a perspective of travel to heavily polluted sites as a form of environmentalist advocacy. Significantly, for this paper, she also addresses the concept of the imaginary of space and place in relation to activist events and tourism, which I will consider shortly. More recently the debate has tended to focus on academic activism within Tourism Studies (For example: Hales, Dredge, Higgins-Desbiolles, & Jamal, 2018; Ooi, 2019), and less so on the imaginary of space and place associated with activist events and mobilities.

Whilst there has been discussion around LGBT + activism and travel, this has been quite limited. Reflection tends to be concerned with what Puar (2002) calls the "selling of liberation" (pp. 123), where the discourse of activism becomes negotiated as the original act of protest is colonised by globalised economic interests (Markwell & Waitt,

2009; Rojek, 2013). Others have drawn on the work of Urry (such as Urry, 1990, 2001) to consider the gaze of the activist tourist in regions of conflict or space of tragedy (Brin, 2006; Hercbergs, 2012; Buda & McIntosh, 2013). But the activist tourist, whose mobility is associated with embodied engagement in protest, does not fit this mould comfortably. Unlike most other forms of tourism, the activist tourist, carrying a placard or joining in a mass chant as they process through some of the primary transit routes of a city, may be more interested in being part of that upon which the gaze falls. They thus become both *visitor* and, providing the protest is successful in its disruption of daily routines and it is mediated well across multiple platforms, some form of *attraction*. Central then, to developing an understanding of the *engaged* activist tourist are the entwined imaginaries of space and place associated with participation in an event of dissent. I use the word 'imaginary' here in a sense derived from that of the *social imaginary* developed by Cornelius Castoriadis (Castoriadis, 1993, also Castoriadis & Ricoeur, 2017), to refer to those practices and discourses assumed to be associated with space and place as articulated by an individual or group engaged in visitor mobility. Grasping that complex and nuanced imaginary of the event of dissent becomes a truly critical project within tourism studies. For, as Bianchi (2009) argues, it endeavours to "…find ways of integrating the study of discourse with ongoing, as well as material forms of power (and, as such) …constitutes a radical departure from the status quo in Tourism Studies". (p498)

The primary purpose of the "Creativity, protest and the city" project, otherwise known as disrUPt, was to draw together a small team of experienced researchers who could explore methodologies appropriate to studying the imaginary of dissent. It adopted a perspective that was closer to individual experience, rather than drawing generalised conclusions based on patterns discernible through the analysis of Big Data (Tsou, 2015). The project ran between October 2016 and October 2017, with most of the methods explored delivered across a number of cities in Yorkshire (England); primarily Leeds and Sheffield. The research team consisted of the author (as Principle Investigator) and four other academics (two from an events/tourism background, one from leisure studies, and two from cultural studies), and several external partners associated with activism in the region.[1] The external partners were self-selected following a general call for collaborators put out through connections already held by the research team. All the academics came from the same University, which was within the same geographical region as the research.

A study of the imaginary of dissent can cut across multiple perspectives in critical research within tourism, leisure and event studies. However, given the proximity of that imaginary to the articulation of identity (Roberts, 2005) and construction of the 'other' (Glynos & Howarth, 2007) it can prove to be problematic when standard social science approaches are used to capture data. In order to address some of those concerns, DisrUPt set out to examine novel techniques that would engage with methods that were disruptive and adaptive. This was done to establish the viability of such approaches as tools for research, rather than as a means of capturing fresh data, or as a means of testing any existing theories or findings: that could be done in later research – if the methods tried were found to be of value.

The topic of this paper will be discussed over four steps. First, it will consider some of the limitations that are associated with the adoption of traditional social science methods when it comes to researching the imaginary of dissent. Following an outline of what CES is, I will argue that it can offer a pathway which can provide fresh opportunities for furthering research in the study of events of protest, activism and activist tourism. Second, after a brief description of the background to adaptive and disruptive methodologies, I will address the issue of what is to be understood by those terms within this paper. Third, the actual methods deployed by the disrUPt project will be described and, where appropriate, how they changed over the course of the project. The were four methods trialled. First, the creation of the kino-cine-bomber as a form of artefact based urban intervention. Second, the use of augmented film screenings as a disruptive *visual* method. Third, film making workshops as a means of articulating empowerment. Finally, the production of safe and structured environments, where participants, usually in a confrontational relationship during periods of protest, were able to meet and share their individual imaginaries of dissent through conversation. Finally, each of the methods described will be reviewed to examine the opportunities and limitations they exhibited, as well as any other issues that emerged which may impact their future viability as adaptive/disruptive methods.

Researching the imaginary of dissent

An encounter with urban events of dissent, either as participant, spectator, or someone whose daily routine becomes disrupted, presents us with a complex and entwined set of imaginaries of space, place and time (Pavoni, 2018). The mainstays of social science research, such as the interview, focus group and survey, may offer a hint of the complexity at work within this diverse social ecosystem, but they tend to be traces, at best. Such fragments, symptoms of the 'event', require sensitive judgments around interpretation and the location of the researcher and the participant within the research process (Feigenbaum, McCurdy, & Frenzel, 2013). As such they require a *reading between the lines* which can open the research up to critiques of researcher insider/outsider bias. Whilst such factors may impact any form of social science research employing standard methods of inquiry (Flyvbjerg, 2001), there are others that are more closely associated with researching eventful protest and dissent, adding more clouds to an already troubled sky.

Both the activists and those that represent forces mobilised to 'contain' manifestations of protest have a vested interest in projecting, and protecting, the world view of which they are a part. For understandable reasons (Monaghan & Walby, 2018; Schlembach, 2018) the groups (such as campaigning organisations and security/policing services), whose physical presence at protests is the most overt articulation of such events, in the mediated imaginary of them, are wary of people asking a lot of questions (Mowatt, Forthcoming). Engagement with the context within which the research is undertaken is problematised by factors the participant and/or the researcher implicitly foreground and background in their apprehension of the other actor. Whilst there may be no paradox in holding a position of activist/academic (Higgins-Desbiolles, 2010) the researcher may brings with them assumptions around norms and behavioural routines that, consciously or otherwise, they may lapse into when articulating their questions or interpreting a participant's response. Such

Goffmanesque foregrounding and backgrounding (Goffman, 1963) operates as a limitation to the rigour of the research. Comparable arguments can be made for participants that might be considered 'bystanders' to the event.

Chesters (2002) has suggested that attempting to adopt a stance of academic neutrality raises several ethical concerns, especially associated with how we establish the locus of knowledge production. Adopting an *ethics of engagement*, he suggests, would move away from considering social movements as an object from which knowledge can be produced; instead it would acknowledge the capacity of such groups to be valued as producers of knowledge in their own right. Such a position resonates with the critical pedagogy of Paulo Freire (1996) and Ernest Boyer's (1997) conceptualisation of *engaged scholarship*; both emphasised co-creation and collaboration over pedagogic and epistemic colonisation (Holland, Powell, Eng, & Drew, 2010). In their approach to researching protest camps, Feigenbaum et al. (2013) attempted to address some of these concerns by adopting a form of photo-elicitation led group discussions, these camp fire chats were set within a context convivial to participants. Using images that had been gathered from a variety of material and digital sources, they deployed a number of game-like activities within the context of chats around a camp's communal fire, utilising their resources to stimulate affective memories and conversation.[2] The use of images also facilitated communication between researcher and activist, where neither shared a common language. Whilst the approach offered opportunities for generating fresh insight into the imaginary of dissent within the setting of protestful occupation, it does require a careful balance being struck between congeniality and a cooler distance, as the researcher is still required to seek some level of impartiality.

Though discussing a different context, one that is aligned to the negotiation of policy change, Teghtsoonian (2016) has argued that social inquiry which adopts a singular framework risks narrowing or closing down possibilities for research and analysis. This is not the same as advocating a mixed-method approach. The limitations that have been mentioned already would not, necessarily be overcome by adopting a mixed-methodology, if anything those limitations would be reproduced and compounded by becoming even more deeply entrenched. What is required, she suggests, is a reflexive, ongoing, dialogue. Echoing Dorothy Smith's feminist activism (2007), she calls "... researchers to a practice of reflexivity in considering the political effects of the methodological choices (they) make" (Op.Cit. p. 331). At this point it is worth considering the perspective critical event studies (CES) can bring, as a conceptual orientation, to this discussion.

CES is an emerging sub-field of event studies which, according to Spracklen and Lamond (2016), begins with a problematisation of the referent - 'event'. In their words;

> "At its most basic level we define event as 'that which intervenes the mundane'. We understand that such interventions will be contested by some of those impacted by them, and that they are never simple, surgical-like incisions; if anything, they are more like a ragged laceration with consequences far beyond those of the location of the wound." (p. 9)

A central characteristic of 'events', CES argues, is disruption and contestation. It is through approaching events from their wider context and grasping their essentially disruptive and contested character that what becomes construed as *the* event is brought into focus (Robertson, Ong, Lockstone-Binney, & Ali-Knight, 2018). Through reflexively examining the multiplicity of the *disturbances* that are drawn together as *the* event, asking how they are being drawn together (by whom and for what purpose), and in what way(s) are they being mediated, what becomes construed as 'the event' emerges (Finkel, Sharp, & Sweeny, 2018). Such multiplicity shifts the emphasis away from conceptualising *event* as singular to what can be more usefully thought of as an evental (Badiou & Corcoran, 2017) spatio-temporal site. It is through confronting those imaginaries, discourses and discursive practices that what comes to be constituted as the 'event' can be studied.

If 'event' is to be constituted through a reflexive exploration of what might be considered the evental site what about the spatio-temporal 'event' that constitutes the site of research? From such a perspective, inquiry becomes an activity that constitutes itself through an iterative and reflexive process. However, it is not simply *an activity*, but one where adopting an evental approach to research acknowledges that it needs to exhibit characteristics that resonate with its contested and disruptive outlook. As such it is consistent with Teghtsoonian's (2016) and Smith's (2007) position; to *consider the political effect of one's methodological choices*. It thus forms part of a family with other critical orientations, which cut across tourism, leisure and events research. Consequently, the division between knowledge gatherer and contributor, between insider and outsider, subject and object (etc.), can be confronted with knowledge being co-created and shared, and thus able to carry the potential to profoundly impact the agents commonly denominated as researcher and participant. Thus, appropriate methodological approaches to investigating the imaginaries of dissent and protest, within a CES, would need to be disruptive (exposing the discursive practices hidden in the mundane and routine of daily life), and adaptive (so that they can engage iteratively and reflexively, whilst challenging the 'researcher', the 'participant' and the presumed boundaries between them). But what can be construed as *disruptive* or *adaptive* methodologies, and how were such terms understood within the disrUPt project?

Adaptive and disruptive methodologies

Adaptive methods

Adaptive methodologies were first proposed by James Kay in the mid-90s (Kay & Schneider, 1994) as a means of developing an understanding of the complexity exhibited in the interplay between human (social) systems and the environment (ecology) in which such systems were located. Working within an arena of environmental systems management, Kay and his colleagues argued that such an eco-social systems approach was required if we are to grasp how "...the nexus in which ecological understanding and sociocultural preferences met and interfaced with policy makers and managers" (Waltner-Toews & Kay, 2005. p.39). This "diamond schematic", as they called it, embraced a holarchic approach (i.e. one where the parts and whole of the system are functionally inseparable and, to an extent, indistinguishable (Koestler, 1969)) that characterised such eco-social systems as resilient, whilst also exhibiting integrity and self-organisation (Kay, Regier, Boyle, & Francis, 1999). Waltner-Toews and

Kay (2005) argue that investigating such systems requires an adaptive rather than fixed methodology because;

> "Within the dynamics of eco-social complexity and the uncertainty this generates, we are faced with finding our way through a foggily perceived landscape rather than charting a scientifically determined course to a known end point." (p.39)

The complexity, or *fogginess*, of eco-social systems emerging from the multiplicity of "... socio-ecological stories, pictures and system descriptions" (ibid, p. 45) articulated by the agents contributing and co-constituting the system itself. In striving to understand such complexity they go on to propose an approach called AMESH (Adaptive Methodologies for Eco-social Sustainability and Health). which can move beyond fixed, singular (or mon-ocular) perspectives by encouraging researchers and practitioners to;

> "... look for and understand system feedback and the emergence of self-organization. The iterative nature of this process and its multi-ocular assessments mean that the unpredictability and uncertainty of eco-social systems, with the potential for dramatic flips, can be more easily accommodated than in more conventional approaches." (Waltner-Toews & Kay, 2005. p.49)

Beyond the field of environmental systems management, adaptive methodologies that derive from the AMESH framework have also been used in broader sociological theory. Armitage, Marschke, and Plummer (2008), argue that incorporating adaptivity is essential if one is to truly consider the impact of an external environment that is in a process of constant change. Cherednichenko and Godlevsky (2015) develop this further. They suggest that adopting adaptive methods, that can adjust as new sources of data become apparent, avoids many of the limitations of fixed methodologies within self-regulatory and open systems where, as Penn (2016) has put it, "... we must be prepared to take action without full system understanding" (p.26).

Because of the complexity of the relations within and between parts and their environment (Koestler's principle of the *holarchic*, 1969) and the constant change of that environment, adaptive methodologies thus emphasis the use of more than one approach (Waltner-Toews, multi-ocular-ism) that can be adjusted as new data is made available. This should be understood as a realistic response to the fluidity of the eco-social systems within which participant and researcher are located, and not as a means of triangulation, as a mixed-methods approach might.

Disruptive methods

As we found in our consideration of *adaptive methods*, disruptive methodologies can, to an extent, be connected to systems theory. Hodgson and Midgley (2015) have suggested that the definition of any system we are investigating is a function of the multiplicity of its stakeholders. Such systems, they maintain, are not a fixed object but mutable, evolving, and adaptive. Through critique (i.e. disruption) of its boundaries, both internally and between it and other systems, the system can confront its ambiguities and assumptions whilst clarifying where power lies; how it is operating, and in whose interest it is working. Analogous to this, Lamond (2018) has argued that agents, whether individual, institutional or otherwise (e.g. a social movement), contributing to the constitution of an event of protest can, with varying levels of ontic and epistemic attachment, be construed as stakeholders within an open system. That is, they can be construed as *advocacy coalitions* (Sabatier & Weible, 2007) associated with policy positions around an issue. Jones and Jenkins-Smith (2009) argue that the mobility of agents, from one coalition to another, depends on the strength of those ontic and epistemic connections. All other factors being equal, if someone has a weak connection to a coalition, they may defer to participation through an online campaign, whilst an agent with strong ontic and epistemic ties will be motivated to travel (as an activist tourist) to a site, or sites, associated with campaigning. Here we find resonances with Pezzullo's (2009) *toxic tourism*. It is by disrupting a coalition that we can, in principle, gain insight into those system factors highlighted by Hodgson and Midgley (2015).

Zielińska, Kowzan, and Prusinowska (2011) and Kowzan, Zielińska, and Prusinowska (2014) have explored the use of disruption as a political pedagogic device by social movements, as well as a pedagogic methodology. Their studies focus on the use of direct intervention, by activists, in lectures halls in Poland between 2009 and 2013; reflecting on the consequences of such disruption and "… on the learning of various actors" (Ibid, p.297). Zielińska et al. (2011) analyse one case study lecture intervention as:

> "… a configuration of three subjects: the speaking one (the lecturer), the listening one (the audience) and the (out)standing one, that is, the protesters. We [argue] … that the activists change the communication between the lecturer and the audience by their protest. They provoke and surprise the other two subjects and point to ambiguities in the world, and in the lecturer's message, showing their resistance to the ideology behind it." (p.257-258)

As such the intervention becomes both a method for the exposure of power and potential pedagogy, as it can present an opportunity for shared learning, growth, and development. At a more local and more overtly pedagogic level, Bastian (2015) has employed disruptive techniques to make visible the regimes at work in choosing texts within the classroom. Such flipped approaches to learning, as they are commonly termed, are gaining increasing traction at all levels of learning (Reidsema, Kavanagh, Hadgraft, & Smith, 2017).

What is common across these and other approaches to disruptive methods is the way they expose discursive routines whilst offering an opportunity, as Walker (2015) puts it, to blur formal and informal space. As such they challenge the boundary between 'insider' and 'outsider' and, ultimately, they can work to dissolve that binary. In her analysis of cultural resistance Sarah Kanouse (2004) suggests that such disruptive approaches can reshape discursive, physical and relational spaces, and facilitate the reimagining of them, i.e. disruptive methods present us with an opening where fresh imaginaries can emerge. Disruption permits a Deleuzian deterritorialization (Deleuze, 2014) that enables the recontextualization and formation of unexpected connections. However, as Halvorsen (2015) has pointed out, disruptive methods cannot offer a fixed approach; to do so would mitigate their capacity to evoke the affects they seek. Instead, any such research approach must operate "… against-and-beyond itself, [in] a constant struggle to overcome any form it takes" (p.467). Consequently, if a method is to be truly disruptive, it must always-already exhibit some form of adaptivity.

Adaptive and disruptive methods in the disrUpt project

Having considered the background to adaptive and disruptive methods it is imperative we now consider how the disrUPt project approached them, and how they emerged within the project. As mentioned at the beginning of the paper, one of the purposes of the disrUPt project was to explore methods that could facilitate a deeper understanding of the imaginaries of dissent associated with events of protest.

Prior to the commencement of the project the author, as principle investigator (PI), instigated a series of meetings that drew together members of the research team, as listed earlier. Through the team's own connections, representatives from a number of regional small-scale and independent cultural agents as well as activists from a range of social movements, based in and around the university's home city of Leeds (West Yorkshire – UK), were also invited. The cultural organisations included an independent arts cinema, an architectural practice with a primarily regional footprint, and an independent film festival curator. The social movements covered a variety of activist interests, including environmental activism; anti-war movements; LGBT + campaigners and others. At the first meeting the conversation focused on the form the research could take. Whilst the core aim, that of trying to find ways of uncovering the diverse imaginaries associated with the practice of events of dissent that took place in an urban context, was maintained, the research approaches to be adopted were openly discussed and allowed to appear organically from the dialogue between all the meeting's participants. Central to the steering of that meeting was a commitment, by the PI, to draw on the experience of non-academic participants in articulating approaches that would resonate with their engagement with the city, both as a physically present space, but also as a spatial imaginary. Having established a loose framework of projects at that first meeting, academics were attached to them depending on their personal strengths and interests. Subsequent meetings took the form of reporting on the progress of the diverse projects initiated.

The idea of disruption as an orientation for the research programme emerged quickly. Whilst central to the PI's construal of 'event', they were careful not pre-impose their own perspective on the group. However, 'disruption' still emerged in the dialogue of that initial meeting. 'Disruption', echoing the formulation of Kanouse (2004), was found to lie at the heart of practice for both the cultural and activist agents present. The framework of methods adopted, which will be outlined shortly, were intended to disrupt the imaginaries of all participants (including those of the researcher) so that they could be more easily articulated and, in a supportive way, something upon which the extended team could reflect. This, to a degree, resonated with the work of Zielińska et al. (2011) and Kowzan et al. (2014), in their suggestion that disruptive methods also represented a form of activist pedagogy. Given this shared orientation, the group agreed to change the project's title from 'Creativity, protests and the city' to 'disrUPt'.

Bearing in mind Halvorsen's (2015) warning that disruptive methods should occur in a constant struggle to overcome whatever form they take, some element of adaptivity was also felt to be required. Koestler's holarchic principle of the complex inseparability of part and whole in an open system, would suggest the research approaches to be explored would also need to incorporate elements of reflexive adaptability.

However, the degree to which this should be incorporated within the project did not receive a consistent level of support. Consequently, the range of methods to be adopted included differing degrees of adaptiveness. The next section will outline the four approaches explored by the disrUPt project and the extent to which they exhibited adaptive and disruptive methods.

The disrUPt *research programme*

The four approaches adopted as part of the disrUPt research programme were: the creation of an artefact as a device for urban intervention; use of augmented film screenings; a film making workshop; and the production of a safe and structured environment, where participants, usually in a confrontational relationship during periods of protest, were able to meet and share their imaginaries of dissent. The order presented here follows that in which they were rolled out during the running of the project, which ran between October 2016 and November 2017. Each will be taken in turn and how they were delivered discussed.

The Kino-Cine bomber: Disrupting urban space

Created by postgraduate architecture students from the University of Sheffield (South Yorkshire – UK), with the support of the Leeds based architecture practice Group Ginger, the Kino-Cine Bomber is a bicycle-based cinema device intended to "... re-imagine disused buildings and obsolete urban infrastructure (as a space) for public leisure" (Group Ginger, 2017, NP.). Drawing on principles inspired by both the praxis of détournement (a form of cultural re-routing, or hijacking), initiated by the Situationists, and the spatial philosophy of Henri Lefebvre, the use of the projected moving image is deployed by the device to reimagine urban space through subtraction. For example, in Coventry (West Midlands – UK) it was deployed to 'daylight' the path and flow of a river hidden by present urban development, it thus subtracted from the contemporary context so as to re-inscribe the river into the urban space, thereby revealing a presence through the imaginary of absence.

Based on a 'Christina' (A form of Danish freight carrying bicycle), the Kino-Cine Bomber is comprised of, "... an added wooden tower to elevate a 2000-lumen (digital) projector; a car battery powers the projector, 50-watt sound system, and radio transmitter (for anyone wishing to listen in via radio)." (Ibid-NP) Figure 1 shows one the Kino-Cine Bomber in operation in Coventry, making visible ('daylighting') a former space of public entertainment in what is now an area dominated by new office developments.

The process of subtraction ('daylighting' and re-imagining), produced through the projections of the device, are intended to disrupt the imaginaries of space, creating what Guy Debord might have described as the undermining of the commoditisation of the 'spectacle' through the creation of rebellious artistic situations (Debord, 2011). The intention is to use the kino-cine bomber on a series of sorties (as those that worked on creating it have referred to its deployment) around Leeds, whilst evaluating the impact of its use through the response of those reacting to its intervention. So far,

Figure 1. The Kino-Cine Bomber.
(Source: Simon Baker/Re-Activist Studio. Used with permission)

the artefact has only undertaken trial sorties in Coventry, though steps to secure its use in Leeds are currently being planned.

Augmented cinema: Disruptive film screenings

Between November 2016 and October 2017, the disrUPt project screened around a dozen films in five venues. Screenings took place in a variety of spaces in and around Leeds. Those spaces included venues designed for cinematic presentation (an independent cinema close to the project's home university) to non-traditional and pop-up venues (including a municipal building; a disused church converted into an arts centre; a small event space attached to a city centre café, and an independent venue close to the city centre's railway stations that is used mainly by emergent unsigned bands). The themes covered by the films shown, which were predominantly documentary films or the product of activist film makers, were initially suggested within the pre-project meetings discussed earlier. As regards the specific films screened, these emerged through dialogue between the activists on the team, the venue host, an independent film events coordinator, the PI, and what could be made available. The film selected for each venue was only finalised between one screening and the next, rather than pre-set at the start of the series. A central element of each screening was how it was to be 'augmented' (i.e. disrupted), adapting each screening to resonate with the films theme and possible attendee reaction. How the screening was to be disrupted varied from one presentation to the next. Before outlining some of the key elements of disruption that were used it is worth taking a small detour to establish the background to *augmented cinema*.

Augmented cinema draws on elements of what has become known, in more commercial circles, as 'Live Cinema'; which is defined by Atkinson and Kennedy (2016) as:

> "…a cinema that escapes beyond the boundaries of the auditorium, whereby film screenings are augmented by synchronous live performance, site-specific locations, technological intervention, social media engagement, and all manner of simultaneous interactive moments, including singing, dancing, eating, drinking and smelling" (p. 139).

However, it supplements this with practices akin to those established by 'third cinema' (Solanas and Getino, 1969), which began to appear as a form of underground film making and sharing in Latin America during the 1960s. 'Third cinema' identified films that were overtly political, that carried an explicit political message, and screened them in such a way as provoke debate. They were, as Presence (2013) argues, films made and presented to be 'detonators' of discussion, often in non-traditional spaces, that would often target marginalised issues and communities. As well as post-screening panel-led discussions, which are a common practice in the showing of documentary and 'politically' motivated films, the disrUPt screenings incorporated disruptive elements. They included; an impromptu book readings, apparently spontaneous performances by community choirs, a pay-as-you-feel buffet of rescued food[3], an unannounced demonstration of placard bearing protesters walking through and round the audience prior to the screening. One screening was located in a space where the attendees had to first navigate through an audience for a completely different event. The augmentations were decided between screenings, adapting to the theme and the audience reaction to the previous intervention. Figure 2 was taken outside one of the screening venues (the converted church). It shows members of the Commoners Choir[4] just before they processed through the audience, carrying banners and singing their own brand of political music.

Adaptivity was central to keeping the screenings unique and responsive, whilst disruption served multiple purposes. Similar to Debord's positioning of situationist interventions, mentioned earlier, the playfully rebellious creative disruptions helped establish an environment where those who wished to engage in discussion, following the film, could do so in a supportive, though candid, manner (the disruptions permitting an almost 'anything goes' setting). It is planned to follow up this part of the project with a number of lightly structured interviews with people who self-selected to participate in that part of the project at the end of the screenings. The disruptive elements within the cinematic presentations will, in those instances, also act as a point of shared memory through which rapport, trust and, it is hoped, participant integrity, can be assured.

Film making workshop: Disruptive pedagogy

Working with the Liverpool based new media arts organisation *First Take*, the disrUPt project set up a short film making course for women, from what might be considered

Figure 2. The Commoners Choir prepares to disrupt a film screening.
(Source: The authors)

marginalised communities in West Yorkshire. Participants were recruited on the basis of the provision of an opportunity to learn more about film making Several community organisations in the region that worked with women from BAME[5] communities were approached, including some working with refugees and asylum seekers. Nine women signed up for the course, which ran for two blocks of three days, over a fortnight, crossing July and August 2017. The orientation of the workshops drew on the pedagogic principles of Paulo Freire, who argued "[t}he oppressed must be their own example in the struggle for redemption" (Freire, 1996, p.54). Though each participant learnt how to make and edit film through the direct intervention of one with expertise in those skills, the created film had to be on a topic that all participants agreed was meaningful to them. This flipped the expertise away from the person conventionally designated as 'the teacher'. Whilst that person remained in role as facilitator, imparting technical knowledge, she also acquired the position of student/learner as she grew to discover more about the issues the participants confronted, how they felt about articulating their concerns, and how they managed those tensions in their daily lives. The skills developed were, as a consequence, directly connected to the empowerment of the participants in articulating their concerns. Figure 3 is a still from the film, it is a

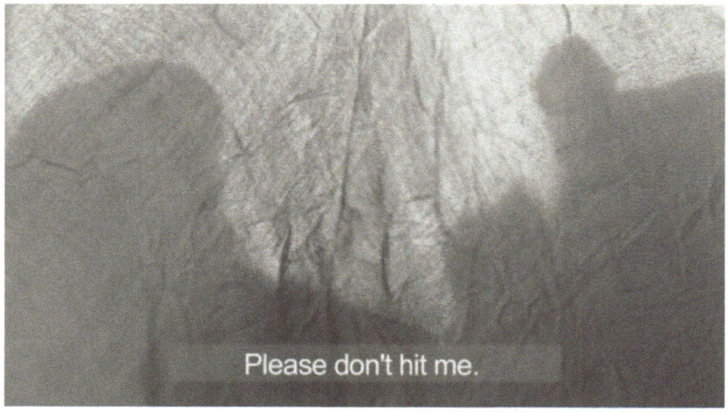

Figure 3. Film still.
(Source: First Take – www.firsttake.org.uk. Used with permission)

representation of some of the domestic violence a number of the participants had experienced.

The finished film, which has no name, runs for a little over 7 minutes and addresses issues of gender equality and domestic violence (In the form of both emotional and physical abuse). In conversation with the PI, after the completion of the film, the workshop facilitator commented on how moved she was by the experience of working with the group, and how much she had learned from the participants. Not just about the violence they faced but how the precarity of their situation (many were seeking asylum in the UK) also impacted that experience, and their capacity to share it with others. Testified by the film's content and anecdotal accounts emergent from conversations with them and the community workers through which they were connected, the female participants also seemed to develop bonding attachments that extended beyond the workshops. With some of them now engaging with other forms of activism, and a few interested in developing their community film making skills further.

Sadly, in part a product of the women's trepidation around the film's content and their appearance in it (Which has arisen from a combination of, for some of them, uncertainty around their current residency status and, for others, concerns associated with the possible repercussion of some family members 'getting wind' of their involvement.), they have stated that the film should not have a public screening. To date the film has had a small, closed, screening, to an invited audience; this was in November 2017. Audience reaction was very positive and supportive. Whilst only one of the group spoke, before the presentation, she spoke with passion about the opportunities the programme had given them all, and the importance of the messages they were trying to communicate through the film. The PI has agreed that future presentations of the film will only be with the agreement of the majority of those people that participated in its making, both in front of, and behind, the camera.

Critical conversations: Disrupting the focus group

The final approach was the creation of a series of three opportunities for conversation. These took the format of focus group meetings; however, though this was the least disruptive of all the approaches, the meetings diverged from those forms of focus group that seek some level of demographic comparability or strive to obtain a degree of consensus around a topic (Barbour, 2018; Krueger, 1988). Demographics was not a consideration, nor was a consensus sort. Given the tension between the camps represented at the meetings contestation was anticipated, not agreement; consequently respecting participants right to their own position, at the meetings, needed to be managed carefully. Whilst actual participants had not been targeted, the character of their connection to events of protest had been. It was organisations and agencies that were approached through their association with protest; participants volunteered through those networks. Each meeting lasted around 2 hours, and the same people were invited to each meeting. Given the work commitments of a number of the invitees, the meetings were scheduled to run at roughly monthly intervals and ran between late April and early July 2017. Despite this, it was not the case that all those contributing could attend every meeting. Over the three sessions we had participants from; activist groups (both young members and those that might be considered veteran campaigners); local councillors; leaders in the regional business community; people with a direct connection to mass media (broadcast and print based); city-based securitisation services (though a representative from the police was not forthcoming), and people with a direct connection to either legal services, or human rights law. As such, the group was composed of people who, whilst connected to events of protest within a city context, did not necessarily share a sympathetic perspective on protest and the practices of social movements. All participants were aware of the composition of the group but were not, at first, aware of the relationship others had to the topic being discussed. Obviously, this became apparent as the first meeting progressed, but it did not form a presupposition attachable to any particular individual, before the conversation began.

In the briefing prior to each group discussion, issues of respecting everyone's right to speak, to have a view that was to be acknowledged as legitimately their own, and that everything said in the room was strictly confidential, were addressed. All participants were given gender neutral names, and only those names were to be used in the conversation. This made disruption and adaptivity central to each participant's participation, while it also formed part of how the conversations developed over subsequent meetings. The first conversation had an open brief, with a loose framework of discussing the relationship between protest and the city. Following the end of session wind-down, options around the structure for the following session were considered. A variation on Feigenbaum et al. (2013) campfire chats approach, mentioned earlier, was approved by the participants. It was agreed that all attending would bring an artefact they associated with dissent and protest, which would provide the opening stimulus for conversation. It was also suggested that the PI would bring a number of images, taken from online sources, to stimulate debate, if 1) nobody brought anything or 2) conversation dried up quickly. It transpired that this was the longest session, with a number of participants carrying on the conversation in a bar after we had been asked by the University's security staff to vacate the building. At the final session it was agreed to share a text that the PI had mentioned in the first meeting, which would be circulated and used as the focus of the final discussion. The text was Chapter 3[6] of Jacques Rancière's book "Hatred of Democracy" (Rancière 2014).

Anecdotally the three meetings proved successful in building bridges between diverse world views. As a framework, they could, in principle, lay the foundations for developing a future programme that seeks a shared vision for how such stakeholders can negotiate the use of city spaces around events of protest. To facilitate data capture, the conversations that took place over the three meetings were audio recorded and transcribed.

Reflections on the disrUPt research programme

Each of the four approaches explored by the disrUPt project contained, to varying degrees, adaptive and disruptive elements. In all cases, those elements were a central part of the research experience for both participant and researcher, and as such permitted articulations of multiplicity which CES characterises as central to its construal of 'event'. Thereby reflecting an examination of the imaginary of dissent, through an orientation that construes the research project as *evental*. Additionally, in acknowledging such multiplicity, the project highlighted hybridity within adaptive and disruptive methods as something that moved beyond mixed methods. Implicit within such methodologies is the desire to achieve some form of convergence on a single world view (Robson & McCartan, 2016). As a form of *militant methodology* (Halvorsen, 2015), disruptive and adaptive methods differ in that they mark a holarchic (Koestler, 1969) turn towards the traditional binaries of subjective/objective; academic/practitioner; or insider/outsider, through a problematisation of the ways in which they are being constructed through the evental site of research: operating, as Halvorsen suggests, "… against-and-beyond itself" (Op.Cit. p.467).

However, as the project developed, it became apparent that the approaches encountered several limitations with regards to the practicalities of their deployment and viability, as well as their conceptual legitimacy.

All four piloted methods required substantial partnership working, which had a significant impact on the scale and reach of the programme. Whilst attempts were made to extend the project beyond the one urban environment, which would have facilitated a greater opportunity to explore potential conversations between culturally diverse imaginaries of dissent, this was only achieved with the Critical Conversations element of the programme. Even in that instance, the partnership was not as straightforward as it had been hoped. Two key factors made partnership cooperation and development difficult. First, was the limited time that trust had to develop between all those involved in the programme's delivery; not just at an individual level, but also institutionally. On a one-to-one basis this, for the most part, worked relatively well, but institutionally it became more problematic. Trust as either a sharing of power or the negotiated give and take within a more nuanced power relationship (Baier, 1986) proved challenging when working with what may be construed as regionally recognised brands. This was not equivalent to the size of the organisation but, to an extent, its visitor/user reputation. Consequently, relating this to the second factor, we found partners reticent around taking, or absorbing, what they encountered as risk. For example; we encountered this in wariness around granting permission for a Kino-Cine Bomber *sorties*. With some potential hosts unsure of how attracting visitors to what might otherwise be considered an *unglamorous* space, would impact their impression of the wider location. Some venues exhibited a degree of caution around offering to host a film to be screened. The film "Just Eat It", about food waste, was rejected by several potential spaces, as they were unsure how attendees would subsequently view the hosts food policy. Such matters made negotiations time consuming, with programme delivery resting, at times, only on a partial indication that the partner was actually as committed to the delivering the output in the format the project had requested. The disparity between the disruptive and adaptive approach being developed by the research team, and some of the lag emergent from some of the careful negotiations with the partner relationships, added, an inertia to how some elements became articulated.

As well as these practical problems there was also a key question around how to extract and analyse data from the approaches that have been trialled by the disrUPt project. Whilst the focus of this paper has been on the research methods used, part of the evaluation of the effectiveness of those methods also considered how data could be extracted from them. However, to date, very little of that work has been undertaken. This is a key limitation in the project as a whole, and one that must be addressed if it is to progress. DisrUPT was established to explore methodologies appropriate for engaging with the diverse imaginaries of protest articulated by agents, whether individual activist tourists; organisations; social movements or institutions, associated with events of dissent. Although the adoption of disruptive and adaptive methodologies were found to be successful in facilitating the articulation of those imaginaries, the proposed routes anticipated for capturing and analysing data from those methods did not exhibit characteristics consistent with the disruptiveness and adaptiveness of the interventions themselves. Both Jackson (2008) and Cashman et al. (2008) in their discussion of community based participatory research (CBPR) point to such limitations within much participatory research. For the most part, the techniques that have been suggested, whether anchored is some form of grounded theory, narrative or thematic analysis still rely on the standard tools of social science research critiqued earlier. As such they undermine one of the ultimate objectives of the project; to examine the imaginary of dissent from a perspective that is closer to that of individual experience. Methodologically therefore, whilst contestation is addressed it does not, as Torre, Fine, Stoudt, and Fox (2012) argue, confront "the deep and knotty relations of privilege and injustice" (p.179) inherent in researching from the *outside – in*. Jackson (2008) and Nind's (2011) approach to participatory qualitative data analysis may, through the seeking of narratives and themes in collaboration with non-academic research partners, offer a way of managing such concerns. A possible limitation is that, in contrast to the groups participating in the disrUPt project, both Jackson and Nind worked with communities where the world view of their participants contributed to their coherence as an identifiable group; many of those that contributed to the disrUPt project were in direct contestation with each other. Even within a loose coalition that could be associated with a specific concern, factional differences in world-view, on approach, language, and intended impact were commonplace. However, the structural embedding of participant engagement in the process of data capture and analysis that Jackson (2008), Nind (2011) and, Torre et al. (2012)

recommend, could support the emergence of a disruptive and adaptive toolkit for participant led data analysis. Such considerations will be borne in mind as the project proceeds. While the detail of how a disruptive and adaptive approach is to reach into those corners of its inquiry is yet to be ascertained, the need to incorporate some form of participatory data analysis is unquestionable.

Conclusion

There is an old English proverb; *the proof of the pudding is in the eating.* To date, the disrUPt project is still at an early stage. If it were a loaf of bread; whilst it may be in the latter stages of proofing, it is yet to be baked. The primary purpose of the disrUPt project was to explore methodologies appropriate to studying the imaginary of dissent, from a perspective that was close to that of individual experience, within activist mobilities and activist tourism. In that it was successful. However, it cannot claim to have succeeded fully if it does not also produce some explicit findings and identifiable conclusions that are drawn from the approaches it has trialled. In order to address that concern the programme has slowed, allowing it time to explore and review participatory qualitative data analytic approaches, ones that can also adopt a disruptive and adaptive perspective on data capture: disrUPt is far from over.

The partnerships that were central to the delivery of the project are still in place and lay a substantial ground upon which disrUPt can grow, and new research programmes can emerge. Over time, the factors of trust and the capacity to take and absorb risk, through the partnering relationships that have been built, are extending and deepening. Such enhanced relationships bode well for future, further, exploration of approaches to the study of the imaginary of protest and dissent. There are plans for the Kino-Cine Bomber to take more *sorties.* Venues and film event programmers are showing increased interest in the use of augmented screenings. Members of the research team have been approached by other community partners who have expressed an interest in film making projects. Conversations with colleagues at universities in Sao Paulo (Brazil) are progressing around how a *critical conversation* framework can have the potential to establish spaces where stakeholders with contested imaginaries of the city can discuss the value and place of dissent within their urban context. Such an approach carries resonances with the findings of de Frantz's study of the planning of the *Museumsquater* in Vienna (De Frantz, 2018), where she concludes that progress was achieved when "…the governing consensus was transformed…-through an open, inclusive and contentious process of discursive interaction" (pp. 502). If fruitful, it could lead to policy changes that result in negotiated moments of dissent within the spatial fabric of the city, welcoming the activist tourist and contributing to the destination's visitor narrative.

Notes

1. Not all activist partners wish to be identified; however, those that are were: Film Fringe (A Yorkshire based film screening group interested in the role film can play in activism. It is also part of the global Radical Film Network and provided links to that network); Leeds Tidal (A cross-*social movements* support group based in Leeds, but supports activist groups

across West Yorkshire); Yorkshire MESMAC (A sexual health and sexuality awareness charity supporting the LGBT + community); though not a formal representative we also had an active member of Yorkshire CND (Campaign for Nuclear Disarmament).
2. An analogous approach, one of participant led photo-elicitation, was adopted by Amsden et al (2010) in their study of the sense of place held by tourism-dependent communities.
3. Rescued food, sometimes referred to as food salvage, is the use of edible food that would otherwise be thrown away from traditional food outlets (such as supermarkets and restaurants). See, for example, the work of https://therealjunkfoodproject.org/ and https://feedbackglobal.org/
4. The Commoners Choir is a West Yorkshire based people's choir. Mainly perform their own music, their work has been presented around the UK and Europe. They have their own YouTube channel, which is worth a look.
5. Black and Minority Ethnic
6. Chapter 3 is entitled, *Democracy, Republic, Representation*; pp. 51-70 in the Verso edition, translated by Steve Corcoran.

Disclosure statement

No potential conflict of interest was reported by the author(s).

Funding

This article was funded by Leeds Beckett University: Research Cluster Award.

References

Amsden, B. L., Stedman, R. C., & Kruger, L. E. (2010). The creation and maintenance of sense of place in a tourism-dependent community. *Leisure Sciences*, *33*(1), 32–51. doi:10.1080/01490400.2011.533105

Armitage, D., Marschke, M., & Plummer, R. (2008). Adaptive co-management and the paradox of learning. *Global Environmental Change*, *18*(1), 86–98. doi:10.1016/j.gloenvcha.2007.07.002

Atkinson, S. A., & Kennedy, H. W. (2016). Inside-the-scenes: The rise of experiential cinema. *The Journal of Audience and Reception Studies*, *13*(1), 139–151.

Badiou, A. (2017). *Conditions*. (S. Corcoran, Trans.) London: Bloomsbury Academic.

Baier, A. (1986). Trust and antitrust. *Ethics*, *96*(2), 231–260. doi:10.1086/292745

Barbour, R. (2018). *Doing focus groups*. London: Sage.

Bastian, H. (2015). Capturing Individual Uptake: Toward a Disruptive Research Methodology

In Composition Forum (Vol. 31). Association of Teachers of Advanced Composition. Retrieved from https://eric.ed.gov/?id=EJ1061545

Bevington, D., & Dixon, C. (2005). Movement-relevant theory: Rethinking social movement scholarship and activism. *Social Movement Studies*, *4*(3), 185–208. doi:10.1080/14742830500329838

Bianchi, R. V. (2009). The 'critical turn'in tourism studies: A radical critique. *Tourism Geographies*, *11*(4), 484–504. doi:10.1080/14616680903262653

Boyer, E. L. (1997). *Scholarship reconsidered: Priorities of the professoriate*. San Francisco: Jossey-Bass.

Brin, E. (2006). Politically-oriented tourism in Jerusalem. *Tourist Studies*, *6*(3), 215–243. doi:10.1177/1468797607076672

Buda, D. M., & McIntosh, A. J. (2013). Dark tourism and voyeurism: Tourist arrested for "spying" in Iran. *International Journal of Culture, Tourism and Hospitality Research*, *7*(3), 214–226. doi:10.1108/IJCTHR-07-2012-0059

Buncombe, A. (2017). March for Science: Thousands march in 500 cities to protest Donald Trump's 'rejection of science'. Published in the Independent 22nd April 2017. Retrieved from https://www.independent.co.uk/news/science/march-for-science-earth-day-march-protest-donald-trump-science-rejection-a7696816.html.

Cashman, S. B., Adeky, S., Allen, A. J., Corburn, J., Israel, B. A., Montaño, J., … Eng, E. (2008). The power and the promise: Working with communities to analyze data, interpret findings, and get to outcomes. *American Journal of Public Health*, *98*(8), 1407–1417. doi:10.2105/AJPH.2007.113571

Castoriadis, C. (1993). *World in fragments: Writings on politics, society, psychoanalysis and the imagination*. California: Stanford University Press.

Castoriadis, C., & Ricoeur, P. (2017). Dialogue on history and the social imaginary. In S. Adams (Ed.) *Ricoeur and Castoriadis in discussion* (pp. 3–20). (S. Davidson, Trans.). London: Rowman & Littlefield International.

Cherednichenko, O., & Godlevsky, M. (2015). A new methodology of complex systems management. Retrieved from http://repository.kpi.kharkov.ua/handle/KhPI-Press/15685.

Chesters, G. (2002). [Review of A. Starr (2000) *Naming the Enemy: Anti-Corporate Movements Confront Globalization*. Zed Books, New York.]. *Organization & Environment*, *15*(2), 214–217.

Debord, G. (2011). *Comments on the society of the spectacle*. (M. Imrie, Trans.). London: Verso.

De Frantz, M. (2018). Tourism marketing and urban politics: Cultural planning in a European capital. *Tourism Geographies*, *20*(3), 481–503. doi:10.1080/14616688.2017.1354392

Deleuze, G. (2014). *Difference and Repetition*. London: Bloomsbury Academic.

Dixson, C. (2001). Finding hope after Seattle: Rethinking activism and building a movement. Retrieved from http://theanarchistlibrary.org.

Feigenbaum, A., McCurdy, P., & Frenzel, F. (2013). Towards a method for studying affect in (micro) politics: The campfire chats project and the occupy movement. *Parallax*, *19*(2), 21–37. doi:10.1080/13534645.2013.778493

Finkel, R., Sharp, B., & Sweeny, M. (2018). *Accessibility, inclusion, and diversity in critical event studies*. Abingdon: Routledge.

Flyvbjerg, B. (2001). *Making social science matter: Why social inquiry fails and how it can succeed again*. Cambridge: Cambridge University Press.

Freire, P. (1996). *Pedagogy of the oppressed*. (M.B. Ramos, Trans.) London: Penguin Books.

Glynos, J., & Howarth, D. (2007). *Logics of critical explanation in social and political theory*. Abingdon: Routledge.

Goffman, E. (1963). *Behaviour in public places: Notes on the social organisation of gatherings*. New York: The Free Press.

Group Ginger. (2017). Urban subversion and mobile cinema. Retrieved from http://www.group-ginger.com/blogs/2018/01/04/urban-subversion-and-mobile-cinema-leisure/.

Hales, R., Dredge, D., Higgins-Desbiolles, F., & Jamal, T. (2018). Academic activism in tourism studies: Critical narratives from four researchers. *Tourism Analysis*, *23*(2), 189–199. doi:10.3727/108354218X15210313504544

Halvorsen, S. (2015). Militant research against-and-beyond itself: Critical perspectives from the university and Occupy London. *Area*, *47*(4), 466–472. doi:10.1111/area.12221

Hercbergs, D. (2012). Narrating instability: Political detouring in Jerusalem. *Mobilities*, *7*(3), 415–438. doi:10.1080/17450101.2012.659469

Higgins-Desbiolles, F. (2010). In the eye of the beholder? Tourism and the activist academic. In P.M. Burns, C.A. Palmer & J.M. Lester (Eds), *Tourism and Visual Culture* (pp. 98–106) Wallingford: CABI.

Hodgson, A., & Midgley, G. (2015). January). Bringing Foresight into Systems Thinking: A Three Horizon Approach. In *Proceedings of the 58th Annual Meeting of the ISSS-2014 United States*. Retrieved from http://journals.isss.org/index.php/proceedings58th/article/view/2278.

Holland, D., Powell, D. E., Eng, E., & Drew, G. (2010). Models of engaged scholarship: An interdisciplinary discussion. *Collaborative Anthropologies*, *3*(1), 1–36.

Jackson, S. F. (2008). A participatory group process to analyze qualitative data. *Progress in Community Health Partnerships: Research, Education, and Action*, *2*(2), 161–170.

Jones, M. D., & Jenkins-Smith, H. C. (2009). Trans-subsystem dynamics: Policy topography, mass opinion, and policy change. *Policy Studies Journal*, *37*(1), 37–58. doi:10.1111/j.1541-0072.2008.00294.x

Kanouse, S. (2004). *Ungrounded Opposition: Cultural Resistance as Virus* (Doctoral dissertation, School of Art and Design, University of Illinois, Urbana-Champaign). Retrieved from http://www.readysubjects.org/writing/kanouse_mfathesis.pdf.

Kay, J. J., Regier, H. A., Boyle, M., & Francis, G. (1999). An ecosystem approach for sustainability: Addressing the challenge of complexity. *Futures*, *31*(7), 721–742. doi:10.1016/S0016-3287(99)00029-4

Kay, J. J., & Schneider, E. (1994). Embracing complexity: The challenge of the ecosystem approach. *Alternatives Journal*, *20*(3), 32–39.

Koestler, A. (1969). Beyond atomism and holism: The concept of the holon. In A. Koestler and J.R. Smythes (Eds) *Beyond reductionism: New perspectives in the life sciences* (pp. 192–232). London: Hutchinson.

Kowzan, P., Zielińska, M., & Prusinowska, M. (2014). Intervention in lectures as a form of social movement pedagogy and a pedagogical method. *Interface: A Journal for and about Social Movements*, *6*, 297–326.

Krueger, R. A. (1988). *Focus groups: A practical guide for applied research*. 1st Ed. London: Sage.

Lamond, I. R. (2018). Conceptualising events of dissent: Understanding the Lava Jato rally in Sao Paulo – 5th December 2016. In R. Finkel; B Sharp; M. Sweeny (Eds) *Accessibility, inclusion, and diversity in critical event studies*. Basingstoke: Routledge.

Markwell, K., & Waitt, G. (2009). Festivals, space and sexuality: Gay pride in Australia. *Tourism Geographies*, *11*(2), 143–168. doi:10.1080/14616680902827092

Monaghan, J., & Walby, K. (2018). Hobocops': Undercover policing's deceptive encounters. *Criminological Encounters*, *1*(1), 7–18.

Mostafanezhad, M. (2018). The geopolitical turn in tourism geographies. *Tourism Geographies*, *20*(2), 343–346. doi:10.1080/14616688.2018.1434820

Mowatt, R. (forthcoming) Events of dissent, Events of the self. (The liminality of protest images). In I.R. Lamond and J. Moss (Eds) *Liminality and critical Event Studies: Borders, Bounderies and Contestation*. London: Palgrave.

Nind, M. (2011). Participatory data analysis: A step too far?. *Qualitative Research*, *11*(4), 349–363. doi:10.1177/1468794111404310

Ooi, C. S. (2019). Indigenisation of scholarship: Go native but don't be naive. Conference paper presented at *CAUTHE 2019: Sustainability of Tourism, Hospitality and Events in a Disruptive Digital Age*. Central Queensland University, Australia. Retrieved from http://ecite.utas.edu.au.

Pavoni, A. (2018). *Controlling urban events: law, ethics and the material*. Abingdon: Routledge.

Penn, A. (2016). July). Artificial life and society: Philosophies and tools for experiencing, interacting with and managing real world complex adaptive systems. In *Proceedings of the Artificial Life Conference 2016*, pp. 26–27. Cambridge, Massachusetts.

Pezzullo, P. C. (2009). *Toxic tourism: Rhetorics of pollution, travel, and environmental justice*. Tuscaloosa: University of Alabama Press.

Presence, S. (2013). *The political avant-garde: Oppositional documentary in Britain since 1990*. (Unpublished PhD thesis). University of the West of England. Retrieved from http://eprints.uwe.ac.uk/20924/

Puar, J. K. (2002). Circuits of queer mobility: Tourism, travel, and globalization. *GLQ: A Journal of Lesbian and Gay Studies*, *8*(1-2), 101–137. doi:10.1215/10642684-8-1-2-101

Rancière, J. (2014). *Hatred of Democracy*. London: Verso.

(2017). Reidsema, C., Kavanagh, L., Hadgraft, R., and Smith, N. (Eds.). *The flipped classroom: Practice and practices in higher education*. Singapore: Springer Nature.

Roberts, J. (2005). The power of the 'imaginary' in disciplinary processes. *Organization*, *12*(5), 619–642. doi:10.1177/1350508405055938

Robertson, M., Ong, F., Lockstone-Binney, L., & Ali-Knight, J. (2018). Critical event studies–issues and perspectives. *Event Management*, *22*(6), 865–874. doi:10.3727/152599518X15346132863193

Robson, C., & McCartan, K. (2016). *Real world research*. 4th Ed. Chichester: John Wiley & Sons.

Rojek, C. (2013). *Event power: How global events manage and manipulate*. London: Sage.

Sabatier, P. A., & Weible, C. M. (2007). The advocacy coalition framework. *Theories of the Policy Process*, *2*, 189–220.

Schlembach, R. (2018). Undercover policing and the spectre of 'domestic extremism': The covert surveillance of environmental activism in Britain. *Social Movement Studies*, *17*(5), 491–506. doi:10.1080/14742837.2018.1480934

Smith, D. E. (2007). Making change from below. *Socialist Studies*, *3*(2), 7–30.

Spracklen, K., & Lamond, I. R. (2016). *Critical event studies*. Abingdon: Routledge.

Teghtsoonian, K. (2016). Methods, discourse, activism: Comparing institutional ethnography and governmentality. *Critical Policy Studies*, *10*(3), 330–347. doi:10.1080/19460171.2015.1050426

Torre, M. E., Fine, M., Stoudt, B. G., & Fox, M. (2012). Critical participatory action research as public science.

Tsou, M. H. (2015). Research challenges and opportunities in mapping social media and Big Data. *Cartography and Geographic Information Science*, *42*(sup1), 70–74. doi:10.1080/15230406.2015.1059251

Urry, J. (1990). *The tourist gaze: Leisure and travel in contemporary societies*. London: Sage.

Urry, J. (2001). Globalising the tourist gaze. In S. Babu; S. Mishra; and B.B. Parida (Eds) *Tourism development revisited: concepts, issues and paradigms* (pp. 150–160). Los Angeles: Sage.

Walgrave, S. and Rucht, D. (Eds). (2010). *The world says no to war: Demonstrations against the war on Iraq*. Minneapolis: University of Minnesota Press.

Walker, M. (2015). The productive and disruptive methodology of doing fieldwork at work. *Area*, *47*(4), 473–478. doi:10.1111/area.12219

Waltner-Toews, D., & Kay, J. (2005). The evolution of an ecosystem approach: The diamond schematic and an adaptive methodology for ecosystem sustainability and health. *Ecology and Society*, *10*(1), 38–53. doi:10.5751/ES-01214-100138

Zielińska, M., Kowzan, P., & Prusinowska, M. (2011). Social movement learning: From radical imagination to disempowerment?. *Studies in the Education of Adults*, *43*(2), 251–267. doi:10.1080/02660830.2011.11661616

The disruptive 'other'? Exploring human-animal relations in tourism through videography

Minni Haanpää, Tarja Salmela, José-Carlos García-Rosell and Mikko Äijälä

ABSTRACT

Although there is a growing body of literature focusing on the use of qualitative research approaches for understanding human-animal relations, surprisingly little attention has been devoted to the topic in tourism studies. In particular, the central role played by animals in the tourism industry and thus in the creation of tourism experiences calls for a more critical reflection on the methodologies through which the relations and encounters between humans and non-humans are studied. Considering this gap, we draw upon videography to explore the potential and challenges of using video as a means to interpret and theorise on multispecies relations in tourism. More precisely, we use videography to better understand the co-constructed relationship between humans and animals engaged in tourism activities, leading us to consider these relationships as multispecies assemblages. Our attempts to explore and document these encounters on video lead us to critically evaluate the ethical and epistemological underpinnings of our study and hence to the genesis of this methodological paper. Although challenged by a post-humanist stance, videography as a 'more-than-representational' approach offers ways to capture non-linguist, sensuous and embodied qualities of the research context. Utilising these possibilities, our study contributes to the development of more inclusive tourism theorising and futures by exploring videography as a vehicle for emergent theorising on human-animal relations through its relational and political nature of engagement.

摘要

虽然越来越多的文献关注于使用定性研究方法来理解人与动物的关系，但令人惊讶的是，在旅游研究中很少有人关注这一主题。特别是，动物在旅游业中以及因此在创造旅游经验方面所起的核心作用呼吁对研究人类与非人类之间的关系和接触的方法进行更批判性的反思。考虑到这一空白，我们利用录像方法探讨了视频作为解释和理论化旅游中多物种关系手段的潜力和挑战。更准确地说，我们使用影像来更好地理解参与旅游活动的人和动物之间共同构建的关系，使我们把这些关系看作是多物种的组合。我们试图通过视频来探索和记录这些遭遇，从而批判性地评估我们研究的伦理和认识论基础，从而形成这篇方法论论文。尽管受到后人文主义立场的挑战，视频摄影作为一种'比表征性更强'的方法，提供了捕捉非语言学、感官和具体研究背景的方法。利用这些可能性，我们的研究通过探索利用视频图像为研究工具研究新兴的人与动物关系理论研究，最终有助于发展更具包容性和前瞻性的旅游理论。

Introduction

This article reflects on the methodological choices, challenges and possibilities of a videography research project on human-animal relations in tourism context. First, we suggest that you to watch the videography on which this text is based (please see the online full text version of this paper to access the video).

Following the ideas of Peter Lugosi (2018), we view videography as a methodology as a *disruptive practice* in academic knowledge production. This means understanding it as a performative act that involves 'human and non-human actors that contradict and/or misperform expected, normalized, and thus patterned ways of thinking, doing, and feeling' in socio-material settings (2018, p. 56). Lugosi refers to disruptive practices as those performed in the research field when producing knowledge. However, we expand the notion from the research setting to additionally encompass the field of academic knowledge production, considering the performative acts of producing, theorising and presenting knowledge (Lugosi, 2018; see also Seregina, 2018).

The research on which we reflect here emerged as part of a larger project focusing on understanding ethical consumers and their relations to animals within a tourism context in Finnish Lapland. Animal-based tourism services such as dog sledding, reindeer sleigh tours and horse riding are among the most popular activities for tourists visiting Lapland (García-Rosell & Äijälä, 2018). There are more than 4,000 dogs, 1,000 reindeer and 150 horses working in the tourism industry in Finnish Lapland, producing more than 15 million euros in annual revenue (García-Rosell & Äijälä, 2018). In 2017, for the first time ever, snowmobile safaris were surpassed in popularity by sled dog safaris. Yet, despite their marketing value and economic impact, these animals have been overlooked in the tourism development context.

Consequently, this article discusses a research process in an animal-based tourism context, which began with a human-focused research agenda, and evolved through the use of videographic methodology (Belk et al., 2018; Rokka & Hietanen, 2018; Seregina, 2018) to a recognition of complex multispecies assemblages (Haraway, 2008) and realisation of powerful non-human[1] agency. The affective and iterative research process made visible how agency became distributed among various agents (Haraway, 2008; Ogden, Hall, & Tanita, 2013; Notzke, 2019) or how agents became co-constituted within socio-material processes (Barad, 2003, 2007) in a tourism research setting.

Inspired by previous methodological videographic and moving image contributions (see Belk et al., 2018; Brown & Banks, 2015; Lorimer, 2010; Rokka & Hietanen, 2018; Seregina, 2018), this article proposes the value of videography in evoking and witnessing multispecies assemblages. With multispecies assemblage we turn to Ogden et al.'s (2013) definition of the term, according to which a multispecies assemblage consists of various subjects and 'considers being as ontologically multiplicitous, in which being is emergent via the present material configuration of multi-being connections' (p. 15). Such theorisation is congruent with recognition of animal agency: while animals may be unaware of consumerist and regulative aspects of tourism activities, this does not mean that they are not part of – and more importantly affecting – tourism experiences (Carr, 2014; Notzke, 2019; see also Dashper, 2019; Koski & Bäcklund, 2017).

We also want to encourage other tourism scholars to evaluate the potentialities of videography as a methodology. The remainder of the article is structured as follows. It begins by describing the research context and the fieldwork undertaken. We then continue with a discussion of the ontological and epistemological principles underlying videography. We proceed by illustrating and reflecting on our videographic research process by focusing on three overlapping themes: *Emergent theorising through affectivity and iteration; Relating agencies in the tourism context; Ethico-politics in making videography*. We conclude by suggesting that videography as a research methodology opens up new avenues for future tourism research, more attuned to the rich variety of multispecies assemblages and their complex nature in tourism settings.

Starting point: filming ethical consumers in the animal-based tourism context

The research process described in this paper originated from a study that examined consumer values in relation to the use of animals in tourism.[2] The original study was part of the very first research project to study animal-based tourism services in Finnish Lapland. Situated in the northernmost part of Europe, Lapland is a wilderness and nature-based tourism destination. Tourists visiting Lapland are particularly attracted by exotic landscapes, natural phenomena (northern lights, polar nights), Santa Claus and encounters with animals. As a winter destination, Lapland attracts tourists from around the world. The number of annual registered overnights in Lapland is nearly three million (Regional Council of Lapland, 2015). The research team of the project consisted of four researchers, the authors of the paper at hand.

Given that the initial aim of our videographic qualitative study was to explore ethical consumption in relation to animal-based tourism (see Lewis & Potter, 2011), our primary data consisted of interviews with tourists who were visiting or planning to visit Lapland. We gained access to the interviewees through the personal and professional connections and networks of the research team. All but one of the six tourists interviewed participated in animal-based tourism activities organised by Lappish tourism companies. These activities were filmed by the research team in order to observe the actual practices of the tourists and to obtain video material, which could be used to illustrate the circumstances and experiences raised in the interviews. During the course of the project, we also conducted separate observation and filming in the animal-based tourism companies to collect data on their activities and operations. Altogether we collected almost seven hours of video material, of which the interviews represented around three hours.

The filming was done with two digital single-lens reflex cameras. One of the cameras was equipped with a directional on-camera microphone and was on some occasions supported by a tripod or handheld stabiliser. The latter was useful to avoid shaky camera shots when moving around outdoors and on the premises of the companies. When the microphone was unavailable, we used phones and/or voice recorders for audio recording. In spite of having the appropriate equipment, we faced some challenges with collecting data due to Lapland's extreme winter conditions. The process of data collection was not completed prior to the analysis – a situation encountered by many academics conducting empirical fieldwork – rather it was part of an interactional and inductive process of data collection, analysis and interpretation. Given that data collection was ongoing, video material was (re-)watched and preliminary interpretations were made simultaneously with collecting further empirical data. Similarly, our field observations and an ongoing review of related literature had a direct impact on the process of data collection and analysis. From this on-going process the premises for the study now being reflected began to develop.

Videography as a methodology

Although there is a growing stream of literature on human-animal relations in tourism studies (e.g. Fennell, 2012; Markwell, 2015; see Äijälä, García-Rosell, & Haanpää, 2016), most studies have relied on traditional research methods such as interviews and observation (e.g. Bertella, 2014). For example, researchers have explored the performance of animals in tourism activities (Bertella, 2014, 2016; Clocke & Perkins, 2005; Picken, 2018; Ratamäki & Peltola, 2015) and how animal subjectivity is socially constructed through touristic representations (Curtin, 2006; Ong, 2017; Yudina & Grimwood, 2016).

An examination of studies focusing on human-animal relations enabled us to discover a manifold and rich ensemble of research, literature and film with various disciplinary backgrounds such as geography and anthropology (e.g. Cahill, 2015; Pick, 2015). Much of the existing work on human-animal studies has explored the potential of moving images in understanding humans' relations with other animals. As Brown and Banks (2015, p. 138) note, video has been used to study animals and their behaviour throughout history. They argue that ethnographic engagement through video contributes to acknowledging (1) animal lives as intertwined with human lives, (2) animal-human relationships as dynamically co-created through the body, senses and emotions, and (3) the agency of the video camera and the imagery produced. Research conducted through video recognises non-human agency and the corporeality, aesthetics and affect embedded in human-animal relations (Brown & Banks, 2015).

In the field of geography, Jamie Lorimer's (2010) work on elephants provides a case in point in developing moving image methodologies 'for witnessing and evoking human-nonhuman interactions' (p. 238). Lorimer additionally highlights how the development of more-than-human methods, whether with the assistance of the moving image or otherwise, seems to be far from easy: the advancement of non-representational and more-than-human methods lag significantly behind theoretical advancements. In the same vein, seven years later Dowling, Lloyd, and Suchet-Pearson (2017) noted the imbalance between theoretical debates and the actual praxis of methodological advancements.

In our case, videography as a methodology contributes to the advancement of more-than-human methods in the field of tourism studies. Videography was originally chosen as a methodology for the research project on ethical consumers due to the aim to understand the perspectives of the people researched, as well as the previous engagement of some members of our research team with videographic methods. According to Petr, Belk, and Decrop (2015), videography can be defined as the process of producing and communicating knowledge through the collection and analysis of visual material. It is also important to note, as Seregina (2018) explains, that the aim of videography is to produce a video or moving visuals. In comparing videography with other moving image methodologies, Rokka and Hietanen (2018) stress that videographic research is relational, affective and performative, enabling researchers to elaborate and express theoretical arguments. Collectively, these characteristics became an important issue of concern to us because they facilitated an alternative perspective on theorising human-animal relations as multispecies assemblages. Indeed, despite the growing body of literature on visual methods in post-human research, propounding theoretical arguments through audiovisual or non-textual means has received little attention in (post-human) social sciences (Wood, 2015; also see Hietanen & Rokka, 2017).

Videography should not be confused with visual ethnography, in which video has traditionally been understood as a tool to capture and represent reality. Rather than documenting and representing ethnographic knowledge, videography opens up new ways of seeing and understanding the world in which we live (see Rokka & Hietanen, 2018). Following this examination, it can be argued that the epistemological underpinnings of videography are non- or more-than-representational in nature. It is in this sense that video is said to be relational, a 'media practice' that is embedded in and produces social relations, and one thus capable of inducing change (Hietanen, Rokka, & Schouten, 2014). Moreover, camera and video material have the potential to intensify the affective qualities of lived events (see Rokka & Hietanen, 2018; Vannini, 2015). When discussing videography in relation to our study, we refer to (1) the actual observations made through video recording in the field, (2) the reiterative process of data, entailing the (re-)watching of video material, and (3) theorising with the video medium by crafting the visual presentation (Hietanen et al., 2014; Rokka & Hietanen, 2018).

Accordingly, this study benefits from the work of marketing scholars on videographic research (e.g. Belk et al., 2018; Rokka & Hietanen, 2018; Seregina, 2018) and social science researchers on video methods (e.g. Vannini, 2015). In spite of the relative lack of attention to post-human theorising in the field of marketing and consumer research, videography has methodological implications for post-human studies. Our investigation suggests such a possibility, as doing research through video and (re-)watching the data collected lead us into an inductive and reflexive process of theorising on human-animal relations (Belk et al., 2018). The methodological potential of video lies in its incessant movement. As Hietanen and Rokka (2017, p. 5) argue, if video 'is taken seriously then what becomes epistemologically important are how it imparts its forces in movement (change in the intensities of our thoughts—how changing thought alters our unfolding material relations)'. Such movement affects us bodily, withdrawing us from plain cognitive observation (Hietanen & Rokka, 2017, p. 5). It is the movement that renders possible the 'more-than-representational' qualities of video: the sensuous, tacit, non-linguistic and embodied, which again gives access to new ways of thinking (Rokka & Hietanen, 2018; Vannini, 2015; see also Buller, 2015). This was what happened in our study when the 'b-reel' footage of the ethical consumer videos began to evoke uneasiness concerning human-animal relations in the researched field: a disruption emerged. In so doing, it offered us new possibilities to explore the collected data. Through this disruption, we commenced an emergent theorising process in which affective and tacit hunches led us to speculative conceptualisations regarding human-animal relations (see Belk et al., 2018).

Emergent theorising on video means to open up ways to think about potential futures. From this perspective, video is not understood as merely a description of past events, but a means to unlock new ways of being, acting and relating in the world (see Hietanen & Andéhn, 2018; Rokka & Hietanen, 2018). Videography engages researchers in theorising, both implicitly and explicitly, by embracing not only linguistic, but also non-linguistic, tacit, sensuous and embodied qualities of video. Theorising on video has also received critique, for 'making up desired realities' (see e.g. Hietanen & Andéhn, 2018; Hietanen & Rokka 2017). However, as Hietanen and Andéhn (2018, p. 13) argue, 'all we express on video is necessarily a fiction'. Nonetheless, when 'understood and crafted into being a form of 'serious fiction' in its own right, a bricolage of edited performances and interpretations, theorizing on video could begin to move from the logic of making an externalized statement of explanation to something that rather attempts to be part of emerging culture itself by animating, attuning, evoking, disrupting, sensitizing or entangling us with affective resonances of lived-through moments and events' (Rokka & Hietanen, 2018, p. 10).

To develop the idea of emergent theorising further, some videographers have drawn attention to the role of editing as a relational and political act (see Rokka & Hietanen, 2018; Vannini, 2015). Following this line of thought, Seregina (2018) points to the role of the audience in editing videographic research. Discussing the role of the audience in videography is not a novel idea. For example, Belk and Kozinets (2005) note that videography not only enables one to reach wider audiences, but also to stimulate enthusiasm, emotions and interest, helping audiences to empathise with the phenomenon at hand.

Yet, as Rokka and Hietanen (2018) show, videography has the potential to engage audiences in powerful affective and reflexive encounters with the videographic screenings and researchers. From this perspective, audiences play a central role in theorising as they become active co-creators of the videography and its meaning (Seregina, 2018). As Seregina explains, videography is a never-ending learning process for the audience and the researchers alike, as meanings exist ephemerally during the performance, and thus continuously develop and change. In this regard, videographic research becomes political, as it works to evoke reflexive thinking and action (Hietanen & Rokka, 2017). Indeed, there is no neutral videography, rather 'all are full of potential for subjectivation and, if treated as such, may just help us towards more reflexive and powerfully critical scholarly contributions in the future' (Hietanen & Andéhn, 2018, p. 14).

Reflecting on the videographic research process

Emergent theorising through affectivity and iteration

This sub-chapter reflects on theorising through video by focusing on the iterative process of watching, editing and showcasing it. The (re-)watching of video material on animal-based activities evoked affective states, thoughts, questions and realisations among our research team that we had not anticipated (Rokka & Hietanen, 2018; Vannini, 2015). It was through moving images (Hietanen & Rokka, 2017), the presence and vitality of multiple bodies (Buller, 2012) in the film, and the iterative process of making a video that recognition of the multispecies assemblage began to shape our agency, together with us shaping that of others. This shaping became reality both in situ in the 'field' during the research process with the video camera and retrospectively when watching the video data and editing the material. Indeed, both represented a process of multispecies engagement and different means to entangle with various agential relations (Ogden et al., 2013; see also Koski & Bäcklund, 2017).

In particular, a video of a reindeer sledge ride – the very first scene of the videography – in which the reindeer turns to look towards the researcher sitting and filming in the sleigh represented a turning point in re-considering the agency of animals in the research setting. It was almost if the reindeer was silently asking 'What are you doing?' Another turning point was the moment when two of the researchers were watching a video clip in which one reindeer was being filmed waiting for a tourist sleigh ride to commence. The reindeer stands still, not looking at the camera or paying attention to it. An insecurity regarding the reindeer's experience in that situation arose, causing a troubling feeling: 'Is it really justified to film this reindeer?'

These moments resonate with Haraway's (2008, p. 22) question: 'And what if the question of how animals engage one another's gaze responsively takes centre stage for people?' In our first example, the reindeer looks back at us, its gaze intersecting with our own (Haraway, 2008). In this moment, confusion and a particular question arose: 'What is the reindeer thinking in this very moment?' In the second example, the reindeer did not look, hence its gaze did not intersect with ours. This disinterest in the camera and us researchers was actually the cause of the troubling effect within us. The lack of interest coupled with the non-active presence of the reindeer in that very situation (the reindeer was not for instance eating and as such its attention was not directed to some other practice/target) stimulated us to develop immediate and varied interpretations. Indeed, some of these interpretations included: 'Is the reindeer not feeling comfortable in this situation? Should we leave it alone? Is this what we are doing, right?' It was as if the reindeer's imagined emotional state was transferred to us researchers: we were nothing but uncertain about the accuracy of our interpretations (see also Lorimer, 2010), and they yet included an ethical orientation and a sense of respect during the multispecies encounter in that particular contact zone (Haraway, 2008).

Thus, videography became a method to deliberate on troubling experiences with a wide range of motives and agendas (the agenda of the project; the agenda of critical tourism researchers; the agenda of making animals visible in a human-centred tourism industry). Through videography, we were able to access the field again and again, retrospectively: numerous aspects of the filming situations appeared only when watching the video material afterwards. The videos acted as a 'link' between everything during the process: research field, people, disciplines and time (see Rokka & Hietanen, 2018). Within the team of researchers they operated as an enabler of multi-disciplinarity. The collective watching and discussing of the videos enabled us to attach our different views to something 'tangible' and elaborate them together through different concepts and during various stages of the project. This point connects the potentialities of videography as a research method to the issues raised by Buller (2015) and Fennell (2013): the importance of co-operation between scholars from different disciplines to more fully understand the subjectivity of animals.

During the moments of filming, our attention was more or less attuned to the filming itself and all of the multiple tools and equipment present during the actual filming procedure, as well as the challenges and even problems that arose during filming. The focus was on finding a good angle, effective lighting and the appropriate timing for recording (especially when filming in a situation with considerable noise and movement, as in the case of sled dog kennels). As such, video data worked as a medium for us to pay attention to the things that went unnoticed in the field. Based on the cues and elaborations from watching the material and relating it to our experiences and theoretical standpoints, a version of the videography was crafted (see Belk et al., 2018). The editing was undertaken by one member of the research team, while another wrote the theoretical framework to the video-format'. The text was then added to the video, a process that demanded more editing in order to ameliorate the viewing experience and readability of the text.

We screened the first ten-minute-long version of the video in two conferences and received feedback on the audience's reactions. From the comments obtained, one might interpret that the videography resonated with the viewers because it evoked considerable discussion and affective response. However, the role of the text in the video was questioned, as it distracted from the viewing experience, drawing attention away from the viewpoint of the non-human subjects. Screening the videography proved a fruitful avenue for reflection, learning and means of communication (see Seregina, 2018). We laid bare our incompleteness in understanding non-human subjects' standpoints and experiences of being part of the tourism industry, but the endeavour also demonstrated our willingness to try to understand and to make space for these subjects. Based on the comments received in conference viewings and the advancing of this paper, one member of the team edited the video further. A joint decision was made to leave most of the textual theorising out of the video and hence 'let it speak for itself' (see Hietanen & Rokka, 2017). With this rationale, the rhythm and the sequences presented in the video were once again edited. Now communicated through the video are the troubling, disruptive moments and interactions that had first evoked affectivity in the research process and viewings. The video does not bring closure or provide answers, but rather leaves certain theorisations open for discussion (Hietanen & Andéhn, 2018).

When starting to write this methodological paper, we were confronted with a question regarding the roles of the written text and the video itself. How to 'write the video into the text'? What is the relationship between the text and the video? Is one given more emphasis than the other? Deviating from (for example) Hietanen and Rokka's (2017) companion essay to their 30-minute videography, our written article is not a companion text but rather one that aims to communicate our methodological process concerning the research in its entirety. Given that the theorising through the video is undertaken in an unconventional manner, the aim of this paper is to document and explore this process.

Relating agencies in the tourism context

Agency is deemed one of the most complex concepts in the social sciences, as it refers to a contradictory space, including questions of self-fulfilment, forms of interaction, political possibilities and restrictions in structural and everyday power relations (Koski & Bäcklund, 2017; see McFarland & Hediger, 2009; Philo & Wilbert, 2000). Animals are perceived as having no or limited possibilities to affect these features and therefore it is debatable whether they possess agency (see McFarland & Hediger, 2009). However, Urbanik (2012) disputes this stance by stating that agency is multi-directional and engaged in relations (see also Räsänen & Syrjämaa, 2017).

Animals are involved in tourism activities without their informed consent; they lack the freedom to choose in a way that humans do, even when their agency can be considered active instead of passive (Notzke, 2019). They are subjects to human values and desires. This is an unequal power relation, which might become even more instrumental under commercial pressures (Carr, 2014; Dashper, 2014, 2019). Therefore, it is difficult to believe that animals have a presupposed agency, as the concept is predominantly connected with notions of free will and moral behaviour. Animals are often said to lack such abilities and are thus separate from human beings. However, beings who adapt to the behavioural standards of their groups act in moral ways. This notion can be broadened to an interspecies sense of morality. For example, humans make daily decisions whether to eat meat and there are tales of heroic animals saving humans (McFarland & Hediger, 2009).

Human-animal relations in tourism and leisure contexts are more complex than a master-slave dynamic, as animals shape tourism encounters through their actions and reactions (Notzke, 2019). Animal agency shapes the material world and multispecies encounters influence the way we relate to the world and ourselves

(Räsänen & Syrjämaa, 2017). For example, as animals are tourism workers in Lapland, their working conditions are developed continuously: dog and reindeer harnesses and sledges are designed to be as light as possible so that they can be easily pulled. Animals can also shape the operations of product suppliers and retailing agencies in tourism, as the values of clients have become more sensitive to animal-based attractions (see Notzke, 2019). The co-evolution of human and animal (Haraway, 2003, 2008; see also Dirke, 2017) has shaped the tourism phenomenon and continues to do so.

The videographic research process enabled the becoming of relating agencies – the multispecies assemblage – to become prominent in the tourism context (cf. Barad, 2003; Haraway, 2008; see also Brown & Banks, 2015). This subsequently became the focus of the final video: to express and discuss multispecies assemblages in the tourism context. As this shaping began to unfold through our methodological choices, we increasingly reflected on the array of relating agencies in the multispecies assemblages. Humans cannot engage in two-way verbal conversation with other animals about their lives, but they can communicate in an embodied way by engaging in non-verbal communication and collaboration, and seeking and receiving responses that transcend human language (Dashper, 2017).

By concentrating on embodied human-animal relationships, other forms of subjectivity are potentially revealed, not through rationalised discourse but through corporeal, haptic and sensory performance (see also Merchant, 2011; Rokka & Hietanen, 2018). In movement lies agency, through which animals (including humans) define not only themselves but also space and time as their own (Buller, 2012; von Uexküll, 2010). Movement is essentially multi-directional and therefore perhaps the most inclusive manifestation of agency (Lulka, 2009). Through movement, animals express, enact and develop their agency and communicate it to others, whether it be through acts of resistance or the active co-construction and co-assembly of the world in presence and vitality (Haraway, 2008; see Buller, 2012). Therefore, movement alongside agency becomes 'communication, interrelation and so on emerging dimensions of this motile ontology' (Buller, 2012, p. 145).

In this process, it may be less important to ascertain whether we understand the other subject 'correctly' than to be curious in the first place (Haraway, 2008). When being curious, we will not miss 'a possible invitation, a possible introduction to other-wordling' (Haraway, 2008, p. 20). Haraway refers to this as 'positive knowledge of and with animals', which she sees as something that 'might just be possible' (2008, p. 21). In our videographic work and this written text, we do not want to present too many interpretations on the behalf of others. This is partly because interpretations within our research group vary, and also because the search for 'correct' interpretations of animal behaviour is not the aim of this paper. Instead, we want to express the value of curiosity in what animal subjects might be experiencing as a manifestation of ethics, 'a practice of recognition' (Ogden et al., 2013, p. 9), which we deem a sustainable path for any (ethnographic or other) fieldwork conducted with animal subjects.

In our study, numerous video sequences were filmed from the tourist perspective in line with the original focus of the videography. The main content of the filmed material follows the tourists participating in animal-based activities. Focusing on these activities, it can be noted how the agencies of animals, humans and other socio-materialities interrelate. For the tourists, participation in the activities requires engagement in versatile and mostly unfamiliar practices (see Rantala, Hallikainen, Ilola, & Tuulentie, 2018). The uncertainty of tourists becomes recognisable from their bodily expressions and posture: they seem a bit reserved and cautious in their actions. They follow the instructions of the guides who are more familiar with the animals: they walk behind the guides, move slowly and approach the animals with caution. Our videography depicts the start of a husky safari, where tourists enter the domain of the animals, specifically the kennel at one of the tourism firms. The reactions of the dogs – very loud barking, jumping and running around – play a key role in constructing the socio-material experiences.

One important notion that became available through watching the video material was acknowledgement of the reactions of the animals when they interacted with tourists. In one sequence, a dog retreats from a situation where a photographing tourist moves slightly too close to its habitat. The dog 'backs out' and barks. We can also see the tourist withdrawing and leaving the situation. It seems that both are 'keen on everyone keeping a polite distance' (Candea, 2010, p. 246). Yet, the interpretation of this very situation highlights the open-endedness of videographic theorising (Rokka & Hietanen, 2018), as another 'reading' is that the tourist leaves the situation for a more human-centred reason: not attaining a desirable picture owing to the dog's behaviour. In another sequence, contrary to retreating, a child carefully approaches and pets a dog. Such encounters – whether successful or not – between humans and animals require (non-linguistic) communication and co-operation between subjects.

During the research process, we became aware that our relations with the animals taking part in the study emerged through or in the immediate presence of an inanimate intermediary: the camera. As Lorimer (2010) notes, the camera has a presence. It provided us with an 'entrance' to the field and offered a sense of safety and justification of our actions: in our case as researchers in a tourism context, we were able to approach the animals that are the key producers of the tourism experience via the presence of the camera. We also recognised the camera's power in the way that the animals were sometimes tentative about us holding it. On some occasions, the guides even mentioned that certain animal subjects were suspicious of the camera equipment.

In the context of the fieldwork, the camera partly granted or communicated our 'official' role as researchers who are allowed to be close to the animals and subjects under scrutiny. This connects the researcher with the tourist, a role with particular privileges, as it is usually a camera or a smartphone that works as an intermediary between the tourist and the animal during the encounter. A tourist with a camera in an inappropriate location/situation might receive sympathy and even gain access to places impossible of a person without such a device. Yet, through the camera lens, our relationship with other subjects alters; our embodied experience of being with others becomes intermediated by material and inanimate agency. On the other hand, the camera was affected by other actors, as was the case when the reindeer whirled snow while running. The video camera held by the researchers sitting in the sleigh was consequently covered with snow, which forced them to stop filming.

Furthermore, recognising the agency of the video camera sets forth a way of approaching visual images as being in continuous transformation rather than as straightforward evidence of truth or reality (see Hietanen & Rokka, 2017; Rose, 2007; Shrum, Duque, & Brown, 2005). When we are open to the fluidity of movements and various modes of relating between humans and animals in a tourism context, we may acknowledge the material connection of the relationship and appreciate the lives and cultural dimensions surrounding animals by viewing them as key agents within these relationships (see also Danby & Hannam, 2016). Embracing the non-human does not mean a rejection of the human; rather, this is an act of de-centring (Buller, 2012, p. 140), while simultaneously being aware of the limitations in our efforts to do so. Instead of valuing some subjects of different species and bodily forms of existence over others, we attend to their relationality, forming a multispecies assemblage as a manifestation of relating agencies in a tourism context.

Ethico-politics in making videography

In our research, the affective nature of videography merged with its ethico-political dimension (Parker, 2003; Pullen & Rhodes, 2014) in two interrelated ways: in the production of video material in the field, and in the choices made when producing the actual video outcome, with videography used as a medium of theorising (Hietanen et al., 2014; Rokka & Hietanen, 2018). Both of these notions pertain to one crucial question: 'What should be included in the final video?'

The non-linguistic co-operation with the animal subjects confronted us with an ethical challenge when documenting their daily lives. The videos made visible the fact that during the fieldwork we had prioritised human agency (owing to the focus of original study) and had largely taken our 'right' to film the animals for granted. Some rules of research conduct only apply for humans (such as asking permission to record a video) and seemingly disappear when applied to other animals. This is a challenge of 'ethnography that takes animals seriously as social beings and ethnographic subjects' (Madden, 2014, p. 279). Reflecting on these choices resulted in us trying not to break completely from anthropocentric research tradition in tourism studies, but at least to discuss it. In practice, this meant recognising and communicating animal agency without striving to speak *for* animals as such (see Bear, Wilkinson, & Holloway, 2017; Lorimer, 2010). Even though our videography is mediated by decisions made by us as researchers of what, who and (partly guided by our will) when to film, it gives a voice and visibility to non-human subjects. With videography we are at least able to make animals present (see Hamilton & Taylor, 2017), even while recognising the power imbalances involved in our efforts to do so (Packard, 2008).

Besides facing and reflecting on the power imbalances between the subjects present in the videography, we were confronted by the ethico-politically charged selection made by us researchers of which particular animal subjects to make visible in the videography. Of the three animal species present in our project (horses, reindeer, sled dogs), our video primarily concentrates on sled dogs. In a later phase of the research process, the material of reindeer was added. Indeed, recognising the importance of permitting space for animal agency in its own right, we aimed to broaden understanding of knowledge-producing subjects in a broader sense by including a sequence on reindeer (see also Brown & Banks, 2015).

The prevalence of sled dogs in the videography was not intentional, but rather a direct consequence of the context and condition of our study. Dog sledding is the most popular animal-based tourism activity in Lapland (García-Rosell & Äijälä, 2018) and hence it is unsurprising that the majority of the companies involved in the project were sled dog kennels. As a result, the discussions on animal welfare within the project tended to revolve around sled dogs. It was through these kennels that we principally attained access to places where we could film the encounters between tourists and animals. We should also note that heated public discussions on the welfare of sled dogs in North America stimulated us to pay more attention to this particular group of animals.

Nonetheless, our choices were also pragmatic. Developing a video that is sufficiently coherent without striving for a single representation of 'reality' requires a practical focus. We were aware that the role of reindeer was crucial for the development of the study. In the previous phases of our research, the involvement of reindeer in tourism work as a whole (acknowledging their nature as semi-wild animals) had instigated ethically charged questions within our research team. It was also from our relation with reindeer – the moment of our gazes intersecting (Haraway, 2008) and the moment when our gazes did not meet – that this study had its origin. These ethical considerations guided us to include video material of reindeer in the outcome despite the possible 'lack of focus' that may have been mitigated by focusing on sled dogs.

By making choices about which animal subjects and scenes to include in the videography, the editing process drew attention to the political nature of videographic research. In editing the video material, we became more aware of our ethical responsibilities considering the effects of the videography on our stakeholders (see Hietanen & Rokka, 2017). Although videography is an open-ended theorising process, inviting diverse interpretations by the audience (Rokka & Hietanen, 2018), we as producers of the film outcome must consider the potential impacts on different stakeholders (tourism companies, public agencies and the university) and their resultant reactions. This becomes even more relevant if we consider our own position as tourism researchers, involved in a project based on close collaboration with the industry and partly funded by tourism companies.

Indeed, we acknowledge the power of our videographic research in producing a particular image of the animal-based tourism industry, and the companies and their employees who make a living from these activities. In particular, our awareness of heated public discussions on the welfare of sled dogs in North America in conjunction with our aim to publicly publish the videography on the Internet influenced the editing process. There were some sequences from the field that would probably require additional explanation to avoid misconceptions among the audience. For example, in one sequence, blood can be seen in a sled dog's paw after a safari ride. This was due to specific weather conditions causing the snow to become exceptionally hard, increasing the strain on the dogs' paws, an explanation provided to the researchers by a safari guide. However, without verbal or textual communication in the video material to explain the reason for the bloody paw and how such circumstances are treated, the scene may have been interpreted in problematic ways.

Editing out this material was thus an ethico-political act undertaken by the researchers accountable for editing the video. This case highlights the ethical decisions posed by videographic material given its powerful affective capabilities. This act stimulated us to reflect on our ethical responsibilities as researchers in a multispecies assemblage. These acts and choices are further connected to the ethico-political nature of the decision to produce the video almost entirely without speech and text. Through this choice, we provided even more space for audiences' interpretations of the video. Thus, viewers are given the opportunity to decipher an interspecies relation that lacks a common verbal language. At the same time, the opportunity challenges the audiences, potentially evoking an affective experience that causes them to reflect on possible ways of being, acting, and relating in the world (Rokka & Hietanen, 2018).

Discussion

Let us now turn to reflect on videography as a disruptive practice in relation to the researched phenomenon, as well as towards academic knowledge production as stated in the introduction of this paper (see Lugosi, 2018). Lugosi argues that performative acts of disruption may emerge either accidentally or purposefully. In the videographic research process discussed here, they appeared in both ways.

During the research process, in the field and through the moving image, we experienced powerful embodied disruption caused by the recognition and manifestation of animal subjectivity, which began to guide our research in unintended ways (Buller, 2012; von Uexküll, 2010; see Lugosi, 2018). We saw videography become an ethical, affective and embodied (but not unproblematic) undertaking that invited us to problematise the human-centricity present in the original focus of our study. Through the research process, we began to appreciate the relational and political potentialities and simultaneous challenges of videography (Hietanen & Rokka, 2017; Rokka & Hietanen, 2018) as a research method and vehicle for emergent theorising when employed in multispecies ethnographic research (Kirksey & Helmreich, 2010; Ogden et al., 2013) in a tourism context.[3] Moreover, videography led us to reconsider our relationship to other animals as research subjects in a particular research context partly guided by third-party motivations, and to imagine new ways of producing knowledge in a multidisciplinary and multispecies context.

In line with Dowling et al. (2017), recognising the multispecies assemblage invited and challenged us to do tourism research differently. Thus, through theorising and presenting through video, a non-human agent, we aimed to purposefully disrupt the epistemological practices of academic knowledge production (Garrett, 2010; Lugosi, 2018; Rakić & Chambers, 2009, Seregina, 2018; see also Lugosi & Quinton, 2018) and to consider how to advance 'methodological invigoration and innovation' inside tourism studies regarding more-than-human and non-representational methodologies (see Lorimer, 2010, p. 239). Building on the epistemological underpinnings of video, movement and its power to affect viewers bodily in sensuous and non-linguistic ways, our aim has been to evoke new ways of thinking on the researched phenomenon, as well as about knowledge inside tourism studies (see Hietanen & Rokka, 2017).

Our experiences cast light on the potential of videography as a research methodology to enable both the in situ and retrospective identification of multispecies assemblages emerging in the research context. Indeed, the disruption to the dominant human-centricity of tourism research experienced resonates with a post-humanist recognition 'that none of us are actually distinct from each other, or the world' (Pepperell, 2003, p. 172), and that our relatedness is much more complex than the clear power inequalities present in the animal-based tourism considered. As we view tourism as an ordering, 'a complex materially heterogeneous assemblage which is both coherent and the same time emergent' (Franklin, 2012, p. 46), it is not a formation of separable elements interacting with one another, with humans the only movers, a sole agent (e.g. Haraway, 2008; see Franklin 2012). Congruent with Haraway's (2008) insights about animals as agents that we 'live with', we gained new tools to theorise and make tangible how animals working in the tourism industry should not be treated as mere objects of our consumption or objects of academic inquiry, but rather as subjects with which to live and encounter with respect. These active subjects additionally shape who we (and therefore what our academic disciplines) become in this process of engagement, that is, 'through relations with other beings' (Ogden et al., 2013, p. 17).

Conclusion

In this paper, we have explored videography as a vehicle for emergent theorising in post-human tourism studies through its relational and political potentialities. Human-animal studies (see e.g. Marvin & McHugh, 2014) and consequently methods to explore multispecies relations represent a growing field of research (see e.g. Bastian, Jones, Moore, & Roe, 2017). This work can be found across various fields, such as geography (e.g. Barua, 2014; Brown & Banks, 2015; Buller, 2015; Lorimer, 2010), anthropology (e.g. Candea, 2010; Fijn, 2012; Grimshaw, 2011; Kohn, 2007; Maurstad, Davis, & Cowles 2013), sociology (e.g. Hamilton & Taylor, 2017; Konecki, 2008; Malone, Selby, & Longo, 2014) and tourism and leisure studies (e.g. Bertella, 2014, 2016; Curtin 2006; Dashper, 2019; Markuksela & Valtonen, 2019; Peltola & Heikkilä, 2015). In this vein, some scholars have explored the potential of moving image methodologies to develop an understanding of our relationships with other animals (e.g. Bear et al., 2017; Brown & Banks, 2015; Lorimer, 2010; Pagel, Scheer, & Lück, 2017). Our videographic study has relatedly sought to disrupt the anthropocentric nature of academic research.

The three themes, *Emergent theorising through affectivity and iteration, Relating agencies in the tourism context, and Ethico-politics of making videography,* have helped unfold the contributions of this study. Through these themes, we have elaborated how videography became a disruptive academic practice in numerous ways (see Lugosi, 2018): first, the affectivity of the methodology led the animal subjects – the disruptive 'others' – to expand our understanding as researchers on the topics to be studied and how and *with whom* we are able to produce knowledge. The videographic methodology forced us to question our conventional ways of doing academic knowledge in practice, as well as at ethico-political level: what counts as knowledge and how we can and should interpret, present and theorise about/on the non-human and the non-representational. Yet, videography does not provide us with definitive answers, but rather leaves us with further questions on (academic) knowledge and its production. Hence, besides the methodology's unfolding of disruptive 'others', it can itself be seen as a disruptive other of academia, its agentic forces challenging the convention whereby linguistic and textual forms of knowledge production are privileged.

Although we conclude by questioning rather than answering, recognition of the challenges of the development of more-than-human methodologies does not preclude us from trying to create space for the co-constitution of knowledge. Leaning on the assumption that we are always limited in our capacity to understand non-human perspectives, to give up trying would in our view be more ethically questionable than acting. Thus, we dare to remain lacking in our efforts. Instead of speaking for or on behalf of animals based on an understanding that we have some kind of special 'right' to do so, we make way for deliberation on the dilemma and struggle arising from us trying, or feeling the need, to do so. Recognising this, we perceive our work as a door opener (Hamilton & Taylor, 2017) that strives to view multispecies assemblages as a path to versatility-embracing knowledge, rather than one privileging a human perspective. We hope that through its relational and political qualities, affective and non-representational videography as a methodology can help us to develop more inclusive tourism theorising and futures.

Notes

1. While being aware of the tendency of the category of 'non-human' to propose human exceptionalism, we use this categorisation with care in those situations where the focus is explicitly to be taken further (or widened) from the human, in order to give room for other types of agencies as well as to bring forth the disruption caused by their realisation.
2. The study was conducted between 2016 and 2018 as part of the project 'Animals and Responsible Tourism: Promoting Business Competitiveness through Animal Welfare'. The project, which was implemented by the University of Lapland, was funded by the European Regional Development Fund (ERDF) through Business Finland, a Finnish funding agency. In addition to public funding, 10 per cent of the project budget was financed by a group of tourism companies, because this is a requirement to receive funding through this particular funding instrument.
3. Even though our research is not to be considered a 'full' ethnography, it correlates with the central notion of multispecies ethnography in pressing (or perhaps) *inviting* 'creatures previously appearing on the margins of anthropology … into the foreground' (Kirksey & Helmreich, 2010, p. 545), and in holding ethnographic elements in its methodological array. Our study takes this 'foregrounding' into practice in the field of tourism studies, and as such builds on existing work exploring human-animal relations in tourism (e.g. Fennell, 2012; Markwell, 2015; see Authors).

Acknowledgements

The authors would like to thank the audiences of the videography screenings in the "2nd Peaceful Coexistence Colloquium - Reimagining Ethics and Politics of Space for the Anthropocene" and the "Living Ethics Seminar" for the inspiring and insightful discussions that helped to develop the videography to its final version. We also thank associate professor Joonas Rokka and the three anonymous reviewers for their encouraging and constructively critical comments received for the earlier versions of this article.

Disclosure statement

No potential conflict of interest was reported by the authors.

References

Äijälä, M., García-Rosell, J.-C., & Haanpää, M. (2016). Kirjallisuuskatsaus: Eläimet osana matkailutoimintaa [Literature review: Animals as part of tourism]. *Matkailututkimus [Finnish Journal of Tourism Research]*, *12*(2), 45–59.

Barad, K. (2003). Posthumanist performativity: Toward an understanding of how matter comes to matter. *Signs: Journal of Women in Culture and Society*, *28*(3), 801–831. doi:10.1086/345321

Barad, K. (2007). *Meeting the universe halfway: Quantum physics and the entanglement of matter and meaning*. Durham: Duke University Press.

Barua, M. (2014). Circulating elephants: Unpacking the geographies of a cosmopolitan animal. *Transactions of the Institute of British Geographers*, *39*(4), 559–573. doi:10.1111/tran.12047

Bastian, M., Jones, O., Moore, N., & Roe, E. (Eds.). (2017). *Participatory research in more-than-human worlds*. Abingdon: Routledge.

Bear, C., Wilkinson, K., & Holloway, L. (2017). Visualizing human-animal-technology relations. Field notes, still photography, and digital video on the robotic dairy farm. *Society & Animals*, *25*(3), 225–256. doi:10.1163/15685306-12341405

Belk, R. W., Caldwell, M., Devinney, T. M., Eckhardt, G. M., Henry, P., Kozinets, R., & Plakoyiannaki, E. (2018). Envisioning consumers: How videography can contribute to marketing knowledge. *Journal of Marketing Management*, *34*(5-6), 432–458. doi:10.1080/0267257X.2017.1377754

Belk, R. W., & Kozinets, R. V. (2005). Introduction to the resonant representations issue of consumption, markets and culture. *Consumption Markets & Culture*, *8*(3), 195–203. doi:10.1080/10253860500160395

Bertella, G. (2014). The co-creation of animal-based tourism experience. *Tourism Recreation Research*, *39*(1), 115–125. doi:10.1080/02508281.2014.11081330

Bertella, G. (2016). Experiencing nature in animal-based tourism. *Journal of Outdoor Recreation and Tourism*, *14*, 22–26. doi:10.1016/j.jort.2016.04.007

Brown, K., & Banks, E. (2015). Close encounters: Using mobile video ethnography to understand human-animal relations. In C. Bates (Ed.), *Video methods: Social science research in motion* (pp. 95–120). Abingdon: Routledge.

Buller, H. (2012). "One slash of light, then gone". Animals as movement. *Études Rurales*, *189*, 139–154. doi:10.4000/etudesrurales.9630

Buller, H. (2015). Animal geographies II: Methods. *Progress in Human Geography*, *39*(3), 374–384. doi:10.1177/0309132514527401

Cahill, J. L. (2015). Animal Photogénie: The wild side of French film theory's first wave. In M. Lawrence & L. McMahon (Eds.), *Animal life and the moving image* (pp. 23–41). London: Palgrave.

Candea, M. (2010). "I fell in love with Carlos the meerkat": Engagement and detachment in human–animal relations. *American Ethnologist*, *37*(2), 241–258. doi:10.1111/j.1548-1425.2010.01253.x

Carr, N. (2014). *Dogs in the leisure experience*. Wallingford: CABI.

Clocke, P., & Perkins, H. C. (2005). Cetacean performance and tourism in Kaikoura, New Zealand. *Environment and Planning D: Society and Space*, *23*(6), 903–924. doi:10.1068/d57j

Curtin, S. (2006). Swimming with dolphins: A phenomenological exploration of tourist recollections. *International Journal of Tourism Research*, *8*(4), 301–315. doi:10.1002/jtr.577

Danby, P., & Hannam, K. (2016). Entrainment. Human-equine leisure mobilities. In J. Rickly, K. Hannam, & M. Mostafanezhad (Eds.), *Tourism and leisure mobilities* (pp. 27–38). Abingdon: Routledge.

Dashper, K. (2014). Tools of the trade or part of the family? Horses in competitive equestrian sport. *Society & Animals*, *22*(4), 352–371. doi:10.1163/15685306-12341343

Dashper, K. (2017). *Human-horse relationships in equestrian sport and leisure*. Abingdon: Routledge.

Dashper, K. (2019). Moving beyond anthropocentrism in leisure research: Multispecies perspectives. *Annals of Leisure Research*, *22*(2), 133–139. doi:10.1080/11745398.2018.1478738

Dowling, R., Lloyd, K., & Suchet-Pearson, S. (2017). Qualitative methods II: 'More-than-human' methodologies and/in praxis. *Progress in Human Geography*, *41*(6), 823–831. doi:10.1177/0309132516664439

Dirke, K. (2017). Changing narratives of human-large carnivore encounters in nineteenth-century Sweden. In T. Räsänen & T. Syrjämaa (Eds.), *Shared lives of humans and animals. Animal agency in the global North* (pp. 163–178). Abingdon: Routledge.

Fennell, D. A. (2012). *Tourism and animal ethics*. Abingdon: Routledge.

Fennell, D. A. (2013). Tourism and animal welfare. *Tourism Recreation Research*, *38*(3), 325–340. doi:10.1080/02508281.2013.11081757

Fijn, N. (2012). A multi-species etho-ethnographic approach to filmmaking. *Humanities Research*, *18*(1), 71–88. doi:10.22459/HR.XVIII.01.2012.05

Franklin, A. (2012). The choreography of a mobile world. Tourism ordering. In R. van der Duim, C. Ren, & G. T. Johannesson (Eds.), *Actor-network theory and tourism. Ordering, materiality and multiplicity* (pp. 43–58). Abingdon: Routledge.

García-Rosell, J.-C., & Äijälä, M. (2018). Animal-based tourism in Lapland. In J. Ojuva (Ed.), *Animal welfare in tourism services: Examples and practical tips for the well-being of animals used for tourism in Lapland* (pp. 10–24). Rovaniemi: Lapland University of Applied Sciences.

Grimshaw, A. (2011). THE BELLWETHER EWE: Recent developments in ethnographic filmmaking and the aesthetics of anthropological inquiry. *Cultural Anthropology*, *26*(2), 247–262. doi:10.1111/j.1548-1360.2011.01098.x

Hamilton, L., & Taylor, N. (2017). *Ethnography after humanism. Power, politics and method in multi-species research*. London: Palgrave Macmillan.

Haraway, D. (2003). *Companion species manifesto*. Chicago, IL: Prickly Paradigm Press.

Haraway, D. (2008). *When species meet*. Minneapolis: University of Minnesota Press.
Hietanen, J., & Andéhn, M. (2018). More than meets the eye: Videography and production of desire in semiocapitalism. *Journal of Marketing Management*, *34*(5-6), 539–556. doi:10.1080/0267257X.2017.1402807
Hietanen, J., & Rokka, J. (2017). Companion for the videography 'Monstrous organizing—The dubstep electronic music scene'. *Organization*, *25*(3), 320–334. doi:10.1177/1350508417738572
Hietanen, J., Rokka, J., & Schouten, J. W. (2014). Commentary on Schembri and Boyle (2013): From representation towards expression in videographic consumer research. *Journal of Business Research*, *67*(9), 2019–2022. doi:10.1016/j.jbusres.2013.10.009
Kirksey, S. E., & Helmreich, S. (2010). The emergence of multispecies ethnography. *Cultural Anthropology*, *25*(4), 545–576. doi:10.1111/j.1548-1360.2010.01069.x
Kohn, E. (2007). How dogs dream: Amazonian natures and the politics of transspecies engagement. *American Ethnologist*, *34*(1), 3–24. doi:10.1525/ae.2007.34.1.3
Konecki, K. T. (2008). Touching and gesture exchange as an element of emotional bond construction. Application of visual sociology in the research on interaction between humans and animals. *Forum: Qualitative Social Research*, *9*(3). Retrieved from http://www.qualitative-research.net/index.php/fqs/article/view/1154
Koski, L., & Bäcklund, P. (2017). Whose agency? Humans and dogs in training. In T. Räsänen & T. Syrjämaa (Eds.), *Shared lives of humans and animals. Animal agency in the global North* (pp. 11–23). Abingdon: Routledge.
Lewis, T., & Potter, E. (2011). Introducing ethical consumption. In T. Lewis & E. Potter (Eds.), *Ethical consumption: A critical introduction* (pp. 3–21). New York, NY: Routledge.
Lorimer, J. (2010). Moving image methodologies for more-than-human geographies. *Cultural Geographies*, *17*(2), 237–258. doi:10.1177/1474474010363853
Lugosi, P. (2018). Disruptive ethnography and knowledge co-creation. In C. Ren, G. Jóhannesson, & R. van der Duim (Eds.), *Co-creating tourism research: Towards collaborative ways of knowing* (pp. 55–72). Abingdon: Routledge.
Lugosi, P., & Quinton, S. (2018). More-than-human netnography. *Journal of Marketing Management*, *34*(3-4), 287–313. doi:10.1080/0267257X.2018.1431303
Lulka, D. (2009). Consuming Timothy Treadwell: Redefining nonhuman agency in light of Herzog's Grizzly Man. In S. E. McFarland & R. Hediger (Eds.), *Animals and agency* (pp. 55–72). Leiden: Brill.
Madden, R. (2014). Animals and the limits of ethnography. *Anthrozoös*, *27*(2), 279–293. doi:10.2752/175303714X13903827487683
Malone, N., Selby, M., & Longo, S. (2014). Political-ecological dimensions of silvery gibbon conservation efforts. *International Journal of Sociology*, *44*(1), 34–53. doi:10.2753/IJS0020-7659440103
Markuksela, V., & Valtonen, A. (2019). Dance with a fish? Sensory human-nonhuman encounters in the waterscape of match fishing. *Leisure Studies*, *38*(3), 353–366. doi:10.1080/02614367.2019.1588353
Markwell, K. (Ed.). (2015). *Animals and tourism: Understanding diverse relationships*. Bristol: Channel View.
Marvin, G. & McHugh, S. (2014). In it together. An introduction to human-animal studies. In G. Marvin & S. McHugh (Eds.), *Routledge handbook of human-animal studies* (pp. 1–9). Abingdon: Routledge.
Maurstad, A., Davis, L. D., & Cowles, S. (2013). Co-being and intra-action in horse-human relationships: A multi-species ethnography of be(com)ing human and be(com)ing horse. *Social Anthropology*, *21*(3), 322–335. doi:10.1111/1469-8676.12029
McFarland, S. E., & Hediger, R. (Eds.). (2009). *Animals and agency: An interdisciplinary exploration*. Leiden: Brill.
Merchant, S. (2011). The body and the senses: Visual methods, videography and the submarine sensorium. *Body & Society*, *17*(1), 53–72. doi:10.1177/1357034X10394670
Notzke, C. (2019). Equestrian tourism: Animal agency observed. *Current Issues in Tourism*, *22*(8), 948–966. doi:10.1080/13683500.2017.1349081

Ogden, L. A., Hall, B., & Tanita, K. (2013). Animals, plants, people, and things. A review of multispecies ethnography. *Environment and Society*, *4*(1), 5–24. doi:10.3167/ares.2013.040102

Ong, C.-E. (2017). Cuteifying' spaces and staging marine animals for Chinese middle-class consumption. *Tourism Geographies*, *19*(2), 188–207. doi:10.1080/14616688.2016.1196237

Packard, J. (2008). I'm gonna show you what it's really like out here': The power and limitation of participatory visual methods. *Visual Studies*, *23*(1), 63–77. doi:10.1080/14725860801908544

Pagel, C. D., Scheer, M., & Lück, M. (2017). Swim encounters with killer whales (Orcinus orca) off Northern Norway: Interactive behaviours directed towards human divers and snorkellers obtained from opportunistic underwater video recordings. *Journal of Ecotourism*, *16*(2), 190–200. doi:10.1080/14724049.2016.1273939

Parker, M. (2003). Introduction: Ethics, politics and organizing. *Organization*, *10*(2), 187–203. doi:10.1177/1350508403010002001

Peltola, T., & Heikkilä, J. (2015). Koira suden hampaissa – Monilajinen yhteiselämä, kuolema ja osallisuus. [Multispecies co-living, death and participation]. In J. Hiedanpää & O. Ratamäki (Eds.), *Suden kanssa* (pp. 120–146). Rovaniemi: LUP.

Pepperell, R. (2003). *The posthuman condition consciousness beyond the brain*. Bristol: Intellect Books.

Petr, C., Belk, R. W., & Decrop, A. (2015). Videography in marketing research. Mixing art and science. *Arts and the Market*, *5*(1), 73–102. doi:10.1108/AM-01-2014-0002

Philo, C. & Wilbert, C. (Eds.). (2000). *Animal spaces, beastly places: New geographies of human-animal relations*. London: Routledge.

Pick, A. (2015). Animal life in the cinematic umwelt. In M. Lawrence & L. McMahon (Eds.), *Animal life and the moving image* (pp. 221–237). London: Palgrave.

Picken, F. (2018). Knowing the aquatic other: Unleashing Blackfish. In C. Ren, G. T. Jóhannesson, & R. van der Duim (Eds.), *Co-creating tourism research. Towards collaborative ways of knowing* (pp. 147–161). Abingdon: Routledge.

Pullen, A., & Rhodes, C. (2014). Corporeal ethics and the politics of resistance in organizations. *Organization*, *21*(6), 782–796. doi:10.1177/1350508413484819

Rakić, T., & Chambers, D. (2009). Researcher with a movie camera: Visual ethnography in the field. *Current Issues in Tourism*, *12*(3), 255–270. doi:10.1080/13683500802401972

Rantala, O., Hallikainen, V., Ilola, H., & Tuulentie, S. (2018). The softening of adventure tourism. *Scandinavian Journal of Hospitality and Tourism*, *18*(4), 343–361.

Ratamäki, O., & Peltola, T. (2015). Drama over large carnivores: Performing wildlife tourism in a controversial space. In K. Markwell (Ed.), *Animals and tourism. Understanding diverse relationships* (pp. 180–193). Bristol: Channel View.

Regional Council of Lapland. (2015). *Lapin matkailustrategia 2015-2018 [Lapland tourism strategy 2015-2018]*. Rovaniemi: Lapin Liitto (Regional Council of Lapland).

Rokka, J., & Hietanen, J. (2018). On positioning videography as a tool for theorizing. *Recherche et Applications en Marketing (English Edition))*, *33*(3), 106–121. doi:10.1177/2051570718754762

Rose, G. (2007). *Visual methodologies. An introduction to the interpretation of visual materials* (2nd ed.). London: Sage.

Räsänen, T., & Syrjämaa, T. (2017). Introduction. In T. Räsänen & T. Syrjämaa (Eds.), *Shared lives of humans and animals. Animal agency in the global North* (pp. 1–7). Abingdon: Routledge.

Seregina, A. (2018). Engaging the audience through videography as performance. *Journal of Marketing Management*, *34*(5-6), 518–535. doi:10.1080/0267257X.2017.1377753

Shrum, W., Duque, R., & Brown, T. (2005). Digital video as research practice: Methodology for the millennium. *Journal of Research Practice*, *1*(1), 1–19.

Urbanik, J. (2012). *Placing animals: An introduction to the geography of human-animal relations*. Lanham, MD: Rowman & Littlefield Publishers.

Vannini, P. (2015). Afterword: Video methods beyond representation: Experimenting with multimodal, sensuous, affective intensities in the 21st Century. In C. Bates (Ed.), *Video methods – Social science research in motion* (pp. 230–240). Abingdon: Routledge.

von Uexküll, J. (2010). *A foray into the worlds of animals and humans*. Minneapolis: Minnesota University Press.

Wood, M. (2015). Audio-visual production and shortcomings of distribution in organisation studies. *Journal of Cultural Economy*, *8*(4), 462–478. doi:10.1080/17530350.2014.942346

Yudina, O., & Grimwood, B. S. R. (2016). Situating the wildlife spectacle: Ecofeminism, representation, and polar bear tourism. *Journal of Sustainable Tourism*, *24*(5), 715–734. doi:10.1080/09669582.2015.1083996

Emplacing non-human voices in tourism research: the role of dissensus as a qualitative method

Abhik Chakraborty

ABSTRACT

Although tourism shares a fundamentally important relationship with the natural environment, non-human voices are easily lost in conventional methods used in tourism research that favor consensus and clear patterns. Even with qualitative methods, researchers tend to seek consensus among peers and often focus disproportionately on the *social* dynamics of tourism, leading to entrenched anthropocentrism and limited efficacy in representing the condition of a multitude of non-human lifeforms, physical formations, and processes. The portrayal of the tourism environment as a passive backdrop for human activities is particularly insufficient for addressing problems at eco or natural tourism destinations where key roles of fellow species and biophysical formations make tourism a more-than-human phenomenon. This situation calls for engaging with *non-human others* in the tourism landscape through novel methods that are creative, fluid, and to some extent free of theoretical or paradigmatic moorings. Informed by a dissensus on the centrality of the human experience for meaning-making, such engagements are disruptive of the dominant anthropocentric tendencies in qualitative research but are nevertheless necessary for addressing wicked problems, the complex nature of which preclude anthropocentric solutions. As an alternative, fluid and speculative inquiries into the inherent plurality and complexity of non-human others have the capacity to grow into an organic toolkit to inform tourism research in a time of pervasive change.

摘要

虽然旅游业与自然环境有着极为重要的关系，但在传统的旅游研究方法中，非人类的声音很容易被忽略，因为传统的研究方法更倾向于共识和清晰的模式。即使采用定性的方法，研究人员也倾向于寻求同行之间的共识，往往不成比例地关注旅游的社会动态，从而导致根深蒂固的人类中心主义，在表现大量非人类生命形式、物理形态和过程的条件方面效力有限。将旅游环境描述为人类活动的被动背景，对于解决生态或自然旅游目的地的问题特别不够，因为在这些地方，同类物种和生物物理结构的关键作用使旅游成为一种远非简单的人类现象。这种情况需要通过创造性的、流动的、在某种程度上不受理论或范例束缚的创新方法研究旅游景观中的非人类他者。通过对意义形成的人类经验的中心地位的异议，这种介入破坏了定性研究中占主导地位的人类中心主义倾向，但对于解决旅游发展的痼疾是必要的，因为这些问题的复杂性排除了人类中心主义的解决方案。作为一种替代方案，对非人类固有的多样性和复杂性进行流动的和推测性的探究，有能力成长为一个有机的工具包，在一个普遍变化的时代为旅游研究贡献新知。

Introduction: the need of acknowledging voices of non-human others in qualitative tourism research

The aim of this article is to argue that the current scope of qualitative tourism research remains inadequate for addressing multiple challenges for sustainability in ecotourism/natural area tourism destinations and should be complemented with fluid and speculative approaches that challenge anthropocentric tendencies and uphold the complexity and multifariousness of non-human life and physical formations (and processes). Eluding sharply defined methodological contours, this engagement is informed by a 'dissensus' on the narratives that seek to promote the centrality of human experience for understanding worldly phenomena, and by the need to emplace *non-human others* in the research narrative.

Dissensus, defined simply, is the lack of consensus, implying the state of divergence of opinions (Oxford English Dictionary Online, n.d.; Merriam-Webster Dictionary Online, n.d.). I use the term for additional and related meaning of causing disruption in the dominant narrative of qualitative research, and for bold, non-mainstream, creative engagements. Specifically, the case argued for is one against entrenched and incipient anthropocentric tendencies in qualitative research that result in 'humanistic' and 'constructionist' narratives placing the autonomous human subject at the center of the research inquiry, process, and outcome. It should also be pointed out for clarity that the terms *humanistic* and *constructionist* themselves have polymorphous origins and characteristics, and that there are multiple constructionist and humanist research agendas. A full analysis of the variants of social constructionism is beyond the scope of this methodological critique, here the aim is to highlight the inherent limitation of the various shades of constructionist and humanist research agendas that privilege human-centered interpretations of worldly phenomena. It will be argued that such tendency to privilege the *human* stymies the development of more creative, apolitical, and reflexive approaches. As the bulk of the social science theories and epistemological underpinnings used in 'critical qualitative research' dwells on the human subject (Denzin & Lincoln, 2018), the term 'dissensus' is particularly appropriate to sum up the open-ended explorations of the 'other' that perennially challenge the epistemological and philosophical moorings of anthropocentric inquiry. It is also an invitation to look beyond what I perceive to be a strong urge to systematize and theorize from our engagements with the material world. The dissensus approach is particularly pertinent for informing tourism research that seeks to analyze natural-or-eco-tourism destinations, where a number of endangered species and sensitive ecosystems may form the core of visitor attractions. However, it should also be mentioned that this critique does not totally negate the important role of critical assessments of the human subject as the 'performer.' Rather, it provides a point of caution that such assessments sometimes fall into the trap of 'linearity' due to their adherence to knowledge paradigms that seek to incorporate worldly phenomena into human subjectivity. In extension this critique constitutes a counterpoint to those tendencies that seek to systematize knowledge, phenomena, and our engagements with the inherently heterogeneous, plural, and chaotic earth solely from a 'human' standpoint.

The argument is developed in the following manner: first a brief 'overview' of linearity in qualitative research (and qualitative tourism research) is presented, followed by arguments for 'rejecting linearity' in its multiple guises, a critique of the paradigm of 'social construction' that is seen as fostering linearity and anthropocentrism, and a call for recognizing peripheral and dissenting voices as well as for 'minding the gap' regarding non-human others in qualitative tourism research. These sections are followed by insights of fluid methodological praxis derived from two case studies conducted in Shiretoko Peninsula and Shirakami Mountains World Natural Heritage Sites (WNHS) in Japan, and a short conclusion.

Qualitative tourism research and linearity: a brief overview

Before presenting the critique of qualitative tourism research as privileging the human and fostering linearity in the process it needs to be acknowledged that tourism research has grown to become a rich, complex, and fertile field over the years. Increasingly more diverse research methods are being used, and the overall trajectory has moved from a predominantly positivist approach of fact-checking or hypothesis testing to more iterative, reflexive inquiries (Tribe & Airey, 2007; Richards & Munsters, 2010). Whereas early tourism research was dominated by economic and management agendas (Tribe, 1997; Jennings, 2009), theoretical inquiries on the nature of tourism and its qualitative aspects became prominent by the end of the twentieth century (Dann & Phillips, 2001; Tribe, 2006; Goodson & Phillimore, 2004). Such a shift implied a turn away from the positivist approach to a more diverse mix of paradigms including postmodernist and constructionist approaches that question 'value-free' research and 'objective truth' (Goodson & Phillimore, 2004). Pritchard, Morgan, and Ateljevic (2011) observed that tourism research is now increasingly reflexive and transformative—a change upheld by the formation of 'critical' and 'hopeful' tourism scholarships.

However several scholars have also identified persistent problems. Some of these problems are: that tourism research is shallow in theoretical engagements (Goeldner, 2005; Tribe, 2010); that it is nevertheless shaped by positivist pressure in funding allocation and publication (Ateljevic et al., 2007; Caton, 2012); and the limited extent to which feminist critiques are employed (Aitchison, 2005; Wilson & Hollinshead, 2015). Furthermore, Botteril (2007, 2014) cautioned against the relativistic turn in tourism research, asking whether the proliferation of voices to enable multiple representations of tourism phenomena in itself can be seen as adequate maturity of the research field. Upon examining the truth claims put forth by constructionist narratives, Botteril (2014) highlighted their limited efficacy in describing the materiality of reality in the following words:

> Entrapped by the insistence of the exploration of social meanings as an end point, the truth claims of constructionist knowledge remains locked within the self-referential transitive domain, validated only by the processes and protocols of the relevant social scientific discipline communities, and defended by the mirrors of reflexivity and situatedness (p. 294).

Botteril's observation underscores the point that even relativist and constructionist positions that stand in apparent direct contrast to positivist claims may induce loop-thinking and linearity.

This point is worth attention because it opens up the debate on how reflexive qualitative tourism research really is. In particular the proliferation of constructionist approaches and their recent domination of the qualitative research narrative make this a pertinent question. It can be argued that while qualitative research has sought to depart with positivist linearity of cause and effect and claims of universal truth, a different type of linearity is present—albeit in a less visible form—in qualitative inquiry paradigms. This linearity stems from, and is in turn bolstered by, the apparent consensus view that renders worldly phenomena into *social artifacts,* or to use a social constructionist term, into 'constructs' that are shaped by the autonomous human agent. It will also be argued below that equating quantitative research with 'linearity' and scientific research with quantitative approaches are both symptomatic of the inertia in qualitative research that originates largely out of relativist and constructionist approaches that disproportionately focus on the autonomous human subject at the expense of the more multifarious non-human stakeholders. While criticizing quantitative and positivist approaches for hegemonic tendencies, qualitative researchers have, perhaps somewhat unwittingly, ended up marshaling like-minded colleagues who are content with criticizing positivist approaches. This implies a buildup of consensus within qualitative—and more specifically social constructionist— research scholarship regarding the centrality of the human experience, which unfortunately narrows down the applicability of qualitative research for exploring the earth (and our place in it) in a more diverse manner. Herein lies the necessity of including a non-human approach that can creatively rupture this dominant narrative and incorporate more fluid ways of engaging with tourism related phenomena.

Rejecting linearity in the research narrative

Whereas quantitative research paradigms are generally underpinned by a strong commitment to linear connection between cause and effect (deductive reasoning), qualitative research attaches import to the crucial role of context, multiple interaction pathways between phenomena (complexity) and the central role of interpretation (inductive reasoning) (Patton, 2002; Creswell, 2012; Silverman, 2016; Bhattacharya, 2017). Qualitative researchers often follow the tenets developed by Guba and Lincoln (1989, 1990) that described ontologically relativist, epistemologically subjectivist, and methodologically hermeneutic/dialectic tendencies in social science inquiries. This guideline was later expanded to interpret the existence of multiple 'truths' influenced by sociopolitical or cultural backgrounds and to embrace ontological relativity (Patton, 2002). However, often insufficiently discussed is the point that scholars like Lincoln and Guba (1989, 1990) were mainly concerned with interpreting 'social' phenomena, and their criticism against quantitative methods appears to be more of a critique of 'positivist' tendencies within the larger and more varied quantitative research landscape. It can also be argued that such tendencies were, and are, more prevalent among quantitative social science disciplines such as economics that seek to apply numerical analysis modeled after the natural sciences in order to understand social dynamics, rather than being a central tenet of natural sciences per se.

Paul Feyerabend, in his provocatively titled book 'Against Method' (2010), claimed that much of the natural sciences lack any consistent devotion to theories and methods, and that science is fundamentally anarchist in nature. Feyerabend based his claim on the examination of Copernican, Galilean, and Newtonian paradigms (which today are seen as 'deterministic'); and by doing so, made the important observation that these dominant narratives of science were once radical departures that 'ruptured' a previous set of dominant narratives. Further along this line we encounter Einstein's Theories of Relativity, which in turn ruptured the Newtonian worldview with the radical proposition that space and time are not different entities but are intertwined, and that all observed phenomena are 'relative' to the position of the observer (for detailed discussions refer to Cooperstock, 2009; Poisson & Will, 2014). While Einstein's theories forced a sweeping rethinking in the way space and time were considered, they still described a deterministic universe where in principle phenomena could be observed and deduced. This line of scientific thought was fundamentally challenged by Quantum Physics which showed that in the world of subatomic particles, linear cause and effect (and in fact any type of 'cause and effect') relations do not hold up. By invoking Schrodinger's famous conjecture that both a 'dead' and 'alive' cat could share the same spacetime, quantum physicists advanced the understanding of the observed universe as a representation of probabilities that are fundamentally irreducible (for a lucid description of the 'quantum turn' in Physics see Gribbin, 2012; New Scientist Instant Expert, 2017). It is perhaps not unjustified to say that social science research has not yet caught up with this level of intense examination of reality.

Apart from the decidedly fuzzy world of Quantum Physics that explicitly rejects linearity and causality in its traditional sense, another major advance regarding the irreducibility of the biophysical world was provided by the environmental sciences during the latter part of the twentieth century through the advancement of the 'organic' wholeness of the earth as a planet (Lovelock, 2000; Wilson, 2016)). This was complemented by the application of the 'chaos' concept that rejects linear causality and upholds complexity as a fundamental component of our world (Gleick, 2008). The insights derived from these inquiries are reflected in the understanding of contemporary environmental problems (such as climate change) as 'wicked problems', implying a solution at one place can compound a problem in another location (Incropera, 2015; Chapman, 2017; Lovelock, 2009, 2015). In addition scholars such as Room (2012) argued that contemporary global systems (including environmental assemblages) are perpetually in motion, which denies any possibility of finding a problem unchanged as soon as some time elapses (and therefore precluding a premeditated solution). The fluid state of the world in this sense denies us any opportunity to identify a problem, analyze it, and solve that *same* problem, and this posits a further challenge for linear thinking in research and management. With respect to environmental decision-making, Stirling (2010) reported of the strong social pressure to make research results clear, politically relevant, and socially acceptable as impoverishing knowledge, and called for the recognition of qualitative research outcomes that uphold ambiguity and uncertainty of knowledge, specifically through employing dissensus in the form of divergence of opinions or opinions outside prevalent social narratives.

The bottomline of understanding worldly phenomena as perennially fluid, irreducibly complex, and chaotic is that it precludes the positioning of the 'human' at the centerstage, by informing us that the inherent, organic complexity in the worldly phenomena perpetually escapes our capacity to *understand* such phenomena. As our understandings of the world, especially those founded on the centrality of the human experience, are found to be partial and incomplete, the need to engage with non-human others in an imaginative and speculative manner so as to understand their irreducible complexity and intrinsic worth, emerges. This is particularly relevant for tourism, especially eco-and other forms of nature-based tourism that enjoy a fundamental relationship with non-human biotic and abiotic assemblages. While there is a need to offer practical solutions, tourism researchers should also examine tourism-related phenomena through imaginative, fluid, and esoteric inquiries. Indeed, some tourism scholars have already provided valuable insights through their engagements with less than certain forms of inquiry about the changing earth (Farrell & Twining-Ward, 2004; Gren & Huijbens, 2012, 2016). Huijbens and Gren (2012) reminded us that we humans are one of the many inhabitants of this planet, and Benediktsson, Lund, and Huijbens (2011), and Lund and Benediktsson (2011) analyzed the tenuous nature of human mobility in the face of a dynamic planet with reference to the volcanic eruption of Eyjafjallajökull that had a disproportionate impact on global air travel in 2010. These engagements inform us that qualitative methods that foster the consensus worldview of natural phenomena as socially constructed have fundamental shortcomings, especially with respect to their inability to provide what Picken (2010) termed as the 'performative definition' of non-human others. Furthermore, insights from such inquiries have also been related to an application of the Actor-Network Theory (hereafter ANT) in tourism studies (Van der Duim, Ren & Johanneson, 2013)—a form of creative engagement with non-human others that will be discussed below. Here I intend to buttress the argument that the equation of quantitative research with positivism/linearity is to present only a partial and simplistic picture; and there are disruptive traditions within the natural environmental sciences that offer less than certain worldviews and nuanced and complex understandings, and that those traditions could be useful for inquiries on tourism related phenomena.

Limitations of the social constructionist approach and (re) assertion of more-than-human approaches

From the preceding discussion it can be posited that it is increasingly necessary to creatively disrupt the dominant qualitative humanist narrative by arguing that knowledge about the earth is only partial if premised on the centrality of the human experience and oriented to 'social justice' (for the centrality of social justice in qualitative research see Denzin & Lincoln, 2018), and should be complemented by embracing more-than-human approaches. The major epistemological claim presented here is that the worldly phenomena, and therefore tourism phenomena, are complex and multifarious, and are often performed by agents of an ancient earth on which we are newcomers—implying that the insistence of human subjectivity for meaning-making makes knowledge partial, immature, and incomplete. The need to question and move beyond anthropocentrism and to embrace fluid, theoretically unmoored methods as an alternative is informed by the position that our knowledge is constantly shaped by non-human others, but qualitative research has so far largely failed to uphold their voices. There are several attempts to disrupt the dominant narrative in social research such as from feminism/ecofeminism based critiques, and here I extend those arguments from the perspective of non-human other(s). It should also be mentioned here that while it is necessary to 'critique' the current state of anthropocentrism in qualitative research, such criticism does not entail an entire negation of humanist tendencies. In extension it also does not imply that tourism researchers should stop asking questions with political connotations; the vital importance of critical inquiries to the research landscape is always acknowledged. The intended message is to 'find the gap' in the critical qualitative research narrative by exposing it to stimuli from non-human other(s).

Limitations of social constructionism

'Social constructionism' is a heterogeneous term that may include different forms of epistemological positions (Demeritt, 2002; see also Pernecky, 2011 for a broad description of varieties of constructionism pertaining to tourism research). Hacking (1999) described two overarching types of social constructionism: one that seeks to refute apparently taken-for-granted stances such as *gender* or *nature* as not naturally given but influenced by social and power structures; and another that claims that all human knowledge is socially mediated. The common factor between these two types of constructionism is the centrality of the human subject (which in extension is reflected in social structures) and while there are differences between different constructionist claims, together they constitute a cornerstone of the postmodern turn in social science (Dear, 1994; Hacking, 1999; Demeritt, 2002).

While social constructionism may appear as liberating due to its interpretation of *truth* as a human artifact (and therefore open to refutation) it typically avoids informing us how non-human materiality shapes our cognition, and the condition of non-human species under the onslaught of anthropogenic change in the geo-biosphere. In addition the tendency of terming everything as socially constructed can be associated with an incipient anthropocentric bias. Hacking (1999) observed that currently the term 'social construction' is used to denote a number of different things (such as gender, nature, and subatomic particles that we cannot directly see or feel), each of which requires a *different kind* of construction metaphor. Demeritt's (2002) critique provided two important conclusions: that as far as all knowledge is constructed, the social construction metaphor may not be the most useful one for refuting certain taken-for-granted claims; and that in the light of the realization that the term 'nature' precisely signifies what is *not socially* contingent, the term 'construction' as an end obfuscates the processes that construct social ideas of nature. This criticism is related to arguments raised by critical realists who observe that by terming problems as social construction, constructionists do not succeed in making those problems less real (for more see Botteril, 2014), and can even make them worse by treating the claims of polluters on equal footing with the claims of their victims. A more recent challenge was put forth by Chakrabarty (2009) who argued that contrary to the social constructionist emphasis that all knowledge is socially continent, in so far as our civilization is dependent on a climatic equilibrium against the broader backdrop of a fluctuating geo-biosphere, all human knowledge can be seen as geologically contingent. Clark and Yusoff (2017) extended this line of argument with the observation that social and political agency is both contingent and resultant of the planetary forces at work, which in an earlier version of his work Clark (2010) had termed 'inhuman' (both in scale and origin). The implication is while social formations require geological formations, the latter does not require the former (see also Yusoff and Clark interview on YouTube, 2017). This line of thought is influenced by the 'Anthropocene' concept initially proposed by geologists and stratigraphers which posits that human modification of the geo-biosphere has changed it to such an extent that anthropogenic mass extinction is now occurring at a similar scale with previous mass extinctions evidenced from fossil records (Zalasiewicz, 2009; Steffen et al., 2015; Kolbert, 2015). Here again, nonlinearity of change is fundamental, for transition of geological epochs is shorthand for cataclysmic events that change the equilibrium of the geo-biosphere, upsetting the 'normal' rules of evolution (Kolbert, 2015). If we are indeed witnessing a change of that magnitude, uncertainty of its future directions is an inherent part of the situation.

If pervasive and 'wicked' environmental change is a hallmark of this new time; then addressing it requires the understanding of both the inherent complexity of the natural (non-human) others and the uncertainty posed by the accelerating transformation of our relationship with other species that share the planet, and even landforms and landscapes. Unfortunately, as argued by Kopnina (2016), the constructionist/postmodernist 'deconstruction' of the human-nature dichotomy can have the unintended effect of entrenching human chauvinism (i.e., in effect it strengthens the human/nonhuman duality), and furthering 'a robust anthropocentric bias' (Kopnina, 2016, p. 421) that denies the welfare of countless other species of animals and plants simply because they are not human and cannot articulate their condition in a human language. Kopnina's work is concerned about the continued importance of protected areas and ecological networks, and replies to the dominant narrative of qualitative research that seeks to subsume worldly phenomena within the human agency. Her works raises the important counterpoint that while it is important to safeguard human interests, ecocentrism and a conceptual dichotomy of humans and 'wild' nature are needed if non-human nature is to be protected outside of economic interests.

De-ontologizing qualitative tourism research: hints from ANT

Based on the foregoing discussion it can be pointed out that describing all worldly phenomena as socially constructed has its obvious limitations, which in turn requires approaches that recognize non-human agency in the shaping of such phenomena. In the field of ecology, biophilic approaches (Wilson, 2006) and the gaia hypothesis (Lovelock, 2009) are notable examples that are devoted to describing the planet as an inherently plural and complex organic system/entity. While these concepts are important for analyzing tourism and environmental issues, a particularly pertinent approach—ANT—is introduced here as it readily aligns with qualitative research. ANT provides an alternative way to explore worldly phenomena because rather than adhering to a specific theoretical understanding, it emphasizes relational descriptions of 'processes'. ANT is popular in environmental studies, especially where environmental problems are analyzed as heterogeneous and polymorphous processes. Instructive examples include Campbell's (2004) analysis of sustainability of rainfed agriculture in Ghana that sought to describe multiple connections between human, social, and environmental factors; and Kishigami's (2010) study of Inupiat whaling that analyzed multiple interactions including local and global level sociopolitical processes and climate change and its impact on sea ice. As these case studies demonstrated, biotic actants such as whales, crops, microbes, and abiotic actants such as local geomorphic processes, pedological characteristics, wind and oceanic circulation have vital performative agency that shapes human responses and institutions. While not explicitly referring to ANT, a recent special issue in this journal (Tourism Geographies) offered complementary insights on how the biophysical environment can shape tourist behavior. Smith, Wilkins, Gayle, and Lamborn (2018) noted how the geophysical characteristics of five national parks in Utah, US, mediated visitor trends, and Jedd et al. (2018) identified precipitation (notably extreme weather events under a changing climate regime) as an important factor mediating visitation. ANT denies a privileged position to humans by conceptualizing them as merely one 'actant' in a complex, enmeshed, and evolving web of interactions with other human and non-human 'actants' (Murdoch, 2001). Further illustrative examples of its fluidity can be seen from multispecies ethnography that seeks to understand how humans become 'human' through their encounters with nonhuman species (Kirksey, 2014), or ethnographic encounters with abiotic elements such as snow (Rautio & Vladimirova, 2017). Bruno Latour (1996) enlightened us on the use of the term 'network' in ANT by noting that rather than describing fixed couplings, it is used to signify multiple, fluid, and sometimes chaotic pathways of interactions that are also perennially varying in configurations. Consistent to its emphasis on plurality of actants and processes, ANT deemphasizes humans and consciously evades theory building (Kanazawa, 2018).

ANT has also been applied to tourism studies in recent years (for some instructive examples on such multi-actant descriptions of tourism phenomena see Cloke & Perkins, 2005; Ren, 2011; Arnaboldi & Spiller, 2011). As van der Duim, Ren, and Johanneson (2013) pointed out, an important feature of ANT-based approaches is their emphasis on materiality and non-human elements which together with humans shape the world as a 'mobile ordering' (van der Duim et al., 2013, p. 5). They also emphasized the scope of ANT in tourism studies due to its characteristic as an unstable, fluid concept. In extension, ANT provides the much-needed 'performative dimension (agency)' to non-human actants, and legitimizes more speculative and fluid inquiries without specific theoretical or paradigmatic moorings, making it a creative tool for de-emphasizing ontological commitments in qualitative inquiries of tourism phenomena.

Dissensus based portrayal of non-human voices as a creative 'rupture

As evidenced from the discussion above, although there are some forays in qualitative research (and qualitative tourism research) that seek to outgrow anthropocentrism it is perhaps fair to say that such engagements are sparse to date. Even in a state-of-the art review of current trends in qualitative research such as Denzin and Lincoln's (2018), most of the focus is on social justice and the need to mobilize people to struggle and to resist oppression, while the non-human other is only mentioned briefly. Although their synthesis framed qualitative research overwhelmingly along the lines of 'human disciplines', Denzin and Lincoln (2018) nevertheless mentioned multiple versions of the non-human turn such as new materialism as representing a vital 'rupture' from an outdated humanism (see also De Freitas et al., 2016; Clough, 2016–2017).

The importance of 'dissensus' and its disruptive credentials come right out of this 'rupture'. I present dissensus as an umbrella term that includes both (i) a methodological critique of prevalent qualitative methods that give centrality of agency to human subjects and the underlying theoretical tenets that converge on the ultimate importance of human subjectivity for interpreting worldly phenomena; and (ii) an opportunity to creatively disrupt this dominant narrative through fluid engagements with non-human others without adherence to specific theories or paradigms.

As far as research methods are concerned, the works of Mirka Koro-Ljungberg (2016, 2017, 2018) represent the most relevant methodological markers for this type pf engagement. Koro-Ljungberg warned against the 'shallow conceptual links to socially accepted qualitative research practices' and voiced the need for moving beyond post-positivistic and anthropocentric discourses and into a 'fluid methodological space where data, theories, methods and research approaches melt, transform, circumvent, infiltrate, appear and disappear' (Koro-Ljungberg, MacLure & Ulmer, 2017, p.3; see also Koro-Ljungberg, 2016). Her works also critique the prevalent linearity and order in qualitative research (as espoused by works like Bogdan & Biklen, 2006), and uphold the need for methodologies to exist beyond clearly defined boundaries. Such 'post-critical' and 'post-anthropocentric' engagements are in direct response to the contemporary environmental crisis that highlights the limitations of critical social science approaches and the necessity of incorporating 'more than human data' and insights on multispecies symbiosis into the discursive space of qualitative research (Koro-Ljungberg, MacLure & Ulmer, 2017). Elsewhere, Koro-Ljungberg (2016) argued for qualitative methodologies 'without methodology', implying fluid examinations of the materiality of the world and our experiences with it that go beyond the static barriers of paradigms or specific ontologies. Koro-Ljungberg, Löytönen, and Tesar (2017) encouraged qualitative researchers to question the very concept of data created by a 'humanist man' and to resist the urge to tame the nonhuman other through human agency. Koro-Ljungberg Löytönen, and Tesar (2017) presented the case for embracing the materiality of 'data' itself, enabling the research process to become an 'immanent critique making thinking to making and making to thinking' (Ibid, p.8). Alongside, the notion of 'complexity' was advanced as a term that highlights the limits of human knowledge (and is therefore fundamentally irreducible). These claims are consistent with works of Savin-Baden and Major (2010) who complained that qualitative research has not adequately addressed the concept of 'uncertainty', Vitek and Jackson (2008) who posited that ignorance is immanent in our knowledge sphere, and Washington (2015) who cautioned against the perceived superiority of human knowledge that he saw as contributing to the current unsustainable resource use patterns.

Here, I extend these arguments with the point that qualitative ecological knowledge can be fruitfully based on the affective association between us and non-human life forms and landscape/landform assemblages to better inform research on eco-and other forms of nature-based-tourism that would help to emplace non- human voices in the landscape of qualitative tourism research. Taking a cue from Koro-Ljungberg and her colleagues (2016, 2017, 2018), I advance the disruptive claim that consensus or socially accepted tourism research questions that seek to unpack eco or natural tourism settings as socially constructed or subjectively enacted space will not suffice for understanding its multiple inhabitants, complex change pathways, and chaotic encounters; instead questions that lead to undoing of the 'humanly constructed' world will be needed. Such questions can take multiple forms, some examples can be: how might a bear in a protected area perceive a sudden increase of tourists in her habitat? Or what does it mean to be a lion in a time when you (the lion) are a prized trophy? How does an ice-free arctic affect the society and tourist attractions from the viewpoint of the materiality of ice? Some may argue that such questions are speculative and we can never find the answers with any degree of certainty for we cannot read the minds of bears and lions. But speculative inquiries are also rich with possibilities, and besides, the point is precisely on avoiding certainty and hedge on complex uncertainty. The point here is not to present a definitive answer on the behalf of the bear or the lion; but to highlight the degree of ignorance that pervades tourism knowledge systems and to discover the value of uncertain, speculative, and fluid knowledge. The challenge for the researcher is neither to observe a bear from a human position, nor to subsume her bear-ness into the human domain, but to understand our inability to feel like a bear, which has obvious ramifications for the position s/he will articulate. It will encourage her to not to speak with a perceived higher ideological authority, but to speak directly for, and possibly with, the non-human other. Such knowledge, unshackled by its commitment to specific ideological positions, can provide a much-needed creative rupture in the current state of qualitative tourism research. Below I provide brief reports on the insights from my two recent case studies in World Natural Heritage Sites to further illustrate these claims.

Insights from Shiretoko Peninsula and Shirakami Mountain Forest

In this section I present some instructive insights gained from the encounters with non-human others from a three-year long (2016–2019) ongoing research project at two natural World Heritage Sites in Japan, during which stakeholders in World Heritage and tourism were interviewed. The main aim of presenting these insights here is to illustrate the necessity of dissensus regarding the status quo of tourism management in these areas—specifically to highlight the need to move beyond anthropocentric framings of non-human others and to embrace their performative agency in shaping landscapes that are termed 'natural heritage'. These insights are also particularly relevant for advocating a less anthropocentric approach for managing natural heritage areas, especially those that are ascribed global-level significance through the UNESCO's World Natural Heritage program.

Shiretoko Peninsula

Shiretoko is a peninsular landmass in Northeast Hokkaido that stretches nearly 70 km into the Pacific Ocean. Geologically it is a part of the Chishima Arc that is shaped by subduction of a cold oceanic plate at the continental margin. Episodic volcanism is an inherent feature of the Chishima Arc in geological time, the very recent manifestations of which are the series of active volcanoes that form a chain at the middle of the peninsula (Chakraborty, 2018). The area is the southernmost location of drift-ice in the Northern Hemisphere—a complex phenomenon that occurs due to the distant Amur River's freshwaters entering the ocean and forming stratified layers of poorly mixed water bodies which in turn freeze and move southward under the influence of oceanic and atmospheric circulations (Ohshima et al., 2001). The Outstanding Universal Value of the World Heritage property stems from the connectivity of marine and terrestrial ecosystems: seasonal drift ice formation helps massive algal blooms that sustain a complex food chain involving many species of fish and cetaceans in the sea and large mammals such as the Brown Bear on the land. Already a multitude of 'actants' can be seen as vigorous performers in this landscape—ranging from lithospheric plates, magma upwelling mechanisms, a distant river that brings in the raw materials of drift-ice, the ocean, winds, and tiny diatoms that sustain bigger and more iconic species as whales and bears. In addition, volcanism was joined by uplift to create a steep land profile towards the tip of the peninsula, precluding consistent human habitation of the area—and making humans occasional visitors who came searching for seasonal fish and seafood. The anomalous relationship between humans as hunters (and therefore possibly plunderers from the viewpoint of numerous non-human species who live in the forests and waters for millennia) and non-human species as residents poses questions on the human position as the voice for local ecosystems, but any ecocentric viewpoint is anathema to current fishermen who operate as a part of the powerful coastal fishing industry and see marine species as *resources* for exploitation. Complicating the situation is the consistent thinning of the drift-ice in response to climate change that has affected ecosystem productivity, creating a further strain of human-non-human relational networks (Figure 1).

However, repeated interviews with key tourism stakeholders and national park managers did not yield much information about the condition of non-human actants from *their* viewpoints. Typically responses presented a reductive view of the natural environment: national park managers sought to highlight a recent rise in the deer (*Cervus nippon*) population as the prime factor behind degradation in native vegetation; and local managers boasted of programs that educate all local schoolchildren on *how to avoid bears*. Humans and wildlife were presented in constant binary terms: even while acknowledging the importance of free roaming space for species like bears, respondents sought to portray the species as *dangerous* and somehow unwelcome. While respondents seemed to have a positive image of ecotourism in general; they consistently sought to portray ecotourism as a 'tool' for economic benefit; there were even suggestions that local interest in World Heritage inscription was at least partially due to a perceived lucrative business of using ice-breakers for winter cruise packages. In a personal communication, a scientist observed that local fishermen were staunchly opposed to the inscription fearing restrictions on their business and only acquiesced

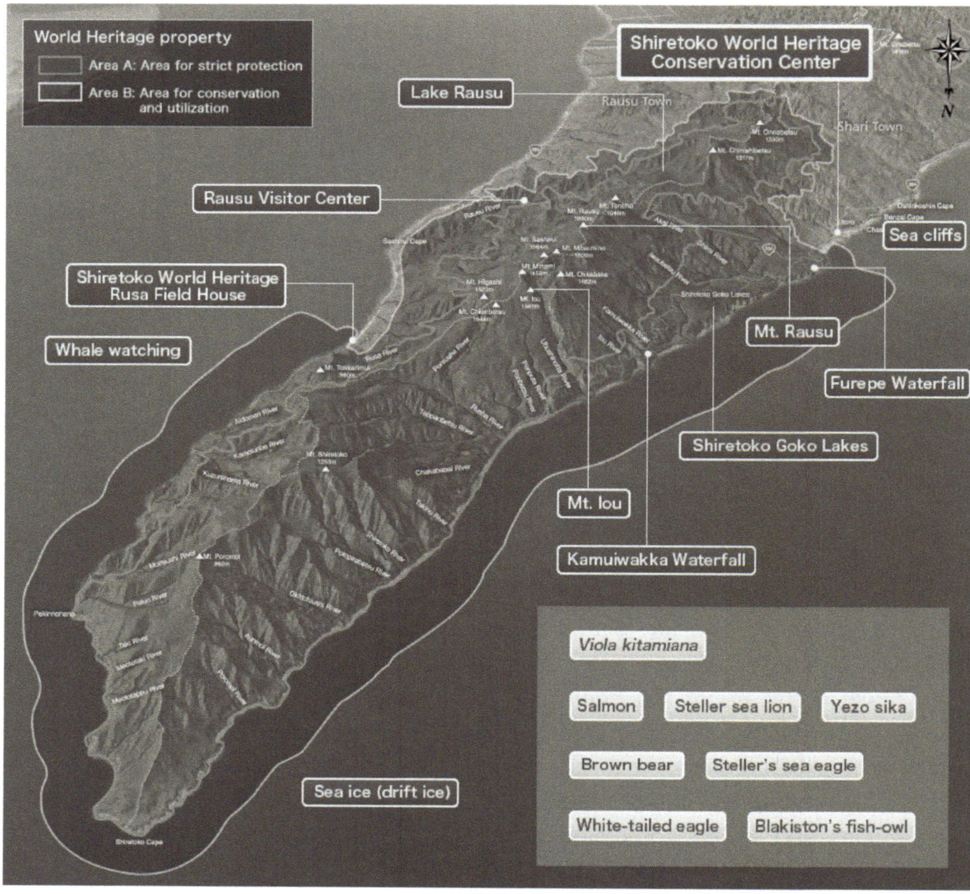

Figure 1. Map of Shiretoko Peninsula World Natural Heritage.
Source: Ministry of the Environment, Japan (n.d.) Website (World Natural Heritage in Japan), Retrieved on 12 Aug 2019.

when they were convinced that the inscription would not restrict their businesses. During the 1990s, the Rausu Town was a hub of commercial Alaskan Walleye Pollock fishing, before the Walleye Pollock stock of the Okhotsk Sea collapsed dramatically, leaving many fishermen without sizable income. The World Heritage inscription in 2005 gave those fishermen a novel opportunity to use fishing vessels for 'marine eco-tours'; and some tour operators expressed satisfaction with the World Heritage because it 'brings a lot of money' now. Local level 'ecotourism planning', carried out under a working group of heritage management and tourism stakeholders, typically favors the approach of increasing tourist access to remote sites. I personally experienced some of these tours: while a large amount of knowledge about marine lifeforms is transmitted in these trips, tour operators frequently used baits to lure endangered seabirds such as the Steller's Sea Eagle (*Haliaeetus pelagicus*) in order to create photo opportunities for birdwatchers. On a typical 'marine ecotour' package, these magnificent birds who can reach wingspans of over two meters were mere 'objects' for enthusiastic tourists; and tour operators—who supposedly are knowledgeable about wildlife—consistently spoke of the birds in terms of their attraction value. Scientists familiar with marine birds of prey have cautioned that Steller's Sea Eagles in Shiretoko are increasingly dependent on artificial feeding from tourist boats and seem to concentrate on artificial breakwaters, a sign that their feeding pattern and behavior is changing (Shiraki, 2010). While there could be an argument that humans have always modified wildlife around them, such arguments do not inform us about the trajectory and quality of the ongoing change. I asked what potential effects could occur to not only the eagles but their ecosystem (these birds are apex predators of their food-chain) but neither scientists nor tourism stakeholders could answer with any degree of confidence.

From participant observations in a number of ecotour packages across summer and winter seasons over two years, I could see that in none of the trips tourists get a sense of the current 'condition' of species they encounter; ecotour guides typically portray the landscape as 'fascinating', full of potential for exploitation and enrichment of the human self, and somehow nearly always desist from explaining the threats faced by non-human others in the process. There is little explanation of multiple connectivities between different species or of interactions between species and the abiotic environment. Justification of exploitative practices such as local industrial fisheries as an integral part of World Heritage and portrayal of local fisheries as an 'ecotourism resource' are further indications of entrenched anthropocentrism at several levels. Even some researchers have claimed that fisheries in Shiretoko are 'sustainable' and operate on the basis of 'adaptive management' (Makino et al., 2009), without providing any reasonable indication of what target at the species level is maintained or being adapted to. Apart from the longstanding tendency to maximize fish catch, and inhibition of salmonid migration pathways even inside the World Natural Heritage territory, fishermen frequently cull endangered species such as the Steller Sea Lions (*Eumetopias jubatus*) due to the species' propensity to raid fishing nets (Figure 2).

While much remains unknown regarding the current population dynamics and threats faced by migratory species such as the Steller's Sea Eagle, useful pointers can be drawn from the plight of the Blakiston's Fish Owl (*Bubo blakistoni*)—a significant non-human other—documented through recent research (Takenaka, 2018; Chakraborty & Takenaka, 2019). The fish owl is a 'resident keystone species', the condition of which provides important hints on the ecological integrity of the environment. Feeding almost exclusively on fish, this endangered and elusive raptor is the largest owl species on the planet. Shiretoko currently forms the largest habitat for this species, and this situation formed an important part of the argument of the property's World Heritage credentials. However, the current condition of the owl is only a reflection of a complex and nuanced process: while the species was found all over Hokkaido Island at the beginning of the century, today it is concentrated in the central highlands and northern areas, with nearly half of the population confined to Shiretoko. This is due to extensive logging of natural forests in Hokkaido during the twentieth century and stream modification through engineering, overfishing in rivers, and pollution through industrial effluents (Takenaka, 2018). The current fish owl population is heavily reliant on conservation efforts such as artificial nest-boxes, as there are currently too few large deciduous broadleaf tree species that allow fish owls to nest and lay eggs in large natural cavities (Takenaka, 2018). Chakraborty and Takenaka

Figure 2. A Blakiston's Fish Owl.
Source: T. Takenaka, reproduced with permission.

(2019) put forth the argument that as the fish owl is an apex predator, its current plight is evidence that the entire ecosystem it fits into is severely jeopardized in terms of ecological integrity, and that without effective protection of its prey species and habitat structure, the predator species is likely to fare poorly in the face of ongoing tourism development and climatic stressors.

In spite of this drastic change in the fish owl habitat and ecology that has taken place within the last hundred years, tourists are not provided adequate information on the current precariousness of the species, and are even lured into believing that *chance encounters* with these elusive raptors are something to be highly valued. Currently several 'ecotour' operators chase the birds with searchlights and take tourists near the nests for possible opportunities to photograph the elusive raptor, and there is even an 'environment-friendly' local inn that lures a pair of owls in its garden with bait to satisfy wildlife photographers. In a somewhat perverse manner, the 'endangered' condition of this species is seen as an economic opportunity that allows exploitation of this status for satisfaction of tourists who mostly come from urban areas and have little firsthand knowledge of wildlife.

As it becomes clear, the task of understanding, planning, and improving ecotourism in an area such as Shiretoko will require disruptive inquiries that depart from the pre-occupation of understanding tourism and the condition of the environment as a social performance, and should embrace the dissenting voices of ecologists, residents who are critical of the current efforts to portray a picture of 'harmony' between humans and other life forms, and dissatisfied visitors (who were identified during the interviews as currently forming an obdurate minority). Ultimately however, the engagements must strive to include the voices of non-human others such as the Blakiston's Fish Owl, the Steller's Sea Eagle, the Steller's Sea Lion, and migrating salmons—even if such engagements have to start with speculative inquiries that have little chance to

Figure 3. Drift ice breaking apart.
Source: author.

be included into the formal fold of management and socially accepted forms of research (Figure 3).

One such stimulus occurred to me on a sunny wintry day while gazing at drift-ice floes lazily breaking apart on the sea as seawater temperature gradually rose, allowing light to penetrate the frozen white layer. This, in ways dimly perceivable to me, fueled an enormously complex web of transfer of energy into tiny diatoms that are gorged upon by small marine animals, which in turn provide the basis of a vast food web that spans out in faraway distances, inviting migratory species and fish who sustain forests far away from the sea. What is more, the appearance of diatoms dates hundreds of million years into the deep time of the Paleozoic; and what I was witnessing was possibly an ancient movement of life; albeit the key actors (species) and their behaviors have changed over the eons. To what extent does human modification of oceans and climate condition these interactions? Is there any change that can be observed on the decadal or sub-decadal scales and to what extent does such change register in the social memory of human stakeholders, and in the memories of non-human counterparts? Such connections are currently poorly explored and inadequately explained in ecotourism literature; yet they form the backbones of many ecotourism systems.

Shirakami Mountains

The Shirakami Mountains World Heritage comprises of 170 km^2 of old growth Siebold's Beech (*Fagus crenata*) forest in the Aomori and Akita Prefectures in the north of the Honshu Island. Although it is the largest remaining old-growth natural forest in the four main islands of Japan, what is protected as a heritage landscape is a fraction of the original extent of the forest that spanned a total of 1300 km^2 as recently as a century back. The forest had remained in a stable condition for nearly 8000 years due to its large size (Chakraborty, 2018; Ishikawa, 2015), before modern machinery

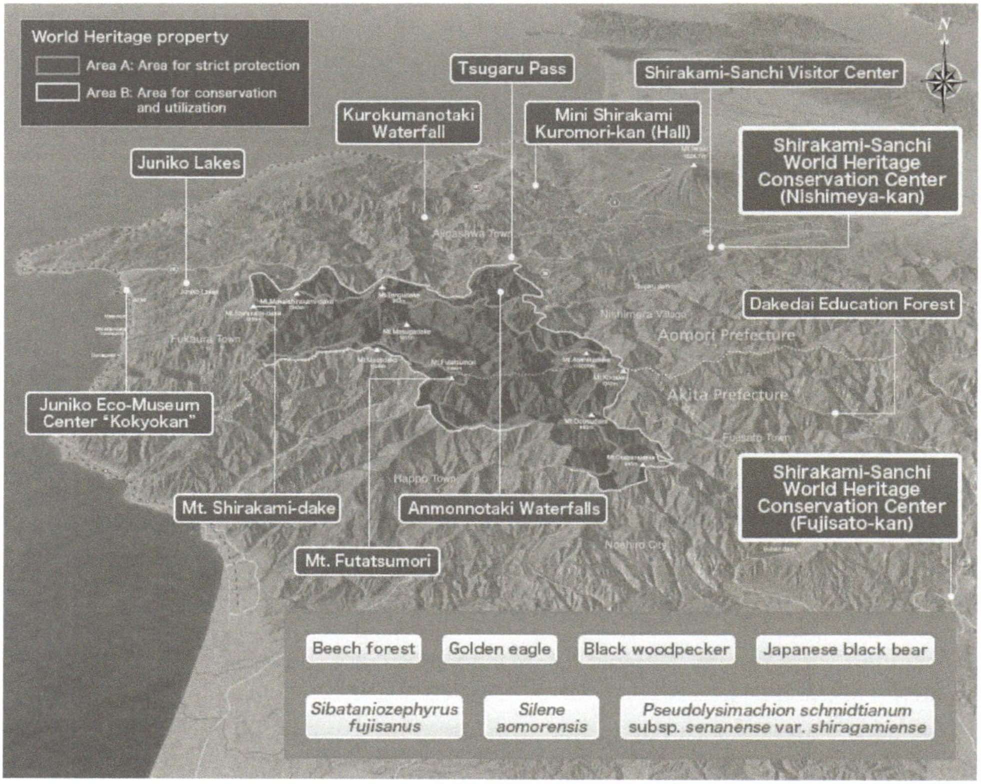

Figure 4. Map of Shirakami Mountains World Natural Heritage.
Source: Ministry of the Environment, Japan (n.d.) Website (World Natural Heritage in Japan), Retrieved on 12 Aug 2019.

depleted it within a few years. Much of that forest was logged off during the economic development of the twentieth century (Sato, 2006), as Siebold's Beech, due to its propensity of storing large volumes of water in its trunk, was seen as 'worthless' as construction wood and thus a prime target for replacement with confer species of high market value (Nebuka, 1992). In 1978 a plan to construct a road that would have traversed what is now the core area of the heritage was proposed for logging and developing mines (Sato, 2006). This led to local protests that eventually succeeded in halting the deforestation and facilitated extensive documentation of the ecological value of the forest. A total of 1500 vascular plant species is known to exist in the area; and nearly 500 of them can be found within the perimeters of the World Heritage property (Ishikawa, 2015). The area is also of the largest natural habitat of the Black Bear (*Ursus thibeatanus japonicas*) and the Japanese serow (*Capricornis crispus*), and is also the home of Japan's last hunter-gatherers known as the matagi (Figure 4).

Swiftly after the World Heritage inscription of 1993, the once worthless Siebold's Beech turned into a tourism resource: the giant trees came to be seen as 'magnificent', with old specimens of several hundred years of age proving to be major attractions. In addition, even as subsistence hunting gathering declined, the matagi tradition became a symbol of attracting tourists who were eager to hear accounts of 'living with the forest.' There are several trails in the peripheral sections of the protected forest, and 'local hunter-gatherer' guides offer explanation of the forest on a number of 'ecotour' packages. Yet what is offered here as ecotourism, and the manner in which the explanations are provided, favor the positioning of humans at the apex of the forest environment. Tourists of course have no way to verify whether this is true, and the myth of human superiority is perpetuated through tales of bravery (mostly featuring able-bodied male protagonists against the 'elements' and 'fierce animals' who are seen as targets that can be justly subdued). This promotes a consensus view of the natural environment as a backdrop for a rich and romanticized human 'culture'—not unlike the tendencies witnessed in the Shiretoko Peninsula.

While there is a number of detailed descriptions on the matagi lifestyle and the relationship between the matagi and the black bear, such descriptions are almost always framed in cultural terms and there is little documentation of the Black Bear's changing fortunes—leading to a situation where the justification of the bear's existence is almost singularly in terms of being the matagi's game. This situation is similar to the one observed by Yudina and Grimwood (2016) with regard to polar bear tourism—where the polar bear is reduced to a spectacle for eager visitors. Further parallels could be drawn with the work of Yudina and Grimwood (2016) with the point that matagis are male, and this cultural justification of hunting bears may actually promote a hidden symbolic justification of human (male) domination of wild forest (feminine nature). The matagi is seen as possessing the skill or knowledge to 'unlock' the secrets of the unruly, unapproachable forest, and is able to kill its most powerful resident—the black bear—and bring the bounties of the forest to people in the process. At the same time, the bear is almost never perceived as an *individual*. However as Nebuka (2017 personal communication)—who has chronicled the forest for nearly four decades—told me: 'Bears distinguish between hunters and non-hunters, I have never been attacked by a bear in the forest although I have seen them up close, very close. But the matter is entirely different if you are a matagi.' This observation may be a personal one but it shows that bears are likely to possess opinions about people they encounter; and such observations on animal cognition and intelligence are corroborated by studies such as Pepperberg's account of a grey parrot Alex, and Safina's (2015) explorations of species like elephants, wolves, and whales. However, affective portrayal of the bear is not mainstreamed either in guidance or management of the area, and bears are nearly always noted by tourism stakeholders for their ferocity. Even as some hunter-gatherers voice their disapproval of guiding urban tourists into the forest citing the incompatibility between their tradition and light-hearted enjoyment of the forest, tourism and its supposedly eco-related variants have strengthened the romanticization of cultural encounters with nature. Tourism over the years has successfully managed to seduce a number of former matagi hunters who now happily guide urban visitors into the forest—resulting in a consensus view of the forest as the backdrop to understand the local lifestyle. This romanticization of the hunter and peripheralization of the hunted pose a challenge for representing voices from the periphery. Even some hunters do not approve of this turn of events, one hunter spoke of the forest today as 'weak, reduced in its vitality', and as 'constrained by human design.' Pressed further, he could not coherently put together the causal links behind the degradation, but nevertheless stressed the degraded condition as the reality and

Figure 5. Ancient Siebold's Beech: conduit of water, habitat for many species.
Source: author.

offered numerous descriptions of *how* the forest has changed during his lifetime. While describing the change in mushrooms, fish, patterns of sandbars in rivers, and windflow patterns in different locations, he lamented that dissenting voices like his that seek to question the romanticization of the matagi lifeway as a touristic artifact are currently consigned to the periphery.

Tourism of course is not merely involved as an act of interpretation, it imparts several direct stressors into the landscape. There are ample signs that tourists here do not simply imagine or perform the touristic place: there are many check dams that purportedly stop 'natural hazards' from affecting tourists, and tourists almost seem like tiptoeing over a sleeping beast when they move around the landscape that is promised to them as 'tamed' through human technology. However the apparent slumber of the beast may be an illusion, a mere movement of the many muscles of the earth can unleash disproportionate impacts on tourism—in the form of landslides and floods that have uprooted roads in the past. The non-human earth in this sense has a palpable agency that influences human actions, but due to the current negation of its performative ability, it is (mis)construed as a background that can be experimented upon, and can be made into a pleasant experience through human design. Ironically this results in a disconnection between the older members of the Siebold's Beech that are held up as 'magnificent specimens' and the larger process of growth of such forests that include the key role of landslides and constant change (Ishikawa, 2015; Chakraborty, 2018). The situation is also not helped by the eagerness of heritage managers to facilitate access to tourists who come from urban centers and their urge to construct facilities that can accommodate all types of visitors. The process of taming and colonizing fierce wilderness continues as a social contract (Figure 5).

Although it is not possible to interview the non-human actants in the forest, observations and speculative questions that disrupt the artificial view of the local culture as the epitome of this landscape can inform ecotourism research and planning. A walk among giant Siebold's Beech trees will reveal that the vast and complex web of life supported by the rise and erosion of the Shirakami Mountains in a sequence that dates back to a time when dinosaurs roamed the earth remains vigorous today. Heavy snowfall blankets much of the area in winter, making roads impassable. When the snow melts in late spring, the giant tree trunks can almost be heard of channeling gushing water within them, and as greenery comes back, species like bears who hibernate during the winter come out to start a new year of their life. Meanwhile abiotic agents such as heavy snowpacks slide off in some locations, trailing large landslides in their wake that duly foster heterogeneity in this old forest as grass-like vegetation sprouts on the bare, exposed soil. Trying to understand how such processes and agents work through the eons to make this forest the way it is requires us to forget the centrality of humans in the earth system. After all, that notion might be the biggest myth that we have constructed through our collective efforts to tame the unfamiliar, wild, non-human, nature.

Concluding remarks

In a time where the earth system is increasingly seen as chaotic and changing without parallel in known history, qualitative tourism research needs to move beyond the centrality of the human subject by accepting non-human others as significant voices. In this article, hints for de-emphasizing ontological commitments (i.e., de-ontologizing) qualitative tourism research were provided from ANT-based approaches, and it was argued that such insights can be combined with affective, speculative, and fluid encounters with non-human others in order to understand tourism landscapes in a more holistic manner. Emplacing non-human voices in the current research landscape is particularly important for ecotourism and natural tourism destinations (such as protected areas) where tourism remains fundamentally dependent on non-human others. As the two case studies demonstrated, recognition of the agency of non-human others requires us to move beyond anthropocentric narratives and to engage in inquiries that challenge the social constructionist interpretation of tourism related phenomena. In line with ANT based approaches, the case studies offered numerous examples of 'actants' such ice, snow, water, animals, landslide, snowmelt, climate shift.

Observations provided in existing ANT-informed tourism research were extended by illustrating both the multiplicity of non-human actants, and the tension between anthropocentric accounts that seek to lump them into relational categories (such as 'fascinating' or 'hazardous'). The analyses also underscore the point that qualitative tourism research can be fruitfully informed by physical and earth science disciplines such as biology, geology, and ecological sciences. As the case studies show, engagements with non-human voices is a challenging, and sometimes messy process. Because it is an emerging field of creative inquiry, there is no reliable time-tested method that can adequately reflect the multitude of non-human voices. However, this situation presents itself as a creative opportunity to think beyond standard qualitative research methods, and the methodological connotations involved therein.

As already mentioned, this methodological critique did not intend to sketch definitive 'methods' as alternatives to humanist inquiries in qualitative tourism research. Means of engaging with non-human others should remain fluid and perennially evolving by necessity. Standard qualitative tools such as interviews, participant observations, and self-reflections can form parts of the toolbox; but the key for upholding the plurality of non-human voices is that they should be used in ways that consciously seek to disrupt socially accepted qualitative research narratives, challenge the centrality of the human perception, and question the role of consensus in the qualitative research community regarding the centrality of the human experience in interpreting worldly phenomena. The arguments developed here stressed the point that post-anthropocentric and (in relation to non-human stakeholders) non-anthropocentric inquiries that diverge from the dominant qualitative research narrative are important because of their potential to inform us about the plurality of the biophysical world. This line of dissensus based research has parallels with ecofeminist critiques of wildlife exploitation in tourism and the two can be fruitfully combined in order to creatively enhance the qualitative tourism research landscape.

Acknowledgment

I wish to thank the three anonymous reviewers for their critical and constructive comments which made the paper better. I am grateful to fellow scholar Adam Doering of Wakayama University for correcting and fine-tuning some of the expressions in the manuscript. I also wish to thank Takeshi Takenaka for the photograph of the Blakiston's Fish Owl.

Disclosure statement

No potential conflict of interest was reported by the authors.

Funding

This work was supported by Wakayama University Center for Tourism Research.

ORCID

Abhik Chakraborty http://orcid.org/0000-0002-3861-3641

References

Aitchison, C. C. (2005). Feminist and gender perspectives in tourism studies. *Tourist Studies*, *5*(3), 207–224. doi:10.1177/1468797605070330

Arnaboldi, M., & Spiller, N. (2011). Actor-network theory and stakeholder collaboration: The case of cultural districts. *Tourism Management*, *32*(3), 641–654. doi:10.1016/j.tourman.2010.05.016

Ateljevic, I., Pritchard, A., & Morgan, N. (Eds.). (2007). *The critical turn in tourism studies: Innovative research methodologies*. Amsterdam: Elsevier.

Benediktsson, K., Lund, K. A., & Huijbens, E. (2011). Inspired by eruptions? Eyjafjallajökull and Icelandic tourism. *Mobilities*, *6*(1), 77–84. doi:10.1080/17450101.2011.532654

Bhattacharya, K. (2017). *Fundamentals of qualitative research: A practical guide*. London: Routledge.

Bogdan, R., & Biklen, S. (2006). *Qualitative research for education: An introduction to theories and methods* (5th ed.). Boston, MA: Pearson.

Botteril, D. (2007). A realist critique of the situated voice in tourism studies. In I. Ateljevic, A. Pritchard, & N. Morgan (Eds.), *The critical turn in tourism studies: Innovative research methodologies* (pp. 121–131). Oxford: Elsevier.

Botteril, D. (2014). Constructionism: A critical realist reply. *Annals of Tourism Research*, *48*, 292–298. doi:10.1016/j.annals.2013.11.004.

Buckley, R. (2003). Ecological indicators of tourist impacts in parks. *Journal of Ecotourism*, *2*(1), 54–66. doi:10.1080/14724040308668133

Campbell, M. (2004). The role of socio-environmental networking in the sustainability of rain-fed agriculture in the coastal savanna of Ghana. *GeoJournal*, *61*(1), 79–88. doi:10.1007/sGEJO-004-1212-2

Caton, K. (2012). Taking the moral turn in tourism studies. *Annals of Tourism Research*, *39*(4), 1906–1928. doi:10.1016/j.annals.2012.05.021

Chakrabarty, D. (2009). Climate change of history: Four theses. *Critical Inquiry*, *35*(2), 197–222. doi:10.1086/596640

Chakraborty, A. (2018). Shirakami mountains: Old-growth forests of Siebold's Beech supporting biodiversity in a dynamic landscape. In A. Chakraborty, K. Mokudai, M. Cooper, M. Watanabe, & S. Chakraborty (Eds), *Natural heritage of Japan: Geological, geomorphological, and ecological aspects* (pp. 49–60).Cham: Springer.

Chakraborty, A. (2018). Shiretoko Peninsula: Dynamic interaction between geology, geomorphology, and ecology at the interface of terrestrial and marine systems. In A. Chakraborty, K. Mokudai, M. Cooper, M. Watanabe, & S. Chakraborty (Eds.), *Natural heritage of Japan: Geological, geomorphological, and ecological aspects* (pp. 31–48).Cham: Springer.

Chakraborty, A., & Takenaka, T. (2019). A qualitative exploratory analysis of ecological integrity for safeguarding World Natural Heritage Sites: Case study of Shiretoko Peninsula, Japan. *Heritage*, *2*(1), 898–919. doi:10.3390/heritage2010060

Chakraborty, A., Mokudai, K., Cooper, M., Watanabe, M., & Chakraborty, S. (Eds.). (2018). *Natural heritage of Japan: Geological, geomorphological, and ecological aspects*. Cham: Springer.

Chapman, K. (2015). *Complexity and Creative Capacity: Rethinking knowledge transfer, adaptive management and wicked environmental problems*. London: Routledge.

Clark, N. (2010). *Inhuman nature: Sociable life on a dynamic planet*. Thousand Oaks, CA: Sage.

Clark, N., & Yusoff, K. (2017). Geosocial formations and the anthropocene. *Theory, Culture & Society*, *34*(2–3), 3–23. doi:10.1177/0263276416688946

Cloke, P., & Perkins, H. C. (2005). Cetacean performance and tourism in Kaikoura, New Zealand. *Environment and Planning D: Society and Space, 23*(6), 903–924. doi:10.1068/d57j

Clough, P. T. (2016–2017). Rethinking race, calculation, quantification and measure. *Culture Studies < = > Critical Methodologies, 16*(6), 423–460. doi:10.1177/1532708616655760.

Cooper, M., Chakraborty, A., & Chakraborty, S. (Eds). (2018). *Rivers and society: Landscape, governance and livelihoods*. New York: Routledge.

Cooperstock, F. I. (2009). *General relativistic dynamics: Einstein's legacy throughout the universe*. Singapore: World Scientific.

Creswell, J. W. (2012). *Qualitative inquiry and research design: Choosing among five approaches*. Thousand Oaks, CA: Sage.

Dann, G., & Phillips, J. G. (2001). Qualitative tourism research in the late twentieth century and beyond. In B. Faulkner, G. Mosacrdo, & E. Lewis (Eds). *Tourism in the twenty first century: Reflections on experience* (pp. 247–265).London: Continuum.

De Freitas, E., Dixon-Roman, E., & Lather, P. (2016). Alternative ontologies of number: Rethinking the qualitative in computational culture. *Culture Studies Critical Methodologies, 16*(6), 1–15. doi:10.1177/1532708616655759.

Dear, M. (1994). Who's afraid of postmodernism? Reflections on Symanski and Cosgrove. *Annals of the Association of American Geographers, 84*(2), 295–300. doi:10.1111/j.1467-8306.1994.tb01739.x

Demeritt, D. (2002). What is the 'social construction' of nature? A typology and sympathetic critique. *Progress in Human Geography, 26*(6), 767–790. doi:10.1191/0309132502ph402oa

Denzin, N. K., & Lincoln, Y. S. (2018). Introduction: The discipline and practice of qualitative research. In N. K. Denzin & Y. S. Lincoln (Eds.), *The sage handbook of qualitative research* (5th ed., pp. 1–26).Thousand Oaks, CA: Sage.

Farrell, B. H., & Twining-Ward, L. (2004). Reconceptualizing tourism. *Annals of Tourism Research, 31*(2), 274–295. doi:10.1016/j.annals.2003.12.002

Feyerabend, P. (2010). *Against method*. New York, NY: Verso.

Gleick, J. (2008). *Chaos: making a new science*. London: Vintage.

Goeldner, R. (2005). Reflections on the historic role of journals in shaping tourism knowledge. *The Journal of Tourism Studies, 16*(2), 44–51.

Goodson, L., & Phillimore, J. (2004). The inquiry paradigm in qualitative tourism research. In J. Phillimore & L. Goodson (Eds.), *Qualitative research in tourism: Ontologies, epistemologies and methodologies* (pp. 30–45).London: Routledge.

Gribbin, J. (2012). *Erwin Schrodinger and the quantum revolution*. London: Bantam Press.

Guba, E. G., & Lincoln, Y. S. (1989). *Fourth generation evaluation*. Newbury Park, CA: Sage.

Guba, E. G., & Lincoln, Y. S. (1990). Can there be a human science?: Constructivism as an alternative. *Person Centered Review, 5*(2), 130–154.

Hacking, I. (1999). *The social construction of what?* Cambridge, MA: Harvard University Press.

Huijbens, E., & Gren, M. (2012). Tourism, ANT and the earth. In V. R. Van der Duim, C. Ren, & G. T. Johannesson (Eds.), *Actor-network theory and tourism: Ordering, materiality and multiplicity* (pp. 146–163).New York, NY: Routledge.

Incropera, F. P. (2015). *Climate change: A wicked problem: Complexity and uncertainty at the intersection of science, economics, politics, and human behavior*. Cambridge, UK: Cambridge University Press.

Ishikawa, Y. (2015). Shirakami ecosystems and plant colonies as a World Heritage. In The Shirakami Institute for Environmental Sciences, Hirosaki University (Ed.), *Introduction to the study of Shirakami Mountains* (pp. 1–10).Hirosaki: Hirosaki University Press [In Japanese].

Jedd, T. M., Hayes, M. J., Carrillo, C. M., Haigh, T., Chizinski, C. J., & Swigart, J. (2018). Measuring park visitation vulnerability to climate extremes in U.S. Rockies National Parks tourism. *Tourism Geographies, 20*(2), 224–249. doi:10.1080/14616688.2017.1377283

Jennings, G. R. (2009). Methodologies and methods. In T. Jamal & M. Robinson (Eds.), *The SAGE handbook of tourism studies* (pp. 672–692).London: Sage.

Kanazawa, M. (2018). *Research methods for environmental studies: A social science approach*. New York, NY: Routledge.

Kirksey, E. (Ed.). (2014). *The multispecies salon*. London: Duke University Press.
Kishigami, N. (2011). Climate change, oil and gas development, and Inupiat whaling in Northwest Alaska. *Études/Inuit/Studies*, *34*(1), 91–107. doi:10.7202/045406ar
Kolbert, E. (2015). *The sixth extinction: An unnatural history*. London: Bloomsbury.
Kopnina, H. (2016). Nobody likes dichotomies (but sometimes you need them). *Anthropological Forum*, *26*(4), 415–429. doi:10.1080/00664677.2016.1243515
Koro-Ljungberg, M., Löytönen, T., & Tesar, M. (Eds). (2017). *Disrupting data in qualitative inquiry: Entanglements with the post-critical and post-anthropocentric (post-anthropocentric inquiry*. New York, NY: Peter Lang.
Koro-Ljungberg, M., MacLure, M., Ulmer, J. (2018). D ... a ... t ... a ... , Data++, Data and some problematics. In N. K. Denzin & Y. S. Lincoln (Eds.), *The sage handbook of qualitative research* (5th ed., pp. 462–484). Thousand Oaks, CA: Sage.
Koro-Lungberg, M. (2016). *Reconceptualizing qualitative research: Methodologies without methodology*. Los Angeles, CA: Sage.
Latour, B. (1996). On actor-network theory: A few clarifications. *Soziale Welt*, *47*, 369–381.
Latour, B. (2005). *Reassembling the social: An introduction to actor-network theory*. Oxford: Oxford University Press.
Lincoln, Y. S., & Guba, E. G. (2000). Paradigmatic controversies, contradictions and emerging confluences. In N. K. Denzin & Y. S. Lincoln (Eds.), *Handbook of qualitative research* (2nd ed., pp. 163–188).Thousand Oaks, CA: Sage.
Lovelock, J. (2000). *Gaia: A new look at life on earth* (3rd ed.). Oxford: Oxford University Press.
Lovelock, J. (2009). *The vanishing face of Gaia: A final warning*. New York, NY: Basic Books.
Lovelock, J. (2015). *A Rough Ride to the Future*. New York: The Overlook Press.
Lund, K. A., & Benediktsson, K. (2011). Inhabiting a risky earth: The Eyjafjallajökull eruption in 2010 and its impacts. *Anthropology Today*, *27*(1), 6–9. doi:10.1111/j.1467-8322.2011.00781.x
Makino, M., Matsuda, H., & Sakurai, Y. (2009). Expanding fisheries co-management to ecosystem based management: A case in the Shiretoko World Natural Heritage area. *Marine Policy*, *33*(2), 207–214. doi:10.1016/j.marpol.2008.05.013
Merriam-Webster Dictionary Online. (2018). Retrieved from https://www.merriam-webster.com
Miles, M. B., & Huberman, A. M. (1994). *Qualitative data analysis: An expanded sourcebook*. Newbury Park, CA: Sage.
Ministry of the Environment, Japan (n.d.). Shiretoko (World Heritage property). Retrieved from https://www.env.go.jp/nature/isan/worldheritage/en/shiretoko/area/index.html.
Ministry of the Environment, Japan (n.d.). Shirakami (World Heritage property). Retrieved from https://www.env.go.jp/nature/isan/worldheritage/en/shirakami/area/index.html
Murdoch, J. (2001). Ecologising sociology: Actor-network theory, co-construction and the problem of human exceptionalism. *Sociology*, *35*(1), 111–133. doi:10.1177/0038038501035001008
Nebuka, M. (1992). *Thinking of the forest: Report from the Siebold's Beech Forest of Shirakami Mountains*. Tokyo: Rippu Shobo Publishing [In Japanese].
Nebuka, M. (2011). *World heritage shirakami mountains: Guidebook for nature experience, field observation and tourism*. Tokyo: Nanatsumori Shuppan [In Japanese].
New Scientist Instant Expert. (2017). *The trouble with reality: Inside the disturbing world of quantum theory*. London: New Scientist.
Ohshima, K., Mizuta, G., Itoh, M., Fukamachi, Y., Watanabe, T., Nabae, Y., … Watatsuchi, M. (2001). Winter oceanographic conditions in the southwestern part of the Okhotsk Sea and their relation to sea ice. *Journal of Oceanography*, *57*(4), 451–460.
Oxford Dictionary Online. (2018). Retrieved from https://www.oxforddictionaries.com
Patton, M. Q. (2002). *Qualitative research and evaluation methods*. Thousand Oaks, CA: Sage.
Pepperberg, I. M. (2009). *Alex & me: How a scientist and a parrot discovered a hidden world of animal intelligence-and formed a deep bond in the process*. New York, NY: Harper Perennial.
Pernecky, T. (2012). Constructionism: Critical pointers for tourism studies. *Annals of Tourism Research*, *39*(2), 1116–1137. doi:10.1016/j.annals.2011.12.010
Picken, F. (2010). Tourism, design, and controversy: Calling on non-humans to explain ourselves. *Tourist Studies*, *10*(3), 245–263. doi:10.1177/1468797611407757

Poisson, E., & Will, C. M. (2014). *Gravity: Newtonian, post-newtonian, relativistic*. Cambridge, UK: Cambridge University Press.

Pritchard, A., Morgan, N., & Ateljevic, I. (2011). Hopeful tourism: A new transformative perspective. *Annals of Tourism Research*, *38*(3), 941–963. doi:10.1016/j.annals.2011.01.004

Rautio, P., & Vladimirova, A. (2017). Befriending snow: On data as an ontologically significant research companion. In M. Koro-Ljungberg, T. Löytönen, & M. Tesar (Eds.), *Disrupting data in qualitative inquiry: Entanglements with the post-critical and post-anthropocentric (post-anthropocentric inquiry* (pp. 23–34). New York, NY: Peter Lang.

Ren, C. (2011). Non-human agency, radical ontology, and tourism realities. *Annals of Tourism Research*, *38*(3), 858–881. doi:10.1016/j.annals.2010.12.007

Richards, G., & Munsters, W. (2010). Developments and perspectives in cultural tourism research. In G. Richards & W. Munsters (Eds.), *Cultural tourism research methods* (pp. 1–12).Wallingford: CABI.

Robinson, M., & Jamal, T. (2009). Conclusions tourism studies—past omissions, emergent challenges. In T. Jamal & M. Robinson (Eds.), *The SAGE handbook of tourism studies* (pp. 693–701).London: Sage.

Room, G. (2012). *Complexity, institutions and public policy: Agile decision-making in a turbulent world*. Cheltenham: Edward Elgar.

Safina, C. (2015). *Beyond words: What animals think and feel*. New York, NY: Henry Holt & Co.

Sato, M. (2006). *Shirakami mountains: Will the forest be reborn?* Tokyo: Ryokufu Shuppan [In Japanese].

Savin-Baden, M., & Major, C. H. (2010). *New approaches to qualitative research: Wisdom and uncertainty*. London: Routledge.

Sayer, A. (2000). *Realism and social science*. London: Sage.

Shiraki, S. (2010). Current conditions and challenges for winter migratory White-Tailed Eagle and Steller's Sea Eagle. In S. Museum (Ed.), *Nature conservation in Shiretoko* (pp. 52–61). Sapporo: Hokkaido Shimbunsha [In Japanese].

Silverman, D. (Ed). (2016). *Qualitative research* (4th ed.). Thousand Oaks, CA: Sage.

Smith, J. W., Wilkins, E., Gayle, R., & Lamborn, C. C. (2018). Climate and visitation to Utah's 'Mighty 5' national parks. *Tourism Geographies*, *20*(2), 250–272. doi:10.1080/14616688.2018.1437767

Soper, K. (1995). *What is nature?* Oxford: Blackwell.

Steffen, W., Broadgate, W., Deutsch, L., Owen, G., & Ludwig, C. (2015). The trajectory of the Anthropocene: The great acceleration. *The Anthropocene Review*, *2*(1), 81–98. doi:10.1177/2053019614564785

Stirling, A. (2010). Keep it complex. *Nature*, *468*(7327), 1029–1031. doi:10.1038/4681029a

Takenaka, T. (2018). Ecology and conservation of Blakiston's Fish Owl in Japan. In F. Nakamura (Ed.), *Biodiversity conservation using umbrella species: Blakiston's fish owl and the red-crowned crane* (pp. 19–46).Singapore: Springer.

Tribe, J. (1997). The indiscipline of tourism. *Annals of Tourism Research*, *24*(3), 638–657. doi:10.1016/S0160-7383. (97)00020-0 doi:10.1016/S0160-7383(97)00020-0

Tribe, J. (2006). The truth about tourism. *Annals of Tourism Research*, *33*(2), 360–381. doi:10.1016/j.annals.2005.11.001

Tribe, J., & Airey, D. (2007). A review of tourism research. In J. Tribe & D. Airey (Eds.), *Developments in tourism research* (pp. 3–14).Oxford, UK: Elsevier.

van der Duim, R., Ren, C., & Thor Johannesson, G. (2013). Ordering, materiality, and multiplicity: Enacting Actor-Network Theory in tourism. *Tourist Studies*, *13*(1), 3–20. doi:10.1177/1468797613476397

Vitek, B., & Jackson, W. (Eds.). (2008). *The virtues of ignorance: Complexity, sustainability, and the limits of knowledge*. Lexington, KY: The University Press of Kentucky.

Washington, H. (2015). *Demystifying sustainability: Towards real solutions*. London: Earthscan Routledge.

Weaver, D. (2008). *Ecotourism*. Milton: John Wiley & Sons.

Wilson, E. O. (2006). *The creation: An appeal to save life on earth*. New York, NY: W.W. Norton.

Wilson, E. O. (2016). *Half-earth: Our planet's fight for life*. London: Liveright.

Wilson, E., & Hollinshead, K. (2015). Qualitative tourism research: Opportunities in the emergent soft sciences. *Annals of Tourism Research*, *54*, 30–47. doi:10.1016/j.annals.2015.06.001

Yudina, O., & Grimwood, B. S. R. (2016). Situating the wildlife spectacle: Ecofeminism, representation, and polar bear tourism. *Journal of Sustainable Tourism*, *24*(5), 715–734. doi:10.1080/09669582.2015.1083996

Yusoff, K., & Clark, N. (2017). Interview on geosocial formations and the anthropocene. Retrieved from https://www.youtube.com

Zalasiewicz, J. (2009). *The earth after us: What legacy will humans leave in the rocks?* Oxford: Oxford University Press.

Hanging out on Snapchat: disrupting passive covert netnography in tourism research

Heather L. Jeffrey, Hamna Ashraf and Cody Morris Paris

ABSTRACT

Tourism research is increasingly turning to online social media conversations as a data source, but seldom assigning space in research outputs for ethical appraisals or reflexive accounts of data collection. Many tourism researchers download conversations after they have occurred, but presence is central to gaining rich insights through netnographic research. We contribute to and disrupt online tourism research by critiquing, extending, and adapting existing methods to Snapchat, which in itself disrupts possibilities for online data collection. Hanging out with tourist Snapchatters highlights how privacy concerns are a primary motive to post on Snapchat. Snapchat is understood by our participants as more private than other social media sites due to the ephemerality of content and the notification of screenshots (when others save your snap). It is no coincidence that these same functionalities render established methods of digital tourism research difficult in the case of Snapchat. Snapchat users also problematise the notion of social media as public space, as they utilise the platforms for much more than communicating with the world. Snapchatters create, augment, and store memories.

摘要

旅游研究越来越多地将在线社交媒体谈话作为数据来源，但在现有研究产出中很少对之进行伦理性评价或者对数据收集进行反思性的描述。许多旅游研究人员会把谈话后的对话下载下来，但在网络民族志研究中，在场感是获取丰富见解的关键。我们通过批评、扩展和调试现有方法进行快拍研究，对在线旅游研究有所贡献，并对其有所颠覆性的发展。快拍本身就破坏了在线收集数据某些可能性。与游客拍照留念的人在一起，突显出隐私问题是人们在快拍上发帖的主要动机。快拍被我们的参与者理解为比其他社交媒体网站更私密，因为内容的短暂性和截屏通知(当其他人保存你的照片时)。同样的功能使得数字旅游研究的既定方法在快拍情形下变得困难，这并非巧合。快拍的用户也对社交媒体作为公共空间的概念提出了质疑，因为他们利用这些平台不仅仅是为了与世界交流。快拍聊天者创建、增强和存储了记忆。

Introduction

The post-television age has seen viral marketing mainstream and the advertising world increasing expenditure on the internet more than any other medium (Pérez-Latre, 2009). As a reflection of this, businesses and academic researchers alike are increasingly turning to online volunteered information as data. One of the main attractions of the internet is tourist created content (TCC) a type of user-generated content (Munar, 2011), which has facilitated the investigation of tourist activities as researchers may never need to leave their desks. It is predicted that this growth in the use of the internet as the 'field' for data collection will only continue in tourism research (Chambers & Rakić, 2015).

For the tourism researcher, 'post-visit narratives provided on these online platforms offer a window into tourists' subjective travel experiences around the world' (Mkono, 2011, p. 255). The sharing of experiences and emotions, in ways improbable or impossible years earlier, within the reflective space provided by online platforms is increasingly becoming the desirable, easily accessible, data for tourism researchers (Tribe, Xiao, & Chambers, 2012). However, this is often only the case when we do not reflect on the ethical implications of digital data collection and carry out passive and covert forms of netnography. The primary purpose of this article is to disrupt tourism netnographers, encouraging discussions of ethics and reflexive reporting on methods as has been promoted elsewhere (Lugosi & Quinton, 2018).

After discussing participation and presence in tourism netnographic research, the article highlights how participation is necessary if netnographies are to deliver rich and thick description and insights (Geertz, 1973). Increasingly accessed via mobile technologies, new social media platforms are themselves disrupting methods of observation, and it is becoming more and more difficult for netnographers to hide behind the computer screen. Issues of privacy and the use of social media for tourists are discussed before introducing the relatively new social media platform, Snapchat.

The article links our critique of passive and especially covert forms of netnography, which we highlight could be synonymous with downloading conversations without watching them emerge, to Snapchatters. Following two months of virtual ethnographic fieldwork, where one of the authors 'hung out' and followed 15 Snapchatters on their international travels, interviews with the 15 participants were carried out in Dubai. The participants were between the ages of 19 and 30, typical ages of avid Snapchatters and belonged to at least three social media networks. Interestingly, the study highlights how, contrary to previous research (Parra-Lopez, Gutierrez-Tano, Diaz-Armas, & Bulchand-Gidumal, 2012), Snapchatters may be concerned by privacy. This case responds to Lugosi and Quinton (2018) call for research to explore the types of interactions afforded by social media through the explication of how our participants use the platform not only to communicate but to create, save, store, and even augment memories, blurring the public and private divide.

Tourism research and online ethnography

The development of technology has revolutionised the traditional methods of sharing travel stories (Dickinson, Hibbert, & Filimonau, 2016; Dinhopl & Gretzel, 2016; White &

White, 2007). From postcard (White & White, 2007) to the present day, the method of communicating travel experiences remains pictorial (Pan, Lee, & Tsai, 2014). Perhaps this continuity explains partly why many academic researchers have sought continuity in methods, attempting to adapt tried and tested methodologies to this new virtual space. Netnography and virtual ethnography, among a whole host of other methodologies adapted from ethnography, have become increasingly commonplace in academic journals (Lugosi & Quinton, 2018).

Netnography is a concept originally coined by Robert Kozinets to describe the appropriation of ethnographic methods for the online environment (Kozinets, 2002). Netnography has been said to excel over traditional qualitative methods (e.g. interviews) due to several factors: it can be less time-consuming, more precise (as there is no need for transcription, comments can be downloaded directly), inexpensive, and honest (Mkono, 2011, 2012b, 2013; Wu & Pearce, 2014). It has been argued that netnography allows uninterrupted access to freely posted public thoughts, potentially providing an 'understanding of the discourse present in the community' (Podoshen, 2013, p. 266).

Similarly, to traditional ethnography, netnography demands the researcher take a role in the online community, and Björk and Kauppinen-Räisänen (2012) highlight four potential roles (see Figure 1). The four roles are dependent on the level of participation and the openness of the researcher in identifying themselves and their role.

Within tourism research, there is a growing trend for researchers to take on the role of a lurker (see, e.g., Björk & Kauppinen-Räisänen, 2012; Mkono, 2011, 2012a, 2013; Tribe & Mkono, 2017; Wu & Pearce, 2014). Lurkers are both passive (not actively participating) and covert (not notifying the community of their presence), a role conceptualised as cyber stealth (Ebo, 1998). Contradictorily, even though many lurking researchers utilise the work of Kozinets to define netnography, Kozinets (2015) stresses that netnography must include participation.

The stated benefits of lurking include the limited interference of the researcher, as tourists decide to write comments and engage in the online environment without being recruited, and the possibilities for open and honest answers afforded by anonymity (Wu & Pearce, 2014). These benefits may come at a cost as researchers find themselves in a time marred by unethical data collection practices that have led to negative publicity towards commercial enterprises.

> A time when social media companies, like Facebook and Instagram, and Big Data processors, like Meltwater and Crimson Hexagon, appear to be offering marketing and research treasures that will reveal everything, betraying secrets long held and opening the pathways to complete monitoring and complex spirals of social control. (Kozinets, Scaraboto, & Parmentier, 2018, p. 231)

Notification	Passive	Active
Overt	Observer	Participant
Covert	Lurker	Spy

Level of Participation

Figure 1. Potential roles for the netnographer (Source: Björk & Kauppinen-Räisänen, 2012).

The ethics of lurking

Active participants (those who comment and share in the online community), who do not disclose their identity, are deemed immoral spies by Kozinets (2002). The disclosure of identity and informed consent are key to Kozinets' (2002, p. 9) guidelines, which adhere closely to conventional ethnography:

> (1) the researcher should fully disclose his/her presence, affiliations and intentions to online community members during any research, (2) the researchers should ensure confidentiality and anonymity to informants, and (3) the researcher should seek and incorporate feedback from members of the online community being researched. [4] [the researcher should] contact community members and obtain their permission (informed consent) to use any specific postings that are to be directly quoted in the research

These ethical guidelines have been previously critiqued for their rigour (Langer & Beckman, 2005). If upheld, they would negate the benefits of netnographic research, such as the limited interference of the researcher, and the affordance of honesty (Wu & Pearce, 2014). Similar to conventional ethnography, participant observation can create an 'observer effect' where participants become aware of the researcher and change their content accordingly (Mkono, 2011).

Mkono (2013) acknowledges that the terming of more passive roles as netnography may be problematic, but still utilises the role of lurker under the term netnography. Informed consent is a basic requirement of most universities' ethics committees, but it poses both a practical and an ethical challenge for netnographers (Lugosi & Quinton, 2018). Björk and Kauppinen-Räisänen (2012) also utilise the role of a lurker, acknowledging there may be ethical issues regarding a lack of informed consent, arguing that as the participants will not be identified, there is little need for consent. Other researchers assume that 'it is virtually impossible to verify the identity of participants who often post content anonymously or pseudonymously' (Tribe & Mkono, 2017, p. 106), but anonymity does not always equate to an inability of the reader to identify participants. It is easy to copy a conversation from a scientific paper into Google and identify the source (Kozinets, 2015). A lack of informed consent highlights how researchers consider personal data freely available for collection regardless of the users' intentions (Janta, Brown, Lugosi, & Ladkin, 2011; Zimmer, 2010).

It might be argued that online space is public space, which would mean that netnographers like ethnographers can ethically capture the lived experience *in situ* (Langer & Beckman, 2005). Geographers have struggled with the problematic concept of virtual space, initially understood as separate and distinct to 'real' physical space, a dichotomy now under scrutiny (Ash, Kitchin, & Leszczynski, 2018; Kinsley, 2014). Issues of privacy and presence may be less important for geographers who fail to conceptualise the digital, the virtual as 'real'. The argument concerning online space as public space might be linked to the way space is conceptualised, and here rather than space, researchers might be understanding the digital as a public archive. This is especially true when researchers are not present in the space when interactions occur (discussed in the next section).

However, public does not equate to consent (Kozinets, 2015), and there are various technological issues with the public/private digital divide. Within certain social media sites, the researcher may have access to data that others do not due to the way internal networks are configured (Zimmer, 2010). Moreover, the possibility of gaining consent has been evidenced in some netnographic research (Kozinets, 2015), which suggests a potential lack of an ethical awareness among some tourism researchers.

Rather than arguments based on the public or the anonymous nature of the data, or the naming of methods as netnography but the utilisation of ethical guidelines that apply to a distinct method such as content analysis (Langer & Beckman, 2005), tourism researchers could base their arguments on moral philosophy. To some extent, Caliandro and Gandini (2017) actor/issue centred divide of digital research is based on a moral philosophical argument. Issue centred research, which deals with large data sets, complicates ethical procedures, such as informed consent, but should not render the researcher outside a discussion of ethics argue the authors. They suggest that one way of upholding ethical standards is to ensure issue centred research returns value to the participants.

Hanging out online

Lurking is not the only role available for tourism netnographers, if participation is viewed as a continuum, lurking could be placed at one extreme and perhaps 'hanging out' at the other (Germann Molz & Paris, 2015; Kendall, 2002). Paradoxically, hanging out enables researchers to formalise their role with participants; they can become members of the virtual space and disclosing their identity, recruit willing participants (Germann Molz, 2013). The disclosure of identity allows researchers to follow up on interesting themes, to obtain informed consent, member check and inform participants of the right to leave the study (Paris, Berger, Rubin, & Casson, 2015). Hanging out allows researchers to develop co-presence with participants, which does not dissipate as the participants cross borders, seas, and continents (Germann Molz & Paris, 2015).

Passive forms of netnography often entail downloading data after it occurred (Costello, McDermott, & Wallace, 2017), not participating at all, and potentially tying the researcher to a desktop computer. Hanging out can be performed anywhere at any time without the need for heavy equipment, as social media advances, how we access data should do too. The multi-temporal, multi-spatial nature of social media and netnography (Lugosi & Quinton, 2018) demands researchers become increasingly able to move with the participants (Hannam, Butler, & Paris, 2014; Paris, 2011). Snapchat is an example of a social media platform that is only accessible via a mobile device. Hanging out has previously been explored utilising mobile virtual ethnography, which 'reflects our attempts to adapt ethnographic techniques to the study of the mobile and virtual social phenomena' (Germann Molz & Paris, 2015, p. 172). Netnographer mobility may occur via a screen but also geographically and spatially if they are to gain rich insights into new platforms and communities.

One of the main differences between hanging out and lurking is not necessarily participation, but presence – hanging out can fall anywhere on the active–passive participation continuum, but the researcher is present. For all disciplines, the effectiveness of netnography (regardless of discipline) as a 'qualitative methodology relies on the need for human presence and personal connections online' (Costello et al., 2017, p. 1).

Non-participant observation does not necessarily reflect actual participation, as even observers participate if they are to notify the community that they are observing. Observers must create rapport to be successfully allowed to observe the group; in contrast to lurking, observation may require an understanding of qualitative capabilities. In much tourism research, there has been a failure to fully explore the levels of participation or the ethics of online research, as researchers appear to solely focus on issues rather than actors (Caliandro & Gandini, 2017), or downloading conversations rather than observing them as they evolve. This can be intimately linked to how the virtual space is conceptualised as a simulated or artificial space (Ash et al., 2018).

Primary or secondary data?

The downloading of conversations without observing them emerge also has methodological limitations. In this instance, netnography can lend little to an understanding of the complex relationships between the actors and between the actors and the context because you were not there (Costello et al., 2017). Lurking netnographers share little with lurking ethnographers, where synchronic presence is key. In netnography, presence is often asynchronistic, occurring after the fact when the conversation has finished rather than synchronic. In addition to the limited understanding of context afforded by lurking, researchers may be naming their methods erroneously, as netnography also alludes to a more qualitative approach to the data that is often absent upon closer scrutiny.

Many researchers who do identify themselves in online communities do so abruptly without building rapport and are subsequently treated with disdain (Hewson, 2014). As identified by researchers utilising email interviews (Martini & Buda, 2018; Paris, 2011) and Skype (Adams-Hutcheson & Longhurst, 2017), qualitative protocols coupled with digital methods can empower participants, which has long been a desire in qualitative geographical research (McDowell, 2010). By simultaneously creating co-presence and distance, 'Power is reconfigured, whereby the participant can turn off, tune out or disengage, choosing to seamlessly reinstate their desired distance and space' (Adams-Hutcheson & Longhurst, 2017). Those who utilise the data as a secondary source may be missing out on the primary benefits of the virtual space, the ability to create co-presence and distance to empower participants to switch off.

Privacy, memory, and ephemerality

As previously highlighted, the online environment poses ethical challenges for tourism researchers, which lack appraisal and consideration in many tourism netnographic studies. This is due to the affordances of technology to share often personal information, including data such as geographical locations, which is a potential threat to privacy (Gretzel, 2011). How as a society, we understand privacy is not certain, and much of what is shared online is done so by users who may lack an awareness of any potential risks (Turkle, 2011). As Urry (2002) emphasised, technology has dramatically shifted corporeal surveillance to digital surveillance, a pervasive disciplinary tool that, for the most part, remains invisible. The invisibility of surveillance and a lack of awareness of risks may explain why it has been found that privacy concerns do not dramatically influence travellers' intentions to share on social media (Berger & Paris, 2013; Parra-Lopez et al., 2012).

In the digital age, information flows and is monitored instantaneously as power is de-territorialised and commercial actors (such as social media sites themselves) become increasingly prominent (Urry, 2002). Mobile technologies are not inherently good or bad; it is the actors who utilise social media for good or bad: the public, the social media organisations, commercial companies, governments, and researchers. As such, the flow of information in social media affords not only digital surveillance but also opportunities that may be considered more positive, such as memory and retrieval (Paris, 2012), as:

> Persons leave traces of their selves in informational space, and can be more readily mobile through space, or simply stay in one place, because of a greater potential for 'self-retrieval', for the retrieval of their personally information at another time or place (Urry, 2002, p. 268)

Within the tourism literature, flows of digitised information are often conceptualised as communicative TCC, especially from the perspective of destination marketing. Social media is thought to have transformed content that would have traditionally been thought of as a souvenir to what is now e-word of mouth:

> Photos and videos were mementos of tourism experiences and their physical manifestations (album, videotape, framed photo print) were important keepsakes. Now they are communication media and in the case of Snapchat they are ephemeral messages.' (Gretzel, 2017)

Tourism research is yet to fully explore the affordance of social media for 'self-retrieval' in general, and there is a paucity of research into the potentialities for 'self-retrieval' specifically when the content has a time limit. Snapchat has dramatically challenged traditional social media by creating a platform where users can share disappearing content. The popularity of the ephemeral feature has led Facebook's photo-sharing application Instagram to launch a 'secondary feed' to share temporary content (Kofoed & Larsen, 2016), and Facebook Messenger and WhatsApp Messenger (also owned by Facebook) have quickly followed suit (Isaac, 2017a, 2017b). Snapchat has emerged as the fastest growing pictorial social media application (Vaterlaus, Barnett, Roche, & Young, 2016) and is particularly popular among Gen Y (Piwek & Joinson, 2016). Little is known about the use of ephemeral content among tourists or their privacy concerns. As the popularity and provision of platforms based on ephemeral content sharing increases, it will also become important to identify appropriate research methods; passive netnography will not capture this disappearing data.

Snapchat

Even though tourism academics are increasingly turning to online content as a data source, they still lag behind the latest developments in social media (Duignan, Everett, Walsh, & Cade, 2018), such as Snapchat. In contrast to other social media sites, Snapchat provides an 'ephemeral' form of visual communication (Piwek & Joinson, 2016). The content shared on Snapchat can be viewed for as long as 10 seconds, and it self-destructs after 24 hours, and this temporality is a primary factor in its popularity (Sashittal, DeMar, & Jassawalla, 2016).

Content shared on social media such as Facebook and Instagram has habitually been 'polished' (Kofoed & Larsen, 2016); for example, pictures are often planned and posed. Tourist photo taking is central to all aspects of tourism, while on holiday, 'other activities are suspended, and people present themselves for future memories by posing' (Larsen, 2006, p. 89). However, the ephemeral nature of Snapchat has led to lesser concerns about the quality of content being posted, and selfies and pictures of everyday activity are the most shared pictures on Snapchat (Kofoed & Larsen, 2016; Piwek & Joinson, 2016). Location-based geo-filters have also facilitated the communication of location on Snapchat since 2014 (Tweedie, 2014), which could pose threats to security (Gretzel, 2011). Ephemerality may have disrupted the ways we take pictures, but there is little understanding of how it may impact the ability of picture taking to 'construct travel memories' (Baerenholdt, Haldrup, Larsen, & Urry, 2004).

Snapchat's unique ephemeral feature does not lend itself to the covert passive forms of netnography common among tourism researchers, as the data (in theory) cannot be saved or stored. If a user does attempt to save the data by utilising a screenshot, the other user is informed, which could blow the cover of the covert netnographer. This creates a specific challenge for companies and researchers interested in monitoring or investigating Snapchat. It has been noted that in the future marketers will have to gain access and permission to join private conversations and learn new rules of engagement (Gretzel, 2017), and it can be logically presumed that researchers will also have to follow suit. The following section discusses how mobile virtual ethnography has been utilised in this study to explore the tourist use of Snapchat, potentially providing a framework for future social media researchers.

Methods

The primary aim of the study was to explore tourist motivations for sharing content on Snapchat, and the study was actor-centred (Caliandro & Gandini, 2017), incorporating mixed methods, and a multi-sited approach. Here, multi-sited refers to the various sites that the participants travelled to (Germann Molz, 2006; Hine, 2011) as well as online and offline researcher/participant interactions. The participants gave consent to being followed from place to place via Snapchat, the researcher delved in and out of the data from various sites in her everyday life and always via a mobile phone. Online 'fieldwork' involved two months of non-participant overt observation at the beginning of 2017, where the second author observed the content posted on Snapchat via a mobile telephone. As participants who Snapchat while travelling were sought, friends and family of the second author were approached before snowballing was utilised to recruit further participants in person.

Researching with family and friends can be critiqued as there is a prior relationship, which may shape the subsequent data, but there are issues of power within all researcher–participant relationships (McDowell, 2010). The intimate nature of netnography made approaching family and friends both essential and advantageous as they would normally share content on Snapchat with the researcher. However, before data

Participant	Age Group	SM Networks	Place Travelled
Respondent 1	19 - 25	4+	Spain
Respondent 2	26 - 30	4	USA & Turkey
Respondent 3	19 - 25	4	Spain & Pakistan
Respondent 4	19 - 25	4+	India
Respondent 5	19 - 25	3	Qatar
Respondent 6	19 - 25	3	Spain
Respondent 7	19 - 25	4	France & India
Respondent 8	19 - 25	4+	India
Respondent 9	19 - 25	4+	Georgia & Hong Kong
Respondent 10	19 - 25	3	Spain
Respondent 11	19 - 25	4+	Spain
Respondent 12	19 - 25	4	Germany & UK
Respondent 13	26 - 30	3	Turkey & USA
Respondent 14	19 - 25	3	Spain
Respondent 15	19 - 25	4+	Spain

Figure 2. Participant profiles.

collection, university ethical protocols were maintained, participants were informed of the aims of the research, and the observation techniques guaranteed anonymity and offered the right to withdraw at any time retaining their data, before being asked for their informed consent in person. The participants also share similar characteristics to typical Snapchat users; for example, they rely heavily on smartphones, and they frequently use three or more social media networks (see Figure 2) and share pictures instantly (Piwek & Joinson, 2016; Vaterlaus et al., 2016).

In contrast to other social media sites that allow the revisiting of data online (Germann Molz & Paris, 2015), the ephemerality of data on Snapchat as previously commented necessitates a variety of methods. An observational logbook assisted in taking notes conveniently and simultaneously (Creswell, 2009). Extensive field notes were taken to assure the depth of observation material. The minor details, for instance, the combination of filters, the audio in videos, the timings, captions, usage of Bitmojis, and Emojis, were noted down for analysis. Additionally, respondents were requested to save their travel snaps and share them with the researcher. Rather than the researcher taking screenshots of the snaps, which might have interrupted and disrupted participants, this method was deemed more respectful of the participants' privacy. After the initial observation period, in-depth semi-structured interviews were conducted in-person with the participants.

These interviews were key to exploring the use of Snapchat from the perspective of the participant (Caliandro & Gandini, 2017). The interview questions were initially shaped to develop a better understanding of motivations behind sharing travel snaps and were informed by the 'Motivation, Opportunity and Ability Model' (MOA) (MacInnis & Jawoeski, 1989, cited in Sigala, Christou, & Gretzel, 2016). The interview began by asking participants about their general usage of Snapchat, before a discussion of the participants' travel snaps, the motivation behind sharing the pictures and challenges they face while travelling and using the application (if any). Questions included: Why do you prefer using Snapchat over other social media networks? How do you feel about the temporality of Snapchat? How do you feel about the screenshot notification? Thematic analysis utilising the initial codes motivation, opportunity, and ability allowed the themes utilised in the article to emerge: privacy and savouring and saving the ephemeral. The privacy afforded by Snapchat became a clear motivation for sharing, and saving and savouring the ephemeral a motivation, and an opportunity.

This article draws heavily on the analysis of the 15 interviews with the participants, but the observation period allowed insights into each participant's use of Snapchat, which shaped the interviews. Interviews after a period of observation can be likened to meeting up with an old friend; due to co-presence, it can feel as though you travelled with the participant. In the era of big data, researchers and corporate organisations are frequently turning to quantitative methods of researching digital spaces (Ash et al., 2018). 'Meaningful, deep, and exhaustive insights' are analysed from small data (Caliandro & Gandini, 2017, p. 17) and are necessary if we are to understand how geographies are 'produced by the digital' (Ash et al., 2018).

Discussion

The primary motive of the study was to explore tourist motivations for sharing content on Snapchat, but the emerging themes are relevant to questions of ethics and online ethnography. Participants chose Snapchat over other social media platforms due to reasons of privacy, which was linked to the ephemerality of content and notification of screenshots. Perhaps unsurprisingly these are the two features that render covert passive netnography redundant on Snapchat. Interestingly, participants also expressed a desire to save this content for themselves, which highlights how often the reasons for posting may not be to make public a private moment. The subsequent sections discuss more fully these emerging themes and how they relate to the extant literature.

Privacy

A key theme that emerged from the interviews was related to the personal and private nature of Snapchat, which may appear to be an oxymoron but was often linked to a comparative evaluation of Snapchat and other social media sites. For example, respondent one contrasted the 'public' nature of Facebook with the privacy afforded by Snapchat:

> I feel Snapchat is more private. On Facebook for example my post could be easily shown on anyone else's newsfeed if one of my friends likes it or anything like that, whereas on Snapchat, as I said I have very limited people on it and those are people that I trust so yeah, I feel more comfortable posting pictures or videos on Snapchat because I feel Facebook is more public and out-there, but Snapchat is more private and limited.

Several other participants considered Facebook to be at the opposite end of the public/private social media spectrum, respondent 12 stated:

> Yeah because I have limited people on there so like closer friends. So, I prefer posting it on there than Instagram and Facebook, especially Facebook, because I have so many people on

> there that I don't even know anymore like from back home and stuff, so I don't like sharing private things with them

Interestingly, the privacy characteristic associated with Snapchat was not always depicted as a positive, and respondent 11 appreciated the public nature of Instagram over privacy afforded by Snapchat:

> Now that I have started a blog I upload more on Instagram, I do post Instagram stories for the blog as well. On Snapchat its just people I know.

All of our participants were aware of risks related to privacy, and for most of them, it was a concern. The nature of Snapchat as a more private social media site was one of their primary motives for sharing content there, which challenges the notion that online users may be unaware of privacy risks (Turkle, 2011). This finding also contradicts previous research that has suggested that travellers are not concerned about their privacy when posting content related to their trips on social media sites (Berger & Paris, 2013; Parra-Lopez et al., 2012).

While tourist photo taking has been linked to the staging of future memories (Larsen, 2006), the ephemeral nature of Snapchat affords the recording of the mundane, the everyday and the silly for our participants. Respondent 14, noted:

> I like the fact that the picture doesn't stay forever so I like it. I don't have to worry about what I am posting cos it's not going to stay forever.

And respondent 17 described a liminal digital space:

> Respondent seven It is really good especially when fooling around with friends and sending them silly pictures, it stays only between you and your friends only for 10 seconds until it disappears forever

The ephemerality of content was also linked to increased privacy in comparison with Facebook, with respondent eight noting:

> Yeah, it's the whole point of that I get to decide when I want to remove it. I get to decide whether I want to post it or not and it's there only for 24 hours and, like now after all these upgrades and all, I can see who screenshots my pictures, who views my pictures, who reads my chats and in the end, it goes away. Otherwise, on Facebook on Instagram it just there.

In addition to ephemeral content, this participant notes how Snapchat allows the visualisation of those who watch your content. Whereas it can be argued that digital surveillance is primarily invisible (Urry, 2002), Snapchat certainly appears to be defying this trend. The identification of who is looking may influence the theorising of the digital gaze, but it will almost certainly impact digital researchers, governments, and businesses. Not only does this impact those who would want to utilise the data, but it also creates an implicit code of conduct among Snapchat users, and respondent three felt that:

> It's kind of creepy. No one does it, except my friend has done it a couple of times just because she was there in the pictures (...), but I can't stalk people. It's good that I can see who screenshots my pictures, but I want to screenshot other people's picture as well.

The ability to screenshot and be screenshotted challenges the belief held by some users that Snapchat is inherently more private than other social networks, but also acts as a surveillance mechanism (Urry, 2002) to ensure that the actors behave accordingly. Many participants confessed their own behaviour was shaped by this surveillance, respondent 15 acknowledged 'Uhhhh I think that it's good for me but when I do it to somebody else not so much I guess'.

Most of the participants confided that only their very close friends screenshotted their snaps and that they were happy with that, but respondent three acknowledged that screenshotting is not the only way to save the snaps:

> initially, I thought that it's so smart that you get a notification when someone screenshots and then I realized if someone's got an extra phone they could just take a picture of it, it's not too safe

Respondent 12 even questioned the role of Snapchat:

> Because they can think that oh it's gone anyways, but then you never know how Snapchat itself saves them, and you know, anyone can take a screenshot it a bit dangerous for younger people.

The participants in the study were aware of the security and privacy issues related to using social media, and it was a concern and a reason to post on Snapchat over other social media sites. The notification of screenshots leads to visible and identifiable surveillance, influencing how Snapchatters interact but also creating an obvious barrier to research for covert netnographers.

Saving and savouring the ephemeral

As previously noted, Snapchat's success lies in its ephemeral functionality. As such perhaps unsurprisingly one participant acknowledged that this was one of the primary motives for sharing on Snapchat, suggesting 'that makes me want to post more because it won't stay there'. However, participants did acknowledge that ephemerality was not always as positive as first thought, and they had to employ strategies to save what they were posting, respondent one noted:

> It has its positives and negatives. Negatives are that if you forget to save, then you don't get to save it you don't get to see it in a few years or sometimes if you forget to save it and send it out you never get to see it unless the person takes a screenshot for you. But yeah, I like the temporary feature because like I said if once you post something, it doesn't have to have a point like you can post your entire day and it doesn't have to have a point like that.

Saving the stories for themselves became a prominent theme from the interviews, even when as previously suggested the content they were sharing was less 'polished' (Kofoed & Larsen, 2016). Some participants believed that ephemerality was a negative feature of Snapchat as it made saving content more difficult, and respondent three stated:

> I don't like it; I have to save all my pictures. It goes away, so I have to save everything, all the pictures I post. Also, I clicked something, and it doesn't directly save to my camera roll. So now I have to do that and save it to my camera roll.

Other more arguably tech-savvy participants had identified a way to save the post as memories within Snapchat. Similarly, to analogue photos before snaps become keepsakes (Dinholp & Gretzel, 2016) and Snapchat affords the construction of travel memories (Bærenholdt, Haldrup, Larsen, & Urry, 2004), respondent four noted:

> a lot of the times I use Snapchat and save, and I don't post things as well because I just like the memory, like now we have the option of just saving on your account randomly.

While we leave traces of ourselves online, which are virtually mobile, being shared across borders, these traces can also remain in one place and facilitate self-retrieval (Urry, 2002). The memory function within Snapchat as respondent six explains affords the saving of these traces in one place:

> Snapchat has the feature now, the memories, so like you can save you pictures to memories, and so you don't necessarily have to save them on your phone, so I can post my pictures using that, in case I don't have internet access.

Participants also highlighted how certain features of Snapchat allowed them to augment their memories with respondent nine commenting:

> I don't always use it but like, for example, I am overseas, and that place is different and obviously out of the normal for me then I put a geo-filter, so I remember and associate in the years to come if I save it, I'll know exactly where I was at that time.

Geo-filters enabled the addition of information as a layer over the picture to include details that would otherwise be forgotten, affording self-retrieval (Urry, 2002) and the promotion of a snap to a keepsake (Gretzel, 2017). Respondent five highlighted:

> you can put the time, you can put the location, you can put the temperature you can put different things like that, so I really think that adds to pictures when you're travelling cause sometimes you don't always remember the names of the places or you don't always remember what the temperature was at that time so having that all saved up as your memory and put up as your story is, I don't know, it's more of a personal touch to your pictures, it is more convenient because, at the end of the day, when I recently travelled I took like maybe 30 pictures and at the end of the day I saved them all.

Interestingly, notwithstanding the ephemeral nature of Snapchat, Snapchatters highlighted how they used Snapchat to create, augment, and store memories.

Conclusion

The contemporary tradition of lurking in netnographic tourism research (passive covert netnography) or the downloading of conversations from social media sites attracts ethical scrutiny. The downloading of conversations in this way can be conceptualised as issue centred netnography (Caliandro & Gandini, 2017), which may involve the collection of mass online conversations. Tourism netnographers often justify their data collection on the nature of the space as public space, but the public/private distinction remains blurry as evidenced by the participants later in this study. A more convincing justification would be that the outcome of the research is more important than the process and may relate to the benefits of the research for participants (Caliandro & Gandini, 2017).

In contrast to ethnographic lurkers, netnographic lurkers may lack presence, and while participation can be viewed as a continuum, presence should be viewed as a prerequisite to rich insights. Snapchat disrupts the ability of passive covert netnography to collect data due to the disappearing nature of that data. As the popularity of ephemeral storytelling continues, both marketers and researchers will have to adapt to very new data collection possibilities. Hanging out online is much more apt for these new platforms as ultimately it demands presence (Germann Molz & Paris, 2015; Kendall, 2002). Social media sites such as Snapchat not only afford the mobility of the researcher but also demand it, making it harder to download conversations and TCC without participating. The research project presented in this article included both online observation and offline interviews; the online observation occurred through a mobile telephone screen as both participants and researcher moved through separate physical spaces but were co-present.

The ethics of online data collection are intimately linked with arguments of public and private space alongside issues of informed consent. The Snapchatters interviewed in this article were aware of privacy issues, and it was one of their primary motives for sharing on Snapchat. Interestingly, the tagging of physical locations in TCC has been identified as a potential threat to privacy and safety (Gretzel, 2011), but most of the participants had not considered the use of geo-filters as a threat. The ephemerality alongside the notification of screenshots of their stories made the platform (at least) feel more private. The noticeable and identifiable acts of surveillance via screenshot meant that they were less likely to 'stalk' or be 'stalked' and has repercussions for how we collect data and theorise the digital gaze.

The article contributes to the extant literature by highlighting how Snapchat can be used for private purposes such as the creation, augmentation, and storing memories. This finding could be furthered by future research on the use of social media for 'self-retrieval' (Urry, 2002). Future netnographic tourism research will be shaped by both the recent shift in EU law and the affordances of new social media networks. Newly introduced GDPR demands all organisations obtain consent from European social media participants in the retrieval and analysis of their data (EUGDPR.ORG., 2018). This change means that non-participant, covert netnography must begin to focus specifically on ethics, and consent to remain within the law.

This article responds to Lugosi and Quinton (2018) call for explorations into the types of interactions afforded by social media and highlights how researcher participation is delimited by Snapchat, which renders covert passive netnography an impossibility. The notification when data is saved using Snapchat and other similar social media platforms will transform the way data can be collected. To negotiate this environment, netnographers can utilise 'small data', engaging with the community on a deeper level, seeking informed consent, and adhere to qualitative research protocols.

While netnographers face challenges in their methods, qualitative researchers can view digital or virtual space as an opportunity to empower their participants. Both co-presence and distance are simultaneously created in online research, which can afford the participant an opportunity to switch off and disengage as and when they desire. While many tourism researchers have utilised sites such as Tripadvisor similarly to archival data to research tourists, generic social media networks such as Snapchat have received less attention. These networks should be understood to offer up an extension to the everyday rather than a simulated or artificial virtual reality; they can provide researchers with meaningful data linked to a variety of issues such as work and employment in the tourism industry which has received less attention from netnographers.

Netnographers have adapted existing methods to take advantage of opportunities afforded by social media and the digital environment, but future netnographies should consider how these methods have traditionally engaged with issues of ethics. The future of netnography should not be dominated by big data and finished downloaded conversations, but by deep contextual understandings afforded by consensual synchronistic co-presence.

Disclosure statement

No potential conflict of interest was reported by the authors.

ORCID

Heather L. Jeffrey http://orcid.org/0000-0001-9097-6928
Cody Morris Paris http://orcid.org/0000-0002-0339-2471

References

Adams-Hutcheson, G., & Longhurst, R. (2017). At least in person there would have been a cup of tea': Interviewing via Skype. *Area, 49*(2), 148–155. doi:10.1111/area.12306

Ash, J., Kitchin, R., & Leszczynski, A. (2018). Digital turn, digital geographies? *Progress in Human Geography, 42*(1), 25–43. doi:10.1177/0309132516664800

Bærenholdt, J., Haldrup, M., Larsen, J., & Urry, J. (2004). *Performing tourist places*. Aldershot: Ashgate.

Berger, E. A., & Paris, C. M. (2013). Exploring the role of Facebook in re-shaping backpacker's social interactions. In Z. Xiang & I. Tussyadiah (Eds.), *Information and communication technologies in tourism 2014* (pp. 299–312). Cham: Springer.

Björk, P., & Kauppinen-Räisänen, H. (2012). A netnographic examination of travelers' online discussions of risks. *Tourism Management Perspectives, 2*, 65–71. doi:10.1016/j.tmp.2012.03.003

Caliandro, A., & Gandini, A. (2017). *Qualitative research in digital environments: A research toolkit*. New York, NY: Taylor & Francis.

Chambers, D., & Rakić, T. (2015). Conclusion: Reflections beyond existing research frontiers. In D. Chambers & T. Rakić (Eds.), *Tourism research frontiers: Beyond the boundaries of knowledge* (pp. 165–176). Bingley: Emerald Group Publishing Limited.

Costello, L., McDermott, M. L., & Wallace, R. (2017). Netnography: Range of practices, misperceptions, and missed opportunities. *International Journal of Qualitative Methods, 16*(1), 160940691770064. doi:10.1177/1609406917700647

Creswell, J., W. (2009). *Research design: Qualitative, quantitative and mixed methods approaches.* Thousand Oaks, CA: Sage.

Dickinson, J. E., Hibbert, J. F., & Filimonau, V. (2016). Mobile technology and the tourist experience: (Dis)connection at the campsite. *Tourism Management, 57*, 193–201. doi:10.1016/j.tourman.2016.06.005

Dinhopl, A., & Gretzel, U. (2016). Selfie-taking as touristic looking. *Annals of Tourism Research, 57*, 126–139. doi:10.1016/j.annals.2015.12.015

Duignan, M., Everett, S., Walsh, L., & Cade, N. (2018). Leveraging physical and digital liminoidal spaces: The case of the #EATCambridge festival. *Tourism Geographies, 20*(5), 858–879. doi:10.1080/14616688.2017.1417472

Ebo, B. (1998). Internet or outernet? In B. Ebo (Ed.), *Cyberghetto or cybertopia? Race, class, and gender on the internet* (pp. 1–12). Westport, CT: Praeger.

EUGDPR.ORG. (2018). Frequently asked questions about GDPR. Retrieved from https://eugdpr.org/the-regulation/gdpr-faqs/

Geertz, C. (1973). *The interpretation of cultures*. New York: Basic Books

Germann Molz, J. (2006). Watch us wander: Mobile surveillance and the surveillance of mobility. *Environment and Planning A: Economy and Space, 38*(2), 377–393. doi:10.1068/a37275

Germann Molz, J. (2013). Social networking technologies and the moral economy of alternative tourism: The case of couchsurfing.org. *Annals of Tourism Research, 43*, 210–230. doi:10.1016/j.annals.2013.08.001

Germann Molz, J., & Paris, C. M. (2015). The social affordances of flashpacking; exploring the mobility nexus of travel and communication. *Mobilities, 10*(2), 173–192. doi:10.1080/17450101.2013.848605

Gretzel, U. (2011). Intelligent systems in tourism a social science perspective. *Annals of Tourism Research, 38*(3), 757–779. doi:10.1016/j.annals.2011.04.014

Gretzel, U. (2017). The visual turn in social media marketing. *Tourismos. 12*(3), 1–18.

Hannam, K., Butler, G., & Paris, C. M. (2014). Developments and key issues in tourism mobilities. *Annals of Tourism Research, 44*, 171–185. doi:10.1016/j.annals.2013.09.010

Hewson, C. (2014). Qualitative approaches in internet-mediated research: Opportunities, issues, possibilities. In P. Leavy (Ed.), *The Oxford handbook of qualitative research methods. Oxford library of psychology series* (pp. 423–452). New York, NY: Oxford University Press USA.

Hine, C. (2011). Towards ethnography of television on the Internet: A mobile strategy for exploring mundane interpretive activities. *Media Culture & Society, 33*(4), 567–582. doi:10.1177/0163443711401940

Isaac, M. (2017a). Facebook mirrors Snapchat (again) with stories feature. *The New York Times.* Retrieved from https://www.nytimes.com/2017/03/28/technology/facebook-mirrorssnapchat-again-with-facebook-stories.html?_r=0

Isaac, M. (2017b, February 20). WhatsApp's new feature continues Facebook's Snapchat Mimicry. The New York Times. Retrieved from https://www.nytimes.com/2017/02/20/technology/whatsapp-statusfacebooksnapchat.html?action=click&contentCollection=Technology&module=RelatedCoverage&region = Marginalia&pgtype = article

Janta, H., Brown, L., Lugosi, P., & Ladkin, A. (2011). Migrant relationships and tourism employment. *Annals of Tourism Research, 38*(4), 1322–1343. doi:10.1016/j.annals.2011.03.004

Kendall, L. (2002). *Hanging out in the virtual pub: Masculinities and relationships online.* Berkeley: University of California Press.

Kinsley, S. (2014). The matter of 'virtual' geographies. *Progress in Human Geography, 38*(3), 364–384. doi:10.1177/0309132513506270

Kofoed, J., & Larsen, M. (2016). A snap of intimacy: Photo sharing among young people on social media. *First Monday, 21*(11), 1–9. doi:10.5210/fm.v21i11.6905

Kozinets, R. V. (2002). The field behind the screen: Using netnography for marketing research in online communities. *Journal of Marketing Research, 39*(1), 61–72. doi:10.1509/jmkr.39.1.61.18935

Kozinets, R. V. (2015). *Netnography redefined.* Los Angeles: Sage.

Kozinets, R. V., Scaraboto, D., & Parmentier, M. A. (2018). Evolving netnography: How brand auto-netnography, a netnographic sensibility, and more-than-human netnography can transform your research. *Journal of Marketing Management, 34*(3–4), 231–242. doi:10.1080/0267257X.2018.1446488

Langer, R., & Beckman, S. (2005). Sensitive research topics: Netnography revisited. *Qualitative Market Research: An International Journal, 8*(2), 189–203. doi:10.1108/13522750510592454

Larsen, J. (2006). Picturing Bornholm: Producing and consuming a tourist place through picturing practices. *Scandinavian Journal of Hospitality and Tourism, 6*(2), 75–94.

Lugosi, P., & Quinton, S. (2018). More-than-human netnography. *Journal of Marketing Management, 34*(3–4), 287–313. doi:10.1080/0267257X.2018.1431303

MacInnis, D. J. and Jaworski B. J. (1989). Information processing from advertisements: Towards an integrative framework. *Journal of Marketing, 53*(October), 1–23.

Martini, A., & Buda, D. M. (2018). Analysing affects and emotions in tourist e-mail interviews: A case in post-disaster Tohoku, Japan. *Current Issues in Tourism*, 1–12. doi:10.1080/13683500.2018.1511693

McDowell, L. (2010). Interviewing: Fear and liking in the field. In Dydia DeLyser, Steve Herbert, Stuart Aitken, Mike Crang, and Linda McDowell (Eds.), *The SAGE Handbook of Qualitative Geography* (pp. 156– 171). Los Angeles: Sage.

Mkono, M. (2011). The othering of food in touristic eatertainment: A netnography. *Tourist Studies, 11*(3), 253–270. doi:10.1177/1468797611431502

Mkono, M. (2012a). A netnographic examination of constructive authenticity in Victoria Falls tourist (restaurant) experiences. *International Journal of Hospitality Management, 31*(2), 387–394. doi:10.1016/j.ijhm.2011.06.013

Mkono, M. (2012b). Netnographic tourist research: The internet as a virtual fieldwork site. *Tourism Analysis, 17*(4), 553–555. doi:10.3727/108354212X13473157390966

Mkono, M. (2013). Hot and cool authentication: A netnographic illustration. *Annals of Tourism Research, 41*, 215–218. doi:10.1016/j.annals.2013.01.014

Munar, A. M. (2011). Tourist-created content: Rethinking destination branding. *International Journal of Culture, Tourism and Hospitality Research, 5*(3), 291–305.

Pan, S., Lee, J., & Tsai, H. (2014). Travel photos: Motivations, image dimensions and affective qualities of place. *Tourism Management, 40*, 59–69. doi:10.1016/j.tourman.2013.05.007

Paris, C. M. (2011). Understanding the Statusphere and Blogosphere: An analysis of virtual backpacker spaces. In R. Law, M. Fuchs, & F. Ricci (Eds.), *Information and communication technologies in tourism 2011* (pp. 443–455) Vienna: Springer.

Paris, C. M. (2012). Flashpacking: A discussion of independent travel in a digital world. In M. Fuchs, F. Ricci, & L. Cantoni (Eds.), *Information and communication technologies in tourism 2012* (pp.191–202). Vienna: Springer.

Paris, C. M., Berger, E. A., Rubin, S., & Casson, M. (2015). Disconnected and unplugged: Experiences of technology induced anxieties and tensions while traveling. In I. Tussyadiah, & A. Inversini (Eds.), *Information and communication technologies in tourism 2015* (pp. 803–816). Cham: Springer.

Parra-Lopez, E., Gutierrez-Tano, D., Diaz-Armas, R., J., & Bulchand-Gidumal, J. (2012). Travellers 2.0: Motivation, opportunity and ability to use social media. In M. Sigala, E. Christou, & U. Gretzel (Eds.), *Social media in travel, tourism and hospitality*. (pp. 171–187). UK: MPG Books.

Pérez-Latre, F. J. (2009). Advertising fragmentation: The beginning of a new paradigm? In H. Powell, J. Hardy, S. Hawkin, & I. MacRury (Eds.), *The advertising handbook* (3rd ed.) (pp. 34–45). Oxon: Routledge.

Piwek, L., & Joinson, A. (2016). "What do they Snapchat about?" Patterns of use in time-limited instant messaging service. *Computers in Human Behavior, 54*, 358–367. doi:10.1016/j.chb.2015.08.026

Podoshen, J. S. (2013). Dark tourism motivations: Simulation, emotional contagion and topographic comparison. *Tourism Management, 35*, 263–271. doi:10.1016/j.tourman.2012.08.002

Sashittal, H., DeMar, M., & Jassawalla, A. (2016). Building acquaintance brands via Snapchat for the college student market. *Business Horizons, 59*(2), 193–204. doi:10.1016/j.bushor.2015.11.004

Sigala, M., Christou, E., & Gretzel, U. (2016). *Social media in travel, tourism and hospitality*. London: Routledge.

Tribe, J., & Mkono, M. (2017). Not such smart tourism? The concept of e-lienation. *Annals of Tourism Research, 66*, 105–115. doi:10.1016/j.annals.2017.07.001

Tribe, J., Xiao, H., & Chambers, D. (2012). The reflexive journal: Inside the black box. *Annals of Tourism Research, 39*(1), 7–35. doi:10.1016/j.annals.2011.11.001

Turkle, S. (2011). *Alone together: Why we expect more from technology and less from each other*. New York: Basic Books

Tweedie, S. (2014, July 15). How to use Snapchat's new geofilters: A feature it kept hidden until now. Business Insider. Retrieved from http://www.businessinsider.com/snapchat-geofilters-2014-7

Urry, J. (2002). Mobility and proximity. *Sociology, 36*(2), 255–274. doi:10.1177/0038038502036002002

Vaterlaus, J., M., Barnett, K., Roche, C., & Young, J., A. (2016). "Snapchat is more personal": An exploratory study on Snapchat behaviors and young adult interpersonal relationships. *Computers in Human Behavior, 62*, 594–601. doi:10.1016/j.chb.2016.04.029

White, R., N., & White, B., P. (2007). Home and away: Tourists in a connected world. *Annals of Tourism Research, 34*(1), 88–104. doi:10.1016/j.annals.2006.07.001

Wu, M., & Pearce, P. L. (2014). Appraising netnography: Towards insights about new markets in the digital tourist era. *Current Issues in Tourism, 17*(5), 463–474. doi:10.1080/13683500.2013.833179

Zimmer, M. (2010). "But the data is already public": On the ethics of research in Facebook. *Ethics and Information Technology, 12*(4), 313–325. doi:10.1007/s10676-010-9227-5

Part II

Re/Creating Methodologies

A critical consideration of LEGO® SERIOUS PLAY® methodology for tourism studies

Yana Wengel, Alison McIntosh and Cheryl Cockburn-Wootten

ABSTRACT

Increasingly, tourism researchers have turned their attention to the use of creative methodologies to gain deeper understandings of tourist phenomena. There is a need to challenge and creatively disrupt conventional methodological approaches as they are criticised for their inability to be participant driven, capture the co-construction of research context or to address the impact of wider social dynamics to knowledge creation in tourism studies. Based on our research focused on host–guests experiences participating in the World Wide Opportunities on Organic Farms (WWOOF) programme in New Zealand we provide a critical consideration of LEGO® SERIOUS PLAY® for tourism studies. LEGO® SERIOUS PLAY® is a communication tool aimed at developing creative thinking through building metaphors around identities and experiences using LEGO® bricks. To demonstrate how the method can be used in tourism studies, we draw on examples from three LEGO® SERIOUS PLAY® workshops to illustrate the benefits and challenges of this methodological approach. LEGO® SERIOUS PLAY® offered a metaphorical way for participants to construct creative artefacts and explain their ideal WWOOFing experience, representing sometimes complex, entrenched and emotional issues, and relationships that may have been difficult to express via traditional methods. The method enables participant driven, co-production of knowledge in a playful, free-flowing way to foster creative thinking, meanings and possible solutions. The method helps participants creatively communicate complex and sensitive issues, especially around their relationships – to objects, landscapes, people and identities – aspects that may otherwise be silenced by traditional research methods. As a novel method LEGO® SERIOUS PLAY® provides opportunities for researchers who want to gain a deeper understanding of the social dimensions of tourism, to co-create spaces for knowledge exchange and develop an in-depth understanding of socially constructed relationships and realities.

摘要

近十年来, 旅游研究将注意力转向创新性质性研究方法, 以期对旅游现象有更深刻的认识。尽管大量的研究强调创新性的方法, 但仍有必要挑战甚至创造性地打破基于传统方法论的研究方法，因为这些传统方法被批评说无法由参与者驱动,无法共同构建研究脉络，或应对广泛的社会动力因素对旅游研究中知识创新的影响。基于我们对新西兰有机农场全球机遇项目(WWOOF)中主客体验的研究，我们为旅游研究提供了乐高严肃游戏方法的批判性思考。乐高严肃游戏（LEGO® SERIOUS PLAY®）是一款沟通工具，旨在通过使用LEGO®砖块围绕身份和体验构建隐喻来发展创新性思维。为了展示该方法如何应用于旅游研究，我们借鉴了三个乐高严肃游戏工作坊的例子来说明这种方法的好处和挑战。乐高严肃游戏为参与者提供了一种隐喻性的方式，让他们构建有创意的人工制品，并解释他们理想的有机农场全球机遇项目的体验，表现有时候是复杂、根深蒂固的情感问题，以及可能通过传统方法难以表达的关系。该方法使参与者驱动，以一种有趣的，自由流动的方式共同生产知识，以培养创造性思维，意义和可能的解决方案。该方法帮助参与者创造性地沟通复杂和敏感的问题，特别是围绕他们与物体、景观、人和身份的关系，这些方面可能会被传统的研究方法所掩盖。乐高严肃游戏作为一种新颖的方法，为研究者提供了很多机会，有望更深入地理解旅游业的社会方面，共同创造知识交流的空间，深入了解社会构建的旅游关系和现实。

Introduction

Previously, tourism scholars have challenged, tourism scholars challenged traditional positivist approaches to tourism research and called for a Critical Turn in tourism studies (Ateljevic, Pritchard, & Morgan, 2007, p. 251). Tourism scholars have critiqued the ontological, epistemological, and methodological limitations dominant within tourism academia instead, offering alternative methods of inquiry which aim to produce inductive, interpretive, reflexive accounts of subjective realities of participants (Decrop, 1999; Phillimore & Goodson, 2004; Riley & Love, 2000). Recently, tourism scholars have highlighted the importance of acknowledging epistemology in determining methodological choices (Ateljevic, Harris, Wilson, & Collins, 2005). They have discussed how this influences endeavours to provide spaces and voice to participants in order to explore their subjective experiences and gain co-created knowledge (Jennings, 2010; Pernecky, 2012). Indeed as Hughes and Sharrock (1997, p. 89) have stated, 'experiences of others can be grasped through the apprehension of their inner meanings' and applications of interpretive, inductive methodological tools of data collection and analysis develop research capacities for this co-production of knowledge to occur.

To contribute to this area of scholarship, we provide a critical consideration of the method and processes of the interpretive, creative methodological tool, LEGO® SERIOUS PLAY® for tourism research. LEGO® SERIOUS PLAY® is a facilitated workshop technique grounded in constructionist epistemology which enables individual participants to depict their understandings in a metaphorical and creative, playful way. This method allows participants to provide meanings about their experiences which might be difficult to articulate through verbal or written processes. By using a creative, playful method, LEGO® SERIOUS PLAY® seeks to reduce tensions involved in research, such as the power-authority dynamic. It also provides opportunities for both individual and collaborative understandings of phenomena through the use of metaphorical explanations. For example, researchers have noted that work across multiple industries and academic sectors creates problems in relation to knowledge management, developing understandings and creating capacities across the seemingly disparate participants (Hart & Wolff, 2006). LEGO® SERIOUS PLAY® attempts to overcome these boundary relationships, potential misunderstandings between people and knowledge by using playful metaphorical depictions. While LEGO® SERIOUS PLAY® is not a panacea for all research that aims to develop participant's voice, it does offer an orientation for developing co-created understandings that attempt to overcome the traditional approaches that rely on a translation of declarative knowledge between participant and researcher.

The article begins by providing some brief background about the development of LEGO® SERIOUS PLAY®. The literature review section will critically evaluate frequently used qualitative tools and provide discussion on the philosophical foundations of LEGO® SERIOUS PLAY® methodology. This section is followed by details of the case study that applied LEGO® SERIOUS PLAY®. Tourism researchers who adopt qualitative, inductive approaches to their research are always seeking tools that allow participant driven understandings and co-creation of knowledge. To address this gap, we then discuss the findings of our case study research and our reflections and conclusions on the use of LEGO® SERIOUS PLAY® methodology for tourism research.

Background

LEGO® SERIOUS PLAY® emerged in the early 2000s and is a method that has initially been used in business, education, community and more recently hospitality and tourism research. The idea of a method which uses a 'playful approach' and LEGO bricks to improve organisational performance originated in 1995 within the LEGO Group. The LEGO Group launched the LEGO® SERIOUS PLAY® method in 2001 (Gauntlett, 2007) and open-sourced a version of the method in 2010. Initially, the LEGO® SERIOUS PLAY® method was considered as a tool for improving organisational performance within the corporate environment (Oliver & Roos, 2007; Peter, Jacobs, & Roos, 2005; Pickard, 2007).

The studies conducted by David Gauntlett pioneered LEGO® SERIOUS PLAY® as a research methodology around the exploration of identities (Gauntlett, 2007; Gauntlett & Holzwarth, 2006). Drawing on his work, LEGO® SERIOUS PLAY® has been applied as a business method to enhance strategic thinking in organisations (Hadida, 2013) and as a brand research tool (Trebbin, 2016). Lately, the LEGO® SERIOUS PLAY® methodology has been used in psychology research (Harn & Hsiao, 2018), and education research as a tool for developing individuals' learning capacities (Barton & James, 2017; Kurkovsky, 2015; Montesa-Andres, Garrigós-Simón, & Narangajavana, 2014). Other studies in education have focused on the application of the LEGO® SERIOUS PLAY® method for providing kinesthetic forms of learning that involve both the body and the physical realm (James, 2013a, 2015; James & Brookfield, 2013b; Peabody & Noyes, 2017). In another example from education, Tseng (2017) found the LEGO® SERIOUS PLAY® method was a useful tool for fostering narrative identity among economically marginalised students. Similarly, within the community studies sector, Fletcher, Greenhill, Griffiths, Holmes, and McLean (2016) illustrated how they used

LEGO® SERIOUS PLAY® to allow the voices of community members and stakeholders in the town planning process.

A recent hospitality study used LEGO® SERIOUS PLAY® as a data collection tool to anticipate future changes in the hotel industry (Tuominen & Ascenção, 2016). In tourism research, LEGO® SERIOUS PLAY® methodology is advocated for its ability to create tourism realities (Wengel, McIntosh, & Cockburn-Wootten, 2016). Wengel, McIntosh & Cockburn-Wootten (2016) argued that LEGO® SERIOUS PLAY® is an effective methodology to gain deeper understandings of complex, dynamic, socially constructed realities in tourism. Although previous studies have provided information on the LEGO® SERIOUS PLAY® methodology, there remain few critical evaluations of this tool for tourism research.

To illustrate both the methodological processes and provide a critical evaluation of LEGO® SERIOUS PLAY® for tourism studies, we draw on a case study that investigated farm tourism research, with a particular focus on the WWOOF programme[1] in New Zealand. In this example, as we will discuss later in the article, LEGO® SERIOUS PLAY® offered a metaphorical way for participants to construct their ideal experiences, representing sometimes complex, entrenched and emotional issues, and relationships that may have been difficult to express via traditional methods. As a qualitative, inductive approach LEGO® SERIOUS PLAY® aims to understand 'how social experience is created and given meaning' (Denzin & Lincoln, 1994, p. 4). To achieve this understanding, LEGO® SERIOUS PLAY® uses metaphorical explanations to demonstrate constructs from the participant's imagination, which helps to illustrate the multiple facets of their complex experiences visually. In our case study, the methodology enabled us to understand how organic farm hosts and their WWOOF volunteers (guests) represented and constructed their sustainability ideals, relationships and the realities that they faced in their social interactions on the farm. Overall, LEGO® SERIOUS PLAY® enabled participants to communicate complex and sensitive issues in a creative way, especially around their relationships – to objects, landscapes, people and identities – aspects that may otherwise be silenced by traditional research methods.

Literature review

Traditional qualitative research methods such as interviews, focus groups, ethnography as well as photo and video documentation, have established themselves as reliable and useful methods in tourism studies (Buda, Martini, & Garcia, 2017). However, challenges still remain for researchers seeking tools that allow participant driven understandings and co-created knowledge. For researchers investigating topics that are sensitive, deal with unethical and/or inhospitable environments, or participants who have historically been excluded and neglected by the wider society, other research tools have been sought (Campbell, Sefl, Wasco Sharon, & Ahrens Courtney, 2004). For instance, tools that can ethically deal with facilitating voice, power, trust and relationship dynamics between the interviewer and the interviewee are crucial. For researchers working within the community, disability and chronic illness sectors, traditional qualitative tools do not do justice to the rich experiences of the participant, or capture hidden illness, and can end up weakening the research relationship – exacerbating participants feelings of exclusion and the academics' reputation as 'epidemics' (Bell, Addy, Madew, & Kainulainen, 2012, p. 95).

Scholars have critiqued qualitative tools for their reliance on the researcher's needs and values rather than enabling an understanding of the research context and a participant's experience of the issue. For example, in-depth interviews have been critiqued for being time consuming, sometimes expensive, and the flexibility of the process which may have negative implications on the quality of data (Bryman, 2016; Esterberg, 2002; Mishler, 1991; Patton, 2015). Bosco and Herman (2010) criticise focus groups for providing relational shared identity at the expense of individual experiences. Other researchers have suggested that the 'charged atmosphere' during a focus group is a risky factor to manage and can negatively impact fruitful discussion (Finn, Walton, & Elliott-White, 2000; Goulding, 1997). Gauntlett (2007) criticises talk-based approaches to data collection (for instance, interviews and focus groups) for their inability to get direct access to individuals' beliefs and mentions that these methods offer limited opportunity for participants 'to express themselves creatively' and 'to significantly affect the research agenda' (Gauntlett, 2005, p. 2). Talk and written based data collection tools can also exclude participants who may have physical disabilities or difficulties communicating in the researcher's language. For example, participants with a disability or those who speak a different language may find it challenging to articulate their experiences using traditional qualitative methods (Cattell, 2001). Within tourism studies, visual methods of data collection are largely employed by using existing visual objects rather than artefacts or images creatively produced by participants (Albers & James, 1988; Hunter, 2012; Nyaupane, Lew, & Tatsugawa, 2014; Pritchard & Morgan, 2003; Scarles, 2004; Uzzell, 1984).

Overall, traditional qualitative methods have been heavily criticised for their inability to be participant driven, capture the co-construction of the participants' realities or to address the impact of wider social dynamics (Liamputtong, 2007). Hence, there are reasons to explore the application of creative research methods tools such as LEGO® SERIOUS PLAY® methodology.

The philosophical foundations of LEGO® SERIOUS PLAY®

Discussing various theoretical positions integrated in LEGO® SERIOUS PLAY®, researchers agree that initially the methodology rooted on four pillars: constructivism/constructionism, the concept of play underpinned by the use of imagination and metaphors, and the theory of flow (Barton & James, 2017; Gauntlett, 2007; James, 2013a; Nolan, 2010).

Constructivism

Historically, the LEGO® SERIOUS PLAY® method draws on many ideas from the fields of psychology and behavioural science. Two core concepts embedded into LEGO® SERIOUS PLAY® are Piaget's constructivism (1955) and Papert's constructionism (1991). According to Piaget's theory, learners are not 'empty vessels into whom we can pour knowledge' (p. 82), but rather, learners could be described as active theory creators who construct and rearrange that knowledge based on their prior knowledge and experience (Kristiansen & Rasmussen, 2014). Piaget (1955) claimed that intelligence increases when the mind creatively interacts with the outer world. This theory was developed further by Papert, a follower of Piaget, in his theory of constructionism which focused on building knowledge through 'hands-on learning' (Papert & Harel, 1991). Central to Papert's theory was that learning is exceptionally productive when people are actively engaged in the creation or construction of something that is external to themselves.

According to Papert, constructionism allows abstract ideas and relationships to become more concrete, more visual, and tangible, and thus more understandable (Papert & Harel, 1991). Papert proposed that learners 'think through fingers' by producing various modes of thoughts, based on their creativity and imagination (Papert & Harel, 1991). Thus, learners are engaged when tangible objects are involved, and their abstract ideas become more concrete, visual, and therefore more understandable (Papert, 1999). LEGO® SERIOUS PLAY® methodology applied this idea of concrete thinking, which is thinking with and through concrete objects as a mode of thinking that can be complementary to more abstract, formal modes of thought. A founding philosophy of LEGO® SERIOUS PLAY® methodology is that participants can unlock their creative thinking through play and 'thinking with objects' or 'thinking through fingers' when constructing their reality with LEGO® bricks.

The concept of play

The second pillar of LEGO® SERIOUS PLAY® is the 'concept of play' which assumes that innovative and creative ideas are most likely to come through the playful free-flowing process (Gee, 2007; Kane, 2004; Terr, 2000). LEGO® SERIOUS PLAY® methodology applies 'play' and encourages learning through exploration, storytelling, and/or metaphors. This method assumes that each play process has a purpose (Rieber, 1996) and defines play as 'a limited, structured, and voluntary activity that involves the imaginary' (LEGO® SERIOUS PLAY®, 2006, p. 4). An advantage of the LEGO® SERIOUS PLAY® method is that when participants play, they play with their sense of who they think they are (identity), and with one or more specific goals in mind, such as social bonding, emotional expression, cognitive development, and constructive competition (LEGO® SERIOUS PLAY®, 2006). The concept of play is linked to identities and imaginations, which are seen as central to the play process (Gauntlett & Holzwarth, 2006).

Researchers argue that the motivational basis for play is primarily emotional and individuals attach emotional meanings to their experiences and objects (Fein, 1987; Vygotskiĭ & Cole, 1978). A further advantage of LEGO® SERIOUS PLAY® is that 'through the use of modelling and metaphor, the objects of play can take on meanings and can embody abstract concepts, thus concretising formal relationships that can otherwise be quite difficult to comprehend' (LEGO® SERIOUS PLAY®, 2006, p. 6).

Imagination and metaphors

LEGO® SERIOUS PLAY® methodology is founded on three aspects of imagination: imagination as a way to describe something; imagination as a way to create something; and, imagination as a way to challenge something. The descriptive imagination is based on our experiences, and its purpose is to evoke images that describe our complex reality. While descriptive imagination allows us to see what is there in a new way, creative imagination enables us to see what is not there, hence allowing the creation of something new and different. In LEGO® SERIOUS PLAY® methodology, creative imagination is associated with innovation and novel thinking. In turn, challenging imagination goes beyond creative imagination; it does not add a new element to what is already there, but starts from scratch and assumes nothing (LEGO® SERIOUS PLAY®, 2006). When these three types of imagination are combined during the LEGO® SERIOUS PLAY® workshop, they create a strategic imagination, which is seen as a source of original ideas.

Imagination is closely linked to story-telling and the use of metaphors. In qualitative research, metaphors help increase the depth of the meaning of understanding (Kangas, Warren, & Byrne, 1998) and 'illuminate the meanings of experiences' (Carpenter, 2008, p. 274). Metaphors represent 'a form of thinking and language through which we understand or experience one thing in terms of another' (Lakoff & Johnson, 2003, p. 7). As a research method based on metaphorical creative exploration, LEGO® SERIOUS PLAY® can reveal underlying thinking, understanding, and meanings of experiences (Carpenter, 2008; Gauntlett, 2007; Kangas et al., 1998). The LEGO® SERIOUS PLAY® methodology aims to foster creative thinking through building metaphors of participants' identities and experiences using LEGO® bricks. LEGO® SERIOUS PLAY® methodology may differ in terms of the content and focus on the workshop process, the role of the facilitator and the engagement of the participants through skills building and flow process. Overall, previous theorists, using similar metaphorical processes, have argued that a metaphorical process is useful for identifying and reflecting on the multiple social discourses that individuals are exposed to, particularly around the tourism experience (e.g. Morgan, 1980). Consequently, LEGO® SERIOUS PLAY® methodology is claimed to stimulate new awareness of 'reality' and provides deeper metaphorical meanings as well as the depth of participants' lived experiences not captured by alternative methods (Wengel et al., 2016).

The theory of flow

Another pillar of LEGO® SERIOUS PLAY® methodology is Csikszentmihalyi's (1991) theory of flow. The theory of flow describes the emotional state of a person while undertaking a focused task or activity when the skills level is matched to the difficulty of the task (Csikszentmihalyi, 1991). LEGO® SERIOUS PLAY® method embraces the theory of flow to the extent that the methodology allows participants to stay in the flow while engaging, enjoying, and concentrating on the process of construction guided by the facilitator regardless of participants' familiarity with LEGO® bricks.

Currently, there is a LEGO® SERIOUS PLAY® 1.0 (an open source version with basic principles released by LEGO Group) and LEGO® SERIOUS PLAY® 2.0 version. The 2.0 version is a complete version of LEGO® SERIOUS PLAY® method developed by the Association of Master Trainers in LEGO® SERIOUS PLAY® method which includes seven application techniques and four core steps of LEGO® SERIOUS PLAY® process. One significant difference of the LEGO® SERIOUS PLAY® 2.0 is that the facilitator ensures that participants are engaged in the process and experience flow. This is achieved by applying step by step the LEGO® SERIOUS PLAY® process and taking participants through the skills-building process where the increasing challenge of the task is balanced by improved skills of the participants (Kristiansen & Rasmussen, 2014). This process illustrates and reflects the theory of flow, as by 'experiencing flow encourages a person to persist at and return to an activity because of the experiential rewards it promises and thereby fosters the growth of skills over time' (Csikszentmihalyi, 2014, p. 249).

The case study

The case study that applied LEGO® SERIOUS PLAY® investigated farm tourism focusing on host-guest experiences in the World Wide Opportunities on Organic Farms (WWOOF) programme in New Zealand. WWOOF is a tourism related phenomenon and is a global exchange programme which connects volunteers (WWOOFers) with organic farmers to support cultural and educational experiences based on trust and nonfinancial exchange, with the aim to help build a sustainable, global community (Federation of WWOOF Organisations, 2016). This research aimed to understand how hosts (farmers) and guests (WWOOFers) construct the 'ideal' WWOOFing experience to ensure a mutually beneficial encounter for both farmers and volunteers. Underpinned by constructionist epistemology, the research adopted LEGO® SERIOUS PLAY® methodology to involve participants in a discussion about their experiences.

The principal author participated in one-week long training in 2013 and is certified by the Association of Master Trainers in the LEGO® SERIOUS PLAY® Method as a facilitator of LEGO® SERIOUS PLAY® method and materials. This research project adopted the 2.0 version and was guided by LEGO® SERIOUS PLAY® etiquette based on the training manual. As such, we applied two application techniques: (1) building individual models and stories, and (2) building shared models and stories (Kristiansen & Rasmussen, 2014). The first application technique allows participants to build an individual model and to share new knowledge with other participants. The second application technique allows collaboration among participants to make decisions and gain a mutual understanding of a given topic by combining some individual models or parts of the individual models into one shared model. We sourced 20 kg of LEGO bricks for the workshops. An individual customised Starter Kit for skills-building exercises was created based on the requirements that had been provided during the LEGO® SERIOUS PLAY® training (Figure 1). In our farm tourism research, a custom starter kit contained solid coloured bricks, transparent bricks, 'multi metaphor bricks (e.g. a fence, a flagpole), 'single metaphor bricks' (e.g. plants, animals), some 'people bricks' (e.g. mini figures, accessories and 'eye bricks') as well as bricks for making movements (e.g. wheels, hinges, and other rotating elements) and a base plate.

The remaining LEGO® bricks were used as an 'Identity and Landscape kit'. This kit was enlarged by various connection elements, green elements, as well as DUPLO[2] elements representing plants and animals to fit the farming theme of the workshops. Another adjustment of LEGO® SERIOUS PLAY® to the current research was that the

Figure 1. Customised Starter Kit. Source: Authors.

Table 1. LSP workshops and participants. Source: Authors.

Participant number	Name*	Country of origin	Farmer/ WWOOFer	Workshop duration
		Workshop 1 at Farm 1 Location: North Island, New Zealand Farm type: private, no commercial activity		
1	Xavier	Spain	WWOOFer	04:12 h
2	Steffen	Germany	WWOOFer	
3	Yana*	Russia	WWOOFer	
4	Mary	New Zealand	Farmer	
5	James	New Zealand	Farmer	
		Workshop 1 at Farm 2 Location: North Island, New Zealand Farm type: private, commercial activity-sheep grazing		
1	Danny	Mexico	WWOOFer	05:03 h
2	Yana**	Russia	WWOOFer	
3	Linley	New Zealand	Farmer	
4	Mike	New Zealand	Farmer	
5	Chloe	New Zealand	Farmer	
6	Caleb	New Zealand	Farmer	
		Workshop 1 at Farm 3 Location: North Island, New Zealand Farm type: private, no commercial activity		
1	Samantha	USA	WWOOFer	03:30 h
2	Lauren	USA	WWOOFer	
3	Yana*	Russia	WWOOFer	
4	Charlotte	New Zealand	Farmer	

*Original names have been changed.

**Principle author facilitated the workshop process and additionally participated as a WWOOFer.

principal author played a dual role during all workshops: the role of the facilitator and the role of a participant. This adjustment was made to gain participants trust and overcome participants' scepticism towards LEGO® SERIOUS PLAY® methodology (as LEGO® bricks are mainly associated with being a children's toy). This adjustment helped to eliminate this limitation and developed a more positive mindset from participants' attitudes towards this creative method.

Three LEGO® SERIOUS PLAY® workshops with a total of 13 participants were conducted on three farms in New Zealand. Each of the workshops employed two application techniques: building individual models and stories and building a shared model. At the end of each workshop participants built a shared model of the 'ideal' WWOOFing experience. All workshops were video- and audio-recorded. As a method, LEGO® SERIOUS PLAY® requires all participants to be available at the same time and location. At times, this might be a limitation as it can be challenging to get participants together in one site. Ethics approval was obtained from the Waikato Management School Human Research Ethics Committee to protect participants' confidentiality and to ensure informed consent. Table 1 provides a summary profile of the participants.

Before starting each workshop, participants were given information about the LEGO® SERIOUS PLAY® process. Additionally, the principal author explained to participants the concept of LEGO® SERIOUS PLAY® and, in particular, noted it as a method of detachment from the person/the issue to instead one of metaphor. Every workshop started with skills building, a series of icebreaker and warm-up exercises. Specifically, participants were asked to build a duck and a tower using a customised starter kit. During all the workshops, the participants were asked to build different models related to WWOOFing by using LEGO® bricks first as an individual and then as a group. During each workshop, the principal author posed questions as a facilitator, explaining to the participants that she was acting as the facilitator. When the participants were being encouraged to provide a metaphorical explanation of their model one after another, she always provided her answers last. She tried to notice what other participants had mentioned and described her models in a neutral way, to reduce any potential bias and so that no new knowledge was minimised by her insights. When she switched roles from a participant to a facilitator, she reminded the other participants that she was doing so.

After a series of LEGO® SERIOUS PLAY® workshop steps, participants were invited to create a joint model consisting of individual models, or parts of an individual model, that embraced the ideas and experiences of farmers (hosts) and WWOOFers (volunteers) at the same time. During this process, participants were encouraged to negotiate what parts of a model or models would constitute the final model, the 'ideal' WWOOFing experience. At this stage, the principal author, as a participant, held back her ideas and let the participants determine the model and the story of the social construction of the WWOOFing experience. By the end of each workshop, the participants had constructed a joint model of an 'ideal' WWOOFing experience, which is further described in the following section.

To create their metaphors, participants used the LEGO® bricks imaginatively, before drawing on their conscious knowledge of what they perceived an 'ideal' WWOOF experience for them to be. Thus, metaphors are used as a filter to uncover the multiple aspects of the WWOOF phenomenon from the participants' subjective points of view. In verbal language, metaphors represent a way to reveal the underlying thinking or understanding of a person. This creative approach to research can be seen to allow a deeper insight into the understanding of participants' lived experiences and helped to explore their personal meanings and attitudes.

Findings

Two core themes were revealed through the LEGO® SERIOUS PLAY® workshops. These themes were people/relationships and sharing/collaboration. WWOOFing participants believed that an 'ideal' WWOOFing experience depended upon the people involved in it sharing things, activities, and experiences. Some of the models created by the participants focused on a mismatch of expectations, communication, and cultural problems in the host–guest experience. For example, James (farmer, farm 1), mentioned that most of the negative cases of WWOOFing experiences on his farm were connected to the lack of engagement from WWOOFers: 'We open our life, and our house and they are not connecting, they are not interested in all of these' (Figure 2). This model used several standard LEGO® elements to describe James' negative experiences metaphorically. In his model, the LEGO® open window element represents a house, the host mini figure is giving a 'rope' and shows his willingness to connect with the WWOOFer mini figure, but this volunteer prefers to have 'his own agenda' (represented by the flag) and is not interested in the hosts' life and farm activities. The volunteer is standing with his side to a fence (which is representing a barrier), and he is also wearing a helmet, which is metaphorically explained as a lack of motivation and interest in WWOOFing. The plants in front of the 'house' represent the farm and WWOOFing activities.

In the LEGO® SERIOUS PLAY® workshops we asked the participants to build models of the positive WWOOFing experience. Many participants reported stories related to the interaction between people. For example, Linley (farmer, farm 2) built a model of a farmer and her WWOOFers working together in the garden (Figure 3). A minifigure with a hammer on the left represents a farmer who is connected with the plants on

Figure 2. Individual Model, James (Farmer, Farm 1): A Negative WWOOFing Experience Model. Source: Authors.

Figure 3. Individual Model, Linley (Farmer, Farm 2): A Positive WWOOFing Experience Model. Source: Authors.

the farm (represented by the rope). On the same side, this connection extends to the figure with many eyes. This figure represents international WWOOFers who come to work on the farm with Linley and to experience New Zealand. The silver pipe connecting WWOOFers (represented by the figure with many eyes) and plants metaphorically describe the interest of some volunteers in their sustainable lifestyle and their willingness to experience the local life by interacting with their farm hosts. Participants used a variety of LEGO® bricks with eyes, smiles, mini figures and elements showing interaction such as ropes, cords, ladders, and tools like spades, hammers, and scissors to describe people sharing work on the farm.

From the workshops, we saw that both shape and also the colour of LEGO® bricks and elements helped participants to explain their experiences and relationships with others, objects and the lifestyle. This was illustrated in the metaphors participants used in their workshops. For instance, in several individual models (see Figure 4), participants used many green LEGO® bricks and elements representing plants. They used these elements to metaphorically describe the farming element of the WWOOFing experience where farmers and volunteers take care of plants and animals. Participants highlighted that being close to nature, working with animals and plants were distinguishing features of WWOOFing and distinguished WWOOFing from other tourism related exchange programmes like Couchsurfing or Help-X.

One participant built a model of his transformational experience by presenting a model of compost (Figure 5). This model visually represented a cubic metre of the compost pile. For Steffen (volunteer, farm 1), the WWOOFing experience had a transformative component as he got to understand a food life cycle by volunteering on the farms. Apart from the plant bricks representing what compost is made of, Steffen explained that the clear, transparent window element represents his 'enlightenment' about the food processes and compost lifecycle as one of the central elements in food production. From one of the hosts, he had learned about the food life cycle and compost as a crucial element in it. Steffen commented that before becoming a WWOOFer he never thought about food and where it comes from; he just used to get it from the supermarket. However, his WWOOFing experience motivated him to learn

Figure 4. Examples of Individual Models by Farmers and Volunteers (Farm 1, 2 and 3). Source: Authors.

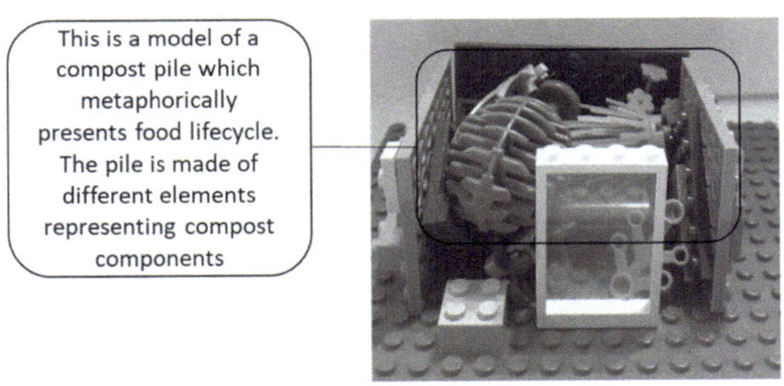

Figure 5. Individual Model, Steffen (Volunteer, Farm 1): Compost Pile Model. Source: Authors.

more about the food cycle and the central role of compost in food growing cycle to keep the soil and the plants healthy and full of nutrients.

The example in Figure 5 shows how participants described their ideas and connected tangible objects and their metaphorical meaning to their experience. Through metaphorical explanations LEGO® SERIOUS PLAY® methodology allows participants to have time (while building a model) to reflect on their experiences and hand-mind connections to draw on their memories bringing their lived experiences to the surface.

An advantage we found of using LEGO® SERIOUS PLAY® methodology is its ability to allow participants to communicate potentially sensitive issues more neutrally. In this research, many participants mentioned intercultural and interpersonal communication as challenging issues. In Figure 6, the left photo is an individual model built by Mary (farmer, farm 1) and the right photo is a model built by Lauren (WWOOFer).

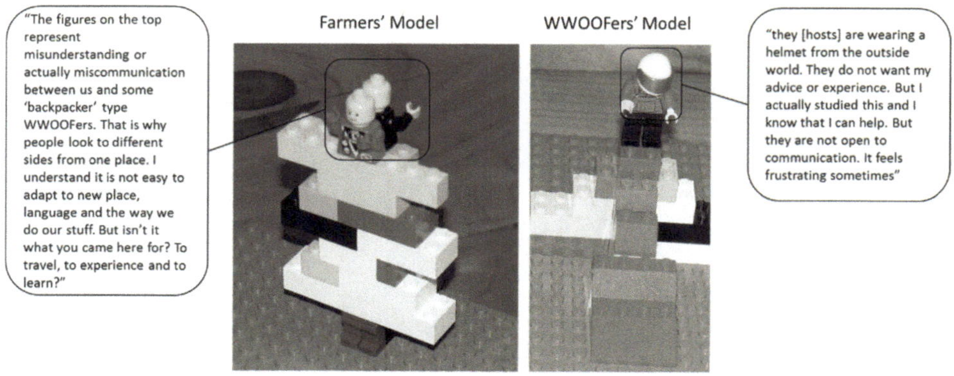

Figure 6. Individual Models, Mary (Farmer, Farm 1) and Lauren (Volunteer, Farm 3): Communication Challenges Models. Source: Authors.

Both models describe the challenges and, at times, frustrating emotions in terms of their communication and mutual understanding about the lifestyle, sustainability values and experiences on the farm. In the left model, the mini figures are shown looking to opposite sides to represent the miscommunication. The WWOOF volunteer described that when a person is 'on top of the tower' or 'in your [their] element', a person tends not to hear what others are trying to communicate. In the model on the right, a helmet was used as a strong metaphor signifying the unwillingness of the farmer to look beyond their own experience and to hear the advice of the WWOOFer. The emotional frustrations of the volunteer are evidenced in the metaphors applied in this model, and her disappointment that communication of advice between WWOOFer and farmer could not be reciprocal.

Another example of a sensitive issue communicated in a workshop was related to the rules. In addition to legal requirements, findings exhibited rules that were more ambiguous and related to the farmer's lifestyle preferences. These rules were open to miscommunication and had the potential of developing negative emotions among either party. These negative emotions arose if the rules were interpreted as unfair, deemed unwarranted or not clearly communicated to the new volunteer. For example, one participant built a model using farm gates with a door to metaphorically describe the importance of following rules on the farm, especially for health and safety and ensuring resource availability. Describing his positive experience with WWOOFers, Mike (farmer, farm 2) mentioned that he enjoyed engaging with international people on his farm because 'they bring the world to us'. Figure 7 below represents Mike's point.

Although the model may appear simplistic as the different shape and colour of bricks are not even attached to each other, it has deep meaning. Mike explained that bricks of different shapes and colours represent WWOOFers coming from various countries, and the central element in this model is his farm and his family (represented by the LEGO® mini figures surrounded by bricks representing the WWOOFers). During the LEGO® SERIOUS PLAY® workshops, other farmers too mentioned that they enjoyed the interaction with WWOOFers and enjoyed seeing them connecting with locals, nature and animals. One of the farmers mentioned that she loved hosting

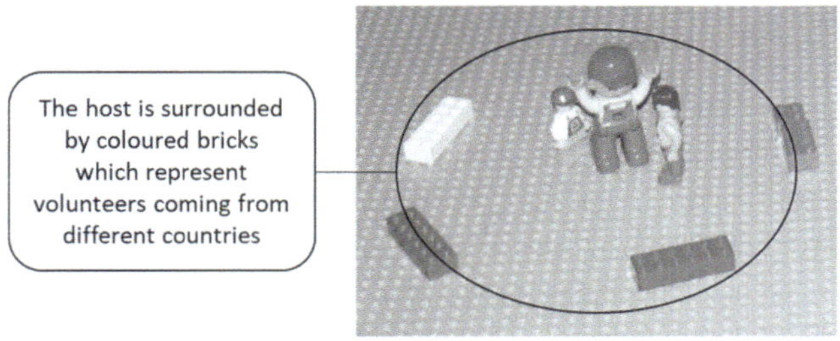

Figure 7. Individual Model, Mike (Farmer, Farm 2): 'The World is Coming to Me' Model. Source: Authors.

Figure 8. Individual Model, Charlotte (Farmer, Farm 3): Positive WWOOFing Experience. Source: Authors.

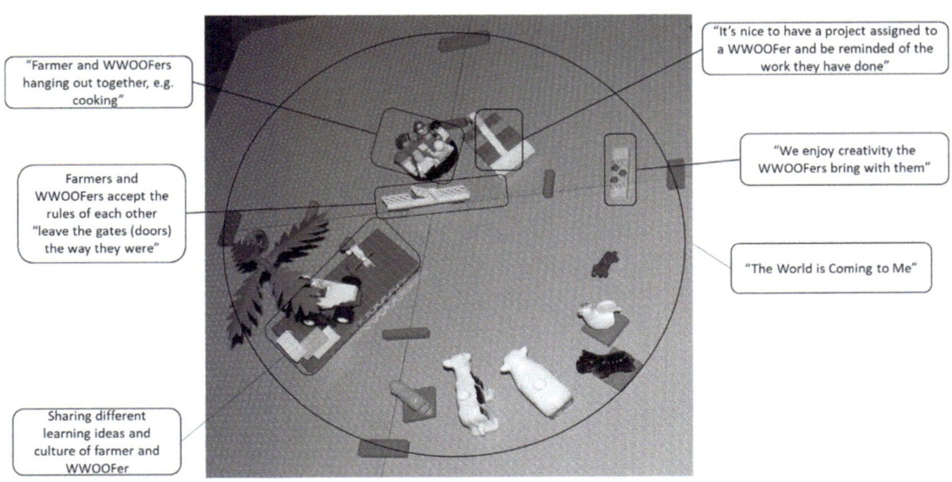

Figure 9. Shared Model (Farm 2): An 'Ideal' WWOOFing Experience Model. Source: Authors.

WWOOFers because she is not able to travel physically but she 'travels' metaphorically by listening to the stories of her WWOOFers. To metaphorically represent this experience Charlotte (farmer, farm 3) built a map and established 'connections' between New Zealand and other parts of the world where the WWOOFers were coming from (Figure 8).

Figure 9 represents an example of a shared model of an 'ideal' WWOOFing experience built by five participants during a LEGO® SERIOUS PLAY® workshop on farm number 2. This model contains several elements of individual models and full individual models built by the participants during the previous stages of the workshop. The whole model is 'embraced' by Mike's individual model explained earlier (see Figure 7). Participants agreed that an 'ideal' WWOOFing experience is one where farmers interact with WWOOFers coming from different countries; WWOOFers are represented by different coloured LEGO® bricks shaping a circle around the model. In the centre of the model, there is a farmer's family and their WWOOFers. In this workshop, hosts mentioned that they enjoyed the creativity of their volunteers, represented by the yellow LEGO® brick with red and green elements on top of it.

Furthermore, this model again mentioned the importance of rules on the farm, which is metaphorically represented by the white gate in the middle of the model. Finally, farmers suggested making this model into a round shape as they predominantly had positive WWOOFing experiences with their volunteers and a circle represented balance for them.

Participants also highlighted the importance of having social time together while sharing and learning about different cultures. For example, in this model (Figure 9), six mini figures are cooking together at the top of the photo. Another important component of the ideal WWOOFing experience co-constructed by the participants related to education and acquiring new skills (for example, represented by the mini figure with a tool next to a tree and a tractor at the bottom left of the photo).

The three models of an 'ideal' WWOOFing experience brought participant's individual experiences together into a co-constructed environment where everyone had a chance to express their voice. The LEGO® SERIOUS PLAY® models revealed the multiple realities within the WWOOF programme. Notably, during the workshops, participants tried not only to co-create their ideal WWOOFing experience metaphorically but also used various green LEGO® bricks, elements and animals to replicate the physical farm appearance. In the workshops, green LEGO® bricks and elements not only represented plants on the farm but also metaphorically described the creativity and energy of the WWOOFers that farmers enjoyed. Participants proposed that an 'ideal' WWOOFing experience could be achieved when participants are mutually interested in each other and farming activities, as well as when they share work, food, social time and understanding the rules.

Reflections on LEGO® SERIOUS PLAY® methodology

The LEGO® SERIOUS PLAY® workshops allowed for a playful and creative process with the freedom to explore the social construction of the New Zealand WWOOFing experience. On reflection of our case study research, we would like to highlight two overall benefits of LEGO® SERIOUS PLAY® for tourism researchers. First, it was clear from our experience that the process allowed a positive, playful experience that enabled participants to (co)creatively represent their experiences metaphorically and with detachment from the person and/or issue. Second, LEGO® SERIOUS PLAY® also facilitated reciprocal learning to occur and allowed tacit 'hidden' experiences and knowledge to be communicated.

In this research, it should be noted that we made two adjustments. First, we created custom LEGO® SERIOUS PLAY® kits using the guidelines provided during the facilitator's training. Secondly, the facilitator also participated in the research process to try and overcome the participants' unfamiliarity with LEGO® bricks as a research tool. In previous LEGO® SERIOUS PLAY® workshops facilitated by the principal author, she found some participants were reluctant to use LEGO® SERIOUS PLAY® as a research method; hence, it is not always easy to predict how participants will feel. LEGO® SERIOUS PLAY® as previous researchers have reported, can have negative associations for participants with creativity and their reluctance towards participation in methods involving creative thinking, painting or producing a collage (Gauntlett, 2005, 2007; Greenwald, Poehlman, Uhlmann, & Banaji, 2009; Mueller, Melwani, & Goncalo, 2012). We feel that the adjustment we made helped facilitate trust and a more positive attitude from participants towards using LEGO® bricks as a new creative method for research. This can be supported in the following quote from one of the participants:

> In the beginning, I did not believe that you are seriously using LEGO® for your research. But now, after we have done it, and it was so much fun, I understand that it is not about bricks but about the stories you tell. Thanks for persuading me to participate. (Mike, Farmer, Farm 2).

Other participants commented that they enjoyed having the opportunity to be involved in a creative process and valued the possibility to share their LEGO® models. Many participants enjoyed being in 'flow' while they were creating their models (Csikszentmihalyi, 1991). This level of enjoyment is certainly found in previous LEGO applications (Wengel et al., 2016). In particular, participants enjoyed sharing and listening to stories of the New Zealand WWOOFing experience. Participants sharing not only their own stories but also seeing and listening to others had provided deeper meanings and understandings of how an 'ideal' WWOOFing experience was socially constructed by the participants of this volunteer farm tourism exchange programme.

We found that LEGO® SERIOUS PLAY® offers participants an opportunity to express themselves creatively, allows researchers to 'dig' deeply into participants experiences to bring to the surface those experiences and facilitated learning. LEGO® SERIOUS PLAY® methodology in this case study proved itself as a good tool to access more complex, sometimes sensitive information from participants' experiences. One advantage of the LEGO® SERIOUS PLAY® is that it offers a variety of pre-formed bricks and elements which are used in the models. This allows participants to overcome any lack of 'creativity' in enabling metaphors to be built and explained. Notably, participants frequently used several elements to metaphorically describe their experiences and relationships with others, objects and lifestyle. For example, windows, gates, and doors were often used to describe interpersonal and intercultural communication aspects, mini figures and eyes were used to describe people and beings, animal and plant elements were used to describe the farm setting.

Furthermore, the colour of the brick was important when participants tried to replicate physical environments. For example, for plants, they used green bricks, and for water blue bricks were used. The window elements and connection elements, such as ropes, cords, and ladders, were frequently used by participants to metaphorically describe networks, connections and disconnections between participants. This proved a more subtle and conflict adverse way to communicate and engage in conversation around the emotional aspects of the experience.

The LEGO® SERIOUS PLAY® process facilitated a learning space for participants through their interactions to reflect on the knowledge gained from their experiences. Traditional data collection tools typically see interactions as merely facilitating the talk based content rather than collaborations for learning (Wibeck, Dahlgren, & Öberg, 2007, p. 251). To achieve these interactions for learning requires the intentional creation of space and activities, and for the researcher to change their expert role to instead use dialogic conversational communication techniques (Hinthorne & Schneider, 2012). The researcher's role alters to one that 'facilitates the learning rather than acting as a knowledge dispenser' (Wibeck et al., 2007, p. 251) Participants collaborating/co-creating over an activity allows interactions that help researchers to learn about how participants develop 'sensemaking' and overcome problems around their experiences. These types of research tools enhance learning for both participant and researcher through the 'act of critical reflection through experiential learning and dialogue' (Hinthorne & Schneider, 2012, p. 2808). Despite these benefits, one of the limitations of LEGO® SERIOUS PLAY® is the level of skills and experience participants have with LEGO® as a toy. For example, participants who had played with the bricks as a child were more experienced and tended to build more complex models. This level of experience may unbalance the flow of the workshop process with people finishing their models at different times. Hence the facilitator needs to be attentive, especially in the skills building stage of the workshop. In our experience, one participant who was at first reluctant to use the methodology (because he perceived it as a child's toy) actually built a complex individual model with deep metaphorical meaning, which was eventually echoed in the models of other participants. Another limitation mentioned earlier is the need for all participants to be at the same time and place, which might be not possible for some research settings.

Although the methodology entails the same critiques and limitations of the constructivist paradigm more generally (including lack of generalisability and replicability of the data), the models from all workshops did share common similarities in the description of the participants' experiences, which means that even in a relatively small sample (such as this study with three LEGO® SERIOUS PLAY® workshops), LEGO® SERIOUS PLAY® allowed rich data saturation to occur and enabled common themes that may be compared to other WWOOF contexts to emerge. We were concerned that constructing and explaining ideas metaphorically might be difficult for participants for whom English is a second language. However, due to its hands-mind connection, implementation of symbols and metaphors, and time given to construct a model, LEGO® SERIOUS PLAY® has proven itself as a particularly useful method for investigating areas where the participants may be using more than one language in order to try to communicate their experiences to the researcher.

Overall, the structured workshop process allowed participants to test ideas without fear of saying something wrong or upsetting relationships, especially when they are living and working together at the farm. While participants assigned meaning to their individual models, they also assigned meaning to their stories, and tacit understandings that were taken-for-granted emerged. In sharing their experiences using this method, participants exchanged advice, understandings and solutions for creating an 'ideal' New Zealand WWOOFing experience.

Conclusion

As a constructionist technique, LEGO® SERIOUS PLAY® offers a valuable methodology for researchers seeking deeper explorations of socially constructed tourism realities that are complex and sometimes sensitive (Wengel et al., 2016). As a visual method, LEGO® SERIOUS PLAY® provides the researcher with insights into participants' imaginations and conceptual understandings of situations and relationships. Also, LEGO® SERIOUS PLAY® offers researchers a tool that can overcome barriers from traditional talked based data collection methods. In our case study, LEGO® SERIOUS PLAY® offered opportunities for bringing different perspectives together to consider an issue with the aim of reciprocal learning, communication and developing solutions. According to Schön (1983), metaphors can play an integral role in answering research questions. In this research, metaphors were used to enrich explanations of what WWOOF means for the participants and to provide a pallet of meanings, definitions and possible solutions.

The LEGO® SERIOUS PLAY® process allows participants to tap into their creativity through the hands-mind connection which provides the researcher with an insight into their complex understandings and descriptions of the investigated tourism phenomena. Also, certain sensitive topics that participants might be reluctant to discuss, especially in front of others, can be explored in this creative, playful process. The process can reduce any possible conflict and negative emotions, for example, if a host is frustrated by guests who do not follow their rules, or if a guest wants to voice that their knowledge might be useful for the farmer to hear. In talk based methodologies this may need careful facilitation but with visuals, metaphors and crucially, through the framework of a 'playful approach' these interpersonal communications can be discussed in a constructive manner.

Finally, for tourism researchers, LEGO® SERIOUS PLAY® facilities a co-creative, reflective thinking and learning framework for tackling understandings around topics that may yield issues of power or sensitivity, or where there might be difficulty in articulating tacit understandings of a tourism phenomenon. The tool enables participants to apply metaphors to 'generate an image for studying subject' (Morgan, 1980, p. 611). By encouraging the use of metaphors to understand how participants make sense of these images, the method could also be applied to look at stakeholder perspectives in the tourism system to engender creative thinking around change for broader action planning for sustainable tourism development.

Notes

1. WWOOF stands for World Wide Opportunities on Organic Farms and represents a network which promotes cultural and educational experiences based on trust and non-monetary exchange. WWOOF connects farmers and volunteers who are interested to exchange labour for food, accommodation and learning about organic farming practices and sustainable ways of living.

2. LEGO DUPLO is a product range created for children between 1.5 and 5 years old. Duplo bricks are two times larger in length, height and width compared to standard LEGO bricks (https://www.lego.com/en-gb/themes/duplo).

Disclosure statement

There is no conflicts of interest to declare.

ORCID

Yana Wengel http://orcid.org/0000-0002-8131-4137
Alison McIntosh http://orcid.org/0000-0003-1593-700X
Cheryl Cockburn-Wootten http://orcid.org/0000-0002-3339-630X

References

Albers, P. C., & James, W. R. (1988). Travel photography: A methodological approach. *Annals of Tourism Research, 15*(1), 134–158. [Mismatch] doi:10.1016/0160-7383(88)90076-X

Ateljevic, I., Harris, C., Wilson, E., & Collins, F. L. (2005). Getting 'entangled': Reflexivity and the 'critical turn' in tourism studies. *Tourism Recreation Research, 30*(2), 9–21. doi:10.1080/02508281.2005.11081469 doi:10.1080/02508281.2005.11081469

Ateljevic, I., Pritchard, A., & Morgan, N. (2007). *The critical turn in tourism studies: Innovative research methodologies*. Amsterdam: Elsevier.

Barton, G., & James, A. (2017). Threshold concepts, LEGO® SERIOUS PLAY® and whole systems thinking: Towards a combined methodology. *Practice and Evidence of the Scholarship of Teaching and Learning in Higher Education, 12*(2), 249–271.

Bell, P., Addy, T., Madew, M., & Kainulainen, S. (2012). Universities as agents in the empowerment of local communities in Germany, Finland and Russia. In L. Goodson & J. Phillimore (Eds.), *Community research for participation: From theory to method* (pp. 89–105). Bristol: Policy Press.

Bosco, F., & Herman, T. (2010). Focus groups as collaborative research performances. In D. DeLyser, S. Herbert, S. Aitken, M. Crang & L. McDowell (Eds.), *The SAGE handbook of qualitative geography* (Vol. 220–236). New York, NY: Sage.

Bryman, A. (2016). *Social research methods* (5th ed.). Oxford: Oxford University Press.

Buda, D., Martini, A., & Garcia, L.-M. (Eds.) (2017). *Qualitative tourism research*. Thousand Oaks, CA: SAGE Publications Inc.

Campbell, R., Sefl, T., Wasco Sharon, M., & Ahrens Courtney, E. (2004). Doing community research without a community: Creating safe space for rape survivors. *American Journal of Community Psychology, 33*(3–4), 253–261. [Mismatch] doi:10.1023/B:AJCP.0000027010.74708.38

Carpenter, J. (2008). Metaphors in qualitative research: Shedding light or casting shadows? *Research in Nursing & Health, 31*(3), 274–282. doi:10.1002/nur.20253 doi:10.1002/nur.20253

Cattell, V. (2001). Poor people, poor places, and poor health: The mediating role of social networks and social capital. *Social Science & Medicine, 52*(10), 1501–1516. doi:10.1016/S0277-9536(00)00259-8. doi:10.1016/S0277-9536(00)00259-8

Csikszentmihalyi, M. (1991). *Flow: The psychology of optimal experience*. New York: Harper Perennial.

Csikszentmihalyi, M. (2014). *Flow and the foundations of positive psychology: The collected works of Mihaly Csikszentmihalyi*. New York: Springer.

Decrop, A. (1999). Triangulation in qualitative tourism research. *Tourism Management, 20*(1), 157–161. doi:10.1016/S0261-5177(98)00102-2. doi:10.1016/S0261-5177(98)00102-2

Denzin, N. K., & Lincoln, Y. S. (1994). *Handbook of qualitative research*. Thousand Oaks, CA: SAGE.

Esterberg, K. G. (2002). *Qualitative methods in social research*. New York: McGraw-Hill.

Fein, G. (1987). Pretend play: Creativity and consciousness. In D. Görlitz & W. J. F. (Eds.), *Curiosity imagination and play*. Hillsdale, N.J.: Lawrence Erlbaum.

Finn, M., Walton, M., & Elliott-White, M. (2000). *Tourism and leisure research methods: Data collection, analysis, and interpretation*. London: Pearson Education.

Fletcher, G., Greenhill, A., Griffiths, M., Holmes, K., & McLean, R. (2016). Creatively prototyping the future high street. *Production Planning & Control, 27*(6), 477–489.[Mismatch] doi:10.1080/09537287.2016.1147094

Gauntlett, D. (2005). Using creative visual research methods to understand media audiences. *Medien Pädagogik, 9*, 1–32. doi:10.21240/mpaed/09/2005.03.29.X

Gauntlett, D. (2007). *Creative explorations: New approaches to identities and audiences*. London: Routledge.

Gauntlett, D., & Holzwarth, P. (2006). Creative and visual methods for exploring identities. *Visual Studies, 21*(1), 82–91. doi:10.1080/14725860600613261 doi:10.1080/14725860600613261

Gee, J. P. (2007). *What video games have to teach us about learning and literacy* (Rev. and updated ed.). New York: Palgrave Macmillan.

Goulding, A. (1997). Joking, being aggressive and shutting people up: The use of focus groups in LIS research. *Education for Information, 15*(4), 331–341. [Mismatch] doi:10.3233/EFI-1997-15406

Greenwald, A. G., Poehlman, T. A., Uhlmann, E. L., & Banaji, M. R. (2009). Understanding and using the implicit association test: III. Meta-analysis of predictive validity. *Journal of Personality and Social Psychology, 97*(1), 17–41. doi:10.1037/a0015575 doi:10.1037/a0015575

Hadida, A. L. (2013). Let your hands do the thinking! *Strategic Direction, 29*(2), 3–5. doi:10.1108/02580541311297976

Harn, P.-L., & Hsiao, C.-C. (2018). A preliminary study on LEGO®-based workplace stress reduction with Six Bricks and LEGO® SERIOUS PLAY® in Taiwan. *World Journal of Research and Review, 6*(1), 64–67.

Hart, A., & Wolff, D. (2006). Developing local 'communities of practice' through local community–university partnerships. *Planning Practice & Research, 21*(1), 121–138. doi:10.1080/02697450600901616

Hinthorne, L. L., & Schneider, K. (2012). Playing with purpose: Using serious play to enhance participatory development communication. *International Journal of Communication, 6*(1), 2801–2824.

Hughes, J. A., & Sharrock, W. W. (1997). *The philosophy of social research*. Harlow: Longman.

Hunter, W. C. (2012). Projected destination image: A visual analysis of Seoul. *Tourism Geographies, 14*(3), 419–443. [Mismatch] doi:10.1080/14616688.2011.613407

James, A. (2013). Lego serious play: A three-dimensional approach to learning development. *Journal of Learning Development in Higher Education, 6*.

James, A., & Brookfield, S. (2013). The serious use of play and metaphor: Legos and labyrinths. *International Journal of Adult Vocational Education and Technology, 4*(3), 1–12.[Mismatch] doi: 10.4018/ijavet.2013070101

James, A. (2015). Learning in three dimensions: Using Lego Serious Play for creative and critical reflection across time and space. In P. C. Layne & P. Lake (Eds.), *Global Innovation of Teaching and Learning in Higher Education: Transgressing Boundaries* (pp. 275–294). Cham: Springer.

Jennings, G. (2010). *Tourism research* (2nd ed.). Milton: John Wiley & Sons.

Kane, P. (2004). *The play ethic: A manifesto for a different way of living*. London: Macmillan.

Kangas, S., Warren, N. A., & Byrne, M. M. (1998). Metaphor: The language of nursing researchers. *Nursing Research, 47*(3), 190–193.

Kristiansen, P., & Rasmussen, R. (2014). *Building a better business using the Lego serious play method*. Hoboken, NJ: Wiley.

Kurkovsky, S. (2015). Teaching software engineering with Lego serious play (pp. 213–218). ACM. doi:10.1145/2729094.2742604

Lakoff, G., & Johnson, M. (2003). *Metaphors we live by*. Chicago, IL: University of Chicago Press.

LEGO® SERIOUS PLAY® (2006). *The science of LEGO® SERIOUS PLAY®*. Enfield, CT: Executive Discovery LLC.

Liamputtong, P. (2007). *Researching the vulnerable: A guide to sensitive research methods*. London: SAGE.

Mishler, E. (1991). *Research Interviewing: Context and narrative*. Cambridge, MA: Harvard University Press.

Montesa-Andres, J., Garrigós-Simón, F., & Narangajavana, Y. (2014). A proposal for using Lego serious play in education. In M. Peris-Ortiz, I. Pechuán & F. Garrigós-Simón (Eds.), *Innovation and Teaching Technologies* (pp. 99–109). New York: Springer.

Morgan, G. (1980). Paradigms, metaphors, and puzzle solving in organization theory. *Administrative Science Quarterly, 25*(4), 605–622. doi:10.2307/2392283. doi:10.2307/2392283

Mueller, J. S., Melwani, S., & Goncalo, J. A. (2012). The bias against creativity. *Psychological Science, 23*(1), 13–17. doi:10.1177/0956797611421018. doi:10.1177/0956797611421018

Nolan, S. (2010). Physical metaphorical modelling with Lego as a technology for collaborative personalised learning. In J. O'Donoghue (Ed.), *Technology-supported environments for personalized learning; Methods and case studies* (pp. 364–385). Hershey, PA: Information Science Reference.

Nyaupane, G. P., Lew, A. A., & Tatsugawa, K. (2014). Perceptions of trekking tourism and social and environmental change in Nepal's Himalayas. *Tourism Geographies, 16*(3), 415–437. [Mismatch] doi:10.1080/14616688.2014.942233

Oliver, D., & Roos, J. (2007). Beyond text: Constructing organizational identity multimodally. *British Journal of Management, 18*(4), 342–358. doi:10.1111/j.1467-8551.2007.00516.x. doi:10.1111/j.1467-8551.2007.00516.x

Papert, S. (1999). Child psychologist Jean Piaget. *Time*, p. 105. Retrieved from http://content.time.com/time/magazine/article/0,9171,990617,00.html.

Papert, S., & Harel, I. (1991). Situating constructionism. In S. Papert & I. Harel (Eds.), Constructionism. http://www.papert.org/articles/SituatingConstructionism.html.

Patton, M. Q. (2015). *Qualitative research & evaluation methods: Integrating theory and practice* (4th ed.). Thousand Oaks, CA: SAGE Publications.

Peabody, M. A., & Noyes, S. (2017). Reflective boot camp: Adapting LEGO® SERIOUS PLAY® in higher education. *Reflective Practice, 18*(2), 232–243. doi:10.1080/14623943.2016.1268117

Pernecky, T. (2012). Constructionism: Critical pointers for tourism studies. *Annals of Tourism Research*, *39*(2), 1116–1137. doi:10.1016/j.annals.2011.12.010. doi:10.1016/j.annals.2011.12.010

Peter, T. B., Jacobs, C. D., & Roos, J. (2005). From metaphor to practice: In the crafting of strategy. *Journal of Management Inquiry*, *14*(1), 78–94. doi:10.1177/1056492604270802

Phillimore, J., & Goodson, L. (2004). *Qualitative research in tourism: Ontologies, epistemologies and methodologies*. Abingdon: Routledge.

Piaget, J. (1955). *The construction of reality in the child*. London: Routledge.

Pickard, A. J. (2007). *Research methods in information*. London: Facet.

Pritchard, A., & Morgan, N. (2003). Mythic geographies of representation and identity: Contemporary postcards of Wales. *Journal of Tourism and Cultural Change*, *1*(2), 111–130. doi:10.1080/14766820308668163

Rieber, L. P. (1996). Seriously considering play: Designing interactive learning environments based on the blending of microworlds, simulations, and games. *Educational Technology Research and Development*, *44*(2), 43–58. doi:10.1007/bf02300540. doi:10.1007/BF02300540

Riley, R. W., & Love, L. L. (2000). The state of qualitative tourism research. *Annals of Tourism Research*, *27*(1), 164–187. doi:10.1016/S0160-7383(99)00068-7. doi:10.1016/S0160-7383(99)00068-7

Scarles, C. (2004). Mediating landscapes: The processes and practices of image construction in tourist brochures of Scotland. *Tourist Studies*, *4*(1), 43–67. [Mismatch] doi:10.1177/1468797604053078

Schön, D. A. (1983). *The reflective practitioner: How professionals think in action*. New York, NY: Basic Books.

Terr, L. (2000). *Beyond love and work: Why adults need to play*. New York, NY: Touchstone Book.

Trebbin, J. (2016). *Let it click with the brick*. Berlin: Berlin School of Economics and Law.

Tseng, W.-C. (2017). An intervention using LEGO® SERIOUS PLAY® on fostering narrative identity among economically disadvantaged college students in Taiwan. *Journal of College Student Development*, *58*(2), 264–282. doi:10.1353/csd.2017.0019

Tuominen, P., & Ascenção, M. (2016). The hotel of tomorrow: A service design approach. *Journal of Vacation Marketing*, *22*(3), 279–292.[Mismatch] doi:10.1177/1356766716637102

Uzzell, D. (1984). An alternative structuralist approach to the psychology of tourism marketing. *Annals of Tourism Research*, *11*(1), 79–99. doi:10.1016/0160-7383(84)90097-5 doi:10.1016/0160-7383(84)90097-5

Vygotskiĭ, L. S., & Cole, M. (1978). *Mind in society: The development of higher psychological processes*. Cambridge, MA: Harvard University Press.

Wengel, Y., McIntosh, A. J., & Cockburn-Wootten, C. (2016). Constructing tourism realities through LEGO® SERIOUS PLAY®. *Annals of Tourism Research*, *56*, 161–163. doi:10.1016/j.annals.2015.11.012. doi:10.1016/j.annals.2015.11.012

Wibeck, V., Dahlgren, M. A., & Öberg, G. (2007). Learning in focus groups: An analytical dimension for enhancing focus group. *Research. Qualitative Research*, *7*(2), 249–267.[Mismatch] doi:10.1177/1468794107076023

Stakeholder engagement in sustainable tourism planning through serious gaming

Lidija Lalicic and Jessika Weber-Sabil

ABSTRACT

Serious gaming is an interdisciplinary and co-creative research methodology, which allows participants to proactively engage in role-playing and co-creating strategies. During the gameplay session, various game mechanisms trigger individual, social, and collective learning outcomes of the players, which in the end can lead to a change in the individual belief system towards the subject. Despite demonstrating its applicability in other complex domains, such as maritime spatial planning or urban development, an investigation into the applicability of serious games within the tourism domain is scarce. In conceiving the process of tourism planning as a strategic plan with dependent actors, resources, objectives, and challenges in a multi-layered decision-making process, creating or re-creating such strategic games can be beneficial to promote the understanding and change of belief systems among stakeholders. A particularly critically question is how serious gaming is able to generate qualitative research inquiries in order to shed more light on the complexities of the tourism destination planning process. Destination planning research can profit from new disruptive methods, such as serious gaming, for a more experimental and explorative approach in understanding the interconnectedness of tourism stakeholders, as well as estimations of short and long-term impact of decisions on a destination. By enabling strategic conversation about tourism planning issues, serious gaming functions as a strategic learning tool that provides opportunities for individual and community learning. Besides practical and conceptual implications, a set of future research avenues are provided that will enhance the qualitative research paradigm.

摘要

严肃博弈是一种跨学科的、协同创新的研究方法, 它允许参与者积极参与角色扮演和协同创新策略。在博弈设置过程中, 各种博弈机制会触发玩家的个体、社会和集体学习结果, 最终导致个体对主体的信仰体系发生改变。尽管证明了它在其他复杂领域的适用性, 如海洋空间规划或城市发展, 但对严肃博弈在旅游领域的适用性的调查很少。将旅游规划过程看作是一个多层次决策过程中的战略规划, 包括相关主体、资源、目标和挑战, 创造或再创造这样的战略博弈, 有利于促进利益相关者对旅游规划理念体系的理解和改变。一个特别关键的问题是, 严肃博弈如何能够产生定性研究调查, 以便更清楚地了解旅游目的地规划过程的复杂性。旅游目的地规划研究可以从新的颠覆性方法中获益, 例如严肃博弈, 以一种更具实验性和探索性的方法来理解旅游利益相关者之间的相互联系, 以及对旅游目的地决策的短期和长期影响的评估。通过促进关于旅游规划问题的战略对话, 严肃博弈作为一种战略学习工具, 为个人和社区学习提供了机会。除了实际应用和概念启示外, 还提供了一套未来的研究途径, 以加强定性研究范式。

1. Introduction

The development of cities is deeply impacted by changing tourism dynamics (Solà & Cooper, 2019). Forces such as global mobilization, disruptive technologies, and digital access create greater access to tourism products and also pressure destinations to respond and re-adjust their governance. The link between tourism management and effective place governance practices is becoming increasingly important in order to achieve better social, cultural, economic, and environmental roles for tourism in urban spaces (Mata, 2019). The critical question of how tourism should or can be incorporated into governance thus remains on the agenda (Gurran & Phibbs, 2017) with the inclusion of questions such as: 'Who should make decisions about the tourism destinations?', 'How do we make inclusive decisions?', and 'How can we best implement them?'. The multiplicity of actors in an interdisciplinary destination network that simultaneously requires stakeholders to have flexible and adaptive capabilities to foster collective governance challenges destinations to answer the aforementioned questions (Koens, Postma, & Papp, 2018). For most, gaining mutual trust is inconvertible as is knowledge exchange and partnerships, however, the mechanisms for doing so are not well understood by practitioners and researchers (Poplin, 2011; Rumore, Schenk, & Susskind, 2016). The shortcomings of the current tourism paradigm combined with a lack of integration into place governance urgently warrants more nuanced contemporary governance arrangements to facilitate the management of tourism as a part of the city agenda (Mata, 2019; Ryan, 2002).

In addition to calling for new governmental arrangements, methodological perspectives on the subject are also called into question by shifting management priorities in the current tourism paradigm. Current qualitative research inquiries fall short in understanding this paradigm shift in destination management and its underlying complexity. Therefore, McCabe, Sharples, and Foster (2012) call for more experimentation and exploration of new disruptive methods in order to better grasp the difficulties of destination planning and the processes that go along with it. According to Wilson and Hollinshead (2015), we need more 'freshly energized' approaches within the broader field of social science that can be re-conceptualized for tourism research and provide new insights into academic and practice-oriented purposes. While observing the wider social science qualitative spectrum, tourism research could greatly benefit from implementing theoretical ideas and debates from other subjects in this matter. As Bramwell (2015) posits, in the end, tourism is not so different from any other field in terms of the distinctiveness of theories dealing with similar issues, such as urban and rural research settings.

This paper introduces the method of *'serious gaming'*. While being recognized as an interdisciplinary and co-creative research methodology that is widely accepted for stakeholder participation in fields like marine spatial planning (Keijser et al., 2018), urban planning (Poplin, 2011), or integrated care (Cheung, Stevens, Evers, & Hiligsmann, 2018), the suitability of serious gaming is discussed in light of tourism in this paper. As other interdisciplinary research fields have shown (Mayer et al., 2013), a serious game is believed to be beneficial for the understanding of the multi-layered, strategic tourism planning process because it challenges individual belief systems among stakeholders.

This paper tackles the inquiry from a methodological perspective, namely, how serious games are able to generate qualitative research inquiries in order to shed more light on the complexities of tourism destination planning processes. First, the paper will illustrate how serious gaming as a method is creative and disruptive at the same time in terms of data generation processes and material production. Second, the paper discusses opportunities for researchers who apply serious gaming to destination planning in order to unravel its complexities and, finally, it identifies future research avenues for further inquiry. Overall, the paper postulates that serious gaming can offer new opportunities to produce methodologically solid data, benefiting qualitative research in light of the growing importance of collective governance in the planning of tourism destinations.

2. Serious gaming: a definition and principles

2.1. A definition

The concept of serious gaming originates mainly from military strategy gaming (Abt, 1970), which has migrated into education and business since the 1950s. For example, in 1969 Abt introduced as one of the first, a computer-assistance simulation game entitled 'Corridor' for the purpose of alternative transportation planning in an urban area (Ampatzidou et al., 2018). Since then, serious gaming has developed into a substantial industry branch and research field (Crookall, 2010). However, one needs to carefully make the distinction between gamification and serious gaming according Deterding, Khaled, Nacke, and Dixon (2011). Where, serious gaming is built on the premised of experiential and behavioral qualities of game play, and in doing so, having the ability to balance entertainment and stimulate learnings of negotiation, deliberation and collaboration among players (Ampatzidou et al., 2018; Ritterfeld, Cody, & Vorderer, 2009, p. 6), whereas gamification only invoke ludic game qualities, such as points and badges to engage with brands (Ampatzidou et al., 2018). Over the years, the learning effects from a serious games, has sparkled interest of scholars given that the learning aspects of serious gaming are associated with a set of benefits for participation and civic engagement (i.e. raising awareness, developing complex problem-solving skills) within a safe environment (Ampatzidou et al., 2018; Helmefalk, 2019). Thus, the main focus of serious gaming is to learn about elements of a complex subject in addition to having 'fun' while playing the game. The engagement and feedback based on the principles of play provide good conditions in which players can learn and change their attitudes (Mayer, 2012). According to Crookall (2010), the method is

Table 1. Serious gaming research frames (Mayer, 2012 used with permission).

Frame	Research	References
Research theory	Game theory as in economics, political science	Leyton-Brown and Shoham (2008)
Research concept	Organization, management, decision-making as a strategic or political game	Scharpf (1997); Steunenberg, Schmidtchen, and Koboldt (1999)
Research object	Studying game cultures, game economics, game politics	Castronova (2006); Ermi and Mäyrä (2003); Salomon and Soudoplatoff (2009)
Design artifact	Game as socio-technical design (artifact)	Wouters, van der Spek, and van Oostendorp (2011); Björk and Holopainen (2006)
Research method	Game as a research method comparable to simulation or experiment	Barnaud, Promburom, Trébuil, and Bousquet (2007); Mayer et al. (2014)
Intervention method	Game as therapy, educational, learning, change, or decision support method	Preschl, Wagner, Forstmeier, and Maercker (2011); Geurts, Duke, and Vermeulen (2007)
Data-gathering method	Game as an environment for observation, group interview, data-modelling	Cooper et al. (2010); Good and Su (2011); Wood, Griffiths, and Eatough (2004)
Foundations and requirements	Design and operationalization	Mayer (2012)

also deemed to be 'serious' in order to avoid misinterpretations about the usage of the method (p. 905).

In a nutshell, a serious gameplay session varies between one and three hours, and it aims to create more awareness and understanding of the underlying topic among its players (Keijser et al., 2018). As the setting is by nature participatory, players have to interact with each other and the presented game elements (i.e. maps, rules, assigned roles, a physical game-board, and/or computer-visualized screen) (Mayer, 2012). Under the facilitation of a game master who moderates and guides the players through their role play and game play, several rounds are conducted until a desired final outcome is achieved. Overall, the game is designed to represent a chaotic experience to some extent, which quite likely reflects the reality of the contextual domain (Keijser et al., 2018; Mayer et al., 2013; Salen, Tekinbaş, & Zimmerman, 2004).

However, depending on the disciplinary background, the approach or framing of serous games has been dealt with differently, ranging from a theory, as is common within the field of political science (Leyton-Brown & Shoham, 2008), to a data-gathering method, see Table 1 for an overview. In this paper, gaming is predominantly framed and discussed as an intervention method and data-gathering method. As this paper questions how serious games are able to generate qualitative research inquiries, the predominant focus is placed on the method from a data-gathering perspective.

The paradigm through which research is framed takes the perspective of social constructionism (Gergen, 1985). Within this stance, it is believed that there is an individual reality that only exists in social interactions. Reality is socially, culturally, historically, and linguistically influenced by the players. Consequently, knowledge is developed and socially constructed through interpretations within the process of gameplay. The use of language, sense-making, role interpretations, decision-making processes, and outcome interpretation are shaped by the player group under the supervision of the game master. The gameplay reality is socially and culturally influenced depending on the composition of the group. Thus, data gathering needs to focus on interpreted intersubjective social realities that emerge from the gameplay interaction (Weibler, 2016).

2.2. The process of the gameplay sessions

At the start of a gameplay session, the game facilitator introduces the players to a narrative that sets them into the context of the gameplay (i.e. managing a tourism destination, strategy planning, etc.). The use of the story and stimulation (SAS) approach (Mendoza & Prabhu, 2005) allows participants to understand the storyline in which key events are described. The responsive dashboard that displays the gameplay also reinforces the narrative based on applied game mechanics and participants' input. Then, the game facilitator provides the players with a set of instructions, rules, and feedback mechanisms to which they have to respond while playing the game.

The gameplay sessions are designed in such a way that players engage in face-to-face negotiation processes with other players. By role-playing real-life circumstances, participants are required to make appropriate actions and choose from a range of policies to manage a specific context (i.e. a destination) in order to further advance in the game and ideally come closer to the final outcome. Furthermore, this kind of role-playing forces players to gain perspective of their given role in the gameplay and create empathy for others players while negotiating. As participants 'play' these processes, the game allows them to jointly explore deep complex issues and relationships without impacting real-life structures and processes (Rumore et al., 2016; Salen et al., 2004). During the gameplay, the game facilitator moderates and steers the gameplay session while also constructively responding to emerging challenges and providing feedback to the players so that the game does not stagnate. While the game evolves, more concrete challenges, viewpoints, and possible solutions will be identified.

The players are in charge of the decisions in the game. In doing so, the co-produced game scenarios are more likely to be accepted by everyone who is playing the game (Haatanen et al., 2014; Mayer, 2012). However, the differing or opposing perspectives of various players can bring tension and conflict to the game. The process of forming coalitions and findings consensus in order to progress on a given challenge can be timely and intense, and it can also be the most interesting part of the game (Mayer, 2012; Salen et al., 2004). This goes hand in hand with negotiations, alignments, discussions, and disagreements in which the ambiguous outcome depends solely on the players' behavior and decisions. The game facilitator supports this process (referring to negotiations) and concludes the gameplay with the solution proposed by the player groups. In the end, a gameplay debriefing takes place, which allows participants to reflect on the game experience and translate and connect the learning from game events to real-life situations (Keijser et al., 2018; Rumore et al., 2016).

2.3. Applicability in other domains

Predominantly, gameplay lends itself to situations where the stakeholders are national and international levels (Gopnik et al., 2012), but also horizontal and vertically integrated at the various governance levels (Meiner, 2010), the process of a project is often procedural and iterative (Mayer et al., 2013). Depending on the context, which may involve governing complex policy arenas (Mayer et al., 2013), the game is designed upon the principles led by the domain, such as unclear system boundaries or ambiguity. Of course, the game sessions can also be applied to less governance-led complexities. However, the game is most effective with the essential knowledge and expertise of players on tourism planning processes, given the complexity of system variables offered by the interactive dashboard, the collaborative mechanisms to evolve the game as well as its associated role playing creates great conditions in which to unravel said complexities, promote understanding among players, and elicit change in such situations (Mayer et al., 2013). Popular domains in which serious gaming has been applied include maritime spatial planning, strategic planning, management, care, education, urban planning, and climate change (Cheung et al., 2018; Hauge et al., 2013; Keijser et al., 2018; Romero, 2013, Rumore et al., 2016).

2.4. Learning outcomes for the players

The gameplay offers various learning outcomes for the players. Different ways of learning can occur during three significant moments (*before, during, and after the game*). Before the game, players become more aware of the topic under discussion and get

Table 2. Participants' learning outcomes of serious gameplay sessions (based on Arnab et al., 2015; Suttie et al., 2012, used with permission).

Learning dimension	Game mechanisms	Gameplay experience	Learning mechanism
Creating	Design/editing, infinite gameplay, ownership, status, strategy/planning, titles/grids	The players have to create solutions, new policies, and approaches to manage the situation and develop a final outcome.	Accountability, ownership, planning, responsibility
Understanding	Appointment, role-play, tutorial, cascading information, questions, and answers	The players have to understand the complexity of the gameplay and link it to real-life scenarios. The players have to understand the other players' reasoning.	Objectivity, participation, question and answer, tutorial
Analyzing	Feedback, meta-game, realism	The players have to analyze a given problem, situation, and/or conflict.	Analyze, experimentation, feedback, identify, observation, shadowing
Applying	Capture, competition, cooperation, movement, progression, selecting, stimulate/response, time pressure	The players have to translate and connect the learnings from game events to real-life situations.	Action/task, competition, cooperation, demonstration, imitation, stimulation
Retention	Cut scenes/story, tokens, virality, behavioral momentum, interactions, goods/information.	The players have to be able to withhold their opinions to avoid stagnation and be able to reflect upon their peers' opinions in order to form solutions together.	Discover, explore, generalization, guidance, instruction, and repetition
Evaluating	Action points, assessment, collaboration, communal discovery, resource management, game turns, rewards/penalties, urgent optimism	The players have to critically evaluate their own as well as other players' knowledge in addition to the development of the game outcomes.	Assessment, collaboration, hypothesis, incentive, motivation, and reflection

to know the other players and their motivations to participate. During the gameplay, players have to build up their social knowledge while playing and interpreting the signals of the gameplay. After the gameplay, players share collective reflections during the debriefing, which is important for the individual player as well as for the group of players (Keijser et al., 2018). This can have a significant influence on the participants' perceived effectiveness of the game overall (Gopnik et al., 2012; Mayer, 2012).

Arnab et al. (2015) divide the learning skills into several dimensions, displaying the range of skills players are exposed to during gameplay (creating, understanding, analyzing, applying, and retention). Often, the learning skills occur simultaneously. Suttie et al. (2012) extend this line of inquiry and refer to how the game mechanics facilitate learning mechanisms (Table 2 provides an overview). For example, the game requires participants to create a strategy, which leads to high levels of responsibility and ownership among the stakeholders towards the developed strategy. Game mechanisms such as role-play lead to active participation and also a certain degree of objectivity towards the subject to some extent. The various mechanisms allow players to go beyond simply interacting with other players.

A collaborative tool like serious gameplay allows the players to explore issues outside of their usual parameters. This implies that the players have to make better sense of the changes in their real-life environments and adjust to them. Therefore, serious gaming allows for experimentation and learning about participants' attitudes and knowledge of a specific subject (Helmefalk, 2019; Mayer, 2012). The fact that players have to play an unfamiliar role in the game leads to new insights into a given phenomenon under discussion (Green, 2002). In doing so, the gameplay allows for social construction (Mayer, 2012; Wright & Goodwin, 2009) as players have to understand their peers in order to proceed with the gameplay, readjust their perceptions of current issues and strategies of a given situation, and also be able to spot early warning signals in order to plan for the future (Haatanen et al., 2014; Volkery & Ribeiro, 2009; Wright & Goodwin, 2009). Furthermore, the gameplay requires players to collaborate with at least one other party to collectively understand the research problem. Players may form associations with each other so that the gameplay can proceed and more information can be shared (Mendoza & Prabhu, 2005). The engaging processes of serious gameplay offer a broad spectrum of understanding at the local level of all stakeholders while players simultaneously improve their learning and adaptive skills as they begin to understand how the future can unfold (Mayer, 2012). Additionally, participants have the chance to enhance the credibility of their decisions as they are the ones in charge and the ones providing input into gameplay (Bellotti et al., 2012). Based on the suggested changes they make and discussions with other players, the dynamics offered by the game triggers much more than the final outcome. It leads to high levels of effectiveness of the players' adaptive changes in their real-life attitudes and behavior. Most importantly, the individual and collective learning (Sheppard et al., 2011) leads players to become more adaptive to external forces in real-life situations (Volkery & Ribeiro, 2009). Lastly, a successful game is one in which all of the participants are satisfied with the outcomes, including the learning outcomes (Gopnik et al., 2012; Mayer et al., 2014). Naturally, all of this is of great interest on its own for the research team (Zarb, 1992).

2.5. Challenges and limitations

As with every method, serious gaming encounters challenges. The primary challenge to serious gaming is the design and use of the game. The research team needs to build a close-to-realistic model of the system, which relies on real scientific claims and perhaps statistics. The procedure of validating these designs is of the highest importance in order to create a trustworthy and believable environment for the players. Although players agree to voluntarily enter a playful system which has some similarities to reality, game designers undergo a balancing act to design the system in a realistic way as opposed to creating a replica. Furthermore, the game needs to be visually appealing to its players, easy to use, and understandable in order to promote effective learning. In previous gameplay sessions in the field of maritime spatial planning, stakeholders' satisfaction with the game experience showed a direct correlation to their behavioral attitudes towards the topic under discussion (Keijser et al., 2018).

Furthermore, the research team needs to critically reflect upon the facilitator's role as moderator and observer. For example, the question of how a moderator can successfully notice and address individual cognitive style differences during the sessions (Grübler et al., 2007) without intervening in the game process is critical. There is also the challenge of measuring the success or failure of a serious game session. The assessment of the perceived success in alliance with the different degrees of stakeholder participation and their input into the game also needs to be clarified. In addition, one needs to carefully consider the players' reasons for playing the game as this can have a direct influence on measuring the effectiveness of the gameplay and players' perceived learning outcomes. Therefore, the selection of the players can also be a critical process for the research team in and of itself in order to ensure a 'generalizable' sample as often other parties are involved in the invitation process and make preliminary or unconscious decisions about the suitability of participants. This can indirectly influence the game dynamics and the outcome of the data-gathering as the composition of the player group might not be consistent between game sessions. Similar to other participatory procedures, the research team needs to pay attention to the status of the players in real life and the numbers of players in a gameplay (Sheppard et al., 2011). For example, Sheppard et al. (2011) demonstrate that if a group exceeds over five players, the member with the highest status tends to influence the decision process more regardless of his/her role in the game. This scenario emphasizes the importance of the role of a game facilitator who needs to sense such power issues and aim to focus strictly on the problem so that an efficient decision-making process that allows for negotiation and consensus-building within the given context can occur during the gameplay (Regan, Colyvan, & Marckovchick-Nicholls, 2006). After all, the gameplay context should be a safe environment where players can equally act upon a given role and implement rules and decisions to 'hypothetically' work towards a solution (Mayer, 2012).

Researchers have to recognize and address a few limitations when considering the implementation of serious gaming for research purposes. First, there is no standardized data collection method or procedures (Smith, Blackmore, & Nesbitt, 2015), nor is the type of data collection clear. Besides evaluating the game experience, which is of interest to the game designer for further improvements on player interaction and usability, social science researchers may primarily be interested in collecting data on the interrelations of player roles, changing power relations, decision-making processes, and stakeholder involvement. Researchers have to be aware that the serious game is not only an informal setting to collect data, rather it is first and foremost an intervention in which participants are presented with a fictive setting in which they are expected to interact as professionals. This can also mean that power relations in real life may change in the game session, and that the use of unconventional tools or methods in the game is not reflective of real-life interactions. All interactions, including the use of language, interpretations, and decisions in the gameplay are collaboratively shaped by co-players and the facilitator and are thus socially constructed (Foko & Amory, 2008; Mayer et al., 2014; Smith et al., 2015). In the serious board game, the game master is needed to structure the round-based game session, enable player discussions in a balanced manner between the groups, and support interpretations presented by the game board and digitalized dashboard. To evaluate the socially constructed gameplay situation, researchers primarily apply a triangulation of different research methods that depend on the evaluation of moments before, during, or after the gameplay. This can yield some difficulties as the researcher might find her/himself soon presented by a wealth of data that needs to be analyzed.

Pitfalls may also occur during data collection moments when the researcher interferes in the gameplay setting. Data collection methods are advised to be unobtrusive and covert in order to be the least disruptive as possible in light of the game flow and engagement process (Mayer et al., 2014). There are several moments during the game session where data collection is advised. Pre-game evaluations often cover demographic measures and previous experience with technology or games. Although this might be valuable data, it may be more important to ask about participants' attitudes, motivation, and personality style to find indications for game effectiveness (Smith et al., 2015). Although most data can be elicited from in-process game event logs (quantitative data), rich data can also be derived from situated case studies, research in the field, and participative methods, which are still scarce in the context of tourism planning processes. Due to interdisciplinary participants, rich data evaluations that reconstruct different perspectives on the topic and roles should be possible

Practical implications for data collection on serious games should be concerned with allowing for enough time and resources for participants, facilitators, and researchers. Often, extended and longitudinal data collection is needed, and stakeholders need to be convinced to participate in repetitive play sessions or remain otherwise engaged in the study (Mayer et al., 2014).

3. Serious gaming and tourism research

3.1. Context

Before placing serious gaming into tourism research, we need to briefly discuss what makes destinations unique settings. Tourism destinations can be defined as a collection of interdependent stakeholders in which new initiatives rely on stakeholder partnerships. Research stresses the great significance of stakeholder participation required by destinations to succeed with planning purposes (Del Chiappa & Baggio, 2015; Jamal & Getz, 1995; Scott, Baggio, & Cooper, 2008; Waligo, Clarke, & Hawkins, 2013). As a result, the importance of stakeholder collaboration in tourism planning is well documented in research. The planning of destinations is perceived as: (i) multi-objective, since it includes social, economic, and governance objectives (Farsari, Butler, & Szivas, 2011; Gopnik et al., 2012) and (ii) spatially-oriented given that results or forecasts apply only to a specific area. In general, stakeholders can be defined as individuals or groups who are interested in, involved with, or otherwise affected by particular tourism developments in their city (McCabe et al., 2012). Popular theories such as resource-dependency, network-approach, principal agent theory, property rights theory, and transaction cost theory have been applied to frame the complexities of stakeholders in a destination. Furthermore, various qualitative approaches are used and analyzed to understand the integration of all stakeholders (Chen, 2015), workshops, scenario-based planning, or open innovation platforms (i.e. Lalicic, 2018; McCabe et al., 2012; Vernon, Essex, Pinder, & Curry, 2005) in a destination. These theories and methods demonstrate that stakeholders need to share a common goal and purpose in order to succeed with collaborative planning approaches. More important is that they understand the value of working together for making the destination a better place for everyone (Ramayah, Lee, & In, 2011; Volgger & Pechlaner, 2014; Zehrer & Hallmann, 2015). Also, an early integration of all stakeholders, both continuously and sustained throughout the whole process (planning phase, plan evaluation, implementation, and post-implementation phase), is beneficial (Pomeroy & Douvere, 2008). This can differ in format, such as simple communications to intensify negotiations about specific details (Pomeroy & Douvere, 2008) and intense collaborations as seen in recent smart-related projects. However, contemporary governmental approaches that can facilitate the changing tourism paradigm and, at the same time, integrate tourism into urban city planning are scarce.

An attempt to fill this deficit was made in 1983 by Louikissas who introduced the Tourism Activity Game (AGT) as a simulation approach that integrates a board game in order to facilitate exchange and cooperation between stakeholders. The participants in his game simulation perceived it to be an excellent setting in which to interact with others. However, up until now, no new significant examples of serious gaming have since been introduced within the field of tourism planning and stakeholder engagement to the authors' knowledge. So far, the concept of gaming has gained significant attention in fields related to enhancing visitors' experiences and interactions with a given object, sight, overall destination (Xu, Buhalis, & Weber, 2017), or tourism education (Adukaite, van Zyl, Er, & Cantoni, 2017), but not for the purpose of bringing stakeholders together to collaborate and react to the transformative tourism paradigm.

3.2. Conceptual implications

Serious gaming can be a promising method to tackle the concerns tourism destinations are currently dealing with and, in particular, to integrate tourism into city

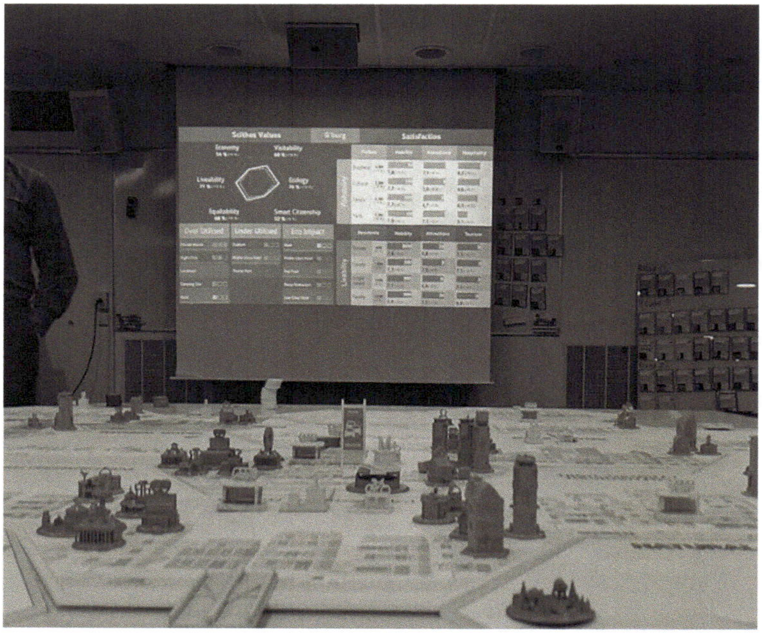

Figure 1. Game Dashboard and Play Board with miniatures (Source: Authors).

governance planning more effectively. The current tourism paradigm shift forces cities to rethink the position of tourism while significant challenges, such as highly overcrowded areas, the commercialization of city centers, and the rising numbers of peer-to-peer accommodations, are influencing various areas of cities. The specific features of a destination, such as the high level of interconnectedness across various domains of urban management (transportation, public spaces, green areas), the temporality of demand and supply, and ownership challenges in urban space as well as the lack of knowledge about tourism and its impacts on the destination from all stakeholders (Getz, 2008; Sheehan & Ritchie, 2005), are excellent conditions to integrate into a serious game play. These complexities can easily be visualized by the game elements, such as the game board, game tokens, and the policies card. See Figure 1 for an example of illustration.

Furthermore, planning sustainable destinations forces planners to focus on specific goals, such as the reduction of Co2 emissions or the number of tourist arrivals, and also requires the adjustments of specific areas in a destination and/or implementation of policies (i.e. taxation regulation), which might impact specific stakeholder groups. During a gameplay, players can focus on a realistic, hypothetically-set goal (i.e. 10% less tourist arrivals by the year 2025) in a safe environment and learn which implementations support them in their work towards this goal without neglecting specific areas, stakeholders, and/or other overseen issues. As the interactive dashboard allows them to receive direct feedback on their suggested changes to city management (i.e. policy 2% increase in tourism taxes) or the various urban dimensions (livability, etc.), players receive direct lessons from their decisions. These interventions in the game sessions allow stakeholders to critically consider the impact of their decision, their role towards tourism and other stakeholders affected by such changes. The dialog among the stakeholders makes it possible to reveal possible conflicts that may hinder effective management of tourism in a city. Thus, the gameplay forces stakeholders to collaborate and design feasible policies in consensus in order to effectively achieve their goals and manage tourism activities across the various dimensions of urban planning.

The inclusiveness of a wide range of stakeholders in the gameplay allows for more a realistic gameplay but also for a more holistic view on the destination and the place tourism will take for future planning initiatives. Often various stakeholder groups are neglected, particularly residents (Lalicic & Önder, 2018). Another benefit of this approach to tourism research is that it emphasizes not only the various views of stakeholders concerning tourism, but also their ideas on the social, environmental, and cultural dimensions of urban spaces. As Mata (2019) states, it is very important for cities to prioritize and settle the role of tourism across various urban management strategies in order to overcome the current tourism-related challenges that destinations are facing. Regardless of whether the game focuses on destination planning in the future, the game also lends itself to the process of re-positioning the destination or short-term planning management. The flexibility of the game design allows for various occasions in a specific context where a large and diverse range of stakeholders are concerned. This paper visualizes the use of serious gaming for the purpose of destination planning while transiting between the tourism paradigms given its current applicability.

3.3. Theoretical implications

From a research perspective, serious gaming has a wide range of opportunities to offer for tourism research and, in particular, the qualitative research paradigm. Seen this way, the role of serious gaming is elaborated on in regards to the most important premises of the qualitative paradigm. In 1997, Bruner (1997) pinpointed the significance of narratives for qualitative research in understanding how people comprehend and communicate about the world as well as how they make sense of the world and use logic and reasoning, a process referred to by Blackburn (2008 in Upham, Klapper, & Carney, 2016) as 'semiotic structuralist philosophy'. The richness of data lies in the reflexive and recursive relationship between time and talk, which can lead to sense-making of this world (Boden, 1997). In fact, the deconstruction and reconstruction of world views are important new insights for the tourism debate (Chia & King, 1998). Perceiving serious gaming as a disruptive way to engage stakeholders in a strategic conversation about their cities, researchers have the chance to observe the role of discussions, conversations, and narratives between the players. According to Mayer et al. (2014), it is the social-technical complexity and professional mastery of the situation by the players where the richness of theoretical knowledge emerges. As Ramírez and Selin (2014) state, those sense-making processes involve social and discursive elements that work mainly through communication and language. Researchers can interpret those communication processes during gameplay and observe the players attitudes and emotions at the same time.

For the players, the sense-making in the setting of a gameplay also allows them to understand the complexity of their destinations and learn both with and from other players. This is related to the critical question of embodied complexities of reflexivity in which the 'social' is the access to subject-interdependent realities and the possibility to 'unworld' these realities (Wilson & Hollinshead, 2015). Burt, Mackay, van der Heijden, and Verheijdt (2017) follow up by referring to serious gaming as review processes with effects on a performative level that aim to change and transform participants' perceptions and ideas about tourism planning and future developments. Helmefalk (2019) refers to the psychological mediators of gameplay such as the sense of control, fulfillment, reinforcement, and companionship. Subsequently, these observations can feed the social constructionist research paradigm (Gergen, 1985; Weibler, 2016) related to stakeholders' maneuverings and dealings during the game in relation to specific tensions thus providing more insight into the non-linearity of the policy-making process (Helmefalk, 2019; Mayer, 2012). Also, from a practitioner-point of view, this might be the most significant outcome of serious gaming. Other variables that can moderate these processes (i.e. players' background knowledge, socio-demographic variables, etc.) can provide an extra dimension of information with which to understand the gameplay and its effectiveness on changes in attitude (Pereira et al., 2012). A critical element, the status of 'researcher-as-expert', receives substantial attention in the context of serious gaming. However, if well executed, the role of the research team is rather at a constant for serious game plays. Through both direct and video observations as well as transcripts of communication and network structure, the data can be aggregated. Interpretations with regards to players' engagement as well as the effectiveness of the gameplay on their attitudes towards the subjects can also be assessed.

3.4. Practical implications

When combined with the collective governance of cities, a destination management organization may profit from serious gameplay sessions. Previous research has pinpointed a few challenges and barriers to facilitating successful planning processes and collaboration in tourism, such as the lack of coordination among government departments (Bornhorst, Brent Ritchie, & Sheehan, 2010; Dodds & Butler, 2010), the absence of collaborative mind-sets, and the decreased involvement of the private sector (Wong, 2017) and residents (Lalicic, 2018), however, serious gaming has the potential to overcome these obstacles. First of all, interdependent stakeholders who seem to differ in their perspectives have to engage with each other in playful activities, collective problem-solving, and collaboration during the gameplay. Considering these various points of discussion and interactions in addition to the game play, bringing tourism stakeholders together to play 'their destination city' would be a promising next step. Second, besides stimulating engagement among tourism stakeholders, policymakers, industry partners, and citizens, these gameplays provide great insight into the perspectives of others and thus allow actors to jointly identify suitable solutions and improve social inclusiveness for those who would normally not engage in tourism matters, such as local communities, NGOs, and the mobility industry.

Furthermore, serious gaming provides a powerful platform for all types of stakeholders to engage in new forms of collaborative and social learning through simulation, role-playing, and deep system thinking (Ampatzidou et al., 2018; Mayer, 2012). The game helps stakeholders to identify the tensions in their destination, formulate hypotheses, and understand the impact of events and changes in policies. With the differences in perspectives and how to approach public tourism challenges, participants do not only learn from each other, they also learn as a community. Particularly for tourism research, these consensus-building processes are an important key to understanding what priorities stakeholders have and how willing they are to collaborate (Lalicic, 2018).

The safe environment of a gameplay allows participants to overcome the level of skepticism present in a network of stakeholders (Hamilton & Alexander, 2013). The question of whether some stakeholder groups are able to contribute in a meaningful manner is addressed throughout the game play. By doing so, the attitudes might change into positive synergies that are fruitful in planning outcomes. Even though destinations can be hindered by historically developed attitudes of trust and the internal mood among stakeholders (Beritelli, Bieger, & Laesser, 2007), serious gaming makes it possible to break such stigmas as conflicts need to be resolved in order for the game to continue. In particular, the gameplay focuses on tensions and solutions involved in working towards a common goal. While players are doing this, they are required to ask themselves which group of stakeholders they belong to, the kind and level of interest they have in the destination, the importance and influence they have, their position towards the use of the destination, the various hats they wear, and the networks they belong to in order to make the partnership and overall destination governance a success (Beaumont & Dredge, 2010; Beritelli et al., 2007).

3.5. Future research avenues

The introduction of serious gaming and its conceptualization for tourism research with a specific focus on qualitative inquiry facilitates several research avenues. Based on the four research streams from Mayer (2012), various directions and questions are provided (Table 3 provides an overview). As elaborated in this paper, the applicability of serious gaming to tourism research seems appropriate, however, a range of questions needs to be answered by future studies. First of all, we call for more qualitative inquiries in the field of tourism to provide solid proof of its applicability. Real-case studies

Table 3. Future avenues serious gaming in a tourism research context (based on Mayer, 2012, used with permission).

Area	Focus	Main research directions
Design-orientated	Making things better	To asses real-case studies and proof-of-concepts in order to elaborate on game design for the tourism discipline.
Intervention-oriented	Making it work	To explore in which way serious gaming can enhance individual, social, and collective earning on various levels so that stakeholders can effectively work together and plan for their destinations in the future?
Domain-oriented	Making it matter	To understand which elements of the complexities in the tourism network can be visualized using serious gaming.
Disciplinary	Making it understandable	To assess the quality of the gameplay and its effects on evoked discussions, players' levels of creativity, perceived trust, and credibility of the gameplay.

and proof-of-concept in the tourism domain is needed to assess the role of serious gaming for tourism research.

In focusing on making serious gaming work for the tourism domain (intervention perspective), we mainly emphasize the learning effects of the game interventions and the game design (Keijser et al., 2018; Mayer, 2012; Pereira et al., 2012). The overall guiding question would be whether and in which way serious gaming can enhance learning on various levels so that stakeholders can effectively work together and plan for their destinations in the future. This means that future studies should explore the processes of social and individual learning as well as sense-making in order to enrich current theories on stakeholder engagement in destinations. This also implies that research needs to grasp upon participants' soft skills (Pereira et al., 2012) and how each player is able to use strategic thinking and creative and problem-solving skills as well as research and adaptive skills to progress during a gameplay. Thus, from a methodological perspective, future studies need to observe and analyze players' personal skills during the game play (Helmefalk, 2019). Considering the collective learning of the gameplay, research needs to document how players develop interpersonal skills (i.e. team work, conflict management, sensitivity) and their awareness of moral ethics and socially relevant issues (social awareness) in the tourism domain. Critical questions, such as, 'How are specific scenarios negotiated and decided upon in terms of groups' dynamics?' and 'How much time has each subject received, and who and in which way does the player contribute to this specific discussion?' are warranted. Particularly for tourism research, these consensus-building processes are important to understand what kind of (individual) priorities stakeholders have and how willing they are to collaborate and sacrifice to achieve a common goal (Lalicic, 2018). Such qualitative data will allow us to demonstrate the value of a serious game to unravel: (i) the process of changing individual belief systems and its influence on both the game and later in real-life as well as (ii) the problem surrounding policy development, acceptance, and implementation processes (Helmefalk, 2019; Mayer, 2012).

This also means that besides the social, individual, and collective learning perspectives, research also needs to delve into the actor-network configuration and engagement in game play. Important questions to be answered are: (i) who are the stakeholders; (ii) what drives these stakeholders to join the game session; (iii) how much power do they have while playing the game; (iv) what are their interdependencies; and (v) what coalitions are likely to be formed (Mayer, 2012; Pereira et al., 2012)? Such observations can also feed the qualitative research paradigm related to stakeholders' motivations, engagement, and the impact of the gameplay on their belief systems.

One can also understand the role of serious gaming from a domain-perspective with a focus on 'making it matter' for the players (Mayer, 2012). In this case, we need to understand the role of serious gaming and its possibility to unravel its complexity. Questions such as, 'Which elements of the complexities in the tourism network are we able to grasp upon?', How much of an effect are we able to generate', 'What is the multiplier effect of such a playful and inclusive manner of stakeholder engagement on a larger network in the city?', and 'What is the second order learning effect on stakeholders?' are important questions to explore. Such information can be retrieved through interviews and longitudinal studies.

Lastly, when taking a more disciplinary research perspective, we need to focus on questions reflecting the quality of the game-based learning interventions. For example, for the game and play itself, we need to ask questions related to the design, for example, we need to investigate which specific storylines to use and how they can provoke discussion and affect participants' levels of creativity, perceived trust, and credibility. Observations and focus groups (i.e. an extended de-briefing moment of the game) would enable the creation of a large knowledge-base about stakeholders' responses to the game elements, the critical role of the setting, and its effects on the players' adaptive learning outcomes. These streams are an important foundation for understanding the role of serious gameplay in the context of tourism research and the qualitative research paradigm. Of course, in addition to discussing the scope of possibilities within qualitative research, serious gaming enables mixed-method approaches, which would make it possible to test the longitudinal effects of using serious gaming as a participatory tool as well. One can include ex-ante and ex-post evaluations of stakeholders' perceptions and changes in attitude, cognition, and learning at the various points of time (Mayer, 2012). However, this is beyond the scope of this paper.

4. Final remarks

Due to the increased impact that tourism has on society, tourism has an important part on any city's political agenda (Dodds & Butler, 2009; Hall, 2008). However, cities have failed to develop responsive contemporary tourism policies to manage it, nor do they integrate mechanisms that would invite collaborative and inclusive planning practices (Dodds & Butler, 2009; Hall, 2008). To respond to these challenges, this paper introduced serious gaming as a contemporary approach to support cities in promoting stakeholders' understanding and change in approach to tourism in their cities. Until now, playful interventions in tourism planning and policy-making have not been so common. Instead, tourism has merely been a part of an urban planning project, rather than being a stand-alone framework and reversing the role of the discussion of urban planning. This paper holds that serious gaming offers opportunities for future studies in tourism, and that it has the potential to be used as an effective method for deepening and refining stakeholder involvement in tourism planning and policy development. In terms of practical relevance, this creative method can be perceived as a useful tool for stakeholders to learn about their destination, enhance future collaboration, and build consensus in a co-created manner. In doing so, they can create better governance of their destinations. Furthermore, the game generates a baseline future that reveals why there is a need to focus on tourism, the cause of its problems, and how to respond to them (Sheppard et al., 2011). The complex system dynamics of a destination translate to responsive visualizations and engagement techniques offered by the game play. This accessible setting triggers a new dialog about the place of tourism in a destination for academia and practitioners. Thus, as this paper posits, serious gaming serves as a research and development tool that facilitates the connection of different stakeholders and academic institutions (Day, 2008), which also creates positive spill-over effects that we need to start embracing. The various research opportunities created by the introduction of this innovative method paves the way for qualitative insights and discovery about this complex topic of destination planning and, hopefully, the development of new theories that can contribute to this ongoing discussion. We hope that we have inspired academics as well as practitioners to integrate contemporary techniques and work together towards an exciting future of tourism planning.

Disclosure statement

No potential conflict of interest was reported by the authors.

References

Abt, C. C. (1970). *Serious games*. New York, NY: University Press of America.

Adukaite, A., van Zyl, I., Er, Ş., & Cantoni, L. (2017). Teacher perceptions on the use of digital gamified learning in tourism education: The case of South African secondary schools. *Computers & Education*, *111*, 172–190. doi:10.1016/j.compedu.2017.04.008

Ampatzidou, C., Gugerell, K., Constantinescu, T., Devisch, O., Jauschneg, M., & Berger, M. (2018). All work and no play? Facilitating serious games and gamified applications in participatory urban planning and governance. *Urban Planning*, *3*(1), 34–46. doi:10.17645/up.v3i1.1261

Arnab, S., Lim, T., Carvalho, M. B., Bellotti, F., de Freitas, S., Louchart, S., … De Gloria, A. (2015). Mapping learning and game mechanics for serious games analysis. *British Journal of Educational Technology*, *46*(2), 391–411. doi:10.1111/bjet.12113

Barnaud, C., Promburom, T., Trébuil, G., & Bousquet, F. (2007). An evolving simulation/gaming process to facilitate adaptive watershed management in northern mountainous Thailand. *Simulation & Gaming*, *38*(3), 398–420.

Beaumont, N., & Dredge, D. (2010). Local tourism governance: A comparison of three network approaches. *Journal of Sustainable Tourism*, *18*(1), 7–28. doi:10.1080/09669580903215139

Bellotti, F., Berta, R., De Gloria, A., Lavagnino, E., Dagnino, F., Ott, M., … Mayer, I. S. (2012). Designing a course for stimulating entrepreneurship in higher education through serious games. *Procedia Computer Science*, *15*, 174–186. doi:10.1016/j.procs.2012.10.069

Beritelli, P., Bieger, T., & Laesser, C. (2007). Destination governance: Using corporate governance theories as a foundation for effective destination management. *Journal of Travel Research, 46*(1), 96–107. doi:10.1177/0047287507302385

Björk, S., & Holopainen, J. (2006). Games and design patterns. *The Game Design Reader*, 410–437.

Boden, D. (1997). Temporal frames. *Time & Society, 6*(1), 5–33. doi:10.1177/0961463X97006001001

Bornhorst, T., Brent Ritchie, J. R., & Sheehan, L. (2010). Determinants of tourism success for DMOs & destinations: An empirical examination of stakeholders' perspectives. *Tourism Management, 31*(5), 572–589. doi:10.1016/j.tourman.2009.06.008

Bramwell, B. (2015). Theoretical activity in sustainable tourism research. *Annals of Tourism Research, 54*, 204–218. doi:10.1016/j.annals.2015.07.005

Bruner, J. (1997). A narrative model of self-construction. *Annals of the New York Academy of Sciences, 818*(1 Self Across P), 145–161. doi:10.1111/j.1749-6632.1997.tb48253.x

Burt, G., Mackay, D. J., van der Heijden, K., & Verheijdt, C. (2017). Openness disposition: Readiness characteristics that influence participant benefits from scenario planning as strategic conversation. *Technological Forecasting and Social Change, 124*, 16–25. doi:10.1016/j.techfore.2016.11.024

Castronova, E. (2006). On the research value of large games: Natural experiments in Norrath. *Games and Culture, 1*(2), 163–186. doi:10.1177/1555412006286686

Chen, J. S. (2015). Tourism stakeholders' attitudes toward sustainable development: A case in the Arctic. *Journal of Retailing and Consumer Services, 22*(3), 225–230. doi:10.1016/j.jretconser.2014.08.003

Cheung, K. L., Stevens, E., Evers, S. M. A. A., & Hiligsmann, M. (2018). Serious gaming as a method for changing stakeholders' perspectives on integrated care. *Journal of Integrated Care, 26*(1), 54–64. doi:10.1108/JICA-10-2017-0035

Chia, R., & King, I. W. (1998). The organizational structuring of novelty. *Organization, 5*(4), 461–478. doi:10.1177/135050849854002

Cooper, S., Khatib, F., Treuille, A., Barbero, J., Lee, J., Beenen, M., … Players, F. (2010). Predicting protein structures with a multiplayer online game. *Nature, 466*(7307), 756. doi:10.1038/nature09304

Crookall, D. (2010). Serious games, debriefing, and simulation/gaming as a discipline. *Simulation & Gaming, 41*(6), 898–920. doi:10.1177/1046878110390784

Day, J. (2008). The need and practice of monitoring, evaluating and adapting marine planning and management—Lessons from the Great Barrier Reef. *Marine Policy, 32*(5), 823–831. doi:10.1016/j.marpol.2008.03.023

Del Chiappa, G., & Baggio, R. (2015). Knowledge transfer in smart tourism destinations: Analyzing the effects of a network structure. *Journal of Destination Marketing & Management, 4*(3), 145–150. doi:10.1016/j.jdmm.2015.02.001

Deterding, S., Khaled, R., Nacke, L. E., & Dixon, D. (2011). *Gamification: Toward a definition*. Paper presented at the CHI 2011 gamification workshop proceedings (Vol. 12), Vancouver, BC, Canada.

Dodds, R., & Butler, R. (2010). Barriers to implementing sustainable tourism policy in mass tourism destinations. *International Multidisciplinary Journal of Tourism, 5*(1), 35–53. Retrieved from https://mpra.ub.uni-muenchen.de/25162/

Dodds, R., & Butler, R. W. (2009). Inaction more than action: Barriers to the implementation of sustainable tourism policies. In *Sustainable tourism futures* (pp. 63–77).

Ermi, L., & Mäyrä, F. (2003). *Power and control of games: Children as the actors of game cultures*. Paper presented at the DiGRA Conference Proceedings, Utrecht, The Netherlands.

Farsari, I., Butler, R. W., & Szivas, E. (2011). Complexity in tourism policies. *Annals of Tourism Research, 38*(3), 1110–1134. doi:10.1016/j.annals.2011.03.007

Foko, T., & Amory, A. (2008). Social constructivism in games based learning in the South African context. In *ED-MEDIA 2008–World Conference on educational multimedia, hypermedia & telecommunications* (pp. 5757–5764). Waynesville, NC: Association for the Advancement of Computing in Education (AACE). Retrieved from http://www.editlib.org/p/29180

Gergen, K. J. (1985). The social constructionist movement in modern psychology. *American psychologist*, *40*(3), 266.

Getz, D. (2008). Event tourism: Definition, evolution, and research. *Tourism Management*, *29*(3), 403–428. doi:10.1016/j.tourman.2007.07.017

Geurts, J. L., Duke, R. D., & Vermeulen, P. A. (2007). Policy gaming for strategy and change. *Long Range Planning*, *40*(6), 535–558. doi:10.1016/j.lrp.2007.07.004

Good, B. M., & Su, A. I. (2011). Games with a scientific purpose. *Genome Biology*, *12*(12), 135. doi: 10.1186/gb-2011-12-12-135

Gopnik, M., Fieseler, C., Cantral, L., McClellan, K., Pendleton, L., & Crowder, L. (2012). Coming to the table: Early stakeholder engagement in marine spatial planning. *Marine Policy*, *36*(5), 1139–1149. doi:10.1016/j.marpol.2012.02.012

Green, K. C. (2002). Forecasting decisions in conflict situations: A comparison of game theory, role-playing, and unaided judgement. *International Journal of Forecasting*, *18*(3), 321–344. doi: 10.1016/S0169-2070(02)00025-0

Grübler, A., O'Neill, B., Riahi, K., Chirkov, V., Goujon, A., Kolp, P., … Slentoe, E. (2007). Regional, national, and spatially explicit scenarios of demographic and economic change based on SRES. *Technological Forecasting and Social Change*, *74*(7), 980–1029. doi:10.1016/j.techfore.2006.05.023

Gurran, N., & Phibbs, P. (2017). When tourists move in: How should urban planners respond to Airbnb? *Journal of the American Planning Association*, *83*(1), 80–92. doi:10.1080/01944363.2016.1249011

Haatanen, A., den Herder, M., Leskinen, P., Lindner, M., Kurttila, M., & Salminen, O. (2014). Stakeholder engagement in scenario development process–bioenergy production and biodiversity conservation in eastern Finland. *Journal of Environmental Management*, *135*, 45–53. doi:10.1016/j.jenvman.2014.01.009

Hall, C. M. (2008). *Tourism planning: Policies, processes and relationships*. New York: Pearson Education.

Hamilton, K., & Alexander, M. (2013). Organic community tourism: A cocreated approach. *Annals of Tourism Research*, *42*, 169–190. doi:10.1016/j.annals.2013.01.015

Hauge, J. B., Bellotti, F., Nadolski, R., Kickmeier-Rust, M., Berta, R., & Carvalho, M. (2013). Deploying serious games for management in higher education: Lessons learned and good practices. *EAI Endorsed Transactions on Game-Based Learning*, *1*(3), 225–234. doi:10.4108/sg.1.3.e4

Helmefalk, M. (2019). An interdisciplinary perspective on gamification: Mechanics, psychological mediators and outcomes. *International Journal of Serious Games*, *6*(1), 3–26. doi:10.17083/ijsg.v6i1.262

Jamal, T. B., & Getz, D. (1995). Collaboration theory and community tourism planning. *Annals of Tourism Research*, *22*(1), 186–204. doi:10.1016/0160-7383(94)00067-3

Keijser, X., Ripken, M., Mayer, I., Warmelink, H., Abspoel, L., Fairgrieve, R., & Paris, C. (2018). Stakeholder engagement in maritime spatial planning: The efficacy of a serious game approach. *Water*, *10*(6), 1–16. doi:10.3390/w10060724

Koens, K., Postma, A., & Papp, B. (2018). Is overtourism overused? Understanding the impact of tourism in a city context. *Sustainability*, *10*(12), 4384. doi:10.3390/su10124384

Lalicic, L. (2018). Open innovation platforms in tourism: How do stakeholders engage and reach consensus? *International Journal of Contemporary Hospitality Management*, *30*(6), 2517. doi:10.1108/IJCHM-04-2016-0233

Lalicic, L., & Önder, I. (2018). Residents' involvement in urban tourism planning: Opportunities from a smart city perspective. *Sustainability*, *10*(6), 1852. doi:10.3390/su10061852

Leyton-Brown, K., & Shoham, Y. (2008). Essentials of game theory: A concise multidisciplinary introduction. *Synthesis Lectures on Artificial Intelligence and Machine Learning*, *2*(1), 1–88. doi: 10.2200/S00108ED1V01Y200802AIM003

Mata, J. (2019). Intelligence and innovation for city tourism sustainability. In E. F. Sola & C. Cooper (Eds.), *The future of tourism* (pp. 2123–2232). Cambridge: Springer.

Mayer, I. (2012). Towards a comprehensive methodology for the research and evaluation of serious games. *Procedia Computer Science*, *15*, 233–247. doi:10.1016/j.procs.2012.10.075

Mayer, I., Bekebrede, G., Harteveld, C., Warmelink, H., Zhou, Q., van Ruijven, T., ... Wenzler, I. (2014). The research and evaluation of serious games: Toward a comprehensive methodology. *British Journal of Educational Technology*, *45*(3), 502–527. doi:10.1111/bjet.12067

Mayer, I., Zhou, Q., Lo, J., Abspoel, L., Keijser, X., Olsen, E., Nixon, A., ... Annen, A. (2013). Integrated, ecosystem-based marine spatial planning: Design and results of a game-based, quasi-experiment. *Ocean & Coastal Management*, *82*, 7–26. doi:10.1016/j.ocecoaman.2013.04.006

McCabe, S., Sharples, M., & Foster, C. (2012). Stakeholder engagement in the design of scenarios of technology-enhanced tourism services. *Tourism Management Perspectives*, *4*, 36–44. doi:10.1016/j.tmp.2012.04.007

Meiner, A. (2010). Integrated maritime policy for the European Union—consolidating coastal and marine information to support maritime spatial planning. *Journal of Coastal Conservation*, *14*(1), 1–11. doi:10.1007/s11852-009-0077-4

Mendoza, G. A., & Prabhu, R. (2005). Combining participatory modeling and multi-criteria analysis for community-based forest management. *Forest Ecology and Management*, *207*(1-2), 145–156. doi:10.1016/j.foreco.2004.10.024

Pereira, G., Brisson, A., Prada, R., Paiva, A., Bellotti, F., Kravcik, M., & Klamma, R. (2012). Serious games for personal and social learning & ethics: Status and trends. *Procedia Computer Science*, *15*, 53–65. doi:10.1016/j.procs.2012.10.058

Pomeroy, R., & Douvere, F. (2008). The engagement of stakeholders in the marine spatial planning process. *Marine Policy*, *32*(5), 816–822. doi:10.1016/j.marpol.2008.03.017

Poplin, A. (2011). Games and serious games in urban planning: Study cases. In *Proceeding ICCSA'11 proceedings of the 2011 international conference on computational science and its applications - Volume part II* (pp. 1–14). Berlin: Springer-Verlag. doi:10.1007/978-3-642-21887-3_1

Preschl, B., Wagner, B., Forstmeier, S., & Maercker, A. (2011). E-health interventions for depression, anxiety disorder, dementia, and other disorders in old age: A review. *Journal of CyberTherapy and Rehabilitation*, *4*, 371–386.

Ramayah, T., Lee, J. W. C., & In, J. B. C. (2011). Network collaboration and performance in the tourism sector. *Service Business*, *5*(4), 411–428. doi:10.1007/s11628-011-0120-z

Ramírez, R., & Selin, C. (2014). Plausibility and probability in scenario planning. *Foresight*, *16*(1), 54–74. doi:10.1108/FS-08-2012-0061

Regan, H. M., Colyvan, M., & Markovchick-Nicholls, L. (2006). A formal model for consensus and negotiation in environmental management. *Journal of Environmental Management*, *80*(2), 167–176. doi:10.1016/j.jenvman.2005.09.004

Ritterfeld, U., Cody, M., & Vorderer, P. (2009). *Serious games: Mechanisms and effects*. London: Routledge.

Romero, M. (2013). Game based learning MOOC. *Promoting Entrepreneurship Education. ELearning Papers*, *33*(May), 1–5.

Rumore, D., Schenk, T., & Susskind, L. (2016). Role-play simulations for climate change adaptation education and engagement. *Nature Climate Change*, *6*(8), 745–750. doi:10.1038/nclimate3084

Ryan, C. (2002). Equity, management, power sharing and sustainability—Issues of the 'new tourism'. *Tourism Management*, *23*(1), 17–26. doi:10.1016/S0261-5177(01)00064-4

Salen, K., Tekinbaş, K. S., & Zimmerman, E. (2004). *Rules of play: Game design fundamentals*. London, UK: MIT Press.

Salomon, M., & Soudoplatoff, S. (2009). Why virtual-world economies matter. *Journal for Virtual Worlds Research*, *2*(4), 4–14. doi:10.4101/jvwr.v2i4.859

Scharpf, F. W. (1997). Games real actors play. *Actor-Centered Institutionalism in Policy Research*, *55*.

Scott, N., Baggio, R., & Cooper, C. (2008). *Network analysis and tourism: From theory to practice*. Tonawanda, NY: Channel View Publications.

Sheehan, L. R., & Ritchie, J. R. B. (2005). Destination stakeholders: Exploring identity and salience. *Annals of Tourism Research*, *32*(3), 711–734. doi:10.1016/j.annals.2004.10.013

Sheppard, S. R. J., Shaw, A., Flanders, D., Burch, S., Wiek, A., Carmichael, J., ... Cohen, S. (2011). Future visioning of local climate change: A framework for community engagement and planning with scenarios and visualisation. *Futures*, *43*(4), 400–412. doi:10.1016/j.futures.2011.01.009

Smith, S. P., Blackmore, K., & Nesbitt, K. (2015). A meta-analysis of data collection in serious games research. In C. S. Loh, Y. Sheng, & D. Ifenthaler (Eds.), *Serious games analytics. Methodologies for performance measurement, assessment, and improvement* (pp. 31–55). Zurich, Switerland: Springer.

Solà, F. E., & Cooper, C. (2019). Conclusion: The future of tourism—Innovation for Inclusive Sustainable Development. In E. F. Sola & C. Cooper (Eds.), *The future of tourism* (pp. 325–337). Cambridge: Springer.

Steunenberg, B., Schmidtchen, D., & Koboldt, C. (1999). Strategic power in the European Union: Evaluating the distribution of power in policy games. *Journal of Theoretical Politics, 11*(3), 339–366. doi:10.1177/0951692899011003005

Suttie, N., Louchart, S., Lim, T., Macvean, A., Westera, W., Brown, D., & Djaouti, D. (2012). Introducing the "serious games mechanics" a theoretical framework to analyse relationships between "game" and "pedagogical aspects" of serious games. *Procedia Computer Science, 15*, 314–315. doi:10.1016/j.procs.2012.10.091

Upham, P., Klapper, R., & Carney, S. (2016). Participatory energy scenario development as dramatic scripting: A structural narrative analysis. *Technological Forecasting and Social Change, 103*, 47–56. doi:10.1016/j.techfore.2015.10.003

Vernon, J., Essex, S., Pinder, D., & Curry, K. (2005). Collaborative policymaking: Local sustainable projects. *Annals of Tourism Research, 32*(2), 325–345. doi:10.1016/j.annals.2004.06.005

Volgger, M., & Pechlaner, H. (2014). Requirements for destination management organizations in destination governance: Understanding DMO success. *Tourism Management, 41*, 64–75. doi:10.1016/j.tourman.2013.09.001

Volkery, A., & Ribeiro, T. (2009). Scenario planning in public policy: Understanding use, impacts and the role of institutional context factors. *Technological Forecasting and Social Change, 76*(9), 1198–1207. doi:10.1016/j.techfore.2009.07.009

Waligo, V. M., Clarke, J., & Hawkins, R. (2013). Implementing sustainable tourism: A multi-stakeholder involvement management framework. *Tourism Management, 36*, 342–353. doi:10.1016/j.tourman.2012.10.008

Weibler, J. (2016). Towards a three-component model of relational social constructionist leadership: A systematic review and critical interpretive synthesis. *International Journal of Management Reviews, 19*(2), 214–236. doi:10.1111/ijmr.12095

Wilson, E., & Hollinshead, K. (2015). Qualitative tourism research: Opportunities in the emergent soft sciences. *Annals of Tourism Research, 54*, 30–47. doi:10.1016/j.annals.2015.06.001

Wong, F. Y. (2017). Influence of Pokémon Go on physical activity levels of university players: A cross-sectional study. *International Journal of Health Geographics, 16*(1), 8. doi:10.1186/s12942-017-0080-1

Wood, R. T., Griffiths, M. D., & Eatough, V. (2004). Online data collection from video game players: Methodological issues. *CyberPsychology & Behavior, 7*(5), 511–518. doi:10.1089/cpb.2004.7.511

Wouters, P., van der Spek, E. D., & van Oostendorp, H. (2011). Measuring learning in serious games: A case study with structural assessment. *Educational Technology Research and Development, 59*(6), 741–763. doi:10.1007/s11423-010-9183-0

Wright, G., & Goodwin, P. (2009). Decision making and planning under low levels of predictability: Enhancing the scenario method. *International Journal of Forecasting, 25*(4), 813–825. doi:10.1016/j.ijforecast.2009.05.019

Xu, F., Buhalis, D., & Weber, J. (2017). Serious games and the gamification of tourism. *Tourism Management, 60*, 244. doi:10.1016/j.tourman.2016.11.020

Zarb, G. (1992). On the road to Damascus: First steps towards changing the relations of disability research production. *Disability, Handicap & Society, 7*(2), 125–138. doi:10.1080/02674649266780161

Zehrer, A., & Hallmann, K. (2015). A stakeholder perspective on policy indicators of destination competitiveness. *Journal of Destination Marketing and Management, 4*(2), 120–126. doi:10.1016/j.jdmm.2015.03.003

Deep reflexivity in tourism research

Émilie Crossley

ABSTRACT

Reflexivity has come to be accepted as standard methodological practice in qualitative tourism research, enabling researchers to reflect on how their subjectivity inevitably impacts on their work. However, reflexive accounts vary enormously from intimate, in-depth explorations to more modest reflexive disclosure statements, which are usually relegated to 'safe' spaces in the text such as the methodology section. Additionally, there is an implicit assumption in tourism studies that, while reflexivity is to be encouraged, some information is too private or embarrassing to be shared publicly. I introduce the notion of deep reflexivity, a radical approach to positionality and intersectionality developed by psychosocial researchers. Deep reflexivity implies that reflexive practice should be extended to include critical reflections on the researcher's emotions, embodiment and unconscious processes that may impact on research. Such information can be accessed through introspection, the interpretation of fantasies, or even dream analysis. This approach is particularly applicable to psychological interview-based research in tourism, but I argue that it has wider relevance and potential. Deep reflexivity in practice is illustrated by sharing a personal account of conducting fieldwork in a volunteer tourism setting in Kenya and I reflect on the challenges that this form of reflexivity presents in terms of ethics, professional integrity, and methodological accuracy. I conclude by arguing that psychosocial studies and the concept of deep reflexivity present a promising avenue for methodological disruption and advancement in critical tourism research.

摘要

反省作为定性旅游研究的一种标准的方法实践, 已逐渐被人们所接受, 使研究人员能够反思其主观性是如何不可避免地影响其研究工作的。然而, 有关反省的描述有很大的差异, 从亲密的、深入的探索到更温和的反思性的披露声明, 这些声明通常被归入文本中的"安全"空间, 如方法论部分。此外, 在旅游研究中有一个隐含的假设, 即尽管反省是受到鼓励的, 但是有些信息太隐私或太尴尬了, 无法公开分享。我介绍了深度反省的概念, 这是一种由社会心理学研究人员发展起来的关于所处的立场和立场交错的激进方法。深层反省是指反省的实践应该延伸到批判性地反思研究者的情绪情感、亲身体验和一些无意识的过程, 它们可能会影响到研究结果。这些信息可以通过内省、对幻想的解释, 甚至梦的分析来获取。这种方法特别适用于基于心理访谈的旅游研究, 但我认为它具有更广泛的指导意义和应用潜力。本文通过分享在肯尼亚一个志愿旅游情景下进行实地工作的个人经历描述，说明深度反思性在实践中的应用。我思考了这种形式的反省在伦理、职业诚信和方法准确性方面所带来的种种挑战。最后，我认为社会心理研究和深度反省的概念为批判性旅游研究的方法论颠覆和发展提供了一条很有前途的方法。

Introduction

Reflexivity has come to be accepted as standard methodological practice in qualitative research. Indeed, Finlay (2002a, p. 212) argues that '[qualitative] researchers no longer question the need for reflexivity: the question is how to do it.' In its broadest sense, reflexivity can be understood as the practice of reflecting on the conditions that enable and constrain the production of research in order to contextualize and qualify findings (Ateljevic, Harris, Wilson & Collins, 2005). A subset of these conditions relates to the conduct and subjectivity of the researcher and it is this form of reflexivity that is most commonly drawn upon in tourism research. This 'personal' or 'positional' reflexivity can be defined as reflection 'upon the ways in which our own values, experiences, interests, beliefs, political commitments, wider aims in life and social identities have shaped the research' (Willig 2013, p. 10). Published accounts of personal reflexivity vary enormously from intimate, indepth explorations of the researcher's subjectivity to more modest disclosure statements, which are usually relegated to what are considered 'safe' spaces in the text such as the methodology section. Tourism studies has typically lagged behind other fields in the social sciences when it comes to practicing reflexivity (Cohen, 2013; Dupuis, 1999; Feighery, 2006; Hall, 2004; Hollinshead & Jamal, 2001; Johnson, 2015; Westwood, Morgan, & Pritchard, 2006) and it has been argued that the field still produces a significant body of qualitative research written from a position of supposed objectivity, detachment and disembodiment (Glancy, 1993; Cohen, 2013).

Theoretical and methodological developments arising from the field's 'critical turn' have, however, provided a platform for advancing reflexivity, which is considered by some to be one of the hallmarks of critical tourism research (Harris, Wilson & Ateljevic, 2007; Tribe, 2005; Wilson & Hollinshead, 2015). The critical turn has promoted greater understanding of tourism studies' epistemological underpinnings (Ateljevic, Morgan & Pritchard, 2013; Franklin & Crang, 2001; Ren, Pritchard & Morgan, 2010) and has fostered methodological pluralism, particularly in relation to qualitative research (Ateljevic, Pritchard & Morgan, 2007; Hollinshead & Jamal, 2001; Phillimore & Goodson, 2004; Riley & Love, 2000). The current climate is therefore conducive to challenging conventional approaches to reflexivity in the pursuit of disrupting and advancing methodological practice in the field. I introduce the notion of 'deep reflexivity', a radical approach to positional and intersectional reflexivity developed through psychosocial research, which implies that reflexive practice should be extended to include critical reflections on the researcher's emotions, embodiment and unconscious processes that may impact on research. Such information can be accessed through introspection, the interpretation of fantasies, or even dream analysis. Deep reflexivity provides a logical extension to personal reflexivity and challenges tourism researchers to confront aspects of their practice that may not conventionally form part of reflexive analysis. I illustrate deep reflexivity by sharing a personal account of conducting fieldwork in a volunteer tourism setting in Kenya and use this example to reflect on the challenges that this form of reflexivity presents in terms of ethics, professional integrity, and methodological accuracy.

Reflexivity in tourism studies

In one of the most comprehensive reviews of reflexivity in tourism studies to date, Ateljevic et al. (2005) articulate four forces that both enable and constrain the production of tourism research. They divide these dimensions of reflexivity into two 'macro' and two 'micro' forces: 'the 'ideologies and legitimacies' that govern and guide our tourism research outputs; the 'research accountability' environment that decides what is acceptable as tourism research; our 'positionality' as embodied researchers whose lives, experiences and worldviews impact on our studies, and 'intersectionality with the "researched" as we carry out our research relationships' (Ateljevic et al., 2005, p. 10). Positional and intersectional reflexivity are arguably the most common forms of reflexive practice in tourism studies and are used to understand both how the researcher's subjectivity shapes their research and how knowledge is co-constituted through the intersubjective dynamics that exist between researchers and participants. Calls for greater positional reflexivity stem from the 'post-modern fracturing of the naturalist, post-positivist tradition and acknowledge[ment] that it is the agency of the researcher as writer that makes the research' (Westwood et al., 2006, p. 34). This erosion of the assurances of post-positivism is aligned with the crisis of representation, a body of scholarship influenced by feminist and post-structural writing that critiques the notion that researchers can represent 'the object of their inquiry from an objective and value-free position' (Ateljevic et al., 2005, p. 11). It is against this backdrop that Feighery (2006, pp. 270–271) argues for recognition of the 'subject-centred nature of knowledge', defining reflexivity as 'the act of making oneself the object of one's own observation, in an attempt to bring to the fore the assumptions embedded in one's perspectives and descriptions of the world.'

There is an implicit assumption in tourism studies that, while reflexivity is to be encouraged, some information is too private or embarrassing to be shared publicly. It has been suggested that social scientists regularly self-censor and sanitize accounts of their research, omitting details of their emotions and subjectivity (Dickson-Swift, James, Kippen & Liamputtong, 2009; Finlay, 2002b; Johnson, 2009; Kleinman & Copp, 1993). Cohen (2013) describes how he has often only written himself into 'safe spaces' in the text and that this 'has sometimes been no more than a sentence or two that divulges [his] own sociocultural background and reasons for interest in the context'. Similarly, Hall (2004) reveals how in previous writing he has removed accounts of personal reflexivity, or relegated these accounts to forewords and afterwords rather than including them in the main body of the text, due to concerns about the 'appropriateness' of such material in an academic publication. I refer to this limited form of reflexivity as 'reflexive disclosure statements'. These admissions from high profile tourism academics provide insight into the feelings of vulnerability clearly experienced by researchers when confronted with the prospect of revealing too much of themselves in a professional setting. While these feelings are both understandable and relatable, some argue that reflexivity has become predictable, formulaic and stripped of its power to disrupt and enhance analysis as a result of our collective inhibition (Frosh & Saville Young, 2008; Hollway, 2004; Pillow, 2003; Saville Young, 2011); it is this disruptive power that deep reflexivity has the potential to reinject into tourism research.

Pillow (2003, p. 192) contends that mainstream reflexivity in qualitative research risks stagnating into a set of practices that become complicit in reproducing the social structures and power relations that many researchers seek to challenge. Instead, she advocates 'a reflexivity of discomfort' that critically interrogates the histories and relationalities upon which research relationships are predicated. This 'reflexivity of discomfort' resonates with Finlay's (2002b, p. 541) warning that practicing reflexivity can be an 'uncomfortable' process in which researchers may need to confront and reveal mistakes made during the research process as well as their personal thoughts and feelings (see also Elliott, Ryan & Hollway, 2012). An example of a strikingly candid reflexive account that certainly provokes discomfort in the reader and must surely have been difficult to write can be found in Johnson's (2009) work from leisure studies on gay bar culture. Johnson's paper, which has the explicit aim of exploring reflexivity, opens with an excerpt from his field journal in which he describes himself urinating in a gay bar and later also recounts stories about his own sexual arousal, drug taking, and racist behaviour as a child. While sharing these highly personal details might be perceived as gratuitous by some, Johnson (2009, p. 484) succeeds in provoking us to consider 'the complexities and risks involved in representing the "researcher self" amidst competing, and perhaps false, dichotomies of the personal/professional life and the scholar/activist.'

It appears that three levels of positional/intersectional reflexivity can be identified in tourism research: reflexive disclosure statements as described by Cohen (2013) and Hall (2004); ongoing conscious reflexive practice woven through the entirety of the text as exemplified by Ateljevic et al. (2005); and 'deep reflexivity' that builds on the previous level through the inclusion of researchers' embodied, emotional and unconscious meanings as demonstrated by Johnson (2009). Despite qualitative research providing detailed accounts of participants' emotional worlds, the emotions and feelings of researchers often go unreported. This is despite evidence that exploring emotions can help us to understand interactional dynamics in interviews or fieldwork, promote empathy with participants and provide explanatory power in analysis and interpretation (Dickson-Swift et al., 2009; Gray, 2008; Harris & Huntington, 2001; Holland, 2007). Though still rare, some tourism researchers have illustrated the value of drawing on their emotions and embodied feelings through reflexive practice (Ali, 2013; Tucker, 2009). Deep reflexivity disrupts conventional approaches to reflexivity as affective and unconscious elements of the researcher's subjectivity that are not usually considered important become central to the reflexive analysis. This method interrupts the orthodox trajectory of reflexivity in tourism research, resisting the formulaic and predictable format that it often takes. In addition, deep reflexivity disrupts the mainstream ontology of the subject in tourism studies, which is usually theorized as rational and coherent, by positing the existence of intersubjective unconscious dynamics. Finally, this approach disrupts the researcher's certitude and control over the research process, challenging them to confront uncomfortable moments of uncertainty, doubt and failure.

Psychosocial approaches to reflexivity

In this section, I explore a field of research that I believe holds great potential for the advancement of reflexivity in tourism studies. Psychosocial studies is primarily concerned with theorizing the intersection between the personal and the social in a non-reductive and non-dualistic fashion (Frosh, 2003; Taylor, 2017; Walkerdine, 2008). It is a theoretically and methodologically plural field of research that aims to articulate 'a place of "suture" between elements whose contribution to the production of the human subject is normally theorized separately' (Frosh & Baraitser, 2008, p. 348). Such pervasive theoretical dualisms include individual/society, psyche/social, structure/agency and body/mind. For example, we would normally assume that the study of the mind belongs to psychology whereas the study of society as a whole belongs to sociology, but psychosocial studies challenges this assumption by suggesting that 'micro' and 'macro' levels of analysis are co-constituting and inseparable (Taylor, 2017). Psychosocial researchers come from a range of disciplinary backgrounds but there has been a strong correlation between psychosocial studies and critical psychology (Frosh, 2003). Critical psychologists critique mainstream, quantitative psychology for disavowing forms of intersubjectivity and relationality that bind subjects socially, producing a reductive and ideologically distorted theorization of the subject (Blackman, Cromby, Hook, Papadopoulos, & Walkerdine, 2008; Henriques, Hollway, Urwin, Venn, & Walkerdine, 1998). Psychosocial research consequently aims to produce 'analyses of subjectivity which are situated, complex, and dynamic, and which stand in contrast to the simplistic, reductionist psychological theories of self' (Gough, 2015, p. 111).

Taylor (2017) suggests that psychosocial research is characterized by three key concepts: an interface or 'inbetweenness', the 'extra-rational' or irrational, and a related concept of 'affect'. The first concept of 'inbetweenness' captures the productive tension intrinsic to the term 'psychosocial' between its two constitutive elements and concerns how the relationship between them should be theorized. There has been considerable debate concerning whether the term psychosocial should remain unhyphenated, implying a 'single dialectical process' (Taylor, 2015), or should leave open the gap that has conventionally separated approaches studying the 'psychic' and the 'social' (cf. Burman, 2008; Hoggett, 2008; Hollway, 2004). The related concepts of the 'extra-rational' and 'affect' point to the ways in which psychosocial research attempts to go beyond analysis of purely conscious, cognitive, rational phenomena by also exploring emotions, feelings, defences and desires. Affect is a contested term that has 'accrued a sweeping assortment of philosophical/psychological/physiological underpinnings, critical vocabularies, and ontological pathways' (Seigworth & Gregg, 2010, p. 5). Taylor (2017) notes that in psychosocial research affect is often used as an umbrella term for feelings and emotions, which are theorized as social phenomena operating between people rather than contained solely within individuals as commonly conceptualized in mainstream psychology. I use affect in this broad sense to refer to an analytic sensitivity to emotions, feelings and, in psychoanalytic terms, transferences and countertransferences – the 'affective' dynamics of subjectivity.

Psychoanalytic theory is frequently used in psychosocial research to enhance readings of personal narratives due to its capacity to theorize extra-discursive, extra-rational aspects of subjectivity and to articulate the complexity of personal, emotional worlds. Psychoanalysis has been utilized as a strategy for 'enriching' and 'thickening' discursive interpretations of narratives, allowing researchers to look for conscious and unconscious 'reasons' behind subjects' investments in identity positions (Frosh & Saville Young, 2008). By paying attention to biographically derived unconscious dynamics, psychoanalytic approaches propose a way of explaining the 'stickiness' of identities; that is, the unconscious reasons why people adopt and maintain particular identity positions even when they induce internal conflict (Edley, 2006). This has allowed psychoanalytic psychosocial researchers to propose that theirs is a 'why' rather than merely a 'how' approach to the discursive construction of identity and subjectivity, providing insights into why subjects appropriate particular discursive resources out of an array of possibilities (Frosh, Phoenix & Pattman, 2003). Psychosocial researchers subscribing to a psychoanalytic viewpoint seek to understand how such unconscious processes 'structure relations between the researcher, the participants, and the data gathered' (Finlay, 2002b, p. 535).

Frosh and Saville Young (2008, p. 113) argue, from a psychoanalytic psychosocial perspective, that reflexivity in qualitative research is too often treated as 'a relatively simple, technical matter, perhaps an issue of confession or self-revelation', such as Cohen's (2013) and Hall's (2004) reflexive disclosure statements. Saville Young (2011, p. 48) proposes that reflexivity should be conceptualized as 'more than simply describing researchers' investment in their work', pointing to the potential for 'exploring unconscious processes that arise in the research encounter through the use of introspection, interpretation of fantasies and even dream analysis'. Equally, Hollway (2004, p. 8) argues that working with forms of deep reflection in psychosocial research constitutes a type of reflexivity that provides the researcher with ways of going 'beyond the rather mechanistic operationalization of reflexivity in qualitative social science in terms of the main, socially-given identities'. What Hollway implies by this is that it is common in qualitative research to find reflexive discussions of how one's membership of particular social groups, such as gender, class and ethnicity, impacts on the researcher's worldview, preferences and interactions with participants in a very broad sense. However, such accounts can be clinical and detached, lacking more personal revelations of feelings, emotions, memories, or embodiment that are seen as critical within psychosocial analysis. Instead, Hollway (2016, p. 2) advocates conceptualizing reflexivity psychoanalytically as an embodied, affective, 'uncognized' way of knowing derived from the 'relational communication of body-based emotional experience, past and present'.

In clinical psychoanalysis, the unconscious flow of affect transmitted between the psychoanalyst and the analysand is referred to as 'transference' and 'countertransference'. Conventionally, transference refers to the projection of the analysand's unconscious desires onto the analyst and to the ways in which the analysand relates to the analyst as a significant other from their childhood (Giami, 2001). The countertransference is the response of the analyst to this transference, and can be defined in a research context as 'the sum of unconscious and emotional reactions, including anxiety, affecting his/her relation with the observed subject and situation' (Giami, 2001, p. 6). These transference and countertransference dynamics result in a theorization of temporally complex encounters in which the 'past and present of both participants, as well as their mutual reactions to past and present, fuse into a unique emotional position involving both of them' (Kernberg, 1965, cited in Giami, 2001, p. 4). Alluding to these transference/countertransference dynamics in qualitative research interviews, Walkerdine, Lucey and Melody (2001, p. 97) suggest that psychoanalytic psychosocial reflexivity allows researchers to pose the following salient questions: 'to which part or parts of me is the subject speaking? Which part of me is responding? In other words, who do I represent for the subject, and who do they represent for me?' This theorization presents people as complex constellations of previous experiences and relationships with others that have shaped their subjectivity.

Whether or not it is possible to identify transference and countertransference in research interviews is a contentious issue that raises questions about the transposition of psychoanalysis from its clinical parameters and about the mechanization of psychoanalytic concepts (Frosh, 2010; Frosh & Baraitser, 2008; Hook, 2008; Kvale, 1999). Frosh and Saville Young (2008, p. 113) warn that discussions of 'the countertransference' in social research can look 'only schematically like the intense exploration of unconscious material characteristic of psychoanalytic reflection on the countertransference in the clinical situation'. After all, research participants are typically only interviewed a small number of times within a relatively short timeframe whereas clients may have hundreds of psychotherapy sessions over a period of several years in a clinical setting. However, Cartwright (2004, p. 223) in his work on the psychoanalytic interview method argues that 'inchoate transference-countertransference impressions' can be present within research interviews. These impressions may enable the researcher to become attuned and affectively receptive to their interviewee so that they can proceed with greater sensitivity and responsiveness (Jervis, 2009). Indeed, Clarke (2002) emphasizes the 'communicative and constructive' aspects of transferences through the term 'projective communication' whereby 'researchers who pay attention to what is going on inside them ... may discover that a respondent has communicated something of how they feel without actually verbalizing it' (Jervis, 2009, p. 148).

Psychosocial or 'deep' reflexivity offers a unique way of analysing intersubjective dynamics between researchers and participants, provides guidance when interpreting data, and mechanisms for researchers to explore their emotional and unconscious responses. Gilbert (2001, p. 11) suggests that research can be 'enriched by our personal and emotional engagement in the research process' and that paying attention to our feelings and even dreams can provide a safe way of dealing with difficult emotions triggered while conducting research. Hollway and Jefferson's (2000) research on the fear of crime in the United Kingdom provides an example of deep reflexivity in which the authors reflect on feelings arising through in-depth interviews and the analytic process. In an analysis of one of his interviewees, 'Tommy', Jefferson reflects on the similarities between their upbringings – both having been raised in large, poor families with strict and often violent fathers – and uses his own biographical experience to question why Tommy frames his narrative so positively (Hollway & Jefferson, 2000, pp. 65–68). Rather than taking Tommy's account at face value, Hollway and Jefferson's psychosocial analysis proposes that Tommy's memories may have become distorted as a way of protecting him from childhood trauma. While Jefferson remembers fear and pain as the result of his father's beatings, Tommy recalls his own beatings with fondness, saying that he 'appreciated' what his father did to him and that he had some great 'laughs' growing up. This disparity in their experiences strengthened Jefferson's conviction in his interpretation of Tommy's use of unconscious psychological defence mechanisms to protect himself from this childhood trauma.

Another example of deep reflexivity is Saville Young's (2011, p. 49) research on race, class and gender in post-apartheid South Africa, in which she uses a 'concentric' model of reflexivity that moves progressively outwards towards psychoanalytic theory, increasingly reading talk 'for the investment positions of the researcher/researched'. Saville Young explores the 'raced, classed, and gendered entanglements' between herself, as a white middle-class woman, and her participants who were black, working-class men. While this sort of reflexive analysis initially seems characteristic of orthodox reflexivity within qualitative research, in that it examines the intersection between the researcher and researched from a social identity perspective, Saville Young quickly deepens her analysis by describing feelings of 'discomfort' as a result of challenging interview encounters. One participant, 'Liye', presents an example when at the close of the interview he seems to ask Saville Young to be his friend before intimating that he would like her to take him to the United Kingdom. Saville Young deflects Liye's request and later reflects on the feelings of 'fear and panic' that it elicited in her, identifying this as the same emotion she experiences when she is approached in the street by beggars who, in South Africa, are 'almost always Black' (2011, p. 51). Through her deep reflexive practice, Saville Young is able to consciously, retrospectively bring coherence and meaning to interview encounters produced in part by 'extra-rational', unconscious eruptions of affect and psychological defence mechanisms. While deep reflexivity is most commonly used in interview-based research, I hope to demonstrate that the approach also has the potential to enrich research adopting ethnographic or observational methods.

These examples demonstrate ways in which qualitative researchers can draw on more private elements of their subjectivity in order to guide and enhance their interpretations of data. Jefferson uses painful memories of his childhood, while Saville Young interprets her feelings triggered by challenging interview encounters. Both researchers present their subjectivity as interviewers in a way that 'includes unconscious, conflictual forces rather than simply conscious ones' (Hollway & Jefferson, 2000, p. 33). These candid accounts disrupt the status quo by embracing uncertainty and discomfort in research encounters (Finlay, 2002b; Pillow, 2003). They also demonstrate the methodological and ethical challenges presented to researchers attempting to use deep reflexivity in terms of limits to self-awareness, relationships with research participants, and exposition of the (professional) self. Working within a psychosocial model presents the challenge of acknowledging the impossibility of total self-awareness, given the impenetrability of the unconscious. However, deep reflexivity is an opportunity for researchers to make use of what is consciously available to them through critical reflection on their subjectivity, allowing access to memories, dreams, associations, feelings, emotions and fantasies. While accepting that our comprehension of our own subjectivity will always be partial, psychosocial researchers have shown that this process can be facilitated through introspection, retrospection, and working collaboratively in research teams (Hollway & Jefferson, 2000; Walkerdine et al., 2001). In the sections that follow, I reconnect this theoretical discussion about psychosocial research with the study of tourism and work through an example of my own reflexive practice from a psychosocial study of volunteer tourism.

Methodology

I undertook a qualitative, longitudinal study of volunteer tourists from 2009–2013 as part of my doctoral studies at Cardiff University in Wales (Crossley, 2013). This research focused on the subjectivities of young travellers, given the popularity of volunteer tourism amongst young people in the United Kingdom and its association with the pre-university gap year (Heath, 2007; Lyons, Hanley, Wearing & Neil, 2012; Simpson, 2004). The primary purpose of the research was to develop a more nuanced, complex understanding of how volunteer tourists' identities and experiences fluctuate and are actively negotiated over time and space. By attending closely to the narratives of young volunteer tourists and applying a psychosocial lens to foreground affect, desire, defence mechanisms and intersubjective dynamics, I hoped that my research would shed light on what MacCannell (2011, p. 4) refers to as the underinvestigated 'depths and intimate contours of tourist curiosity, subjectivity, and motivation'. The research provided insight into how young volunteer tourists respond to touring poverty, how they are caught between competing discourses of socio-economic development and environmental conservation, and how their investment in consumer identities presents a barrier to ethical transformation through travel.

The research involved conducting repeated semi-structured interviews with ten young volunteer tourists over the course of one year. The volunteer tourists that I interviewed were all British, white, mainly middle-class and aged between 18–24 years. Participants were recruited with the aid of a large commercial volunteer tourism provider based in the United Kingdom that offered medium-term community development and wildlife conservation programmes of between one to three months' duration. As part of the research, I joined the volunteer tourist group for one month's participant observation in Kenya in the summer of 2010. By putting myself in the position of my participants, I hoped to gain a more detailed, first-hand appreciation of what volunteer tourism involves and also to gather supplementary data by observing activities, interactions and conversations that might not have been mentioned in the interviews. Birkeland's (2005, p. 17) account of her positioning as a 'reflexive tourist' in the North Cape of Norway has been influential in this respect; I too wanted to explore the preoccupations and taken-for-granted meanings that I brought to the research by sharing my participants' experiences and reflecting on how this made me feel. I maintained a daily reflexive research diary throughout the duration of the fieldwork in Kenya that captured not only the interactions and activities of the volunteer tourists but also gave an account of my most intimate thoughts, feelings, fantasies and dreams (Brody, 1981; Hunt, 1989). I also recorded reflections systematically after each interview, commenting on how I felt about the participant, whether the interview triggered any feelings or associations for me, and whether the participant appeared uncomfortable or emotional at any point (Cartwright, 2004; Finlay, 2002b; Hollway & Jefferson, 2000).

The starting point for the reflexive analysis documented in this paper was a memorable and troubling dream from my fieldwork, which prompted me to review my research diary for content that could help with its interpretation. Once I had identified fieldwork encounters and experiences that appeared to shed light on the most overt features of the dream, I then analysed these data using Walkerdine et al.'s (2001) 'three levels of analysis' for guidance. I began the analysis by taking the data at face value, exploring themes, discourses and identity positions in the text relating to the overt content of the dream. The second level of analysis involved looking for evidence of unconscious dynamics by identifying narrative breaks, contradictions, metaphors, fixations, connections and absences; here I also noted manifest references to emotionality and affectively laden encounters. The final stage of the reflexive analysis was a reflection on how the research dynamics were shaped by my subjectivity, including feelings, unconscious desires and defences, which I then related to my interpretations of the interview and observational data. I achieved this by drawing on reflections I had written throughout the course of my research, which documented thoughts and 'feeling states' arising during the interviews or fieldwork (Cartwright, 2004).

It is important to note that this reflexive analysis has evolved over time and involved multiple iterations. I first presented a seminar on the topic of 'Using dreams to interpret fieldwork dynamics in psychosocial research' to my doctoral group in 2010 (Crossley, 2010), when the experience of being in the field in Kenya was still fresh and vivid in my mind. I then revisited this analysis in 2017 to develop a conference paper, in which I connected this work more explicitly to scholarship on reflexivity and also made linkages to the conference theme of 'stories' in tourism research (Crossley, 2017). The current paper, written in 2018, is therefore the third formal reflexive analysis I have conducted on these particular events from my fieldwork in Kenya. However, the feelings I describe here are not distant memories but rather re-articulations of reflections that I initially recorded during the first iteration of this analysis back in 2010. Having said this, memory is an unavoidable part of the reflective process and is significant in shaping the meanings that we attribute to research encounters. The way in which I remember my fieldwork now will undoubtedly be different from earlier memories, but I have treated this discrepancy as an analytic strength because the emotional distance I now feel from my earlier memories has allowed me to examine them critically, revealing aspects of my subjectivity that had previously been obscured from my conscious awareness.

Deep reflexivity in practice

At the time of my fieldwork in Kenya, I was in my mid-twenties and pursuing a doctorate on the topic of volunteer tourism. I was drawn to tourism as a subject for my research as I have always loved travelling – something that stems in part from my upbringing as a British/French dual national. The frequent visits my family made to France during my childhood instilled in me a desire to experience other places and cultures, and I quickly came to see travel as a vital part of my identity. I had backpacked and travelled independently around many countries in Europe but had never volunteered while travelling, so volunteer tourism held a certain fascination for me. I hoped to learn what the motivations of the volunteer tourists were and whether the practice presented a more ethical, reciprocal form of tourism. I was excited to be visiting Africa for the first time but also nervous about how the research would go. When I arrived in Kenya, I found that the majority of the volunteer tourists were British, white, middle-class women – as am I – in their late teens or early twenties. We lived communally in dorm-style accommodation located in or close to villages in rural Southern Kenya. The volunteer tourism programme consisted of rotating placements in a variety of locations, which involved building houses and schools, assisting with school activities, and wildlife conservation work. We would typically volunteer for three or four days per week, with the remainder of our time spent either travelling to the next placement or taking part in leisure activities such as snorkelling, safaris and cultural excursions. I now believe that I went into the research with strong views on what constituted 'good' and 'bad' tourism practices, particularly in relation to volunteer tourism's highly moralized discourse.

Volunteer tourist photography

It was the second week of my stay in Kenya and we were about to leave on an excursion to look for herds of elephants in the verdant hills that surrounded our accommodation. As we gathered our rucksacks and cameras, ready to get on the bus with the guide, one of the volunteer tourists announced that her camera had broken and so she would not be joining us. Some of us suggested that she might still enjoy coming along and seeing the elephants, but she explained that there would be 'no point' in going if she could not take pictures of them. It was at this point that I realized how deeply intertwined photography, seeing and experience were for the volunteer tourists and how compelled they were to document, package and take home their travels. In addition to being used on touristic excursions, the volunteer tourists also regularly brought along cameras to the volunteer work sites, lodging them into pockets as we undertook demanding, dirty manual labour. The volunteer tourists would pose in groups to have their picture taken, often with a shovel or wheelbarrow in hand as evidence of the work they had completed. I too succumbed to this impractical urge to document my volunteer efforts and returned with photographs of myself shovelling soil and of my blistered hands.

A recurring feature of the photography that I observed was the failure to ask local people's permission before taking their picture, despite having been advised to do so by our volunteer tourism provider. On one occasion, a woman in traditional dress appeared from behind a house, an infant in her arms and a second child standing shyly by her side. One of the volunteer tourists, Kate[1], went up to the woman and, without addressing her, took her photograph. Moments later, another volunteer tourist approached the woman and asked whether she could take a picture, to which the woman shook her head, raised her hand and said 'no'. Kate laughed nervously and said, 'Oh well, too late!' Similar scenes of assumed consent played out regularly in other settings and, looking back through my diary after the conclusion of my fieldwork, I was able to recognize my growing fixation and sense of unease in relation to the volunteer tourists' photographic practices:

> **Day 2:** I found it strange that people immediately started using their cameras ... taking photos of people's houses and of the children. I did take some photos myself, but felt rather self-conscious brandishing such an expensive bit of kit in such a poor village. I couldn't decide whether photographing the children was a harmless bit of fun that, admittedly, they seemed to enjoy or a potentially damaging display of western materialism and wealth.
>
> **Day 8:** I saw several photo-shoots taking place with the volunteers and children, but to me this seems like such a stereotypical image that I wouldn't feel comfortable [doing the same]; it's the image of smiling black children in the arms of adoring white youths that you find plastered all over the volunteer tourism brochures.
>
> **Day 13:** I was appalled at how the others just whipped out their cameras [at the school] and started photographing the children without asking their permission. Two of the gappers even went up to a child at his desk and had their photo taken with him. They were grinning from ear to ear but the little boy looked glum and less than impressed. Afterwards they joked that he would've told them to 'fuck off' if he'd known how. ... I was fuming.
>
> **Day 17:** Driving between the dwellings I was again offended by those who photographed the people, and especially the children, as if they were simply objects to be observed and documented. I reflected on my own photographic practice. I have taken photos with local people in them, but only when these people have been part of a particular landscape or scene that is the focus of my picture. In contrast, I have seen the others zooming right in on groups of waving children. I wonder whether I am being too uptight about it.
>
> **Day 18:** A tiny girl with cute braids ... ran up to us, smiling. Her father held her up so that she could shake our hands. He carried her along the side of the bus, but when they reached Katy she just took a photo of her rather than greeting her. I found it so awful. Jack, sitting on the other side of the bus, asked whether it was a white girl. We said that it wasn't and he said that that made more sense: 'It's a black kid! Get the camera!' he exclaimed. It was such a relief to hear someone else echoing my thoughts.

In these extracts from my field diary, I repeatedly contrast my behaviour to that of the other volunteer tourists in order to position myself as morally superior. I also use emotive language, such as 'appalled', 'offended', 'awful' and 'fuming', to express the strength of my sentiments. However, this moral superiority and criticism of the volunteer tourists' behaviour is tinged with self-doubt, for example when wondering

[1]All names in the research diary and interview extracts have been changed to pseudonyms.

Figure 1. Me posing with two members of the Maasai tribe. *Source:* Author.

'whether I am being too uptight' or when 'I couldn't decide' whether taking photos of the children was harmless fun or potentially detrimental.

Nearly 3 weeks into the trip, we took part in a cultural excursion to a Maasai village that I would later recognize as a significant moment in my fieldwork:

> **Day 18:** Going to the Maasai village was one of the most surreal and fascinating experiences of my life. We pulled up at the group of dwellings ... just at the side of the main road. The Maasai people were dressed in traditional gowns hung loosely on them, although they were not as decorated as images I had previously seen. One man spoke very good English and welcomed us, collected the money and told us that we would be shown traditional dances and have the opportunity to walk around the village. ... I questioned the ethics of going into people's homes and photographing them and their children. It seemed like a strange form of prostitution in which they were selling their bodies to the tourist gaze. I asked people's permission before taking their pictures, but in a situation where we had paid them how could they refuse? Having said this, some of the Maasai practically dragged us beside them to have our pictures taken. Despite my intellectualizing of the situation, I have to admit that when the family lined up and started singing and dancing I had chills running down my spine. Those piercing, wailing voices, the gruff roars and grunts of the men, the astounding straight-legged leaps into the air, the long sticks and flowing robes. I was in awe. ... There was one young man in the village who was more resplendent than the rest, with long, red hair clipped with beads, earrings jutting out at rightangles from his lobes and white and silver necklaces around his neck. He wore a red robe and his face was painted red. His name was Julian and I felt a strange attraction to him. ... Maybe it was the explanation that he was decorated because he is not yet married and has to first fight with a lion to fully become a man; this created associations of potency, sexuality and availability.

This excursion provoked in me a strong desire to photograph the Maasai, which I had not experienced up until that point. It was as if I had been liberated from my anxieties, reassured perhaps by the apparent agency of the Maasai or by the structure of the commodified encounter. Figure 1 shows me posing with two members of the tribe, holding some of the traditional beaded jewellery that I had bought at the village. Unlike many of the other photographs that I have from the visit, in which it was the Maasai who had encouraged us to pose for pictures, this is one that I had requested. I asked for my photograph to be taken with Julian, but the man on the right leaned in for the photograph at the last minute, which made me feel embarrassed for not having wanted him to be included. Today this image is difficult for me to look at and arouses feelings of shame, as it reminds me of the way in which I had objectified Julian and had valued his image above that of the less colourfully dressed man.

The dream

Despite enjoying our visit to the Maasai village, afterwards I had a vivid, disturbing dream that seemed to evoke imagery of the village together with feelings of anxiety and fear:

> **Day 20:** Last night I dreamt that I had been bitten by a snake. The snake was long and broad with a teal body and burnt orange head; just like a lizard I had seen a few days previously. At first, I was unsure whether the snake had actually bitten me, as we were caught in a struggle, which ended in the snake's decapitation. But as the dream progressed the graze on my skin became more and more apparent. I remember the challenge of trying to get to hospital using the dysfunctional ambulance system, which was slow and, like the buses around here, dropped other people off at stops along the way. The route meandered along small roads, and there seemed to be no quicker alternative. I felt that the urgency of the situation was not being appreciated and that the systemic failures were so severe that they threatened my life. Later in the dream I came across a group of men and women on the grassy outskirts of a town who were standing amongst a multitude of tethered snakes. There were snakes of every type and colour imaginable, and I feared that many could possess a lethal bite. The people wouldn't let me leave, saying that I had to have my picture taken with the snakes. I was afraid. Someone had told me that the snake that had bitten me was not dangerous, and yet now I was in renewed peril. I finally managed to get away by promising that I would return the following day with another twenty people and that we would all pay to be photographed with the snakes.

The dream opens with a scenario that many of the volunteer tourists were afraid of – being bitten by a snake. This sets into motion a chaotic series of events that reference the Kenyan landscape and which give rise to unconscious feelings of fear, threat, anxiety and confusion. I am still unsure whether decapitating the snake at the beginning of the dream holds any particular significance; if it does, I am not currently able to consciously access or articulate this meaning. I believe that the snakes at the end of the dream most likely represent the Maasai – their colourful skins evoking the brightly coloured clothes of the Maasai and the elongated way that the men appeared as they performed the *Adumu* (their iconic jumping dance). Being under pressure to have my photograph taken with these creatures, the obligation to pay, and the large number of people that I promise to bring are all clearly references to the commodified relation that had been established between the Maasai and the volunteer tourists. The snakes being 'tethered' symbolizes the oppressive, objectifying effects of this commodified relation. The Maasai being represented by snakes in the dream implies that on an unconscious level I was processing the visit as an encounter with a dangerous Other; the cultural difference that seemed exotic and alluring at the time had become transformed into something alien and threatening.

This dream drew attention to a fantasy relation in which the local people were beckoning us to go and enjoy their authenticity, papering over a far more traumatic realization of the potentially exploitative practice we were participating in and enjoying. I now believe that part of the reason I was so frustrated by the actions of the other volunteer tourists was not just because I found them unethical or because I was attempting to position myself as morally superior, but precisely because I shared their desire to enjoy the authenticity and exoticness of the Kenyan people. The Maasai village appeared to present a safe space in which to fulfil this desire in what I consciously experienced as a non-exploitative, nonthreatening commodified exchange. The volunteer tourists were able to capture and retain part of the exotic Maasai to enjoy again later in the form of souvenirs and photographs, and in return the Maasai received money. It also felt as if the Maasai were in control of this exchange: they welcomed us, took our money, beckoned us to take photographs with/of them, showed us around their village, and invited us to purchase their jewellery. It was the Maasai that appeared to have the agency, which created a reassuring feeling not only that this encounter was not exploitative but, on an unconscious level, that we were pleasing the Maasai by fulfilling their requests; in this fantasy, we were fulfilling their desires as much as they were fulfilling ours. After the dream, I did not record any further criticisms of the other volunteer tourists photographing the local people for the remainder of our stay, indicating that it had helped to process and bring into cognizance a previously unrecognized conflict.

Photo-elicitation interviews

During the final retrospective interviews with the volunteer tourists, conducted in the United Kingdom, I carried out a photo-elicitation exercise using a selection of the participants' own images in order to get them to reflect on why they had taken particular photographs in Kenya. I asked my participants whether there was anything they felt they could not take pictures of and the almost unanimous response was 'no':

> *Sarah*: Not really, I think they were way too, like all the people were way too welcoming for that? There was nothing that you couldn't take photos of 'cause if you didn't they'd be like "Take photos! Take photos!" Wouldn't they? And they loved cameras and stuff like that so there was nothing really that, you kinda think oh I can't take pictures of that. Because I didn't take any pictures until they were all like "Take pictures! Take pictures!" And they did it every single place you went. They all wanted you to take pictures. So no, there's nothing that really, that I was like oh I can't take a picture of that because if I didn't they'd ask me to. 'Cause they all wanted their picture taken? … I just liked the reaction when you took a picture of them and then showed it to them and they were like "Uh! That's me!" Like they're all laughing. So, no I don't think there was anything that

> I couldn't really take pictures of. They were all too welcoming I think. Made you feel like you could do whatever. Made you one of the locals.

Sarah emphatically constructs her photography as invited or coerced by the local people and as a way of fulfilling *their* desire to be photographed rather than *her* desire to take pictures. She repeatedly uses extreme case formulations to suggest that the Kenyans 'all wanted their picture taken' in 'every single place you went' and were 'all laughing'. This presents an exaggeratedly positive picture that omits examples of the locals refusing to be photographed or being visibly unhappy with the volunteer tourists' behaviour, which I know that she witnessed. Sarah's rhetoric is constructed to absolve her of any wrongdoing; not only did the locals enjoy being photographed but she asserts that she 'didn't take any pictures' until she was asked to by the locals. Finally, Sarah tells me that the Kenyans' warm welcome made her feel like 'one of the locals', discursively presenting herself as an insider and again sanctioning her behaviour.

After the interview with Sarah, I recorded the following reflection in my research diary:

> I felt a bit uncomfortable at a couple of points during the photo elicitation exercise. The first was when Sarah showed me a photo of one of the Australian girls playing music through headphones to a child, with others queuing up to listen. I had found this girl the epitome of everything that is wrong with Western tourists in the Third World; she was constantly thrusting her camera into the faces of the local children, never respected the dress code, and was always dripping with technology. Beyond that, I'm not entirely sure why I felt such unease. The second moment was when Sarah picked out the group photo from the hill. I had suspected that at least one participant would choose this photo, and she commented that I had been sitting on my own on another rock. I desperately wanted to defend myself and explain how sad I had felt that evening, but it just wasn't appropriate. I felt very upset at this point, which reinforced the sense that I had found Kenya incredibly difficult socially and that even now feelings of vulnerability and separateness from the other volunteers remain.

When I wrote this reflection, I felt that these two moments of discomfort were unrelated, but now I am not so sure. Initially, I thought that the photograph of the Australian girl disturbed me because it represented something colonial and disrespectful that I found offensive in the context of supposedly ethical tourism. There is still some truth to this interpretation, but I believe there is a deeper meaning that can be accessed by exploring the second moment of discomfort, in which I was reminded of the strong sense of exclusion that I experienced within the volunteer tourist group. Sarah mentions the evening we went to watch the sun set from a picturesque hilltop:

> **Day 27:** The boys started drinking on the bus; I should have known then that we were in for a rowdy evening. The bumpy ride through [the reserve] brought us to a hill which we had to mount in our flip flops. At the top were clusters of large boulders and the most extraordinary 360° panorama across the reserve. … It was such a beautiful moment … and one that moved in me emotions and memories. I thought again of my father and other people whom I had lost; of how they were lost forever to the world and how they would never see all of this. I was overcome with loss and sadness. The others, on the other hand, seemed to be having a frivolously good time. They had started to play tinny rock music on the iPod speakers, were drinking, shouting and taking pictures of one another. I distanced myself as far as possible from them, and declined their invitation for

> a group photo. I didn't feel part of this group and didn't want to be faking a smile in some tacky photoshoot in this wonderful place that had affected me so deeply. They were ruining this moment and as they approached my rock and tried to engage me in conversation all I could think was how awful they were and how I longed to be on my own.

At the time, I experienced only the depth and distress of these emotions. Writing about this episode now, I recognize the parallels between my interactions with the other volunteer tourists and earlier experiences of being unpopular and bullied at school. I had felt rejected by Sarah in Kenya; we had socialized with different groups and she had seemed reluctant to do her interview, to the point that I was worried she would withdraw from the research. Sarah had been close to the Australian volunteer tourists that I had fallen out with, which I believe contributed to this sense of distance. Could the unease I felt when confronted by the image of this popular, confident Australian girl have been generated by temporally layered feelings of exclusion in Kenya combined with my previous experiences at school?

The feelings that I experienced during the interview with Sarah can be conceptualized as inchoate countertransference impressions – unconscious and emotional reactions affecting how I related to my participants and the interview situation (Cartwright, 2004; Giami, 2001). In terms of Walkerdine et al.'s (2001, p. 97) question: 'who do I represent for the subject, and who do they represent for me?', on an unconscious level I believe that I responded to Sarah as an amalgamated Other representing popularity and exclusion, which stemmed in part from my difficult social experiences at school. This would explain my strength of feeling during the interview and subsequent recollections. Reflecting on the dream about the Maasai village allowed me to further unpack these feelings arising in relation to the interview, helping me to realize that my hostility and sense of 'unease' may have also been a defensive response to my own complicity with the volunteer tourists' photographic practices in Kenya. Returning to Sarah's interview, it is possible that her excessive characterization of the locals was a defensive response protecting her from unconscious anxiety regarding her conduct as a tourist in Kenya. However, it is also possible that she felt I was being judgemental and implying that she should not have been so liberal in her photography. Given the strained relationship that I had with Sarah and her friends, it would not surprise me if she was a little on edge.

Collectively, the volunteer tourists' narratives of their photographic practices pointed to a more complex picture than Sarah's account implies. Some participants suggested that it was inappropriate to photograph beggars and that adults required more caution and respect than the children, who seemed to love playing with cameras and having their picture taken. I concluded that photography was acknowledged as having a potentially objectifying effect on the visited communities but was also perceived as a means of creating intimate encounters with the children, constituting a form of connection across the linguistic and cultural divide. Some of the volunteer tourists were also aware of how photography could be used as a way of negotiating moral boundaries, such as that demarcating observation from intrusion. These findings were significant given that volunteer tourism is often advocated as an antidote to mainstream tourism's objectifying gaze, instead promoting meaningful interactions with visited communities. Although I had known that the volunteer tourists would still take photographs, I had not anticipated that these photographic practices would form a part of my analysis and would shed so much light on broader dynamics of enjoyment, objectification and intimacy in the volunteer tourism experience.

Evaluating deep reflexivity

Using deep reflexivity in this research drew my attention to parts of the volunteer tourism experience that I might otherwise not have considered as being analytically important. I did not originally intend to write about the volunteer tourists' photographic practices but thinking about the dream drew my attention to this part of the dataset and helped me to reevaluate its significance. The approach also allowed me to explore my experience of the volunteer tourists' desire to consume exotic authenticity in Kenya, providing a detailed account that I was able to use as supplementary qualitative data by effectively treating myself as an additional participant. Accepting that I experienced many of the same feelings as the other volunteer tourists allowed me to develop greater empathy with my participants, which in turn helped to build rapport that facilitated the interviews. The approach prompted me to analyse these interviews for possible unconscious, intersubjective dynamics, which provided valuable information on how the data were co-constructed. In these ways, deep reflexivity answers Ateljevic et al.'s (2005, p. 17) call for positional reflexivity to 'move beyond the transparent reflexive practice of just being autobiographical and writing oneself 'in' to our research', towards 'uncover[ing] the complexities, negotiations, fissures and tensions that entangle the researcher with his/her research'.

It could be argued that deep reflexivity dissolves the boundary between researchers' professional and personal lives, leaving us vulnerable to scrutiny from peers and students. Through the inclusion of extracts from my field diary, a photograph of myself and a dream, I have tried to provide a candid account of how I felt and acted throughout this research, even when it cast me in a negative light. However, perhaps inevitably, I exercised a certain degree of self-censorship and engaged in an extensive curatorial process when writing this paper. Although I revealed a great deal of my subjectivity, I was always conscious of the control that I had over this process and my ability to withhold certain information about myself in order to retain some privacy and prevent my disclosures from becoming excessive or self-indulgent (Finlay, 2002a, 2002b). It is therefore important to remember that researchers are ultimately always in control of what they publicly reveal of themselves, even for those adopting a deep reflexivity approach. Going through the process of considering what information I felt comfortable sharing about myself enabled me to proceed more cautiously and sensitively when it came to writing about aspects of my participants' lives.

I found that in practice it was difficult to confidently apply psychoanalytic terminology such as 'transference' and 'countertransference' to the data I was analysing, echoing Frosh and Saville Young's (2008, p. 113) concerns that psychosocial research pales in comparison to the 'intense exploration of unconscious material characteristic of psychoanalytic reflection on the countertransference in the clinical situation'. This is not to say that the deep reflexivity method did not allow me to explore affect, subjectivity and the unconscious in deep and meaningful ways, but that assigning specific labels to these affective flows was more problematic than I had originally anticipated. Writing this reflexive analysis challenged me to confront parts of my subjectivity that I did not like or understand. There were times when I felt more confident in my interpretations of the other volunteer tourists than of myself. This introspective process was a reminder of my theoretical conviction that total self-awareness is impossible and that the comprehension we have of our own subjectivity will always be partial and conflicted. However, I believe that revisiting my reflections eight years after the original fieldwork in Kenya provided the emotional distance that I needed to view these experiences with a fresh perspective.

Conclusion

I have argued that deep reflexivity presents a creative opportunity for qualitative tourism researchers to disrupt and extend their reflexive practice by attending to aspects of their subjectivity conventionally considered too private to be shared publicly. Deep reflexivity implies that reflexive practice should include critical reflections on the researcher's emotions, embodiment and unconscious processes (Hollway, 2004; Saville Young, 2011); it goes beyond more common reflexive disclosure statements typically written into 'safe spaces' in the text such as the methodology (Cohen, 2013; Hall, 2004) and exemplifies Ateljevic et al.'s (2005) recommendation for researchers to get 'entangled' in their research. This approach enabled me to reflect on how my subjectivity as a researcher/tourist and my complicity in potentially detrimental photographic practices produced unconscious defensive responses during research on volunteer tourism in Kenya. Reflecting on my subjectivity in this way enabled me to develop greater empathy with my participants, provided additional qualitative data, revealed intersubjective interview dynamics, and drew my attention to parts of the volunteer tourism experience that I might otherwise not have considered as being analytically significant. In conclusion, deep reflexivity presents a promising avenue for challenging and creatively disrupting conventional approaches to reflexivity in pursuit of advancing methodological practice in tourism research.

Disclosure statement

No potential conflict of interest was reported by the author.

References

Ali, N. (2013). Researcher reflexivity in tourism studies research: Dynamical dances with emotions. In I. Ateljevic, N. Morgan, & A. Pritchard (Eds.), *The critical turn in tourism studies: Creating an academy of hope* (pp.13–26). Abingdon, UK: Routledge.

Ateljevic, I., Harris, C., Wilson, E., & Collins, F. L. (2005). Getting 'entangled': Reflexivity and the 'critical turn' in tourism studies. *Tourism Recreation Research, 30*, 9–21. doi:10.1080/02508281.2005.11081469

Ateljevic, I., Pritchard, A., & Morgan, N. (Eds.). (2007). *The critical turn in tourism studies: Innovative research methodologies*. Oxford, UK: Elsevier.

Ateljevic, I., Morgan, N., & Pritchard, A. (Eds.). (2013). *The critical turn in tourism studies: Creating an academy of hope*. Abingdon, UK: Routledge.

Birkeland, I. (2005). *Making place, making self: Travel, subjectivity and sexual difference*. Aldershot, UK: Ashgate.

Blackman, L., Cromby, J., Hook, D., Papadopoulos, D., & Walkerdine, V. (2008). Creating subjectivities. *Subjectivity, 22*(1), 1–27. doi:10.1057/sub.2008.8

Brody, E. B. (1981). The clinician as ethnographer: A psychoanalytic perspective on the epistemology of fieldwork. *Culture, Medicine and Psychiatry, 5*, 273–301. doi:10.1007/BF00050772

Burman, E. (2008). Resisting the de-radicalization of psychosocial analyses. *Psychoanalysis, Culture & Society, 13*, 374–378. doi:10.1057/pcs.2008.25

Cartwright, D. (2004). The psychoanalytic research interview: Preliminary suggestions. *Journal of the American Psychoanalytic Association, 52*, 209–242. doi:10.1177/00030651040520010501

Clarke, S. (2002). Learning from experience: Psycho-social research methods in the social sciences. *Qualitative Research, 2*, 173–194. doi:10.1177/146879410200200203

Cohen, S. A. (2013). Reflections on reflexivity in leisure and tourism studies. *Leisure Studies, 32*, 333–337. doi:10.1080/02614367.2012.662522

Crossley, É. (2013). Volunteer tourism, subjectivity and the psychosocial (Doctoral thesis). Cardiff Univerity, UK. Retrieved from http://orca-mwe.cf.ac.uk/61298/

Crossley, É. (2010, October). *Using dreams to interpret fieldwork dynamics in psychosocial research*. Paper presented at PostGrad Café, Cardiff University, UK.

Crossley, É. (2017). Psychosocial stories: Narrative and deep reflexivity in tourism research. In D. Dredge & S. Gyimóthy (Eds.), *Euro-TEFI 2017. Building our stories: Co-creating tourism futures in research and practice*. (Tourism Education Futures Initiative Annual Conference) (pp. 8–19). Copenhagen, Denmark: TEFI.

Dickson-Swift, V., James, E. L., Kippen, S., & Liamputtong, P. (2009). Researching sensitive topics: Qualitative research as emotion work. *Qualitative Research, 9*, 61–79. doi:10.1177/1468794108098031

Dupuis, S. L. (1999). Naked truths. Towards a reflexive methodology in leisure research. *Leisure Sciences, 21*, 43–64. doi:10.1080/014904099273282

Edley, N. (2006). Never the twain shall meet: A critical appraisal of the combination of discourse and psychoanalytic theory in studies of men and masculinity. *Sex Roles, 55*(9-10), 601–608. doi:10.1007/s11199-006-9116-x

Elliott, H., Ryan, J., & Hollway, W. (2012). Research encounters, reflexivity and supervision. *International Journal of Social Research Methodology, 15*, 433–444. doi:10.1080/13645579.2011.610157

Feighery, W. (2006). Reflexivity and tourism research: Telling an(other) story. *Current Issues in Tourism, 9*, 269–282. doi:10.2167/cit/mp006.0

Finlay, L. (2002a). Negotiating the swamp: The opportunity and challenge of reflexivity in research practice. *Qualitative Research, 2*, 209–230. doi:10.1177/146879410200200205

Finlay, L. (2002b). "Outing" the researcher: The provenance, process, and practice of reflexivity. *Qualitative Health Research, 12*, 531–545. doi:10.1177/104973202129120052

Franklin, A., & Crang, M. (2001). The trouble with tourism and travel theory. *Tourist Studies, 1*, 5–22. doi:10.1177/146879760100100101

Frosh, S. (2010). *Psychoanalysis outside the clinic: Interventions in psychosocial studies*. New York, NY: Palgrave MacMillan.
Frosh, S. (2003). Psychosocial studies and psychology: Is a critical approach emerging?*Human Relations*, *56*, 1547–1567. doi:10.1177/00187267035612005
Frosh, S., & Baraitser, L. (2008). Psychoanalysis and psychosocial studies. *Psychoanalysis, Culture & Society*, *13*, 346–365. doi:10.1057/pcs.2008.8
Frosh, S., & Saville Young, L. (2008). Psychoanalytic approaches to qualitative psychology. In C. Willig & W. Stainton-Rogers (Eds.), *The Sage handbook of qualitative research in psychology* (pp.109–126). London, UK: Sage.
Frosh, S., Phoenix, A., & Pattman, R. (2003). Taking a stand: Using psychoanalysis to explore the positioning of subjects in discourse. *British Journal of Social Psychology*, *42*, 39–53. doi: 10.1348/014466603763276117
Giami, A. (2001). Counter-transference in social research: Beyond Georges Devereux. In M. W. Bauer (Ed.), *Papers in social research methods – Qualitative series, 7*. London, UK: London School of Economics, Methodology Institute.
Gilbert, K. R. (2001). Introduction: Why are we interested in emotions? In: K. R. Gilbert (Ed.), *The emotional nature of qualitative research* (pp.3–16). London, UK: CRC Press.
Glancy, M. (1993). Achieving intersubjectivity: The process of becoming the subject in leisure research. *Leisure Studies*, *12*, 45–59. doi:10.1080/02614369300390041
Gough, B. (2015). Qualitative methods: Critical practices and prospects from a diverse field. In I. Parker (Ed.), *Handbook of critical psychology* (pp.107–116). Hove, UK: Routledge.
Gray, B. (2008). Putting emotion and reflectivity to work in researching migration. *Sociology*, *42*, 935–952. doi:10.1177/0038038508094571
Hall, M. (2004). Reflexivity and tourism research. In L. Goodson & J. Phillimore (Eds.), *Qualitative research in tourism: Ontologies, epistemologies and methodologies* (pp.137–155). London, UK: Routledge.
Harris, C., Wilson, E., & Ateljevic, I. (2007). Structural entanglements and the strategy of audiencing as a reflexive technique. In I. Ateljevic, A. Pritchard, & N. Morgan (Eds.), *The critical turn in tourism studies: Innovative research methodologies* (pp.41–56). Oxford, UK: Elsevier.
Harris, J., & Huntington, A. (2001). Emotions as analytic tools: Qualitative research, feelings, and psychotherapeutic insight. In K. R. Gilbert (Ed.), *The emotional nature of qualitative research* (pp.129–145). London, UK: CRC Press.
Heath, S. (2007). Widening the gap: Pre-university gap years and the 'economy of experience'. *British Journal of Sociology of Education*, *28*, 89–103. doi:10.1080/01425690600996717
Henriques, J., Hollway, W., Urwin, C., Venn, C., & Walkerdine, V. (1998). *Changing the subject: Psychology, social regulation and subjectivity*. London: Routledge.
Hoggett, P. (2008). What's in a hyphen? Reconstructing psychosocial studies. *Psychoanalysis, Culture & Society*, *13*, 379–384. doi:10.1057/pcs.2008.26
Holland, J. (2007). Emotions and research. *International Journal of Social Research Methodology*, *10*, 195–209. doi:10.1080/13645570701541894
Hollinshead, K., & Jamal, T. B. (2001). Delving into discourse: Excavating the inbuilt powerlogic(s) of tourism. *Tourism Analysis*, *6*, 63–73. doi:10.3727/108354201108749746
Hollway, W. (2004). Editorial. *International Journal of Critical Psychology*, *10*, 5–12.
Hollway, W. (2016). Emotional experience plus reflection: Countertransference and reflexivity in research. *The Psychotherapist*, *62*, 19–21.
Hollway, W., & Jefferson, T. (2000). *Doing qualitative research differently: Free association, narrative and the interview method*. London, UK: Sage.
Hook, D. (2008). Articulating psychoanalysis and psychosocial studies: Limitations and possibilities. *Psychoanalysis, Culture & Society*, *13*, 397–405. doi:10.1057/pcs.2008.23
Hunt, J. C. (1989). *Psychoanalytic aspects of fieldwork*. London, UK: Sage.
Jervis, S. (2009). The use of self as a research tool. In S. Clarke & P. Hoggett (Eds.), *Researching beneath the surface: Psycho-social research methods in practice* (pp.145–166). London, UK: Karnac.

Johnson, C. W. (2009). Writing ourselves at risk: Using self-narrative in working for social justice. *Leisure Sciences*, *31*, 483–489. doi:10.1080/01490400903199815

Johnson, L. (2015). Why are you here? Toward reflexivity, positionality, and multivocality at the intersections of research and tourism. In A. Holden, S. Slocum, & C. Kline (Eds.), *Scientific tourism: Researchers as travellers* (pp.27–39). London, UK: Routledge.

Kleinman, S., & Copp, M. (1993). *Emotions and fieldwork. Qualitative Research Methods* (Vol. 28). Newbury Park, CA: Sage.

Kvale, S. (1999). The psychoanalytic interview as qualitative research. *Qualitative Inquiry*, *5*, 87–113. doi:10.1177/107780049900500105

Lyons, K. D., Hanley, J., Wearing, S., & Neil, S. (2012). Gap year volunteer tourism: Myths. *Of Global Citizenship? Annals of Tourism Research*, *39*, 361–378. doi:10.1016/j.annals.2011.04.016

MacCannell, D. (2011). *The ethics of sightseeing*. Berkeley, CA: University of California Press.

Phillimore, J., & Goodson, L. (Eds.). (2004). *Qualitative research in tourism: Ontologies, epistemologies and methodologies*. New York, NY: Routledge.

Pillow, W. (2003). Confession, catharsis, or cure? Rethinking the uses of reflexivity as methodological power in qualitative research. *International Journal of Qualitative Studies in Education*, *16*, 175–196. doi:10.1080/0951839032000060635

Ren, C., Pritchard, A., & Morgan, N. (2010). Constructing tourism research: A critical inquiry. *Annals of Tourism Research*, *37*, 885–904. doi:10.1016/j.annals.2009.11.006

Riley, R. W., & Love, L. L. (2000). The state of qualitative tourism research. *Annals of Tourism Research*, *27*, 164–187. doi:10.1016/S0160-7383(99)00068-7

Saville Young, L. (2011). Research entanglements, race and recognisability: A psychosocial reading of interview encounters in (post-)colonial, (post-)apartheid South Africa. *Qualitative Inquiry*, *17*, 45–55. doi:10.1177/1077800410389443

Seigworth, G. J., & Gregg, M. (2010). An inventory of shimmers. In M. Gregg & G. J. Seigworth (Eds.), *The affect theory reader* (pp.1–25). London, UK: Duke University Press.

Simpson, K. (2004). 'Doing development': The gap year, volunteer-tourists and a popular practice of development. *Journal of International Development*, *16*, 681–692. doi:10.1002/jid.1120

Taylor, S. (2017). Psychosocial research. In B. Gough (Ed.), *The Palgrave handbook of critical social psychology* (pp.225–241). London, UK: Palgrave Macmillan.

Taylor, S. (2015). Discursive and psychosocial? Theorising a complex contemporary subject. *Qualitative Research in Psychology*, *12*, 8–21. doi:10.1080/14780887.2014.958340

Tribe, J. (2005). Editorial: New tourism research. *Tourism Recreation Research*, *30*, 5–8. doi: 10.1080/02508281.2005.11081468

Tucker, H. (2009). Recognizing emotion and its postcolonial potentialities: Discomfort and shame in a tourism encounter in Turkey. *Tourism Geographies*, *11*, 444–461. doi:10.1080/14616680903262612

Walkerdine, V. (2008). Contextualizing debates about psychosocial studies. *Psychoanalysis, Culture & Society*, *13*, 341–345. doi:10.1057/pcs.2008.31

Walkerdine, V., Lucey, H., & Melody, J. (2001). *Growing up girl: Psychosocial explorations of gender and class*. Houndmills, UK: Palgrave.

Westwood, S., Morgan, N., & Pritchard, A. (2006). Situation, participation and reflexivity in tourism research: furthering interpretive approaches to tourism enquiry. *Tourism Recreation Research*, *31*, 33–41. doi:10.1080/02508281.2006.11081260

Willig, C. (2013). *Introducing qualitative research in psychology* (3rd ed.). New York, NY: Open University Press.

Wilson, E., & Hollinshead, K. (2015). Qualitative tourism research: Opportunities in the emergent soft sciences. *Annals of Tourism Research*, *54*, 30–47. doi:10.1016/j.annals.2015.06.001

Challenges in outdoor tourism explorations: an embodied approach

Jelena Farkic

ABSTRACT

Methodological challenges are rarely discussed in depth among outdoor adventure tourism scholars. Despite the prevailing qualitative approaches in this field, and the recognition that the fleetingness of the human experience and the non-linearity and unpredictability of the more-than-human world have the power to influence the research process, the messy, negotiated and often contested researcher's role has been less considered. In addressing this, the aim here is to critically discuss the methodological approach to explorations of the outdoor experiences through deconstructing the researcher's role. Through renderings of the existentialist propositions of being in the world and a post-structuralist philosophy of fluidity and flux, the attention is granted to embodied experiences as a way of generating knowledges. Being situated in the research setting, space is created for interrogation of the processual dimensions of commodified outdoor journeys from an emic, researcher-as-tourist perspective. Research in the outdoor scenaria is by no means a linear process but rather a messy, complex and often ruptured journey, further complicated by the ethical concerns, struggles and idiosyncrasies of the researcher. I thus discuss the nuances and complexities of doing the embodied research and the haphazard ways of data collection. In shifting attention to more existential aspects of being in the outdoors through the process of post-experiential reflections, discomfort emerged as a critical quality of the outdoor experience. I thus illuminate the significance of embodied research and epiphenomenal discoveries in the production of new knowledges, to which greater attention, both in theoretical and methodological conversations, should be paid in the future.

摘要

户外探险旅游学者很少深入讨论方法上的挑战。尽管该领域质性方法盛行，也认识到飘忽无常的人类经验以及非线性和不可预测的超人类世界有能力影响研究过程，但是研究者自身混乱的、待协商的以及经常有争议的角色很少被考虑到。为解决这个问题，该文目的是通过解构研究者自身的角色，批判性地讨论探索户外体验的方法论问题。通过对存在主义关于存在于世界的命题和后结构主义关于流动性的哲学阐释，我们将注意力放在了亲身体验上，并将其作为一种产生知识的方式。置身于研究环境中，从一个主位的、研究者即游客的视角，仔细思考了商品化户外旅行的过程维度。 户外场景的研究绝不是一个线性的过程，反而是一个混乱、复杂、经常破裂的旅程，研究者的伦理顾虑、挣扎和个人喜好使之更为复杂。因此，我将讨论做亲身研究的细微差别和复杂性，以及数据收集的随意性。通过经验后反思的过程，将注意力转移到身在户外的更为存在主义的方面，不适成为户外体验的一个关键品质。因此，我阐明了亲身性研究和超现象发现在产生新知识方面的重要性，未来更应该多多关注理论和方法论方面的对话。

Introduction

It is ever more common that modern humans search respite in nature as compensation for their accelerated pace of life. In the contemporary, mainly western, world 'the contrast between the being of outdoors and the being of indoors could hardly be more pronounced' (Hay, 2015, p. vii). Indoor beings, aiming to compensate for their highly digitalised everyday lives, and re-establish the ontological connection with their authentic being, tend to travel to wild and nature-rich areas and thus they increasingly purchase so-called 'adventure holidays'.

Spaces that such holidays are usually embedded in are often termed as 'liminal', for their otherworldly character, geographical remoteness and relative inaccessibility (Varley, 2011). Marine and coastal environments, for example, described as the 'penultimate frontier for tourism' (Cater & Richardson, 2017), have become particularly attractive to people seeking novel and adventurous experiences. An increased number of tourism products have also been designed to bring the otherwise inaccessible landscapes and elusive wildlife closer to consumers. People are now enabled to go, for example, white water rafting, kayaking, scuba diving, swimming with dolphins, long-distance cycling or chasing the northern lights (Fox et al., 2014; Heimtun, 2016; Beedie & Hudson, 2003).

When going on holidays, however, people take a great part of everydayness along with them, entering tourism spaces in which they dwell and move (Edensor & Falconer, 2011). Likewise, as part of their outdoor pursuits, they are greatly 'buying' convenience. Unlike before when some of the essences of the adventure tourism product were risk, danger and uncertainty, adventure holidays have moved towards more immersive, slower and safer experiences in the outdoors (Cater, 2006; Easto & Warburton, 2010, Farkic, Taylor & Bellshaw, 2020; Varley & Semple, 2015). Since the experiences sought for 'have shifted from the search for the utmost challenging and dangerous to that of safety and comfort' (Trauer, 2006, p. 185), people require services of activity providers, not only to access places but also to ensure their safe and convenient consumption. Within the context of commercial tours, activities are delivered and managed by outdoor guides that are negotiating outdoor environments, providing access to wild places, choreographing people's experiences, maintaining high levels of security and ensuring the overall hospitality and wellbeing of tourists (Farkic, 2018; Farkić & Taylor, 2019; Varley et al., 2018; Beedie & Hudson, 2003). However, despite the commodification of adventure which assumes highly managed and sanitised experiences (Beedie, 2008), entering a liminal arena in search of otherworldly landscapes and extraordinary experiences may mean stepping out of the comfort zone for some. Being in alien environs is not necessarily pleasant, and the sense of comfort and security may be reduced. Insecurities, anxieties and other unpleasant feelings and conditions do not only apply to tourists partaking in adventure holidays, but also to researchers who set off exploring the tourism experience in the Great Outdoors.

This paper therefore aims to voice the challenges in the field work process and the knowledges that were produced through the corporeal engagement with the outdoor research setting. The overarching aim of the broader study was to phenomenologically explore how tourists construct their experiences in the outdoors and the ways in which hospitality works on commercial multi-day adventure holidays. To come closer to the phenomena in question, I conducted ethnographically informed research. I aimed to embody myself in a research setting and conduct fieldwork, which required my body to get attuned to a hitherto alien environment. I joined two multi-day tours on the Isle of Coll, the Inner Hebridean island off the west coast of Scotland. One of the well-established activity providers "Basking Shark Scotland" operates from both Oban on the mainland and the Isle of Coll, taking tourists to experience the extraordinary, liminoid event – close encounters with Scotland's biggest fish, basking sharks, as well as other wildlife, such as white-tailed eagles in the Sound of Mull, seals, porpoises, minke whales and numerous sea birds.

The transition from an 'indoor body' into an 'outdoor body' and into a 'researching body' was far from smooth. Adventurers are usually depicted in a rather masculinist way, as strong, capable, risk-taking individuals. Similarly, perhaps, the adventurer-researcher's role is constructed on the assumption that they are able-bodied, well-versed or experienced in a certain activity, which may not always be the case (Nairn, 1999). Moreover, amongst adventure tourism scholarship, few studies focus on the role of a researcher and the methodological challenges faced in the field (Kennedy et al., 2018; Houge Mackenzie & Kerr, 2012). Thus, my aim here is to extend this conversation by voicing my personal experience of the field work journey and disclose some challenges I had in simultaneously taking roles of both a tourist and a researcher. I also aim to highlight what qualities of the embodied research in the outdoors emerged from the process, juxtaposed to well-trodden ethnographic techniques normally used in tourism explorations. In doing so, I illuminate the knowledges that such an approach has produced, and the potential it has to advance outdoor tourism scholarship. Ultimately, the paper adds to the emerging methodological conversation building on pheno-existentialist philosophy and poststructuralist ideas on embodiment, through disrupting the dominant approaches to studying lived experiences in the outdoors.

Dwelling and embodiment as knowing the world

Building on early existentialist ideas, I more prominently situate the study within the philosophical discourse. My thinking about the human experience and its construction in relation to context (in this case the outdoor spaces) takes departure from existential phenomenology. Its founder Martin Heidegger (2010 [1972]), in his thesis on being and time, posited that 'spaces receive their essential being from particular localities, and not from space itself' (p. 332). By this he meant that human beings *dwell*; they co-exist in relationships that they maintain with places and others, as objects of awareness, through which they acquire meanings (1993). He also expressed the idea of directedness towards others, both humans and non-humans, through the concept he termed *Mitdasein*, or being-with, as an extension of Being (*Dasein*). The 'others' are encountered via a medium of common ground; through things that we use and the processes of exchange. Although he stressed the ontological significance of place, it was, however, untypical of him, as well as other phenomenologists, to unpack the human body in its process of being. In placing an emphasis on *Dasein* and dwelling as a form of being, Heidegger failed to address in more depth the role of the body in experiencing places. In confronting this omission in his work, Casey (1997, 1993) attempted to provide a comprehensive account of the role of place in human experience by building on the notion of dwelling. He philosophically explored the pervasiveness of place in people's everyday lives and the ways we engage with it, positing that the human body always 'takes place' in this world. He explored how Heidegger understood place and the essential part it plays in defining the life-world of experience, highlighting that the human body, perception and habitation are key aspects of being in the world. Casey (1993) suggested that 'bodies do not only perceive but *know* places' (p. 34, original emphasis). The experiential aspects of the place-world and questions of inhabitation of both built and wild places, positing that humans find their place within them, give them meanings and make sense of them. Spaces and places are made and remade through bodily practices, they come into presence through the experience of bodily sensations, being further heightened through affect and emotions.

The paradigmatic shift across the disciplines brought with itself new theories. Body metaphors, like the concept of embodiment, for example, suggest that the body is used as a convenient way of experiencing and thinking about the world. This is understood as an existential condition of human life and cancels any possibility of Cartesian dualism, challenging the objectification of the body and arguing for temporality and fluidity of the human experience (Swain, 2004). As Weiss and Haber (1999) remarked, 'the very notion of embodiment suggests an experience that is constantly in the making... being constituted and reconstituted from one moment to the next... changing in significance and form' (p. 43).

French philosopher Michel Foucault (1979) takes the credit for conceptualising the embodiment theory. He is claimed to have greatly influenced social theory in thinking how the human body is subject to power, which in the early 1990s caught the attention of feminist scholarship. In attempts to understand how power works at an individual level, feminists turned to Foucault's poststructuralism and his examination of the multiple ways in which the power relations are deployed within society. They, therefore, unpack the body by questioning personal body politics and power relations, such as gender, class, race, age or sexuality, and argue for positionality as an agent in constructions of meanings. Wolf (1982) suggested that 'meanings are not imprinted into things by nature; they are developed and imposed by human beings. The ability to bestow meanings – to 'name' things, acts, and ideas – is a source of power' (p. 388).

A phenomenological approach to thinking about being in the world was later supported by poststructuralist theorists of embodiment and non-representational theorists. Hence, the original Heideggerian meaning of the notion of dwelling as being in the world has been extended to embrace both geographical and social aspects of bodily practices and relations it has with the place. Drawing on the concept of dwelling, a number of scholars have described being in the world as an embodied engagement with the environment, and accentuated everyday practices as the basis for how people dwell and construct their worlds (Ingold, 2000; Casey, 1993; Dreyfus, 1993; Macnaghten & Urry, 2000; Thrift, 1997, 1999). For example, following a phenomenological train of thought, which holds that people are in the world in as much as they dwell in it, Ingold (2000) proposed the 'dwelling perspective' which holds that 'it is through being inhabited, rather than through assimilation to formal design specification, that the world becomes a meaningful environment for people' (p. 173). People are not only physically present in the place, they dwell in it - they are aware of it, they are sensually engaged with it through various embodied practices (Ingold, 2005, 1995). Similarly, drawing on Ingold's (1995) understanding of landscape as a milieu of embodied, quotidian dwelling, Wylie (2005) argued for 'placing the self in the body and embedding the body in landscape' (p. 240). In this vein, perceptions, memories and bodily movements are understood as being both in and of landscape and gather landscape together as a lived milieu.

Thrift's (2004, 1999, 1997) work on the body and nature through his non-representational theory provided valuable insights. He not only suggested why contemporary people may seek more embodied nature-based activities but also offered insights into understanding how the human body engages with the natural world. The human body and other natural objects do not stand as separable entities but exist in assemblages, multiplicities and co-evolving 'blocks of becoming' (Thrift, 2004). Thus, the landscape might best be described in terms of the entwined materialities and sensibilities with which humans act, move and sense the world. Kinetic empathy may stimulate and engender the cultural turn to the affective and sensuous (Thrift, 2008). Edensor (2006, 2000a, 2000b) suggested that in the process of embodiment, the materiality of places and spaces should not be neglected. Spatial affordances, as he termed them, greatly constrain and enable a range of actions; 'the surfaces, textures, atmospheres, smells, sounds, contours, gradients, and pathways of places encourage humans to follow particular courses of action, producing an everyday practical orientation dependent upon a multisensory apprehension of place and space' (Edensor, 2006, p. 30). Humberstone (2015, 2013) takes the credit for elaborating the embodiment theory in the context of nature-based physical culture in focusing on the ways in which humans learn through their bodies. Further understandings of the relationships between body, emotions and the elements in relation to adventurous activities are the focus of her work, which advances the knowledge on sensorial experiences of participants in outdoor activities.

Embodied tourism experience

For a long time, little attention was paid to phenomenological explorations of tourists as individual, 'corporeal subjects' that know the world and gain experiences through their bodies. The body has become of great interest to tourism scholarship since Veijola and Jokinen (1994) and Johnston (2001) challenged the disembodied, hegemonic and masculinist research in tourism. Desmond (1999) proposed that 'we must have a more fully embodied concept of the tourist, expanding the notion of the 'tourist gaze' to include other embodied aspects of experience (movement, sound, touch and so forth), both in the physical and imaginary realms' (p. xxi). Since then, the human body and the senses have been significantly considered within tourism theory.

Although Heidegger's concept of dwelling was not initially conceived to embrace tourist practices and mobilities, some authors have used this notion to describe embodiment in the world through tourism (Varley, Carnicelli and Farkic, 2018). For example, Obrador Pons (2003) articulated tourists' dwelling as their engagement with and involvement in a physical environment as a way of creating knowledge. Recounting Heidegger's ideas, he made a point in suggesting that physical presence in a certain space is not sufficient, but one needs to be bodily immersed in it, in order to gain experience. A place like a beach for example 'needs to be apprehended as a place where the body lives, experiments and desires, a place of embodied utopias and nondiscursive pleasures' (p. 55). He drew attention to numerous non-visual sensations that influence the tourism experience, such as 'the feeling of the sun caressing the skin, the sensual movement of the naked body into the seawater and the unpleasant infiltration of sand into body orifices' (2007: 134). There is also much to thank cultural geographers for their theorising of the construction of spaces through the concept of performativity as a form of dwelling. For example, Edensor (2000b) analysed a number of embodied tourist performances (walking, gazing and photographing in particular) that play an important role in constructing tourist spaces. In his account, he decentred visual sense and shed light on the importance of other senses, proposing that via immersed tourism practices tourism spaces are enacted and performed.

Engagement of and experiencing through the body has also been acknowledged by adventure tourism scholars, positing that the body is a medium for negotiation of experiences while undertaking adventurous activities. Adventure tourism has been traditionally claimed to provide challenging activities in natural environs that test the human body and trigger intensive experiences. It seems more obvious that embodied experiences are easier to capture in cases of more intense activities, where the body is challenged and tested (Cater & Cloke, 2007; Cloke & Perkins, 1998). Cater (2006) explained that adventure tourism is 'clearly about active participation with the body' (p. 66). Illustration of this can be found in Fox et al. (2014) example of cycling, where they claim that 'part of the attraction to cycling begins with the bodily experience of being on, melting into, and flowing with the bicycle as it moves through spacetime: that is, the embodied experience of cycling (p. 74).

The body, understood as a medium through which experiences and knowledges are created, has obviously received a more pronounced recognition across disciplines in the past few decades, which laid the foundations for the further phenomenological understanding of lived experiences and embodied knowledges in the context of tourism in the outdoors. To unpack the body as the site of knowing, I, therefore, build on early pheno-existentialist ideas on being in the world, and complement them with poststructuralist propositions of dwelling, embodiment and multisensory engagement in outdoor tourism spaces. The ensuing section positions the body as a credible instrument for research.

Body as an epistemic medium

Tourism researchers have taken an increased interest in understanding people's experiences and social dynamics during commercial journeys in wild nature (Cater &

Richardson, 2017; Carnicelli-Filho, 2013; Varley, 2011; Curtin, 2010; Scarles, 2009). To embody themselves into a research setting and gain insights into human interactions, they take an ethnographic approach, as it allows them 'to describe the lives of people other than ourselves, with an accuracy and sensitivity honed by detailed observation and prolonged first-hand experience' (Ingold, 2008, p. 69). However, ethnography is not always characterised by long-term engagement with other people's lives, as in traditional anthropological studies (Malinowski, 1922; Marcus, 1998). Rather, as Pink and Morgan (2013) explain, it involves intensive excursions into their lives with the aim of creating context through which to explore what is of the researcher's interest. In the short-term research encounter, the intensity of interaction with participants is pronounced, through which knowledge is generated . To that end, spaces of outdoor adventure lend themselves well to ethnographic practice (Cater & Cloke, 2007). Although it has gained momentum in social sciences, in outdoor adventure tourism it has only recently started to be increasingly utilised (Kennedy et al., 2018; Humberstone, 2015; Houge Mackenzie & Kerr, 2012; Carnicelli-Filho, 2013; Rantala, 2011; Rantala et al., 2011; Kerr & Houge Mackenzie 2012; Sharpe, 2005; Beedie & Hudson, 2003).

If we want to think about elusive social realities, however, we need 'to teach ourselves to think, to practice, to relate, and to know in new ways' (Law, 2004, p. 2). With the critical turn in tourism, the 'orthodox methodologies' (Wilson & Hollinshead, 2015) were disregarded and more disruptive approaches have been adopted. Researchers have been increasingly encouraged to embrace unusual and innovative methods in order to advance qualitative inquiry in tourism studies (Buda et al., 2014; Bianchi, 2009; Ateljevic et al., 2005; Botterill, 2001). Therefore, more humanistic approaches have been adopted, which 'embody the researcher and humanise the research process' (Wilson & Hollinshead, 2015, p. 30). Rethinking tourism as an embodied experience and the tourist as an affective, embodied subject, has largely helped stimulate their thinking as contextual, fluid and dynamic, created as a result of embodied, affective and performative practices.

In tourism research, the attention has been greatly shifted to the agency of the individual, as a way of deconstructing and deepening understanding of the processes and contexts within which the meanings of experiences emerge. However, being a social scientist is a messy condition. We bring into research our idiosyncrasies and subjectivities, which have a major influence on the research process (Bourke, 2014; Ateljevic et al., 2005). Feminist scholars have long disrupted 'the idealised image of the lone, ungendered, unbiased researcher, going into the field like a neutral, empty vessel simply waiting to be filled with data' (Billo & Hiemstra, 2013, p. 321). It is 'reasonable to expect that the researcher's beliefs, political stance, cultural background (gender, race, class, socio-economic status, educational background) are important variables that may affect the research process' (Bourke, 2014, p. 2).

The researcher's positionality therefore greatly matters. Hall (1990) emphasised that 'there is no enunciation without positionality. You have to position yourself somewhere in order to say anything at all' (p. 18). At the time of writing, I identify as a middle-class female of Serbian background, currently in my late thirties. Prior to setting off to do the field research, I had possessed the pre-understanding of what Heidegger

(1993) termed fore-structures, that is, pre-knowledge of concepts that influence our understanding of the world. For example, in my explorations of human experiences and social dynamics in the outdoors, I had pre-knowledge of relevant concepts such as tourism, adventure, embodiment, lived experiences or guiding. It was impossible to bracket what I had previously known, and what I strived to find out, and enter the field entirely tabula rasa. Yet I was an inexperienced researcher. Having a background in the positivist research paradigm, doing research in a foreign land, and in a different language, all brought tensions and insecurities both prior to and during the research process. Things were further complicated by having to conduct it in a non-native language. I had to acquire new terminology and become comfortable with it, as I was what Stebbins (1998) termed an 'ethnic outsider'. I also found it difficult to switch between my insider-outsider roles (Kennedy et al., 2018) – being me and me being a researcher, doing things for the purpose of my research I would not have done otherwise (for example chatting to the skipper in the cockpit while getting seasick).

However, writing about knowing through our bodily responses, that is, on the notions of haptic knowledge, of proprioception, is scant among adventure tourism scholarship. Crang (2003) asserted that 'the body quite often ends up as providing a sort of inescapable positioning of the researcher – through race, dis-ability or gender – but less often is the body the instrument of research'. In confronting this, I attempted to sense the field by way of my body and became an 'observant participant' (Walsh, 2009). I was doing what humans do on holidays anyway – chatting, laughing, cooking, walking, snorkelling, eating, taking photos. This approach enabled me to corporeally enter various social and physical spaces, more easily take part in activities and become more personal with others (Dowler, 2001).

I utilised a patchwork of ethnographic methods to complement, extend or replace one another, depending on the given circumstances. To document the experiences, however, I kept a field diary, eventually assembling a large amount of textual material. Initially, such knowledge was heterogeneous, disconnected and disorganised. Allowing time for passiveness and idleness during and post-fieldwork, some themes crystalized, assembled, and became more potent in generating theory than others. I came to realise that I was paying considerable attention to my bodily reactions and emotional journeys whilst doing the research that equally populated my diary pages alongside observations of the reactions and experiences of others. In the coding process I initially grouped the data into broader themes. Rather frequently, I noted the moments of stress, fear, hesitancies, anxiety, awkwardness or exhaustion, to which I eventually assigned a common denominator and termed them collectively as *discomfort*. Across the following sections, through the presentation of diary excerpts, I disclose these moments and subsequently discuss the knowledges and opportunities that emerged from them.

Voicing moments of discomfort

The first tensions about conducting fieldwork arose alongside methodological considerations. I was certain about doing the ethnographically informed research, and yet I had no previous experience in being away for days, researching. I also realised that there is no formula for how to do the research, as it is by no means a straightforward enterprise. I recall the feeling of a lack of confidence before I set off on my first tour. Being introduced as a researcher, although I was also a tourist who purchased the tour, I could not help feeling like an intruder, as an embodied subject observing the interactions of others. I had ethical concerns in relation to formalising the encounters and declaring myself as a researcher and explaining my intentions. I faced my first challenge when I was about to hand out consent forms at the start of the first tour, which is something that is supposedly a common practice while commencing a field work. Nonetheless, negotiation of the time and space to give out consent forms was somewhat problematic:

> I knew I had to find the right moment to tell the group that I was the one doing my research, briefly explain what it was about and kindly ask them to sign the consent forms. I was a bit nervous as the time and place didn't really look right (actually, it looked ideal – white beaches all around and infinity views, far away from everything and everyone!). I waved at the group and they soon gathered around me. I introduced myself again and said that I was in fact on 'busman's holiday'. Having explained that I'd need their signed consent to be able to proceed with the research, they all agreed and nodded. While I was reaching out to grab the forms from the bag, someone asked: "Do we need to sign them NOW?!", moving his eyes like he was pointing to the weather conditions. It was dreary and it was blowing a hooley there at the cliff and it was of course absurd to do it right there and just then. Someone else was witty and said he'd sign the form and I could wait for the paper at the other side of the beach, suggesting the wind would sweep it off his hands. Everyone laughed and simply walked away, scattering across the beach again as if nothing happened. I looked at the folder with the pile of unsigned forms, turned around as if I needed to confirm the poor weather conditions and thought to myself: "Yeah, why not take them to the boat tomorrow?".

I always thought handing out consent forms was out of place – whilst sitting in the bunkhouse, standing at the beach of Coll being exposed to the sweeping wind, or sailing on the rib rocking on the choppy sea. It was never a good moment for it, as depicted below:

> We were on the way back to Oban, the tour was nearing its end. I still needed to give out the consent forms to the guides and John and Y.G. so I asked if it would be okay to sign just then. I gave them the papers, and they started reading through the text. I knew it wasn't the right moment, the rain was hammering, the rib was at its maximum speed and the papers were soaking and constantly folding.

Due to its instability and unpredictability, the weather is usually a major challenge in outdoor tourism. Its materialities are critical to framing the actions throughout the trips (Rantala et al., 2011). Therefore, anticipation and coping with the weather was crucial throughout all stages of the trips: from sailing out on a rib, to doing the activities, to doing the research. There were episodes of very poor weather conditions, during which it seemed that nothing was going on for hours, we did not spot any sharks nor swam with them, we were only sitting at the deck, very obviously disconnected from one another, just going with the flow and bearing with the conditions. And yet, these experiences were memorised and later recorded in my diary:

> For the next couple of hours, we were just sailing. Our energy dropped and everyone looked miserable and soaked to the bones. Some of us took out a packed lunch and were eating with wet hands. Matt's bread was soaking and falling apart. I was completely

> unaware of the time, all I could see around me was water, water, endless water. Everyone was silent. I was eating my tuna sandwich, completely hypnotised by the movements of the waves all around me. I was hallucinating I was seeing whales, but these were just the waves, appearing, disappearing, changing their shapes. This lasted for a while. I couldn't take it any longer, I felt exhausted and I wanted to go back and feel the solid ground under my feet.

In situations like these, I simply had to 'work with' the weather and allow for more flexibility in managing my research process, putting off the conversations or notetaking or any other research activity. I was feeling seasick and mentally exhausted, like everyone else was, yet in these moments the research, in its conventional shape and form, was hardly possible.

> I took my motion sickness pill, it turned out that it was a wise decision since other people on the boat were getting sick on the choppy sea and kept throwing up. Emily didn't even want to change into a wetsuit on the second day. She said she saw the sharks and that was enough for her. Y.G. was occasionally throwing up from the boat, bending over the edge and splashing her face. Vincent had his antacid pills which he was offering to everyone. I couldn't stand on my feet, I sat down and gripped the rail. Most of the group were pale in the face and impatient to jump into the water.

Getting attuned to the immediate environment assumed our gradual familiarisation with the boat, as a space that was shared most often, knowing how it moves, how it is operated, where to sit in order to avoid the splashing of the waves, or how to get in and out of it. We spent most of our time sailing on the rough sea, and it seemed like no one could find their cosy, personal place:

> The boat seemed very uncomfortable in the beginning, and whatever corner I found myself in, I felt I was always in someone's way. It was wet, and small and claustrophobic and our stuff was all around it.

It was interesting to observe how our relationship with the vessel developed over time – from complete strangeness to almost complete familiarity. Technologies, such as boats, are described as spaces through which individual and inter-subjective practices may be extended, making the encounters possible (Crouch & Desforges, 2003). There was a lot of social activity going on on the boat: watching wildlife, changing into wetsuits/dry clothes, zooplankton sampling, conversing, eating, or jumping into the water after sharks. Humberstone (2013) suggested that in these situations 'the participant loses track of time and becomes one with their equipment, the elements and the natural environment' (p. 568). At times, overwhelmed and engrossed in the activity, taking notes was not my priority. Very little time and space was allowed for it, thus I mostly relied on taking photos or making mental notes. I first needed to make sure that I was comfortable on the boat and in the water, that I was well and rested and present in the moment.

Personal belongings, too, played a significant role in the process of familiarisation, or 'becoming comfortable' (Varley & Semple, 2015) with the space we occupied. For example, while standing at the deck of the rib, pressing my hands hard against the cup of tea to warm up, I was watching others, or doing the participant observation, appearing between the cairns and playing with seals just in front of the boat. Unpleasant bodily sensations overwhelmed me after several hours of being in the wetsuit, buffeted by the winds, being seasick and learning how to snorkel in cold waters. I had changed into dry clothes but was constantly shaking in the cold weather. In my later reflections on this event, I noted:

> Now I was properly cold. I started shivering and didn't dare to go further. I felt my skin was all wrinkly and salty and the wetsuit carved a new texture on its surface. Thousands of midges were hovering above us in clouds. I decided to go back to the boat, to escape the midges, to get rid of the wetsuit, to warm up and forget about this nonsense. I just wanted to be wrapped up in a blanket, that's all I wanted.

Instead of noting down my observations, what I initially intended to do, I kept standing at the deck, wrapped up in a towel and waiting for my body to relax from a spasm. My gaze into the distance was interrupted by Vincent's spontaneous move. He swiftly put his hat on my head, saying: "You'll catch a cold, your hair is completely wet!". Having his hat on, and not feeling the cooling breeze anymore, I immediately felt tingling sensations all over my body. The blood was returning into my head and I could feel the warmth around my neck.

Adventurous activities rarely go without a certain amount of hardships and unpleasantness (Beard et al., 2012), even on commercial outdoor tours where comfort and safety are a priori paid for. Clothing and other specialist equipment, as well as its efficient use and briefing on the kit (in this case wetsuits, fins, masks and snorkels) is crucial in guided outdoor adventures (Beedie & Hudson, 2003). Familiarisation with equipment is important as it becomes part of people's bodies, through which they negotiate environment (Merchant, 2011). When it produces confusion, it detaches the body from the environment and the experience of undertaking activities is diminished. Small omissions on behalf of the activity provider, such as providing inappropriate gear, can affect the overall experience of clients. There were many occasions when our bodies suffered due to the influence of inclement nature and the unfamiliarity with the environment that was not our normal habitat. For example, on the first tour I was accidentally given a larger size of wetsuit, which made my body less capable of doing the research in/under water. I wrote about this particular experience:

> Trying to snorkel in the cold water I could entirely identify with Cain's experience, one of the Knausgaard's characters in his novel "A time for everything", when he attempted to run across the fields and ran out of breath. My throat was constricting and my lungs began to get all knotted after only a few metres and I felt as if I was breathing through a straw. The cold water was flushing into my wetsuit, straight down my spine. I won't last long, was the only thought I had as I was trying to make a move towards the guide who was floating with ease in the distance. My fingers and toes went numb after a few moments, the blood had already retreated from my extremities. I then forced my legs to make a quick move forward, but the more I tried, the further from the guide I seemed to be. I then tried to use my hands and make wide strokes, but instead, my hands were flapping back and forth in the water and I think I didn't make a single breathe while I was trying to pull myself together. When in despair I raised my hand to attract attention, the air flooded back into my lungs and I all of a sudden felt at ease. I had no energy to wobble my legs, nor my arms, and I involuntarily relaxed. To my great surprise, I realised I was floating in my neoprene suit. In the contact with freezing cold water I simply sank into complete oblivion and fought for bare life.

It was incredibly difficult, at times impossible, to collect data, particularly in the moments of panic, or whilst sailing in a small dinghy on the choppy sea, standing at the deck tightly gripping the rail whilst being splashed by giant waves, soaking in the rain for many hours during the day, getting seasick, shivering in a wetsuit, or being underwater. The body would simply 'get in the way' of enjoyment in activities, constrain and define its movement through spaces (Markwell, 2001). It was difficult for me to learn the technical skills of snorkelling, and at the same time be alert to what was going on – to focus on conversations, events and social interactions, as well as manage my own emotions and observe the behaviour of others. However, through the states of discomfort, which occurred rather frequently, some new knowledges emerged.

Lessons learnt

Embodied fieldwork and the process of reflexivity helped me understand the complexities and nuances of research in the outdoors, and the ways in which new knowledge was produced (Billo & Hiemstra, 2013; England, 1994; Kobayashi, 1994). To become both a tourist and a researcher, it was necessary to accommodate myself to 'the affordances that were dwelt within and passed through, and out of these adaptions sensuous apprehensions and practical epistemologies emerged [sic]' (Edensor, 2006, pp. 31-32). Due to the interplay of shifting sensualities and emotions in the field encounter, I largely put an emphasis on the body as an instrument of research. Although adventurers are normally depicted as strong, confident, courageous and daring, an adventurer-researcher has by all means a challenging task – to simultaneously undertake activities, manage emotions, control shifting mental states and collect data, all within a very short timeframe. I was frequently switching between my various roles, learning from attachment to and detachment from the research setting, active participation in activities, shifting mental states, negotiations, observations, conversations and reflections. In the research process my idiosyncrasies and subjectivities came to the surface and influenced the way I was doing the research, even though I was part of the commodified, convenient and to an extent safe setting. Nonetheless, I learned what it felt like to be physically and mentally exhausted from undertaking the activities, accompanied by adrenaline rush, exaltations, hesitations and anxieties, while at the same time making sure I was maintaining the focus on my research. This, however, produced new knowledges, which led me to think of the existential aspects of being in wild nature, the creation of a sense of security and psychological wellbeing in the outdoors.

Benefits of the intensive encounter

Intensive short-term ethnographically informed research may be greatly insightful. I learnt how the anticipation, reading and modification of lived experience was influenced by entering various social and physical spaces, and how they were affected by the more-than-human world. Participation in outdoor activities presupposes attunement, where both the researcher's and participants' movements are grounded in shared circumstances (Lee & Ingold, 2006). It is suggested that 'both researcher and respondent become vulnerable as they expose their selves; each finding solace within the visual as they come together in sharing the intensities of emotions, somatic knowledges and haptic spaces of encounter' (Pink, 2010, p. 923). Senses, emotions and affects are relational and emerge from the space between sensing and sensed, that is, 'to sense is always to sense with' (Anderson & Wylie, 2009, p. 326). However, sensing-with is not always reduced to a single human being. It is 'an aesthetic sensibility, a sensation of being at one with the immediate environment that might include the presence of others, together with a combination of memory and anticipation of specific events' (Bissell, 2008, p. 1700). Such 'intersubjective exchange materialises as shared corporeal expression as the immanence of encounter opens intimacies of the self' (Pink, 2010, p. 923).

In particular, the act of sharing (spaces, time or things) as a manifestation of *Midasein*, that is the ontological 'withness' proposed by Heidegger (2010 [1972]), is a prerequisite for 'communitesque' (Lugosi, 2008) encounters in commercial tourism settings. In either sharing the confined spaces of the boat, bad weather, moments of fear and exhaustion, both materiality and sociality of the outdoor experience were bound together. Through the consumption of various 'scapes', we became attuned to the ever-changing and unpredictable nature of the outdoor environment. Furthermore, in sharing the confined space of the boat most of the days, our personal spaces seemed to have collapsed, which allowed for closer insights into one another's bodily reactions and mental states, which in 'normal' circumstances would most likely not be the case. Furthermore, the imagined void I felt between myself and others was colonised by our interactions, and spaces were created for getting more personal with others. I began to realise the significance of the space in-between and I soon became more comfortable with myself, as a participant in activities and as a researcher. Enforced proximity triggered our more frequent communication and prompted reflections and stories surrounding the towel, pills, flip flops, hats, flapjack, GoPro screws and all the things lent and borrowed during the trips. The temporary sharing of space and time not only broke down formal barriers and enabled me not only to collect richer data, but also to create intimacy and fellowship and build trust with others (Still, 2011). The feeling of belonging to a small group is considered to be one of people's existential needs, which may be fulfilled through sharing adventurous spaces loaded with diverse, often contrasting, emotions.

Agency of the more-than-human world

The 'social' is a more heterogeneous mix of not only humans, but also other-than-human 'things' (Latour, 2005). Being embodied in the research setting, it was necessary to allow my body to sense the hitherto alien environment and learn its rhythms, taking into account various non-human actants as agents in constructing the outdoor experience, as well as in the process of collecting data. Outdoor spaces are not fixed, but rather a continual process of interactions between materialities and immaterialities within them, of which humans are only a part and do not control. Materiality is multiple: it involves natural objects, phenomena and processes that are beyond human control, defined in terms of movement and process rather than stasis. Outdoor and social spaces were thus negotiated through the act of sharing of both material and immaterial aspects of a tourist habitus. Edensor (2006) termed these elements 'spatial and material affordances' through which people explore, engage, and understand their experience, and which have power and agency in constraining or enabling their practices, and more or less influence the intensity of the lived experience as well as the research process.

Perceiving, sensing and becoming attuned to natural rhythms was highly relational 'insofar as these materialities/mobilities are the very sparks which ignite passages of perception and sensation, and concordances or dissonances, of bodies and things' (Anderson & Wylie, 2009, p. 326). The weather conditions influence the colouring and shading of the surrounding landscapes, creating and providing a unique backdrop for outdoor activities. Rare and unusual natural spectacles are particularly appreciated by tourists as they are perceived as more affective encounters (Edensor, 2006). Rather unsurprisingly, however, bad weather can obstruct activities and influence people's psychological wellbeing. I felt how the materialities of the weather had a forceful influence not only on my sense of comfort but also on the process of my research. Being outdoors and doing the activities, most of the time exposed to strong winds and rain, I was free of audio recorder, pen and paper, as technologies that would normally assume research. This, however, led me to rely on my body as the only available instrument of research in most situations.

Alternative data collection methods

As previously mentioned, I found the ethnographic approach and participation in guided tours an appropriate way to gain embodied, situational, and practice-related knowledge around guiding and hospitality in the outdoor scenarios. However, I did not have any a priori defined methods nor a rigid field research design. Rather, I had a loose framework and kept my mind open and alert to various 'imponderabilia' (Malinowski, 1922) throughout the research process. Chang (2008) claimed that 'the flexibility of research design should not be misconstrued as a lack of diligence or indecisiveness in planning' (p. 67), and therefore a less structured approach, although much more challenging, allowed for generating richer and more nuanced data.

I realised that much could be gained through spontaneous, entirely unstructured, leisurely conversations, free from disturbances and anxieties that could have arisen from the formalities of the interview process. I thus privileged conversations on the move, which allowed myself, both as a researcher and informant, to be more exposed to the multisensorial stimulation by being embodied in the surrounding environment (Adams & Guy, 2007). This was particularly valuable as it allowed the context, both social and environmental, to frame the conversations (Evans & Jones, 2011). This way, any conversation we had during tours was not interrupted by formalities, allowing for naturally flowing, unconstrained, insightful conversations. Such impromptu conversations occurred, for example, standing in the rain in the pier, shivering at the deck or struggling to learn how to snorkel. Each datum collected during the research encounter inevitably had an agency in layering, shaping and constructing the knowledge around the lived experiences.

Most of the time, however, I had to rely solely on my memory, as a method for 'storing' and 'processing' the data. I was mindful of other people's conversations as invaluable empirical data. I found it particularly difficult to memorise conversations and reproduce them afterwards as accurately as possible. For this reason, whenever possible I was turning to my phone and writing notes in the form of direct quotes, or noting snippets while chatting to others. In cultural studies, memory is a key method through which lived experiences are explored, usually through memory work. However, its utilisation as a method of generating and eliciting data, which is the case in this study, this may appear problematic as the credibility of the method might be questioned. For this reason, Pickering (2008) suggested that elicited memories should be triangulated with other documentary evidence or narrative data, in order to achieve a more methodologically sound account. Writing my field diary, based on memories, was thus supported by a large number of brief notes, various memorabilia, images, sketches, text messages, emails or videos, that not only helped me in the reconstruction of events but also serve as their evidence. Eventually, an account loaded with descriptions of the bodily sensations and moments of discomfort emerged.

From voicing emotions to epihenomenal discoveries

I also learned much from the moments when methods did not quite work and the periods in which no data were collected. There was a number of situations, some of which I illustrated earlier through the diary excerpts, in which standard ethnographic methods were of no use. No usual research technologies assisted me in this process, and the only instrument for collecting data was my body. Therefore, to aid this, post hoc expressive writing of a field diary with sincere, deep, emotion-rich accounts turned out to be a valuable resource in my reflecting on the past events, from which new knowledges emerged.

Various affective and emotional encounters provided me with 'analytical clues' (Davies, 2010; Koning & Ooi, 2013) and offered an opportunity to enrich my understanding of the field as well as my role in the production of knowledge. This process is regarded as highly emotional, transformative and revelatory (Moriarty, 2016). However, Punch (2012) argued that voicing emotions and personal challenges into discussions of the research process is still not fully accepted as a scholarly endeavour. Researchers tend to silence emotions either not to reveal their weakness or to achieve objectivity and get closer to the 'truth' (Widdowfield, 2000). They still struggle with 'transparent reflexivity' articulated through more explicit writing of their accounts. Arguably however, acknowledging, writing and speaking about emotional experiences during the field research can add considerable value to the research, both during its process and its outcomes. The field diary may thus greatly assist in reflecting on hidden struggles in the production of knowledge, reveal not so obvious aspects of the experience and reveal some of its aspects that might not have been considered otherwise (Corbetta, 2003). For example, moments of hesitancies, anxieties, pleasures, security, belonging or not belonging, were later considered in relation to humans' existential being; through our striving for ontological security and existential comfort whilst journeying in the outdoors (Farkic, 2018; Varley et al., 2018).

Conclusions

This paper sought to answer the call for disrupting the entrenched ways of doing the qualitative research and gaining novel knowledges within tourism studies through expanding the epistemological, methodological, and ontological approaches (Buda, 2015; Burrai et al., 2017; Hollinshead & Ivanova, 2013, Farkic, 2020). In so doing, it discussed the methodological journey of the neophyte snorkeler who found herself in the 'liminal playground' to do research (Gyimothy & Mykletun, 2004). However, doing the research as well as being a tourist on the outdoor tours was far from an easy task. Stepping out of the comfort zone carried certain constraints due to embodied reactions to the immediate, alien and at times harsh environment. There was a series of uncomfortable occurrences at work. I therefore aimed to voice these moments of discomfort to show that doing research in the outdoors, particularly on commercial tours on which affordances are a priori paid for and risks and thrills are adequately managed, is certainly not devoid of potential subjective in-situ constraints (Dimmock & Wilson, 2009, 2011). They may diminish the overall outdoor experience and may in many ways constrain the research process. However, this has brought invaluable insights which may in their own right contribute to the extant tourism literature.

In deconstructing the experience, I aimed to provide the philosophical underpinnings for this study and to more prominently acknowledge the inseparable relationship of the body and space, which further led into discussions of the body as a medium through which experiences are created. In so doing, however, I invited existential philosophers into the conversation. I privileged works by the early phenomenologists. In particular, I saw strength in the ontology of Martin Heidegger, and his proposition of the existentialist concepts of being and dwelling. To supplement this, I introduced the poststructuralist theory of embodiment, which suggests that embodied approaches may provide more nuanced, contextualised knowledges around interactions and negotiations of subjective experiences in outdoor tourism explorations.

Much can be gained from active participation in activities and observing how the body dwells and listening what it has to say. This is a kinetic and sensual process which is in most cases unpredictable and uncontrollable yet has the huge potential to disrupt the conventional ways of knowing and produce new knowledges. In moving away from thinking of the body as something bounded, fixed and finite (as boundedness makes us think anthropocentrically), I argued for the agency of the individual, however in relation to the social. I disrupted existing conceptualisations of, for example, the researcher-adventurer who is strong, masculine and able-bodied and whose feelings and emotions are in most cases removed from discussions in the context of doing research in the outdoors. Thus, through exploring a more affective and emotional response to being on outdoor tours, I made sense of the relational nature in meaning-making processes and co-production of experiences and knowledge. I embraced the rhizomatic, relational, more than human thinking which made more space for alternative ontologies in explorations of outdoor adventure tourism.

The paper further adds to the debate around more flexible and less rigid qualitative methodologies in tourism explorations. To that end, embodied research approaches may deepen our understanding of tourist experiences and provide nuanced knowledges of the consumption and production of tourist spaces, particularly in the context of outdoor adventure. Pink (2015) suggested that more sensory approaches may bring 'to the fore the tacit, normally unspoken (about) ways of knowing and doing that are part of our everyday life. These activities underpin much of what we do and means to us; they are often mundane and feel so 'normal' to our participants that they might not even think that they are worth mentioning, but they are also part of the way we feel and sense our futures' (p. 193). Subsequent reflections on bodily reactions provided me with more nuanced, contextualised knowledges around interactions and negotiations of the lived experience in the outdoors.

In summary, the paper disrupts the dominant, masculinist ways of doing research in the field of adventure tourism studies. Firstly, it deconstructs the researcher's role and voices the experiences less spoken of. Secondly, it argues for the credibility of idiographic studies, particularly from the researcher-as-tourist perspective. Thirdly, it suggests that autoethnographic, immersive, and reflective approaches have the potential to convey idiosyncrasies and reveal hidden aspects of the experience which can make meaningful contributions to theory. In this particular study, the sense of discomfort came out as an epiphenomenon, and a critical quality of the whole research process, as well as the overall outdoor experience. This finding may extend the scholarship concerned with tourist experiences in nature-based tourism and extend the debate on the notion of (dis)comfort on outdoor guided tours and its management. Ultimately, the study aims to encourage philosophical underpinnings of and pheno-existentialist approaches to adventure tourism research through partly addressing the famous Heidegger's question – what it means to *be* (a human, a researcher, a tourist).

Disclosure statement

No potential conflict of interest was reported by the author(s).

ORCID

Jelena Farkic http://orcid.org/0000-0002-2135-2254

References

Adams, M., & Guy, S. (2007). Senses and the city. *The Senses and Society*, *2*(2), 133–136. doi:10.2752/174589307X203047

Anderson, B., & Wylie, J. (2009). On geography and materiality. *Environment and Planning A: Economy and Space*, *41*(2), 318–335. doi:10.1068/a3940

Ateljevic, I., Harris, C., Wilson, E., & Collins, F. L. (2005). Getting 'entangled: Reflexivity and the 'critical turn' in tourism studies'. *Tourism Recreation Research*, *30*(2), 9–21. doi:10.1080/02508281.2005.11081469

Beard, C., Swarbrooke, J., Leckie, S., & Pomfret, G. (2012). *Adventure tourism*. Routledge.

Beedie, P. (2008). Adventure tourism as a 'new frontier' in leisure. *World Leisure Journal*, *50*(3), 173–183. doi:10.1080/04419057.2008.9674551

Beedie, P., & Hudson, S. (2003). Emergence of mountain-based adventure tourism. *Annals of Tourism Research, 30*(3), 625–643. doi:10.1016/S0160-7383(03)00043-4

Bianchi, R. V. (2009). The 'critical turn' in tourism studies: A radical critique. *Tourism Geographies, 11*(4), 484–504. doi:10.1080/14616680903262653

Billo, E., & Hiemstra, N. (2013). Mediating messiness: expanding ideas of flexibility, reflexivity, and embodiment in fieldwork. *Gender, Place & Culture, 20*(3), 313–328. doi:10.1080/0966369X.2012.674929

Bissell, D. (2008). Comfortable bodies: sedentary affects. *Environment and Planning A: Economy and Space, 40*(7), 1697–1712. doi:10.1068/a39380

Botterill, D. (2001). The epistemology of a set of tourism studies. *Leisure Studies, 20*(3), 199–214. doi:10.1080/02614360127084

Bourke, B. (2014). Positionality: Reflecting on the research process. *The Qualitative Report, 19*(33), 1–9.

Buda, D. M., d'Hauteserre, A. M., & Johnston, L. (2014). Feeling and tourism studies. *Annals of Tourism Research, 46*, 102–114. doi:10.1016/j.annals.2014.03.005

Buda, D. M. (2015). *Affective tourism: Dark routes in conflict*. Routledge.

Burrai, E., Mostafanezhad, M., & Hannam, K. (2017). Moral assemblages of volunteer tourism development in Cusco, Peru. *Tourism Geographies, 19*(3), 362–377. doi:10.1080/14616688.2016.1236145

Carnicelli-Filho, S. (2013). The emotional life of adventure guides. *Annals of Tourism Research, 43*, 192–209. doi:10.1016/j.annals.2013.05.003

Casey, E. (1997). Smooth spaces and rough-edged places: The hidden history of place. *The Review of Metaphysics, 51*(2), 267–296.

Casey, E. (1993). *Getting back into place: Toward a renewed understanding of the place-world.* Indiana University Press.

Cater, C. I., & Richardson, S. (2017). Marine and coastal tourism. In *Marine and coastal resource management: principles and practice* (p. 18). Taylor & Francis.

Cater, C., & Cloke, P. (2007). Bodies in action. *Anthropology Today, 23*(6), 13–16. doi:10.1111/j.1467-8322.2007.00548.x

Cater, C. (2006). Playing with risk? Participant perceptions of risk and management implications in adventure tourism. *Tourism Management, 27*(2), 317–325. doi:10.1016/j.tourman.2004.10.005

Chang, H. (2008). *Autoethnography as method (developing qualitative inquiry)*, Walnut Creek, CA: Left Coast Press, Inc.

Cloke, P., & Perkins, H. C. (1998). Cracking the canyon with the awesome foursome: representations of adventure tourism in New Zealand. *Environment and Planning D: Society and Space, 16*(2), 185–218. doi:10.1068/d160185

Corbetta, P. (2003). *Social Research. theory, methods and techniques*. Sage.

Crang, M. (2003). Qualitative methods: touchy, feely, look-see? *Progress in Human Geography, 27*(4), 494–504. doi:10.1191/0309132503ph445pr

Crouch, D., & Desforges, L. (2003). The sensuous in the tourist encounter: Introduction: The power of the body in tourist studies. *Tourist Studies, 3*(1), 5–22. doi:10.1177/1468797603040528

Curtin, S. (2010). Managing the wildlife tourism experience: The importance of tour leaders. *International Journal of Tourism Research, 12*(3), 219–236. doi:10.1002/jtr.747

Davies, J. (2010). Introduction: emotions in the field. In J. Davies and D. Spencer (Eds.), *Emotions in the field: The psychology and anthropology of fieldwork experience* (pp. 1–31). Stanford University Press.

Desmond, J. (1999). *Staging tourism: Bodies on display from Waikiki to Sea World.* University of Chicago Press.

Dimmock, K., & Wilson, E. (2011). Take a deep breath: How recreational scuba divers negotiate in-water constraints. *Leisure/Loisir, 35*(3), 283–297. doi:10.1080/14927713.2011.614839

Dimmock, K., & Wilson, E. (2009). Risking comfort? The impact of in-water constraints on recreational scuba diving. *Annals of Leisure Research, 12*(2), 173–194. doi:10.1080/11745398.2009.9686817

Dowler, L. (2001). The four-square laundry: Participant observation in a war zone. *Geographical Review*, *91*(1-2), 414–422. doi:10.2307/3250844

Dreyfus, H. L. (1993). Heidegger's critique of the Husserl/Searle account of intentionality. *Social Research*, *60*, 17–38.

Edensor, T., & Falconer, E. (2011). Sensuous geographies of tourism. *In J. Wilson (Ed.), The Routledge Handbook of Tourism Geographies*. New York: Routledge. 74–81.

Edensor, T. (2006). Sensing tourist spaces. In Claudio Minca and Tim Oakes (Eds.), *Travels in paradox: Remapping tourism* (pp. 23–45). Rowman & Littlefield Publishers.

Edensor, T. (2000a). Staging tourism: Tourists as performers. *Annals of Tourism Research*, *27*(2), 322–344. doi:10.1016/S0160-7383(99)00082-1

Edensor, T. (2000b). Walking in the British countryside: Reflexivity, embodied practices and ways to escape, *Body & Society*, 6(3-4), pp. 81–106.

Easto, P., Warburton, C. (2010). *Adventure tourism in Scotland: Market analysis report for the tourism innovation group*. Wild Scotland. Retrieved September 28, 2016, from http://wild-scotland.org.uk/wp-content/uploads/2010/12/TIG-report.pdf

England, K. (1994). Getting personal: reflexivity, positionality and feminist research. *The Professional Geographer*, *46*(1), 80–89. doi:10.1111/j.0033-0124.1994.00080.x

Evans, J. & Jones, P. (2011). The walking interview: Methodology, mobility and place. *Applied Geography*, 31 (2), 849–858. doi:10.1016/j.apgeog.2010.09.005

Farkic, J. (2020). Consuming dystopic places: What answers are we looking for? *Tourism Management Perspectives*, *33*, 100633. doi:10.1016/j.tmp.2019.100633

Farkic, J. (2018). Outdoor guiding as hospitality work. *Annals of Tourism Research*, *73*(C), 197–199. doi:10.1016/j.annals.2018.04.004

Farkić, J., & Taylor, S. (2019). Rethinking tourist wellbeing through the concept of slow adventure. *Sports*, *7*(8), 190. doi:10.3390/sports7080190

Farkic, J., Taylor, S., & Bellshaw, S. (2020). Slow adventure in peripheral areas. In O'Rourke, T. & Koscak, M. (Eds.), *Ethical & responsible tourism: managing sustainability in local tourism destinations*. Routledge.

Foucault, M. (1979). *The History of Sexuality Volume 1: An Introduction*, London, AllenLane.

Fox, K. M., Humberstone, B., & Dubnewick, M. (2014). Cycling into sensoria: Embodiment, leisure, and tourism. *Tourism Review International*, *18*(1), 71–85. doi:10.3727/154427214X13990420684563

Gyimóthy, S., & Mykletun, R. J. (2004). Play in adventure tourism: The case of Arctic trekking. *Annals of Tourism Research*, *31*(4), 855–878. doi:10.1016/j.annals.2004.03.005

Hall, S. (1990). Cultural identity and diaspora. In J. Rutherford (ed.), *Identity: Community, culture, difference*, (p. 2–27). Lawrence & Wishart.

Hay, P. (2015). Outdoors. Not, therefore, indoors. In Margaret Robertson, Ruth Lawrence and Gregory Heath (Eds.), *Experiencing the outdoors. Enhancing strategies for wellbeing*. Sense Publishers.

Heidegger, M. (2010 [1972]). *Being and time*. Suny Press.

Heidegger, M. (1993). Building, dwelling, thinking. In M. Heidegger (ed.), *Basic writings, revised and expanded edition* (pp. 344–363). Routledge.

Heimtun, B. (2016). Emotions and affects at work on northern lights tours. *Hospitality & Society*, *6*(3), 223–241. doi:10.1386/hosp.6.3.223_1

Hollinshead, K., & Ivanova, M. (2013). The multilogical imagination: Tourism studies and the imperative for postdisciplinary knowing. In *The Routledge* Handbook of Cultural Tourism (80–89). Routledge.

Houge Mackenzie, S., & Kerr, J. H. (2012). A (mis) guided adventure tourism experience: An autoethnographic analysis of mountaineering in Bolivia. *Journal of Sport & Tourism*, *17*(2), 125–144. doi:10.1080/14775085.2012.729901

Humberstone, B. (2015). Embodiment, nature and wellbeing: More than the senses? In *Experiencing the Outdoors* (pp. 61–72). Brill Sense.

Humberstone, B. (2013). Adventurous activities, embodiment and nature: Spiritual, sensual and sustainable? Embodying environmental justice. *Motriz: Revista de Educação Física, 19*(3), 565–571. doi:10.1590/S1980-65742013000300006

Ingold, T. (2008). Bindings against boundaries: entanglements of life in an open world. *Environment and Planning A: Economy and Space, 40*(8), 1796–1810. doi:10.1068/a40156

Ingold, T. (2005). Epilogue: Towards a politics of dwelling. *Conservation and Society, 3*(2), 501.

Ingold, T. (2000). *The perception of the environment: Essays on livelihood, dwelling and skill.* Psychology Press.

Ingold, T. (1995). Building, dwelling, living: how animals and people make themselves at home in the world. *In M. Strathern (ed.) Shifting Contexts: Transformations in Anthropological Knowledge*, London: Routledge, 57–80.

Johnston, L. (2001). (Other) bodies and tourism studies. *Annals of Tourism Research, 28*(1), 180–201. doi:10.1016/S0160-7383(00)00012-8

Kennedy, S., MacPhail, A., & Varley, P. J. (2018). Expedition (auto) ethnography: an adventurer-researcher's journey. *Journal of Adventure Education and Outdoor Learning, 19*(3), 187–201. doi:10.1080/14729679.2018.1451757

Kerr, J. H., & Houge Mackenzie, S. H. (2012). Multiple motives for participating in adventure sports. *Psychology of Sport and Exercise, 13*(5), 649–657. doi:10.1016/j.psychsport.2012.04.002

Kobayashi, A. (1994). Coloring the field: Gender, "race" and the politics of fieldwork. *The Professional Geographer, 46*(1), 73–80. doi:10.1111/j.0033-0124.1994.00073.x

Koning, J., & Ooi, C. S. (2013). Awkward encounters and ethnography, *Qualitative Research in Organizations and Management. Qualitative Research in Organizations and Management: An International Journal, 8*(1), 16–32. doi:10.1108/17465641311327496

Latour, B. (2005). *Reassembling the social: An introduction to actor-network-theory.* Oxford University Press.

Law, J. (2004). *After method: Mess in social science research.* Routledge.

Lee, J., & Ingold, T. (2006). Fieldwork on foot: Perceiving, routing, socializing. *Locating the Field: Space, Place and Context in Anthropology, 42*, 67–85.

Lugosi, P. (2008). Hospitality spaces, hospitable moments: Consumer encounters and affective experiences in commercial settings. *Journal of Foodservice, 19*(2), 139–149. doi:10.1111/j.1745-4506.2008.00092.x

Macnaghten, P., & Urry, J. (2000). Bodies in the woods. *Body & Society, 6*(3-4), 166–182. doi:10.1177/1357034X00006003009

Malinowski, B. (1922). *Argonauts of the Pacific.* Holt, Rinehart and Winston, New York, NY.

Marcus, G. E. (1998). *Ethnography through thick and thin.* Princeton University Press.

Markwell, K. (2001). 'An intimate rendezvous with nature'? Mediating the tourist-nature experience at three tourist sites in Borneo. *Tourist Studies, 1*(1), 39–57. doi:10.1177/146879760100100103

Merchant, S. (2011). Negotiating underwater space: the sensorium, the body and the practice of scuba-diving. *Tourist Studies, 11*(3), 215–234. doi:10.1177/1468797611432040

Moriarty, J. (2016). Autobiographical and researched experiences with academic writing: an analytical autoethnodrama. *Writing in Practice, 2*(1).

Nairn, K. (1999). Embodied fieldwork. *Journal of Geography, 98*(6), 272–282. doi:10.1080/00221349908978941

Pickering, M. (2008). *Research methods for cultural studies.* Edinburgh University Press.

Pink, S. (2015). *Doing sensory ethnography.* Sage.

Pink, S. (2010). The future of sensory anthropology/the anthropology of the senses. *Social Anthropology, 18*(3), 331–333. doi:10.1111/j.1469-8676.2010.00119_1.x

Pink, S., & Morgan, J. (2013). Short-term ethnography: Intense routes to knowing. *Symbolic Interaction, 36*(3), 351–361.

Pons, P. O. (2003). Being-on-holiday tourist dwelling, bodies and place. *Tourist Studies*, 3 (1), 47–66. doi:10.1177/1468797603040530

Punch, S. (2012). Hidden struggles of fieldwork: Exploring the role and use of field diaries. *Emotion, Space and Society, 5*(2), 86–93. doi:10.1016/j.emospa.2010.09.005

Rantala, O. (2011). An ethnographic approach to nature-based tourism. *Scandinavian Journal of Hospitality and Tourism, 11*(2), 150–165. doi:10.1080/15022250.2011.576829
Rantala, O., Valtonen, A., & Markuksela, V. (2011). Materializing tourist weather: ethnography on weather-wise wilderness guiding practices. *Journal of Material Culture, 16*(3), 285–300. doi:10.1177/1359183511413646
Scarles, C. (2009). Becoming tourist: Renegotiating the visual in the tourist experience. *Environment and Planning D: Society and Space, 27*(3), 465–488. doi:10.1068/d1707
Sharpe, E. K. (2005). Delivering communitas: Wilderness adventure and the making of community. *Journal of Leisure Research, 37*(3), 255–280. doi:10.1080/00222216.2005.11950053
Still, J. (2011). *Enlightenment hospitality: Cannibals, harems and adoption*. Voltaire foundation.
Stebbins, R. (1998). The ethnic outsider: The hurdles of Anglophone field research on North American Francophones. *Doing Ethnographic Research*, 65–75.
Swain, M. (2004). (Dis)embodied experience and power. In J. Phillimore & L. Goodson (Eds.), *Qualitative research in tourism: Ontologies, epistemologies, methodologies*. Routledge.
Thrift, N. (2008). *Non-representational theory: Space, politics, affect*. Routledge.
Thrift, N. (2004). Intensities of feeling: Towards a spatial politics of affect. *Geografiska Annaler, 86*(1), 57–78. doi:10.1111/j.0435-3684.2004.00154.x
Thrift, N. (1999). The place of complexity. *Theory, Culture & Society, 16*(3), 31–69. doi:10.1177/02632769922050610
Thrift, N. (1997). The still point: resistance, expressive embodiment and dance. In S. Pile & M. Keith (Eds.), *Geographies of resistance* (pp. 124–151). Routledge.
Trauer, B. (2006). Conceptualizing special interest tourism—frameworks for analysis. *Tourism Management, 27*(2), 183–200. doi:10.1016/j.tourman.2004.10.004
Varley, P., Farkic, J., & Carnicelli, S. (2018). Hospitality in wild places. *Hospitality & Society, 8*(2), 137–157. doi:10.1386/hosp.8.2.137_1
Varley, P., & Semple, T. (2015). Nordic slow adventure: Explorations in time and nature. *Scandinavian Journal of Hospitality and Tourism, 15*(1-2), 73–90. doi:10.1080/15022250.2015.1028142
Varley, P. J. (2011). Sea kayakers at the margins: the liminoid character of contemporary adventures. *Leisure Studies, 30*(1), 85–98. doi:10.1080/02614361003749801
Veijola, S., & Jokinen, E. (1994). The body in tourism. *Theory, Culture & Society, 11*(3), 125–151. doi:10.1177/026327694011003006
Walsh, K. (2009). Geographies of the heart in transnational spaces: love and the intimate lives of British migrants in Dubai. *Mobilities, 4*(3), 427–445. doi:10.1080/17450100903195656
Weiss, G., & Haber, H. F. (1999). *Perspectives on embodiment: The intersections of nature and culture*. Psychology Press.
Widdowfield, R. (2000). The place of emotions in academic research. *Area, 32*(2), 199–208. doi:10.1111/j.1475-4762.2000.tb00130.x
Wilson, E., & Hollinshead, K. (2015). Qualitative tourism research: Opportunities in the emergent soft sciences. *Annals of Tourism Research, 54*, 30–47.
Wolf, E. (1982). *Europe and the people without history*. University of California Press.
Wylie, J. (2005). A single day's walking: narrating self and landscape on the South West Coast Path. *Transactions of the Institute of British Geographers, 30*(2), 234–247. doi:10.1111/j.1475-5661.2005.00163.x

Leveraging digital and physical spaces to 'de-risk' and access Rio's favela communities

Nicola Cade, Sally Everett and Michael Duignan

ABSTRACT

Building relationships of trust and mutual understanding between researcher(s), local stakeholder(s) and gatekeeper(s) is widely regarded as a critical factor in successful research. Methodological strategies and tactics are often based on contextual variability and accessing some communities presents a harder and riskier proposition than others. Here we propose an empirically driven and holistic methodological approach for accessing high-risk communities whereby deprivation and criminality characterises everyday living. Following the 'legacy' of both the 2014 FIFA World Cup and Rio 2016 Olympics, this paper charts a research journey by detailing how local perspectives were accessed at the heart of one urban favela in Rio: *Morro dos Prazeres*. The methodological framework is underpinned by leveraging social networks to aid the identification of key stakeholders required to access and bridge the void between researcher(s) and community. Furthermore, creative digital and physical access routes were also employed (including social messaging platforms such as 'WhatsApp') which helped build and maintain trust with highly respected community leaders before, during and after the research. We suggest that our proposed 'Digi-cal model' (reflecting the digital and physical nexus) approach is transferable to similar tourism projects that require sensitive approaches and complex stakeholder navigation in 'high-risk' community settings.

摘要

在研究人员、当地利益相关者和守门人之间建立信任和相互理解的关系被广泛认为是成功研究的关键因素。选择研究方法的策略和方法通常基于研究情景的可变性，存在一个命题认为访问一些社区比其他社区更困难，风险更大。在这里，我们提出了一种经验驱动的、整体性的方法论途径用来访问高风险社区，这里贫困和犯罪是该社区日常生活的特征。继2014年国际足协世界杯和2016年里约热内卢奥运会的"遗产"之后，本文通过详细描述如何在里约热内卢的一个城市贫民窟Morro dos Prazeres（欢乐山）社区的核心地带接触当地人的视角，描绘了一段研究旅程。方法框架的基础是利用社会网络来帮助识别关键的利益相关者，从而在研究人员和社区之间建立联系和桥梁。此外，研究还采用了创造性的数字和实体空间访问路径(包括网络信使等社交消息平台)，这有助于在研究前、研究中和研究后与受人尊敬的社区领袖建立和保持信任。我们认为，我们提出的"数字-实体模式"(反映数字和实体空间的联系)方法可以应用于类似的旅游项目，这些项目需要在高风险社区环境中采用敏感的方法接触复杂的利益相关者。

Introduction

This paper presents a case study analysis conducted by the lead author (referred to as the 'researcher') and supported by two supervisors (the 'co-authors'). As this is a co-authored paper, the majority of the paper is written in the third-person, however, to reflect the deeply personal nature of the researcher's experiences, the methodological reflection section is written in the first-person alongside field notes and personal reflection. We track the researcher's journey from initial methodological planning to execution, outlining the digital and physical mechanisms leveraged to build relationships with what are often considered hard to reach and closed communities – in this case, Rio de Janeiro's favelas. Specifically, the researcher sought to access perspectives across the socially-excluded and marginalised favela of *Morro dos Prazeres – 'Hill of Pleasures'* (*Prazeres*). Through analysis of the researcher's fieldwork diary notes, and correspondence between stakeholders and gatekeepers, we explore some of the issues, challenges and access methods experienced by the researcher in the field.

The original aim of the investigation was to explore how favelas respond to economic, social and cultural impacts as a result of Rio's mega event and associated tourism through adopting social and community resilience systems and processes. An undertaking of this kind of research, particularly given the context under investigation, often involves engagement with potentially dangerous, risky, unpredictable and unfamiliar environments. Accessing high-risk research settings in any form poses a series of methodological challenges. Moreover, a critical aspect of existing research methodologies into how researcher–gatekeeper relationships are formed, developed and maintained are limited (Clark, 2011; Emmel, Hughes, Greenhalgh, & Sales, 2007), with little work offering useful practical strategies to overcome such challenges. Ateljevic, Pritchard, and Morgan (2007) argue that there is too much emphasis on identifying the issues and problems in critical tourism research without proposing multi-dimensional and disruptive solutions. Therefore, researchers wishing to voyage into challenging environments must consider strategies and tactics to 'de-risk' the research setting and develop solutions to achieve this.

Shefner and Gay (2002) found that gaining access into favela communities was challenging – taking over two months to access neighbourhood associations to build trust. Furthermore, other studies on Rio's favelas fail to explicate the processes in *how* they gained access into such dangerous and risky environments. For example, Alves and Evanson's (2011) study investigated favela residents' perceptions on the warfare between police and drug gangs. Although they provide a methodological 'note', there is no insight into *how* they did so, but instead briefly mention that the process required long and careful preparation. Notably, Cataldo's (2008) study of the effects from HIV treatment chose the *Morro dos Reis* favela in the *Santa Teresa* neighbourhood as their case study for easy accessibility and guidance from a local health centre team. Although perilous conditions impeded their progress, violence and gang activity evidently characterised their chosen community, highlighting that many researchers seek easier and fundamentally, safer options when accessing favelas.

Favelas, often derided as *'slums'*, are informal settlements spread throughout the city of Rio de Janeiro; a contrasting landscape to the well-known *'marvellous city'* of Brazil. Drug trafficking has plagued the city through two rival and extremely violent gangs – the *Red Command* (*Commando Vermelho*) and *The Third Command* – who have monopolised favelas since the late 1980s (Ramos, 2017; Steinbrink, 2013), bringing crime and violence across the city and dominating power within many of Rio's favelas due to a lack of state presence (Freire-Medeiros, Vilarouca, & Menezes, 2013). Perlman's (2010) study discovered almost one in five families had been a victim of homicide, finding that youth homicide is almost seven times higher within favelas than the rest of Rio. Violence further increased in Rio's favelas in 2017 between gangs and police with shootouts becoming a regular and unpredictable occurrence. In June 2017, 181 shootings took place in Rio in just one week, highlighting that extreme violence still plagues the city, reportedly exacerbated by its recent mega events (see Strobl, 2017). Since Rio 2016 Olympic Games, 632 people were hit by stray bullets across Rio state – 480 of these died and many were sadly young children (Strobl, 2017). Foreign Commonwealth Office (FCO, 2018) advise travellers that the security situation is unpredictable and visits to favelas can be dangerous, with specific advice to avoid such areas. This is evidenced in *Morro dos Prazeres* in late 2016, when two Italian tourists accidently strayed into the favela on their motorcycle, one was shot in the head and killed (The Rio Times, 2016).

Concern about safe and appropriate access to communities for research is not unique to favela research. Such approaches also characterise much tourism research, where marginalised, rural and hard to reach areas are often avoided or deemed inaccessible. For example, Nyaupane and Poudel's (2012) study into rural communities and tourism development was compromised by initial reluctant research participants who had previously experienced research where information was simply collected without building relationships. Therefore, they adopted 'rapport' through informal meetings with stakeholders to facilitate participant identification and selection, an often-overlooked approach, as an initial tactic to gain trust and developed an appreciative inquiry in their methodological approach. Accessing vulnerable, sensitive and marginalised populations is a recognised physical and emotional challenge for researchers (see Elliott, Watson, & Harries, 2002) – with many time-consuming and challenging barriers to overcome (e.g. identifying and accessing gatekeepers in high-risk places previously only afforded through 1-1 interaction and longitudinal study (before the development of real-time communicative platforms; e.g. McCann, 2014; Perlman, 2010). Yet, if we hope to gain a better understanding of the impact of tourism and mega-events, de-risking hard to reach and dangerous research settings is an increasingly valuable dimension of the research process if approached sensitively and ethically.

Given that research must approach tourism as a 'multidisciplinary phenomenon' (Decroup, 2004:163), we need to pursue interdisciplinary methodologies (Janesick, 1994), by tightly integrating different research methods that provide clarity and generate richer data for lost and hidden narratives. In looking beyond tourism studies to challenges in gaining access to research participants such as drug-users, the academy can build new knowledge and confidence in its methodological approaches. Therefore, this paper tackles the challenges of researcher access by drawing on multi-disciplinary research from disciplines including health studies and sociology, to identify issues associated with gaining access to high-risk participants and communities.

We outline how researchers can leverage social messaging communications as platforms to identify the key stakeholders required to access and bridge the void between researcher(s) and community. This is then overlaid with a series of creative digital and physical access approaches which include social messaging platforms (SMPs) – mechanisms helping to build and maintain trust with highly respected community leaders before, during and after data generation. The overarching objectives of this paper are to:

1. Make the case for *why* more research is required relating to how researchers can reach risky, often invisibilised host communities to amplify local narratives;
2. Deconstruct the practical digital and physical methods required to build networks, trust and relationships with key gatekeepers and locals within high-risk communities;
3. Serve as a platform for reflection of the researcher's journey, positionality, and personal safety;
4. Synthesise good research practice with empirical insights to develop general and specific principles and practices that disrupt methodological approaches to stakeholder and gatekeeper relationships.

The article is structured into four key sections aligned to our objectives. Firstly, we examine *why* building relationships of trust and mutual understanding is central to gaining access and important for impact studies that tend to exclude and/or marginalise local voices, especially in deprived communities. Secondly, we examine *how* other studies have leveraged digital and physical methods to gain access and de-risk research settings. Thirdly, we examine these questions of *why* and *how* in our context of our research site, *Prazeres*. This section synthesises (i) *a priori* propositions, (ii) *good practices* identified by other case analyses, and (iii) *a posteriori* themes generated from the researcher's empirical analysis, experiences in the field, and subsequent intellectual discussion(s) with co-authors. We present practical methodological considerations required to access high-risk communities whereby deprivation and criminality characterises everyday living in the context of a new model we have termed 'Digi-cal'. This model offers a practical approach to the digital and physical nexus of creative methods used throughout the research process for gaining access into marginalised communities. We conclude by reflecting on the wider generalisability of a more disruptive model for future researchers to utilise and mould to their own unique contexts and situations. We argue that the creative use of digital methods before entering the field helps develop strong relationships early in the research process. Therefore, this paper seeks to bridge the gap in addressing how such relationships can be fostered to enable quicker access into hard to reach areas, extending past the data generation stage of field research.

Literature review

Qualitative research methodologies within the tourism domain remain contested in their ability to provide dependable, credible and interpretive validity due to its exploratory nature (Wilson & Hollinshead, 2015). Johnson (1997) explicates that interpretive validity is the way in which the researcher portrays and accurately attaches meaning to the '... participants' viewpoints, thoughts, feelings, intentions, and experiences ...'. Harris, Wilson, and Ateljevic (2007) further claim that researchers must embrace more reflexive and critical forms of academic inquiry in the 'new' postmodern phase of tourism research (Tribe, 2005), seeking more in-depth and complex understandings of what lies beneath the data, e.g. the stories behind them. However, Tribe (2006) notes that the terms 'critical' and 'new' in tourism research could be (mis)construed labels that may cause limitations, often associated with Marxist theory when explaining social processes. Consequently, Ateljevic et al. (2007:3) clarify that '... it [critical tourism] is more than simply a *way of knowing*, an ontology, it is a *way of being*, a commitment to tourism enquiry which is pro-social justice and equality and anti-oppression: it is an academy of hope'. Extending this, we explicate that '*a way of being*' is how the researcher positions themselves within the process and how that influences the outcome, e.g. 'inquiry from the inside' (see Ospina, 2004:4).

Gaining access: building relationships and establishing trust

Low and Everett (2014) found that access issues are not often explicated for new researchers and are instead presented as strategic issues rather than challenges. Emmel et al. (2007) found researchers apply strategic ways of accessing hard to reach communities but concluded that little research existed in understanding implications of building trust and maintaining relationships between researchers and their participants. Furthermore, they argued that attributes of credibility and trust can be built by the researcher's commitment to understand (and immerse themselves) into the research setting. They base this idea on Kuebler and Hausser (1997), Elliott et al. (2002), and Sixsmith, Boneham, and Golding (2003), who fully immersed themselves within health-based research environments, employing multi-method ethnographic techniques as a strategy of discernibility to build credibility and rapport between them and the subjects of study that they claim would otherwise be invisible and problematic to access. Elliott et al. (2002) employed 'peer interviewers' (established mediators between the drug-users and community drug team), or 'Privileged Access Interviewers' (PAI) (Kuebler & Hausser, 1997), to access their participants (drug-users) that they would have otherwise found difficult to access such 'hidden population' groups. Similarly, Sixsmith et al. (2003) employed multi-ethnographic methods that helped build credibility and maintain trust between them and their research participants within a socially-deprived community. The multi-participatory techniques included activities, walking on site and appearing visible to residents via conversations. Sixsmith et al. (2003) highlighted that access to their participants was time-consuming, with it taking up to twenty-four months to build relationships.

Much research has suggested that the first principle with any social research is the development of trust (e.g. Denzin & Lincoln, 2005). Harrison-McKnight and Chervany

(1996) argue that although trust is an important element within social science research, there is little agreement of what 'trust' means despite extensive studies on the topic (Rousseau, Sitkin, Burt, & Camerer, 1998; Schoorman, Mayer, & Davis, 2007). However, a clear definition of this elusive concept of 'trust' is missing, with many empirical studies defining it in multitudinous ways across multi-disciplines (Harrison-McKnight & Chervany, 1996; Kerasidou, 2017). Yet, without it, we face barriers to research such as obtaining consent from participants and without this, research becomes difficult, if not impossible. Denzin and Lincoln (2005) argue that building a relationship of trust and mutual understanding between the researcher(s) and gate-keeper(s) is a critical success factor for achieving research aims. However, trust may not be mutual or reciprocal, it can only be developed over time (Schoorman et al., 2007). Various scholars identify 'credibility' – alongside an empathetic disposition – as equally central for building strong, reciprocal relationships of trust between the researcher and those within the research setting to gain access (Sixsmith et al., 2003). Yet, building trust and applying such attributes are highly dynamic: situational and contextually dependent, requiring differing strategies to access those residing within unique social and economic environments.

According to Kerasidou (2017), trust can take many forms and is situational and contextual. Kerasidou (2017) explains that there are two different forms of trust: (i) personal trust which is developed between two people and (ii) institutional/a personal/impersonal trust which lies between an individual and institutions; or between two institutions, companies, governments or other large social systems. However, institutional/personal trust is questioned by some, as trust is regarded as a *relationship* that is developed between *two* individual agents over time (Mayer, Davis, & Schoorman, 1995; Rousseau et al., 1998). Schoorman et al. (2007) argue that trust is an aspect of *relationships* rather than a characteristic and can vary with individuals and relationships. Additionally, they argue that reciprocal linkages in empirical research are rarely referred to in how one's trust may affect another's in return, thereby hoping that future research addresses this area. Accessing any collective unit, from a residential community right through to a business enterprise, requires a level of trust. However, the issue remains in *how* trust is gained.

Trustworthiness: writing oneself into the research

Decroup (2004) claims that the issue of trustworthiness needs to be better considered within qualitative tourism research. Trustworthiness can make tourism studies more rigorous and acceptable to positivist researchers (Decroup, 2004). Schnackenberg and Tomlinson (2016) link trust to the concept of 'transparency' which is presented as a three-dimensional (information disclosure; clarity; accuracy) precursor to trustworthiness, in increasing stakeholder confidence. Transparent approaches in the research process allows the researcher to subjectively report their individual experience as an 'insider' (Finlay, 2002:210). The reflection of 'self' during fieldwork stages of research have increased considerably. For example, Everett (2010) extends the concept of reflexivity into the research field, whereby one must transcribe themselves into their work to ensure the researchers' '*physicality*', '*performativity*' and '*positionality*' can be enunciated. Reflexivity is the relationship between the researcher and research participants and can occur in all forms of research to some degree (Davies, 2007), the deeper the researcher is involved, the more relevant reflexivity becomes (Veal, 2017). However, we argue that reflexivity should also consider the relationship between the researcher and gatekeeper(s)/key informant(s) and *not* solely research participants, as they [gatekeeper/informant] are the *actors* at the *core* of access, consequently bridging the gap between researchers and participants. Additionally, Everett (2010) asserts that publishing one's own research experiences from the field is an important element of our own epistemological journeys. These aspects are central in the process of knowledge construction, yet are so often concealed (Feighery, 2006; Spivak, 1988).

Gatekeepers: facilitating access to socially excluded groups

Emmel et al. (2007) claim that gatekeepers play a significant role in facilitating access to socially excluded groups. Their work highlights the range of relationships that exist between gatekeepers and participants which may be either trusting or distrusting depending on whether the gatekeeper is formal, comprehensive or informal. Emmel et al. (2007) propose that formal, comprehensive and informal gatekeepers exist when attempting to access socially excluded groups. Focusing on formal gatekeepers who Emmel et al. (2007:5) characterise as 'those working with the community to achieve a particular end'. Although these authors discuss the relationship between the researcher and gatekeeper, they acknowledge that they do not address *how* this is accomplished and retained. This gap in methodological knowledge is further substantiated by Clark (2011) who asserts that little research exists into researcher–gatekeeper relationships.

It is evident from the available literature that trust is key in gaining access, but the issue remains in *how* trust is built – the academy is lacking in empirical research. Additionally, researcher–gatekeeper relationships are highlighted as playing a significant role in accessing any group (especially socially excluded and marginalised groups), yet there is little evidence on approaches to building such relationships and nurturing reciprocity of trust within them.

Accessing unpredictable and chaotic communities

Accessing chaotic communities like favelas can be risky for researchers. A notable investigation into accessing favelas is Perlman's (2010) reflective follow-up study to her first study of Favelas in the 1960s. She realised that violence and drug gangs had become a barrier in accessing her intended participants (a limitation not faced before) and acknowledged it to be the greatest challenge facing her and her research team. However, Perlman overcame this challenge through careful negotiated access with the Residents' Associations (mostly run by drug gang leaders) and by raising the visibility of the team with an individual 'kit' consisting of an identification lanyard, T-shirts, and a letter explaining the research to make them visible to residents (Perlman, 2010). Additionally, Perlman (2010) 'de-risked' the research setting by contacting participants on the day of interview to confirm the area was safe for the research team to enter.

Perlman (2010) recommended that future studies within favela communities should be conducted in pairs with constant contact between them and the field supervisors to ensure researcher safety (obviously made easier now with mobile phones). Although, mobile phones may be a useful tool for safety, this present study also suggests they provide an effective methodological tool in accessing hard to reach communities.

Leveraging digital and physical methods for access

Traditional approaches to generating qualitative data often involve lengthy *physical access* processes of face-to-face, telephone conversations, or postal correspondence (Saunders, Lewis, & Thornhill, 2016). Furthermore, building rapport and obtaining mutual trust and respect between the researcher and participants is also time-consuming, particularly through these rather dated communicative methods (DiCicco-Bloom & Crabtree, 2006). Prior to the internet the world appeared seemingly widespread and socially divided, subsequently intensifying distances between researchers and their intended participants. However, social messaging platforms (SMPs) such as Facebook Messenger, WhatsApp, and direct message functions on Instagram and Twitter, have increased and improved interactions – reducing the time-space distance between people from different geographical areas, permitting digital access into the lives and spaces of others (Côté, 2013; Liu, Ainsworth, & Baumeister, 2016).

Given technological changes, gaining access to data can now be provided through internet-mediated access involving the use of different online technologies (e.g. Web, Email, instant messaging, (Saunders et al., 2016). Seemingly, the concept of SMPs as methodological tools is rarely explored for leveraging access into research settings (Côté, 2013) and initiating contact between researchers and gatekeeper(s)/key informant(s) and/or participants. Much of the literature surrounding qualitative research methodologies discusses SMPs as useful mechanisms to facilitate relationships for gathering data, e.g. online surveys (e.g. Henderson, 2011; Kapoor et al., 2018). Furthermore, Côté (2013) highlights that such platforms may help to build and maintain trust with participants. However, the use of online instant messaging services such as WhatsApp, have rarely been adopted and utilised within any methodological frameworks for obtaining access into research groups.

Adopting social messaging platforms (SMPs) to negotiate access

The adoption of SMPs as useful tools in accessing participants and data within qualitative (tourism) research studies is under-explored despite them offering an opportunity to reduce time and accessibility constraints that may not have been an option in earlier longitudinal studies (e.g. Kuebler & Hausser, 1997; Perlman, 2010). However, Saunders et al. (2016) suggest that physical access is still necessary before using internet-mediated access routes for negotiating access. This approach is supported by Henderson (2011), who claims that social media platforms can be useful as a primary tool to advance beyond traditional qualitative research methods; helping to leverage [physical] access into hard to reach places. However, Henderson (2011) argues that such platforms may create issues for qualitative research when recruiting participants, by being less rigorous like traditional forms, such as 'screening' suitable participants, potentially running the risk of [SMPs] 'users' not being genuine.

Maramwidze-Merrison's (2016) study promoted the use of social media platforms (mainly Facebook and Instagram) as an innovative methodological strategy for accessing elite organisational participants, considered difficult to access via traditional routes. Social media platforms have been found to enhance a researcher's ability to access participants for both data generation and in building a foundation of mutual trust before entering the field. However, Côté (2013) emphasises that this form of approach carries risk and may inadvertently expose participants and thus breach pre-approved ethical assessments. We recognise that digital approaches may pose a risk in terms of compromising participant anonymity and data confidentiality, requiring rigorous ethical considerations when adopting such methods. Furthermore, Maramwidze-Merrison (2016) delineated that adopting social media as a methodological strategy remains relatively unrecognised within traditional conformed methods of gaining access such as physical contact; already established relationships; or physical intermediaries (those who know the intended participants) and so should attract more reflective discussion within studies that choose to adopt it.

Methodological approach: the case of Prazeres

In terms of context for this paper, the researcher's main overarching project is an investigation of how favela residents respond to socio-cultural 'shocks' from mega events and associated tourism through adopting social and community resilience systems and processes. The researcher sought to provide *voice* and *narrative* for socially excluded and marginalised groups, rarely considered within the context of mega event research. Field research was conducted across two weeks in the favela of *Morro dos Prazeres* (see Figure 1) in the touristic neighbourhood of *Santa Teresa* in Rio and adopted multi-method techniques for data generation. Although *Prazeres* is located within a rich and popular touristic neighbourhood, it is still *excluded* from the rest of Rio's society. The police vehicle that sits parked at the entrance to Prazeres warns off any outsiders – making the statement that the favela is not safe to enter. Favelas are chaotic, unpredictable, diverse and culturally rich communities, with many differing in terms of their socio-economic status (Freire-Medeiros et al., 2013). The socio-economic demographics of *Prazeres* is different to other favelas such as *Maré* (north zone) and *Vidigal* (west zone) as many can access the city to work or sell self-made products (e.g. yoghurt drinks). South zone favelas have suffered less gentrification to other favelas (World Bank, 2012). Additionally, south zone favelas can access central Rio easier than other peripheral favelas, and although *Prazeres* may appear less marginalised, it still suffers socio-economic exclusion and stigmatism, lacking integration with the rest of society (ISSC, IDS and UNESCO, 2016), as well as domination from the *Red Command* gang and unpredictable violence/shootings.

In the early stages of research, purposive-sampling was used to establish contact with Theresa Williamson (permission given to be named) who recommended Prazeres as a fascinating case study. Theresa is the founder of Catalytic Communities (http://catcomm.org) – a non-governmental organisation working for Rio's favelas, and Rio On

Figure 1. Location of Morro dos Prazeres, Rio.

(Olympic News) Watch (http://www.rioonwatch.org) – an online news blog that was established to report the Olympic impacts on Rio's favelas; a community reporting platform for favela communities. Theresa explained: '[...] if there's an opportunity to do some research [on Rio's mega event host community impacts], I highly recommend [*Prazeres*], it's a really fascinating case where a lot of them were building [...] opportunities around tourism and in an instant it's over.' *Prazeres* was impacted by Rio's recent mega-events, from UPP (Unidade de Polícia Pacificadora – Police Pacification Unit) interventions – a strategy for partnership between local-residents and law enforcement (UPP, n.d.) aimed to 'create conditions for social, economic, and political integration of favelas into the city' (World Bank, 2012:22) – and drug gang/police conflicts that subsequently destroyed many self-made community projects (e.g. *Caminho do Grafite*) and successful tourism initiatives (e.g. *Prazeres Tour (n.d.) – see* https://prazerestur.weebly.com) – destroying the *visibility* they were trying to self-create in a fight against marginalisation.

The qualitative methods employed included twenty-eight semi-structured face-to-face interviews, and unstructured conversational interviews. Research participants included *Prazeres* leaders/activists; rap singer/songwriters (within *Prazeres*); food/drink/creative entrepreneurs and restaurant/café owners (within *Prazeres*); a hostel manager (next-door to *Prazeres*); a local photographer; and UPP commander of *Prazeres*. The informants were based on a purposive-sample identified by a key gatekeeper, as who would best support the research – it was noted that some residents would not understand the effects of mega events and would associate them with current issues, therefore, unlikely to provide answers to interview questions. Furthermore, they were mainly residents of *Prazeres* who live among daily uncontrollable conflict, potential danger and challenges of social exclusion. Other data generation methods included walking methodologies; observations from daily and touristic activities; photographic imagery; and daily field notes/voice recordings (sections are quoted below to illustrate key research moments).

As identified earlier, researchers have leveraged a range of creative tools, techniques and methods to access high-risk environments (LeBaron, 2015). Favelas are difficult to access social [chaotic] organisations, requiring researchers seeking to undertake empirical and ethnographic methods to potentially risk personal safety (Perlman, 2010). Favela communities can appear on the surface as pleasant and safe, but we argue there are strong hidden contrasts to this. For example, warfare between police and drug gangs often results in innocent local people and tourists being killed and caught up in crossfire or shot for accidently entering a favela (see Phillips, 2017; Ramos, 2017). *Prazeres* is no exception. A daily life occurrence for many favelas is that anyone can find themselves in the middle of such a *'war'*. As the researcher's initial focus was based upon Rio's marginalised favelas and its recent mega-events, *Prazeres* seemed like a worthy case study. However, there were issues to consider: *How* would the researcher access such a community due to language barriers and safety implications? *How* would the researcher remain safe as a solo-female researcher?

Leveraging digital and physical spaces in Prazeres to access local voices

In this section, we present the digital and physical methods leveraged by the researcher to secure access to local voices and simultaneously de-risk research spaces within the heart of *Prazeres*. The use of gatekeepers and digital methods helped leverage physical access for interviews to take place, allowing the researcher to access individuals more quickly and safely. A central element of this paper is to present some findings on an approach and methodology that proved highly effective and outline personal access techniques that helped cultivate what we have termed a *'digi-cal'* model of access.

At an elementary level, a researcher must identify specific actors existing 'within' a collective unit or 'connected into' an organisational construct (like a community, or a business). In a tourism research context, the main actors are usually regarded as either: (i) the *'gatekeeper'*, and/or (ii) the *'key informant'*. The gatekeeper, as the name suggests is often situated between the researcher and the researched (Ren, Pritchard, & Morgan, 2010) and serves to connect the researcher into a new social group: a facilitative role. The key informant is often (but not necessarily) a gatekeeper and retains a specific role in empirical research: as someone who has current – or – potential access to knowledge which is required of those invoking a purposive sampling technique. Central to purposive sampling is the strategic targeting of individuals or collective groups with access to knowledge that helps the researcher (in part) fulfil their research aim(s). Recruitment of such an individual is usually expected at the 'ethics approval' stage of research: requiring the researcher to engage in strategic thinking (especially in high-risk environments) and ensure the gatekeeper(s) are appropriate.

The provision of additional facilitators was vital in terms of language, and project leads. As found in Browne (2005) in her adoption of snowball sampling to recruit 'hidden' stories, it is crucial the researcher is able to access participants' social networks to access specific populations. We applied a similar approach and Theresa

If you could provide me with some links and contacts that you think can help me with what we were talking about e.g. FUNAI (I think I contacted them and they stated they wont help individual researchers (Survival rings a bell?)) and the Prazeres favela community, anyone who can help here.

Do you think that the [redacted] would be a good focus too? Or perhaps use them as a case study for a paper...we could perhaps work on together? I can always collect extra data during my stay - although this will likely be a maximum of two weeks. Just some ideas!

It was great talking to you and I hope that we can meet up when I come to Rio! I shall keep you informed of this anyway.

Please keep in touch. Thank you again.

Kind regards

Nikki

Dear Nikki,

Great talking today and looking forward to seeing your research expand!

I think Prazeres and Museu do Índio would both be very interesting cases, for different reasons.

Contacts there:

- Prazeres - community photographer Saulo Nicolai who will connect you with Prazeres Tur: [redacted] WhatsApp [redacted] email [redacted] I'd start with him!
- Museu do Índio: [redacted] and [redacted] I believe both speak English. Here's our latest on them: http://www.rioonwatch.org/?p=35133

Good luck and let me know if you don't get far with these folks!
Theresa

Figure 2. An example of the email exchange between the researcher and Theresa Williamson.

offered this access (as the exchanges in Figure 2 illustrate). She recommended that '... we've got people [in *Prazeres*] we can introduce you to who were affected by the tourism "boom" and "bust" ... ', subsequently providing an English-speaking contact, a self-made photographer from *Prazeres* and co-director of *CriaAtivo Film School*, who then provided a contact number – it was the number of a man called Charles.

Securing a key informant or gatekeeper usually marks the beginning of a long and enduring journey between the researcher and subjects of study. In this case, the researcher invoked a purposive-come-snowballing sampling technique (Browne, 2005), starting with one highly influential gatekeeper. In this case, this was a man called Charles Siqueira (permission to be named obtained). Charles, originally a professional dancer, became a social entrepreneur and community leader/activist after moving to Prazeres in 2002. He is also a director of the *CriaAtivo Film School* which addresses social exclusion and helps prevent young people entering the violent underworld of drugs, through instilling self-respect and running grassroots programmes which help create better opportunities for those living in Rio's peripheries. Charles, colloquially described as the 'mayor' of the favela by many, provided not only in-depth insight into the community and its dynamics, but fundamentally served to 'physically' connect the researcher to participants by helping the researcher establish 'virtual' empathy and build trust with them prior to field research.

The reliance upon one key gatekeeper can sometimes be problematic (Clark, 2011), but the in-depth insights and connections that Charles was able to provide meant that the *quality* was more crucial rather than the *quantity*. Charles' well-known reputation in Rio allowed for connections to be made quickly and effectively, that may have otherwise been hindered if using additional gatekeepers. The continual communication between Charles and the researcher, pre- and post-research, meant that the relationship built between them was incredibly strong – a key preparatory tool in accessing less accessible places as identified earlier in the literature (e.g. Alves & Evanson, 2011).

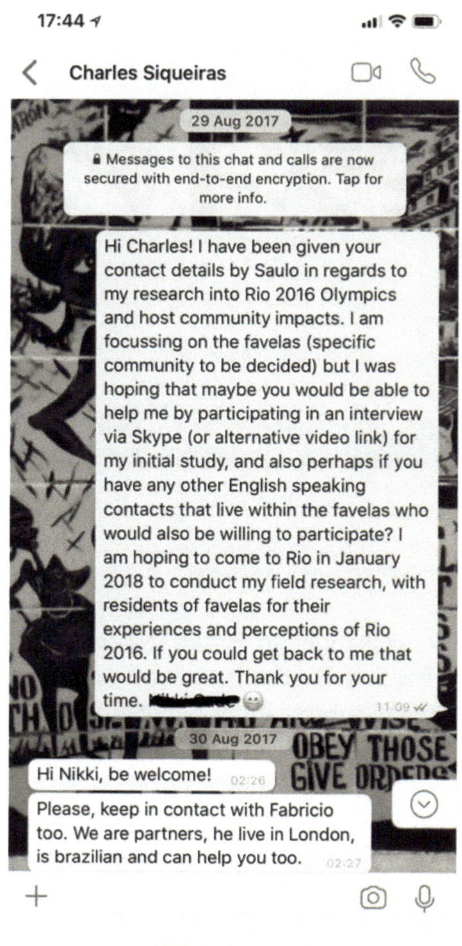

Figure 3. Initial WhatsApp messages and Internet-mediated access to Charles.

Charles opened access and de-risked the research setting by providing a plethora of participants which included Prazeres locals, UPP commander, Brazil's British Council, social legacy project *CriaAtivo* Film School students and staff, and other community leaders e.g. Morro da Providência favela. The initial contact was through WhatsApp (https://www.whatsapp.com – a key digital tool). This digital communication was maintained over a six-month period which helped to nurture trust and develop a sound relationship between the researcher and gatekeeper. This digital engagement is certainly a transferable method that researchers should look to adopt if they wish to access hard to reach communities. An example of how this online method was utilised via WhatsApp to build an initial relationship and generate a network is presented in Figure 3 (screen shot of the WhatsApp messages).

Access to hard to reach groups may be simplified and accelerated by adopting an 'Internet-mediated access' approach via SMPs to access gatekeepers and research participants. An additional benefit is that this access gateway is free, but only *if* people have access to the internet. Access to the internet may be a limitation for favelas as they are less likely to have access to the internet as they often lack basic resources, especially those in more marginalised, peripheral favelas. Fortunately, this was not the case in Prazeres where communal internet access is available. Additionally, owning mobile phones is common across favelas, Fabricio (director of *CriaAtivo Film School*) stated that '[...] everyone has a smart phone, the poorest guy has a smart phone, and OK it might not be the best iPhone, but it's a smart phone that takes pictures, does WhatsApp, [he] watches it [...]'. Essentially emphasising that marginalisation is not always a complete isolation or restriction to technology. By utilising digital platforms, researchers cannot only practically develop early relationships, but generate deeper understandings of local dynamics, challenges, and inter-dependencies. Fundamentally bypassing boundaries and providing ways in which we can access less accessible areas more quickly. Of course, some settings represent a harder and riskier proposition than others, complete with cultural contexts that may greatly differ with the everyday practices and value systems of the researcher. In other words, environments like favelas, represent unchartered and unfamiliar territory that must be navigated otherwise the success of the research and the researcher's safety may be placed at risk.

It was felt that digital platforms offered immediate, personal and convenient vehicles to build relationships. Social networks allow for relationships to be leveraged more quickly. Contact with Charles was established via WhatsApp, who subsequently provided the link to Fabricio who helped finalise plans between the researcher and Charles via email and WhatsApp (Figures 4 and 5). Over five- to six-months, in the lead up to the fieldwork stage (January 2018), the research was discussed and planned to ensure that the fieldwork would be successful. Digital approaches were essential in the set up, but of course, it had to sit with the in-person engagement and interaction.

At this point, we must also reflect on the positionality of the researcher: a young, female PhD researcher who is monolingual with limited understanding of Portuguese language, wider Brazilian and local favela customs. Sitting in a bland-scape of a beige doctoral study room: the very thought of entering what is a very different, dangerous world was both daunting and unnerving. Furthermore, significant socio-epistemological differences (Freire-Medeiros, 2012) between the researcher and the residents of *Prazeres* existed in terms of the researcher's different social class coming from the UK, and significant cultural, lifestyle, and language differences. Yet, the desire to engage in critical scholarship to amplify a 'lost' – and often invisibilised – narrative remained powerful. Early engagement of local interest in her work through the digital communications platforms was therefore reassuring and set the study up well before physically entering the field. It is important to note, the monoglot anglophone nature of the researcher may have hindered some local narrative and varieties of the native language in translation, however, steps were taken to ensure that minimal context was lost. An English-speaking Brazilian local was recommended by the gatekeepers (a close friend) to help with the interviews and translate verbatim, subsequently reducing the language barrier between the researcher and participants, additionally providing interview notes and explaining any areas that seemed confusing to the researcher. This paper now adopts the first person to emphasise the personal journey taken by the researcher.

Figure 4. The snowball effect of WhatsApp: Charles digitally linking the contact details of his partner to the messages.

Physical access: making the researcher and researched visible

It was vital that I [researcher] was not seen as the '*enemy*' by those who may not welcome unwanted external attention, especially from someone who could have been mistaken as an undercover police officer or journalist, essentially risking my own personal safety. I found myself immersed into an unknown world regardless of being in the safe hands of my gatekeeper. The research could have gone one way or another. This anxiety is evidenced in my field notes from the first day in *Prazeres* (Sunday 14 January 2017):

> '[T]wo young men, no older than sixteen, passed us by, but strangely Charles did not acknowledge them fully, other than a nod in their direction. I found this odd as he had literally thrown his arms around everyone else we had come across on our guided tour until now. I quickly realised that they were gang members, marked by their reasonably large handguns that they had tucked into the waistband of their trousers. Remarkably, I felt no fear or concern that we were in the passing company of these individuals and I

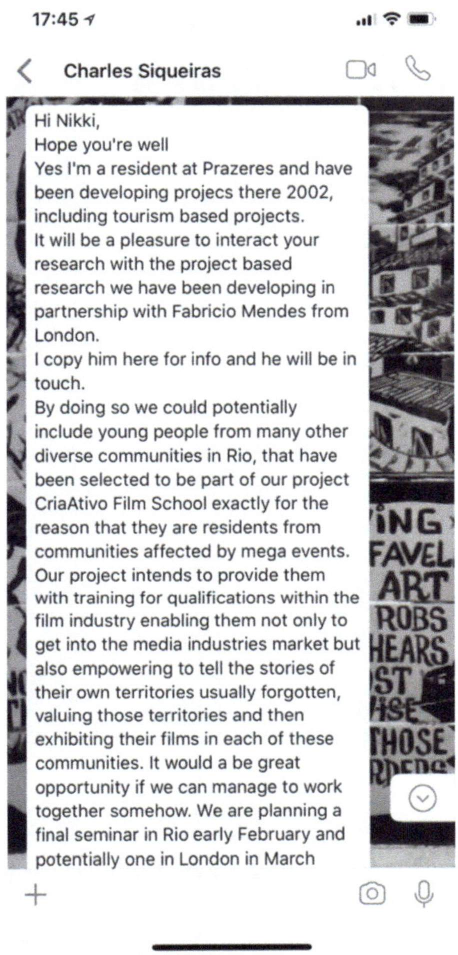

Figure 5. Leveraging digital methods to prepare the field work research in Prazeres: WhatsApp from researcher.

> smiled to acknowledge them with one nodding back in acceptance but his eyes not leaving us until we were out of their sight. I thought to myself, they are so young, something that I found more surprising than fearing them. Anyway, what were they going to do? Shoot me I guess.'

Consequently, the key gatekeeper played a key role in negotiating spaces, and personalities. He offered a form of 'protection' and reassurance from the outset (Figure 6). Charles facilitated access to many parts of *Prazeres* that would otherwise be inaccessible to people from the 'outside'. I engaged with many people sitting outside their homes (even invited to look inside), interacted with gang members, and got a *real* feel for favela life – engagement was not hindered by language barriers – I could see everyone who lived within *Prazeres* was happy and extremely welcoming – no matter who they were.

Charles managed, recruited and prepped suitable research participants on my pre- and post-arrival in Rio. Essentially, direct contact with the research participants was

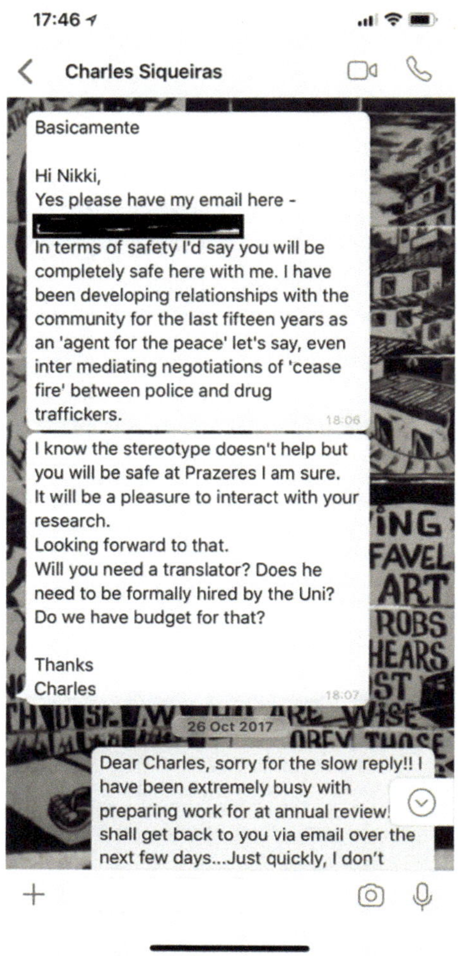

Figure 6. Reassurance of safety from Charles to the researcher.

not made until I physically arrived in *Prazeres*, and sometimes not until the point of interview. I relied on the gatekeeper to act as a mediator and bridge the trust between myself and participants before conducting interviews as I had no means of contacting them beforehand due to language barriers. I adopted Schnackenberg and Tomlinson's (2016) three-dimensional approach (outlined above) to foster transparency and presented each participant with a Portuguese translated (*clarity*) document that consisted of a Participant Information Form (PIF) (*disclosure of information*) and a Participant Enquiry Sheet (PES – *accuracy*). Both forms clearly provided the purpose of the research, with a consent form, as well as the researchers, supervisor and the University's contact details, for them to agree and sign that they were willing to participate in the study.

As trust was built prior to my arrival in Rio through Charles, I was slightly concerned that the interviews were planned so late into the field research, however, Charles was constantly reassuring me that everything was ready for my arrival. Charles and I planned back-to-back interview and focus group schedules for the coming week on the first working day of research. Charles provided names and occupations of potential participants and their association/position within *Prazeres* that he felt would provide the richest and most useful data. The increased uncertainty of recruiting willing participants who can help generate enough insightful data for a doctoral thesis is a major challenge in locating and accessing research subjects that you believe can provide insights to help address research questions (Low & Everett, 2014).

Indeed, participants can make or break, and even alter the entire direction of research (which can be illuminating, but also unnerving for novice researchers). Ritchie, Burns, and Palmer (2005) rightly point out that research is rarely linear and does not always go to plan, no matter how well designed. I had to be prepared for whatever challenges may have arisen in such an unpredictable setting. For example, on day five of the interviews, the last interview of the day was cancelled, I recall (Thursday 18 January 2018):

> 'The last interview with [participant] was cancelled as it was getting late and the general agreement was to leave, as we would not be with a representative when we left the restaurant in *Prazeres*, which is situated at the top of the favela – and it was starting to get dark, so we had to ensure our safety.'

However, this was not a problem, the interview was re-scheduled to the following week. Although the schedule spilled over into the second week, initially intended for touristic observations in Rio, I felt this time was better spent interviewing participants to give voice and narrative for the favela locals – the intended aim of the research. Both myself and Charles were flexible and able to respond to participant needs rather than follow a structured schedule – illustrating the unpredictable research setting increasingly reflected in ethnographic research (Bajc, 2012). We were able to adopt this approach to the planning of interviews as we could stay in contact with each other via WhatsApp remaining flexible and adaptable to change – with accessible transport links to navigate the city easily.

Additionally, I was welcomed and greeted by most residents and perhaps more significantly, the drug gangs and the UPP. By day six, I diarised:

> 'Day in Prazeres – I am starting to feel like a resident myself! Most of my time has been spent here [*Prazeres*], or, if out of the favela, I am with the residents (mainly Charles), the only thing I am yet to do is sleep there!'

The UPP mediator offered his support to me during my time within *Prazeres*. I felt a strong 'sense of belonging' and 'safety', with one of the locals commenting that I was 'like a local'. Similarly, I was warmly greeted by one of the gang leaders in *Prazeres* and told '*to be welcome*' and '*feel safe*' – which came as a relief! The earlier 'investment' of time and (digital) social engagement through the gatekeeper had clearly paid off. Having such a proactive gatekeeper seems to contradict many traditional forms of research where the researcher is mainly in control of their research. By placing the 'power' with the locals in this instance suggests a form of 'risk-taking' from myself in 'trusting' the gatekeeper to deliver what was effectively 'promised' from over 5000 miles away via online digital communications. Conversely, it is intriguing how one man's [Charles] trust and respect for everyone, no matter who they are, what they do, or where they are from, *infused* into my approach. His positionality secured 'permission' for the me to access *Prazeres* and then entrust me with one of Rio's many dangerous favelas.

Charles encapsulates the kind of approaches tourism researchers need to be cognisant of, i.e. the time and immersion required to build trust in our field work sites. Charles placed me in a safety bubble by raising visual awareness to other people in the favela of my existence and research intentions. The sustainability and depth of this link was perhaps best demonstrated when Charles sent a Facebook friend request before my arrival in Rio which, along with WhatsApp, have remained communicative vehicles between us since. To emphasise the relationship we built, when asking permission to refer to him as 'Mr Siqueira', he responded with *'Even better you call me Charles. You are my friend, I'm not Mr Siqueira for you.'* Essentially, the significance of our relationship is more than just one of access (Crowhurst, 2013), it was a privileged insight into unique and hidden realities that may never have been exposed without the digital formation of mutual respect and reciprocal trust and the beginning of a lifelong cross-cultural friendship.

Discussion and reflections: building relationships across a digital and physical nexus

Punch (1994) argued that to benefit from the richness of empirical work, a researcher must find ways to fully immerse oneself within the research setting to build trust. As outlined earlier, much literature theorises the trust relationship between the researcher and gatekeeper from the perspective of gaining trust from the gatekeeper(s) and/or key informant(s) and it is rarely presented as a reciprocal relationship. We argue here that the tourism researchers' approach in the favelas highlights that trust must be reciprocal between both the researcher and the gatekeeper(s)/key informant(s); and can (and should aim to) extend beyond the fieldwork stage of research (Jobbins, 2003; O'Gorman, Maclaren, & Bryce, 2014).

It was noted above that Shefner and Gay (2002) attempted to access favelas but found it incredibly challenging – they had to participate in events to 'persuade' the neighbourhood associations to permit access. Trust clearly had to be nurtured over some time through showing willingness and commitment before they could fully immerse themselves in their research setting. Similarly, and with already [digitally] agreed access, our researcher was taken on an introductory tour of the favela prior to conducting interviews and meeting the participants. The researcher was introduced by Charles to many residents along the way, being made *visible*, in what we consider to be a *safety tactic* that was used, but not initially disclosed to the researcher. This was an indirect way to identify her within the community, so she was known to the locals – and importantly, drug gangs.

Furthermore, transparency was crucial in gaining acceptance within Prazeres (Lincoln & Guba, 1985). According to Punch (1994) in the first instance, the researcher must gain social acceptance from the group they wish to research, referred to as 'social access' by Clark (2011). Clark argues that gatekeepers allow us physical access, but this does not always translate into social access. However, by Charles being able to introduce the researcher as more of a long-term *friend* than an *external alien* meant social engagement was accepted more easily. This certainly echoes Emmel et al. (2007) who found that a researcher's external 'position' to their participants is important in devising suitable methods to facilitate trust between them. The social separation that exists prior to building trust can be characterised as *distrust* or mistrust.

We found that 'distrust' is often dependent upon different circumstances. For example, during the fieldwork, most people were very accepting and trusting, allowing the researcher to eat in their café and not pay for the next three to four days, after meeting them for the first-time. Perhaps this was down to Charles' already existing relationships between them that emitted *that* trust to the researcher. Social networks (e.g. personal contacts) can be useful channels for gaining access to individuals in hard to reach communities and can 'open-up' 'hidden populations' (Browne, 2005:53). Everett's (2010) 'solo female researcher' (also see Jordan & Gibson, 2005) experience was certainly experienced in this research and there was a mutual experience in terms of interactions between the researcher, not just with participants, but also other locals of Prazeres and Rio. This researcher diary extract from Tuesday 16 January illustrates this:

> 'During a trip to AquaRio today, a tourist attraction in Porto Maravilha, with underprivileged children from the social housing development in north Rio, a young girl of 11 years seemed to be fascinated by me and followed me and showed me various sea life as we walked around the attraction. Although we could not communicate orally due to language barriers, we instead communicated through actions and I felt a close 'bond' between us, the girl seemed trusting of my being. At the end of the day trip, she gave me a big hug before we left leaving me with a sense of love and admiration, and more importantly acceptance'.

It is evident that the researchers' early and carefully planned social interaction with her participants, as well as other locals in their daily natural settings, enhanced the information generation. One important dimension was gender – a young solo female researcher entering an economically poorer setting compared to her own – as Everett (2010:171) indicates in her reflections as a female researcher, this 'may have differed had [she] been male'. On reflection, the researcher may have appeared less confrontational than a man may have appeared to community residents – specifically drug gang members. Furthermore, Gurney (1991:379) suggests that female researchers may appear less threatening and 'warmer' than their male counterparts, consequently aiding such access and acceptance. Sinha's (2017) reflections of ethical and safety issues in sex research, discusses how working as a solo-female researcher proved beneficial allowing women to openly share their views with her rather than NGOs. As researchers we should take time to reflect on our own identities and positionalities, and ensure they strengthen (rather than detract from) our work and research relationships. After all, as Schnackenberg and Tomlinson (2016) find, the three-dimensions of *information disclosure*, *clarity* and *accuracy* must always be considered when recruiting research participants and accessing research environments.

The 'Digi-cal model'

As the methodological planning developed, it became clear that this project could contribute new perspectives to similar studies which seek to advance research beyond traditional (and safe) practices of (tourism) research design (e.g. Phillimore & Goodson, 2004). It was through the development of both digital and physical methods that the research design seemed to situate itself on a 'Digi-cal' (digital and physical) nexus. This model values both digital and physical methods working together and works by finding the most appropriate ways to build trust, transparency and personal insight in potentially high-risk research environments.

On reflection, if it was not for the early digital platforms informing later physical 'in situ' field methods in a complex nexus of methodological design, this research indisputably would have failed to progress which often happens with many academics opting for easier and more accessible research topics, particularly within mega-event studies, marginalising the *voice* of those who are mostly affected. This simple yet effective process is summarised in the 'Digi-cal' model below (Figure 7). At its heart (centre) lies the process of 'digital access' which is necessary to underpin the entire process. The identification of an initial relationship with gatekeepers is initially triggered and then galvanised with SMPs (e.g. WhatsApp). This 'soft' introduction was also important in addressing any gender or cultural complexity or misunderstandings that may have emerged in the design phase. This initial phase of work is then overlaid with the use of digital platforms to develop trust over time and build a deeper relationship of reciprocity and mutual understanding. It is only once the digital spaces of research are in place and set, that it may then be possible to access the physical (and potentially higher risk) spaces of research with reassurance and a degree of confidence and insight. As the researcher progresses from the digital to the physical (in person), a relationship of mutual trust and respect may be established over time in advance of fieldwork. In this study, these relationships extended well beyond the lifetime of the research.

Although this approach would benefit researcher beyond the field of tourism and events research, we argue here that digital platforms offer a more disruptive and under-utilised methodological vehicle for reducing research costs as well as reducing the spatial (and cultural) distances often cited between the researcher, gatekeeper(s),

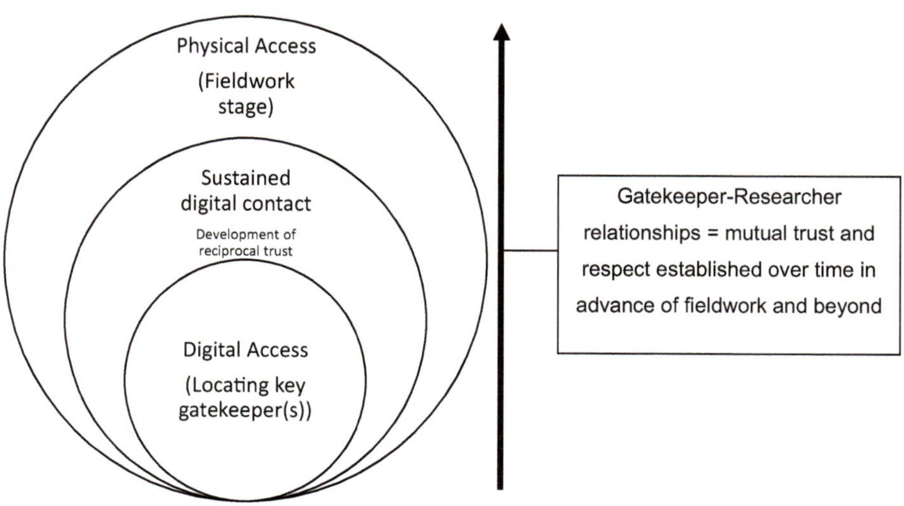

Figure 7. The 'Digi-cal' model: traversing the digital and physical nexus of research (Authors' own).

key informant(s) and participants. By completing key advance stages of fieldwork in the digital space, researchers place themselves in a position to be able to leverage quicker access into inaccessible communities, simultaneously building the all-important element of trust between all stakeholders involved.

Conclusions

This paper has sought to make visible the complexity and 'messiness' of the research process in hard to reach environments and highlight new ways that tourism researchers might look to ethically and sensitively embrace difficult to access or 'risky' research contexts (if valuable to the expansion of knowledge and theory). Through the lens of a tourism research study in the favelas of Rio, we have sought to draw on field notes, pre-research planning approaches and internet-mediated communication methods to articulate a creative way to navigate access to participants and places that are usually hard-to-reach, invisible and deemed risky and dangerous.

Although social media is a relatively recent phenomenon and more studies are using it as a data generation vehicle (see analysis by Lu, 2018), it is rarely considered a useful tool for attaining access to key gatekeepers and attracts even less attention as a way of fostering constructive relationships with key informants in qualitative research studies. Our proposed approach based on a real-life favela research experience has identified that without support of such digital and technological advancements, this kind of in-depth tourism research may have been costlier and most definitely more time consuming as relationships would not have been built had it not been for this methodological process being adopted. Aside from qualitative, in-field research we argue that there is also value in utilising digital platforms to advance other methodological approaches in quantitative and mixed methods research which are equally reliant on access to gatekeepers and participants.

In returning to our original objectives, we began by highlighting the limited body of literature on how researchers might seek to access risky spaces and invisibilised host communities; secondly, it has outlined how physical and digital methods can build networks, trust and relationships with key gatekeepers and locals in the context of a favela. Thirdly, with field diaries and social media excerpts we have reflected one researcher's journey, positionality and how personal safety can be reassured by working with well-positioned gatekeepers. Finally, through the adoption of the 'Digi-cal' model, we have aimed to synthesise good practices and offer a simple framework which summarises the principles and practical steps that worked to both accelerate and 'open up' the research process.

Disclosure statement

No potential conflict of interest was reported by the authors.

ORCID

Sally Everett https://orcid.org/0000-0001-8395-339X
Michael Duignan https://orcid.org/0000-0003-1539-2310

References

Alves, M. H. M., & Evanson, P. (2011). Living in the crossfire: Favela residents, drug dealers, and police violence in Rio de Janeiro. Philadelphia: Temple University Press.

Ateljevic, I., Pritchard, A., & Morgan, N. (Eds.). (2007). The critical turn in tourism studies: Innovative research methodologies. Oxford: Elsevier.

Bajc, V. (2012). Abductive ethnography of practice in highly uncertain conditions. In E.Anderson, D.Ashbury, D.Austin, T.Kim, & V.Kulkarni (Eds.), Bringing fieldwork back in: Contemporary urban ethnographic research. Abductive ethnography of practice in highly uncertain conditions. Annals of the American Academy of Political and Social Science (pp.72–85). Thousand Oaks: Sage. doi:10.1177/0002716212438197

Browne, K. (2005). Snowball sampling: Using social networks to research non-heterosexual women. International Journal of Social Research Methodology, *8*(1),47–60. doi:10.1080/1364557032000081663

Cataldo, F. (2008). New forms of citizenship and socio-political inclusion: Accessing antiretroviral therapy in a Rio de Janeiro *favela*. Sociology of Health & Illness, *30*(6),900–912. doi:10.1111/j.1467-9566.2008.01124.x

Clark, T. (2011). Gaining and maintaining access. Qualitative Social Work: Research and Practice, *10*(4),485–502. doi:10.1177/1473325009358228

Côté, I. (2013). Fieldwork in the era of social media: Opportunities and challenges. PS: Political Science & Politics, *46*(3),615–619. doi:10.1017/S1049096513000565

Crowhurst, I. (2013). The fallacy of the instrumental gate? Contextualising the process of gaining access through gatekeepers. International Journal of Social Research Methodology, *16*(6),463–475. doi:10.1080/13645579.2013.823282

Davies, C. A. (2007). Reflexive ethnography: A guide to researching selves and others (2nd ed.). London: Routledge.

Decroup, A. (2004). Trustworthiness in qualitative tourism research. In: L.Goodson & J.Phillimore (Eds.), Qualitative research in tourism: Ontologies, epistemologies and methodologies. London: Routledge.

Denzin, N. K., & Lincoln, Y. S. (2005). Handbook of qualitative research (3rd ed.). London: Sage.

DiCicco-Bloom, B., & Crabtree, B. F. (2006). Making sense of qualitative research. The qualitative research interview. Medical Education, *40*(4),314–321. doi:10.1111/j.1365-2929.2006.02418.x

Elliott, E., Watson, A. J., & Harries, U. (2002). Harnessing expertise: Involving peer interviewers in qualitative research with hard-to-reach populations. Health Expectations, *5*(2),172–178. doi: 10.1046/j.1369-6513.2002.00158.x

Emmel, N., Hughes, K., Greenhalgh, J., & Sales, A. (2007). Accessing socially excluded people – Trust and the gatekeeper in the researcher-participant relationship. Sociological Research Online, *12*(2),1–13. doi:10.5153/sro.1512

Everett, S. (2010). Lessons from the field: Reflecting on a tourism research journey around the 'Celtic' periphery. Current Issues in Tourism, *13*(2),161–175. doi:10.1080/13683500902853502

Feighery, W. (2006). Reflexivity and tourism research: Telling an (other) story. Current Issues in Tourism, *9*(3),269–282. doi:10.2167/cit/mp006.0

Freire-Medeiros, B. (2012). Favela tourism: Listening to local voices. In F.Frenzel, K.Koens, and M.Steinbrink (Eds.), Slum tourism: Poverty, power and ethics (chapter 10, pp.175–192). Oxon: Routledge.

Freire-Medeiros, B., Vilarouca, M. G., & Menezes, P. (2013). International tourists in a 'pacified' favela: profiles and attitudes. The case of Santa Marta, Rio de Janeiro. Journal of the Geographical Society of Berlin, *144*(2),147–159.

Finlay, L. (2002). Negotiating the swamp: The opportunity and challenge of reflexivity in research practice. Qualitative Research, *2*(2),209–230. doi:10.1177/146879410200200205

Foreign Commonwealth Office (FCO). (2018). *Brazil*. Retrieved from https://www.gov.uk/foreign-travel-advice/brazil/safety-and-security.

Gurney, J. N. (1991). Female researchers in male-dominated settings. In M.Pogrebin (Ed.) Qualitative approaches to criminal justice: Perspectives from the field (2002). London: Sage.

Harris, C., Wilson, E., & Ateljevic, I. (2007). Structural entanglements and the strategy of audiencing as a reflexive technique. In I.Ateljevic, A.Pritchard, & N.Morgan (Eds.), The critical turn in tourism studies: Innovative research methodologies (pp.41–56) Oxford: Elsevier.

Harrison-McKnight, D., & Chervany, N. L. (1996). The meanings of trust. Citeseer,1–86. Retrieved from http://citeseerx.ist.psu.edu/viewdoc/summary?doi=10.1.1.155.1213

Henderson, N. R. (2011). Is there a role for social media in qualitative research? The two can work together for better results. Marketing Research, *23*(3),36–37.

ISSC, IDS and UNESCO (2016). World social science report 2016. Challenging inequalities: Pathways to a just world. Paris: UNESCO Publishing. Retrieved from https://unesdoc.unesco.org/ark:/48223/pf0000245825

Janesick, V. J. (1994). The dance of qualitative research design: metaphor, methodolatry, and meaning. In N. K.Denzin & Y. S.Lincoln (Eds.), Handbook of qualitative research (pp.209–219) Thousand Oaks, CA: Sage.

Jobbins, G. (2003). Translators, trust and truth: Cross-cultural issues in sustainable tourism research. In J.Phillimore & L.Goodson (Eds.), Qualitative research in tourism: Ontologies, epistemologies and methodologies (pp.313–323). London: Routledge.

Johnson, R. B. (1997). Examining the validity structure of qualitative research. Education, *118*(2),282–292. Retrieved from https://www.researchgate.net/publication/246126534

Jordan, F., & Gibson, H. (2005). "We're not stupid … But we'll not stay home either": Experiences of solo women travelers. Tourism Review International, *9*(2),195–211. doi:10.3727/154427205774791663

Kapoor, K., Tamilmani, K., Rana, N., Patil, P., Dwivedi, Y., & Nerur, S. (2018). Advances in social media research: Past, present and future. Information Systems Frontiers, *20*(3),531–558. doi: 10.1007/s10796-017-9810-y

Kerasidou, A. (2017). Trust me, I'm a researcher!: The role of trust in biomedical research. Netherlands: Kluwer Academic Publishers.

Kuebler, D., & Hausser, D. (1997). The Swiss hidden population study: Practical and methodological aspects of data collection by privileged access interviewers. Addiction, *92*(3),325–334. doi:10.1111/j.1360-0443.1997.tb03202.x

LeBaron, E. (2015). Reimagining the geography of the Favelas: Pacification, tourism, and transformation in Complexo do Alemão, Rio de Janeiro. Tourism Review International, *18*(4),269–282. doi:10.3727/154427215X14230549904099

Lincoln, Y. S., & Guba, E. G. (1985). Naturalistic inquiry. Newbury Park, CA: Sage.

Liu, D., Ainsworth, S. E., & Baumeister, R. F. (2016). A meta-analysis of social networking online and social capital. Review of General Psychology, *20*(4),369–391. doi:10.1037/gpr0000091

Low, T., & Everett, S. (2014). Industry engagement with tourism research: The impacts of social control and emotional labour. Tourism Management Perspectives, *12*,134–143. doi:10.1016/j.tmp.2014.09.006

Lu, Y. (2018). Mapping the progress of social media research in hospitality and tourism management from 2004 to 2014. Journal of Travel & Tourism Marketing, *35*(2),102–119. doi:10.1080/10548408.2017.1350249

Maramwidze-Merrison, E. (2016). Innovative methodologies in qualitative research: Social media window for accessing organisational elites for interviews. Electronic Journal of Business Research Methods, *14*(2),157–167.

Mayer, R. C., Davis, J. H., & Schoorman, F. D. (1995). An integrative model of organizational trust. Academy of Management Review, *20*(3),709–734. doi:10.5465/amr.1995.9508080335

McCann, B. (2014). Hard times in the marvelous city: From dictatorship to democracy in the favelas of Rio de Janeiro. Durham, NC: Duke University Press.

Nyaupane, G. P., & Poudel, S. (2012). Application of appreciative inquiry in tourism research in rural communities. Tourism Management, *33*(4),978–987. doi:10.1016/j.tourman.2011.10.009

O'Gorman, K., Maclaren, A., & Bryce, D. (2014). A call for renewal in tourism ethnographic research: The researcher as both the subject and object of knowledge. Current Issues in Tourism, *17*(1),46–59. doi:10.1080/13683500.2012.718321

Ospina, S. (2004). Qualitative research. In G.Goethals, G.Sorenson, & J.MacGregor (Eds.), Encyclopaedia of leadership (pp.1279–1284). London: Sage.

Perlman, J. (2010). Favela: Four decades of living on the edge in Rio de Janeiro. New York: Oxford University Press.

Phillimore, J., & Goodson, L. (2004). Qualitative research in tourism ontologies, epistemologies and methodologies. London; New York: Routledge.

Phillips, D. (2017). British woman shot in Brazil favela named as Eloise Dixon. *Guardian News*. Retrieved from https://www.theguardian.com/world/2017/aug/07/british-woman-shot-after-family-stray-into-brazil-favela

PrazeresTur. (n.d.). *Prazeres tour*. Retrieved from https://prazerestur.weebly.com/#

Punch, M. (1994). Politics and ethics in qualitative research. In N. K.Denzin & Y. S.Lincoln (Eds.), Handbook of qualitative research (pp.83–98) London: Sage.

Ramos, S. (2017). Death toll mounts in Rio de Janeiro as police lose control of the city – And of themselves. *The Conversation*. Retrived from https://theconversation.com/death-toll-mounts-in-rio-de-janeiro-as-police-lose-control-of-the-city-and-of-themselves-80862

Ren, C., Pritchard, A., & Morgan, N. (2010). Constructing tourism research: A critical inquiry. Annals of Tourism Research, *37*(4),885–904. doi:10.1016/j.annals.2009.11.006

Ritchie, B. W., Burns, P., & Palmer, C. (2005). Tourism research methods: Integrating theory with practice. Wallingford: CABI Publishing.

Rousseau, D. M., Sitkin, S. B., Burt, R. S., & Camerer, C. (1998). Not so different after all: A cross-discipline view of trust. Academy of Management Review, *23*(3),393–404. doi:10.5465/amr.1998.926617

Saunders, M., Lewis, P., & Thornhill, A. (2016). Research methods for business students (6th ed.). Harlow: Pearson.

Schnackenberg, A. K., & Tomlinson, E. C. (2016). Organizational transparency. Journal of Management, *42*(7),1784–1810. doi:10.1177/0149206314525202

Schoorman, F. D., Mayer, R. C., & Davis, J. H. (2007). An integrative model of organizational trust: Past, present, and future. Academy of Management Review, *32*(2),344–364. doi:10.5465/amr.2007.24348410

Shefner, J., & Gay, R. (2002). The politics of field research: Lessons from Brazil and Mexico. Canadian Journal of Latin American and Caribbean Studies, *27*(54),199–213. doi:10.1080/08263663.2002.10816823

Sinha, S. (2017). Ethical and safety issues in doing sex work research: Reflections from a field-based ethnographic study in Kolkata, India. Qualitative Health Research, *27*(6),893–908. doi:10.1177/1049732316669338

Sixsmith, J., Boneham, M., & Goldring, J. E. (2003). Accessing the community: Gaining insider perspectives from the outside. Qualitative Health Research, *13*(4),578–589. doi:10.1177/1049732302250759

Spivak, G. C. (1988). Can the subaltern speak? In C.Nelson and L.Grossberg (Eds.), Marxism and the interpretation of culture (pp.217–313). London: MacMillan.

Steinbrink, M. (2013). Festifavelisation: Mega-events, slums and strategic city-staging – The example of Rio de Janeiro. Journal of the Geographical Society of Berlin, *144*,129–145.

Strobl, T. (2017). 8 root causes of violence that will be overlooked in coverage of Rio's Olympic Anniversary. Rio on Watch. Retrieved from http://www.rioonwatch.org/?p=37723

The Rio Times. 2016. *Italian tourist killed in Rio after accidentally entering favela*. Retrieved from https://riotimesonline.com/brazil-news/rio-politics/local-politics-rio-politics/italian-tourist-killed-in-rio-after-accidentally-entering-favela/

Tribe, J. (2005). New tourism research. Tourism Recreation Research, *30*(2),5–8. doi:10.1080/02508281.2005.11081468

Tribe, J. (2006). The truth about tourism. Annals of Tourism Research, *33*(2),360–381. doi:10.1016/j.annals.2005.11.001

Unidade de Polícia Pacificadora (UPP). (n.d.). *About*. Retrieved from http://www.upprj.com/index.php/o_que_e_upp_us

Veal, A. J. (2017). Research methods for leisure and tourism. Harlow: Pearson Education Limited.

Wilson, E., & Hollinshead, K. (2015). Qualitative tourism research: Opportunities in the emergent soft sciences. Annals of Tourism Research, *54*(1),30–47. doi:10.1016/j.annals.2015.06.001

World Bank. (2012). Bringing the state back into the favelas of Rio de Janeiro: Understanding changes in community life after the UPP pacification process (English). Washington, DC: The World Bank. Retrieved from http://documents.worldbank.org/curated/en/255231468230108733/Bringing-the-state-back-into-the-favelas-of-Rio-de-Janeiro-understanding-changes-in-community-life-after-the-UPP-pacification-process

Walking methodologies, digital platforms and the interrogation of Olympic spaces: the '#RioZones-Approach'

Michael B. Duignan and David McGillivray

ABSTRACT

Mega-sport events (MSEs) target, sequestrate and territorialise a 'host' (city – or nation) to exist, sustain themselves, survive and deliver on organisational objectives. Yet, little systematic and empirical evidence examines such urban processes, specifically in real-time during 'live staging' periods, including related implications for host community inclusion (and exclusion). In this article, we present the *#RioZones* project; a series of physical-embodied and digitally enabled primary data collection methods focused around embodied walking and participatory digital methods aimed at: (i) providing a deeper – and disruptive – interrogation of the spatial, social and economic implications of MSE organisation, (ii) developing creative ways to share empirical insights via microblogging, vlogging, media and public engagement platforms, and (iii) bringing together academic, industry, and policy stakeholders to debate and influence change. Embedded physically and digitally (albeit temporarily) into Rio's Olympic city, we refer to this strategically planned amalgam of methods as the '*#RioZones-Approach*'. Our findings reveal that utilising a set of post-positivist methods and approaches can be highly effective for creating new lines of (disruptive) inquiry and qualitative knowledge insights, currently latent in this field of analysis, and in related critical tourism and event studies). Additionally, we illustrate how projects with a longitudinal ethos present a unique opportunity to maintain longer-term scrutiny on spatial-urban processes and involve key stakeholders to critique contested MSE legacies. Our article provides a detailed, but single snapshot into the life of an ever evolving, live and stakeholder engaged research project. In terms of significance and contribution, we argue that *#RioZones* – and the *#RioZones-Approach* – presents an ideal platform to bring together new, existing, creative and disruptive research projects striving to offer a deeper level of analysis and interrogation toward the study and intersection of events, urban space, and host community inclusion and exclusion.

摘要

大型体育赛事(MSEs)旨在让一个东道主城市或国家脱颖而出，形成特色区域，使其得以存续并最终实现组织目标。然而，很少有系统的经验研究考察这种城市过程，尤其是在大型体育赛事实时举办期间的城市化过程，包括对东道主社区包容(和排斥)的相关影响。在本文中，我们介绍了里约区项目，该项目采用了一系列亲身体验与数字化驱动的主要数据收集方法，旨在研究： 1)对大型体育赛事组织的空间、社会和经济影响进行更深入且具有颠覆性的审视;2)通过微博、视频博客、媒体和公众参与平台开发创新性的方法来分享经验见解;3）汇集学术界、实业界与政策实施的利益相关者讨论与影响城市的变化。通过亲身沉浸在里约热内卢奥运城并且采用数字化的方法(尽管是暂时的)研究该城市，我们将这种经过战略规划的方法组合称为“里约区方法”（#RioZones-Approach）。我们的发现表明，利用一套后实证主义的方法和取向可以非常有效地创建新的(甚至是颠覆性)调研路线和定性的知识洞见，目前这种方法和取向在分析领域以及批判性旅游和事件研究领域尚不多见。此外，我们展示了一个具有长期特点的项目是如何提供一个独特的机会，得以持续地长期地审视城市空间过程，并且引导其关键利益相关者参与到评判争议性的大型体育赛事遗产中来。我们的文章提供了一个详细的“生动写照”，以了解一个不断发展的涉众研究项目实况。就其意义和贡献而言,我们认为里约区项目或里约区方法提出了一个理想的研究平台,汇集了现有崭新的、创造性的甚至是颠覆性的研究尝试，对事件、城市空间以及目的地社区包容与排斥的研究及交叉领域提供更深层次的分析与审视。

Introduction

Mega-sport events (MSEs), in order to exist, sustain themselves, survive and deliver on organisational objectives, *must* depend on a 'host' city – or even whole nation – abundant in both physical-infrastructural and symbolic-cultural resources. Territorialising urban space is deemed necessary to (re)shape and (re)configure existing geographies to temporarily stage official elite sport, cultural and commercial activity. The most visible aspects of spatial territorialisation are witnessed in the build-up to and live, physical staging of MSEs. Yet, while there is an abundance of critical studies focusing on the social and economic-geographic consequences of pre-Games planning (i.e. displacement) and post-Games outcomes (i.e. gentrification) (Giulianotti, Armstrong, Hales, & Hobbs, 2015, Kirby, Duignan, & McGillivray, 2018; Broudehoux & Sanchez, 2015; Cade, Everett, & Duignan, 2019; Smith, 2016; Duignan, Pappalepore, & Everett, 2019), fewer studies interrogate the way MSEs effectively sequestrate, territorialise and takeover public space to make way for live staging. As a result, McGillivray and Frew (2015) have called for a deeper interrogation of such spatial-urban processes, and the broader relationship between MSE demands, urban space and public-civic spaces turned into private commercial enclaves – and who is included and excluded from these spaces (Duignan et al., 2019) before, during and after live staging periods. This is of importance as MSEs continue to extend their reach into both core and peripheral urban and rural geographies of their host cities. Examples of this spatial creep include the creation of non-sports, commercial and festive zones across the host city (Osborn & Smith, 2016; Giulianotti et al., 2015) and occupying public spaces like parks and squares in the process – often referred to as 'Live Sites' (McGillivray & Frew, 2015).

Throughout this article, we respond to McGillivray and Frew's (2015) call by making the case for a deeper interrogation of such spatial processes, how this can be achieved methodologically, and why this remains important for the field of critical event and tourism studies.

We argue this is an increasingly pertinent issue as MSEs continually find themselves subject to significant scholarly, policy and public critique for their exclusionary tendencies, including locking-out local interests, while securing exclusive access for 'official' partners like the 'Olympic family' of sponsors, supporters and suppliers (Weber-Neuth, 2016). Most critical commentary to date has focused on the wider-exclusionary socio-economic conditions residing at the heart of the host city and/or nation with stakeholders often targeted for event-led regeneration or (re)development policy (Duignan & Pappalepore, 2019; McGillivray, Duignan, & Mielke, 2019). By focusing on touristic-event spaces, places and environments our study served to unpack spatial imperatives, arrangements and organisational phenomena (re)shaping geographical contours, everyday public space, and the temporary inclusion and exclusions that simultaneously enable access to some and preclude access to others. There is a need for researchers studying MSEs to make use of real-time, embodied research methods to more effectively capture local nuances and idiosyncrasies relating to each host context. It is with these factors in mind that we purposively draw on an amalgam of embodied walking methodologies (i.e. real-time walking observation, photo and video diaries) and participatory digital methods (i.e. utilising vlogging and micro-blogging sites like Twitter) that come together to comprise the *'RioZones-Approach'*. In response to the aforementioned challenges, and opportunities surrounding novel methods utilised and presented across the *RioZones-Approach*, our study was guided by the following research questions:

1. In what way can an amalgam of both embodied and digital practices be utilised by researchers to better understand, and more deeply interrogate, the spatial arrangements and impacts of MSE spaces?
2. How can these practices serve to disseminate findings, foster collaboration, and stimulate debate between academic and non-academic audiences?
3. Beyond the event itself, what are the benefits of both physical and digital platforms for scrutinising (highly contested) legacies?

Structurally, we begin by justifying the need to interrogate the organisational and spatial implications of MSEs, through examining how and why MSEs sequestrate, territorialise and privatise public-civic space. Second, we identify a number of creative and post-positivist methodological practices that can enable deeper and more disruptive insight(s) into the spatial measures and implications rarely identified in the context of MSE impact analysis (Pappalepore & Duignan, 2016). Third, we introduce the *#RioZones-Approach*: focusing on the benefits for researchers in applying mobile methods (embodied and digital) to examine the Rio 2016 Olympic Games. Throughout, we illustrate how a strategic and creative amalgam of physical and digital methods has the power to disrupt linear, sometimes entrenched, approaches to generating knowledge – producing new insights, sharing them more effectively (with academic and non-academic actors) and providing guidance for how others might make use of these techniques in pursuing critical tourism and event studies.

Interrogating [mega] event spaces: situating the study

As highlighted in the introduction, MSEs require a 'host' to support the organisational requirements of official sports, cultural and commercial activities. Large, complex organisational projects like the Olympics require experienced personnel to execute project objectives at both the *national* and *local* level. Successful project planning and delivery is highly dependent on the availability of both internal *and* external resources (Pfeffer & Salancik, 1978). This is because the Olympics is one of the world's most complex organisational projects, requiring significant financial and human resources in order to fuel and sustain itself. Following a successful bid, the Olympics influences (some would argue, determines) urban planning and development in the host city, requiring the reconfiguration of public and private space to enable the event project to be delivered, on time and to the specification of the Host City Contract (HCC) signed with the International Olympic Committee (IOC) (e.g. Müller, 2015). In the process, urban public spaces, including parks, squares, streets, rivers and beaches are opened up to commodification processes, offered to external agents as 'assets' to be exploited for extrinsic gain (Osborn & Smith, 2016; McGillivray & Frew, 2015; Gogishvili, 2018). This involves real estate grabs (Gaffney, 2016), the creation of exclusive commercial enclaves (Broudehoux & Sanchez, 2015) and the development of complex and sophisticated human and non-human security and policing operations (Coaffee, 2015). Legislative exceptions are passed (Broudehoux & Sanchez, 2015), new (temporary) advertising and trading regulations introduced, and existing business activity restricted (Pappalepore & Duignan, 2016) to comply with the contractual obligations of the HCC. As devices to reimagine cities-as-destinations, MSEs enable the local entrepreneurial state to engineer an urban facelift, prioritising developments that spearhead highly aestheticized images and narratives (by and large) for external consumption (Steinbrink, 2013; McGillivray, 2018).

An emergent group of scholars in the tourism, (critical) events and urban geography fields focus on examining the complex spatial arrangements influenced and demanded by hosting MSEs (Hagemann, 2010; Klauser, 2011; Smith, 2013, 2016; Frew & McGillivray, 2008; Broudehoux & Sanchez, 2015; Pappalepore & Duignan, 2016; McGillivray & Frew, 2015; Weber-Neuth, 2016; Pavoni, 2017). This list is not exhaustive, but illustrative of an international grouping of scholars who have a common interest in exploring the relationship between events, urban space, and multivariate effects on both internal and external MSE actors (including residents, businesses and sponsors). Increasingly, there has been a focus on how the reconfiguration of public and private spaces across host cities foregrounds some interests while marginalising others, often on the basis of possession of social and economic capital (Raco & Tunney, 2010). Various studies have shown how MSE planning and delivery arrangements have the power to alienate host communities and local stakeholders (both residents and small businesses) across the Olympic (project) lifecycle – before, during and after the event itself (OECD, 2008; Work Foundation, 2010; House of Lords, 2013; McGillivray & Frew, 2015; Smith, 2016).

Crucially, these scholars have also adopted methods of inquiry that are at odds with conventional thinking in respect of understanding MSE impacts. In the 1990s and early 2000s, the study of MSEs was dominated by economic analyses, quantifying the return on investment host cities could expect from hosting. Adhering to positivist paradigms, scholars relied on quantitative measures of 'success' that could be presented to policy makers and potential event hosts to justify event bids (MacRury &

Poynter, 2009). Additionally, non-economic MSE outcomes like sports participation, health improvement, and infrastructural enhancements were also translated into key performance indicators and played a significant role in ex-ante projections and ex-post event analysis and evaluation (Getz, 2018). Yet, such entrenched epistemological, methodological and evaluative approaches offered limited qualitative insight into the highly situational and contextual factors that influence the spatial impacts on the host city, and within its communities.

Post-positivism and MSE academic-activism

Though occupying diverse fields and disciplines, scholars examining the complex spatial arrangements of MSEs are increasingly drawing on post-positivist epistemological positions. By doing so, they are foregrounding lines of inquiry that seek to avoid reductionism: accepting complexity, contestation and even contradiction (e.g. Müller, 2017) as they seek to observe and document power relations present in the planning and delivery of MSEs. Rather than assess whether MSEs are an 'efficient' or 'effective' means of delivering on a set of prescribed performance indicators, critical scholars are interested in what these mega spectacles afford, for whom and with what short and longer-term consequences – especially for those parts of the host population constituted as 'beneficiaries'. We too adhere to the notion that the mode of inquiry needs to be able to capture how, recalling Foucault, 'individuals constantly escape, evade, and subvert the functioning of discipline' (Miller, 1987, p. 196) and how local resistances, 'fragile relays' and 'contested locales' (Rose, 1999, p. 51) are present in MSE governance arrangements, defined by 'subtleties and micro negotiations of relations of power' (Petersen, 1997, p. 203). Reviewing official pronouncements, documenting policy and legislative actions, observing changes in the physical urban environment and working alongside local activists and social movements affected by MSE developments are just some of the ways that scholars can describe, make visible and, ultimately, disrupt conventional thinking. Many of these values, attributes and foci serve as the foundation for our *#RioZones-Approach*.

In keeping with a focus on disruption and using research to initiate change there is some evidence that action-focused methodological approaches are also being utilized, revealing deeper insights, and acting to subvert or disrupt aspects of MSE planning and delivery. For example, scholars like Gaffney (2016) and Boykoff (2012, 2015) have been making use of immersive, ethnographic, activist-academic methods of inquiry to draw attention to the worst excesses of the MSE 'takeover' (Müller, 2015), particularly on the most vulnerable segments of host city populations. Gaffney (2016) moved to Rio in 2009 and spent the years up to, and including, the Rio 2016 Olympic Games, documenting the changes taking place across the city. This included the passing of exceptional legislation, enabling the displacement of residents from favelas and the creation of spatial exclusions as new housing developments and official venues were erected. Boykoff (2015) has also been a very 'public' activist-academic, drawing attention to the intense militarised securitisation that accompanies MSEs, and translating his research enquiries into public pronouncements that seek to mobilise change within awarding bodies (i.e. the IOC and FIFA), sporting federations and host governments alike. As indicated in the latter sections of this article, we attempted to use similar communicative techniques to politicise similar organisational and spatial challenges identified across *#RioZones*.

More recently, Talbot and Carter (2018) conducted a year-long ethnography of the activities of activist groups in Rio before the Olympic Games, focused on housing evictions and human rights issues. Like Boykoff and Gaffney, by making use of a more 'involved' form of research inquiry, they sought to access deeper insights into complex phenomena, involving attendance at public meetings, observing resident protests, documenting housing demolitions and working closely with activist groups in a translation and journalistic capacity (Talbot & Carter, 2018). Foregrounding activists' local resistances, and contested locales, these authors are more effectively able to uncover the practices of organisers and their external partners (IOC and sponsors) as well as those who are most affected by urban developmental strategies.

We are encouraged by the increasing quantity and quality of research being conducted into the spatial effects of MSEs and the greater emphasis towards (participatory) action research (Reason & Bradbury, 2013; Chevalier & Buckles, 2013) is also to be welcomed. Specifically, researchers' adoption and application of these methods enable them to access deeper insights related to the localised impacts of MSE planning and delivery, for often marginalised or silenced voices (Duignan, Everett, Walsh, & Cade, 2017). In generating new understandings, participatory approaches give greater value to the knowledge, capabilities and experiences of all those involved in or affected by a situation. As a result, these approaches represent powerful means of exposing power relations, and offering pathways to enact positive change for marginalised groups – whether they be residents of an MSE host destination or small businesses that feel sidelined by the trading regulations imposed upon them. The work of Gaffney, Boykoff, and Talbot & Carter is intended to be disruptive, challenging conventional discourses about MSEs (as economic and social panaceas), and adopting modes of inquiry that are unapologetically participatory and action-focused, working *with* research participants to initiate change. Their 'findings' are presented for public consumption, using traditional and new media platforms to extend reach to audiences other than academic contemporaries.

However, there remains a need for more MSE-focused research that utilises other embedded or involved methodological techniques, extending analysis to – and beyond – the event delivery phase itself. Physically immersive, sensory, and affective methods of knowing the urban environment, married to digital methods for capturing, documenting and communicating can be used effectively to access 'what' and 'how' particular spatial arrangements are produced when hosting MSEs. These tools of knowledge generation can also be useful in bridging the gap between the academic and non-academic actors involved in – and affected by – MSE planning and delivery arrangements. Our attempt to develop a method that integrates the physical and the digital is the *#RioZones-Approach*. This approach was designed to produce deeper insights and interrogations into the relations between MSEs and urban space, centred on the Rio 2016 Olympic Games during the pre-event and live staging period. The *#RioZones* project involved fieldwork conducted over three distinct time phases:

Phase 1: immediately before and during 'live staging' – between the 31 July and 8 August 2016;

Table 1. Physical and digital methods and tools.

Physical	Observational data gathered through walking planned routes; Geotagged photos (over 2000 photos generated); Auditory techniques (hours of audio narration of spaces under investigation recorded); Geotagged video capture (over 400 videos captured); Analysis of archival and media reports – before, during and after Rio 2016 (e.g. event sites, official Rio 2016 publications); Interviews (Phase 1 and Phase 2).
Digital	Official RioZones Vlog RioZones.wordpress.com; Microblogging Twitter hashtag: #RioZones Dissemination of findings via special interest groups (e.g. Tri-Net, and JISC lists); International media (e.g. Australian news interview on network 'ABC', through to 'Yahoo News'); National media (e.g. public engagement piece via 'The Conversation'); Regional media (e.g. BBC Radio interview by author live from Copacabana).

Phase 2: immediately after 'live staging' – 25 August – 30 September 2016;

Phase 3: 2 years on legacy analysis between 5 March and 31 August 2018.

For the purpose of this article, our focus is on Phases 1 and 2, though we will briefly mention the importance of creating and sustaining a (digital) network and returning to the MSE host environment to assess (spatial and urban) event legacy outcomes. Phase 1 consisted of two overarching approaches: (i) 'the physical' – a walking methodology and (ii) 'the digital' – an official Twitter hashtag (*#RioZones*) for microblogging, official vlog (riozones.wordpress.com) and a number of other media and public engagement activities (see Table 1 for details of the methods used across #RioZones). We will now attend to each approach in order.

Embodied walking and journeying through the MSE city

Walking methodologies have become popularised in recent years across a variety of disciplines and fields, including anthropology, human geography, sociology, and tourism studies (Edensor, 1998; Fletcher & Platt, 2018; Ingold & Vergunst, 2008). Ingold and Vergunst emphasise that walking can provide access to the entangled relationships that exist between humans, non-humans, natural and social environments. One significant strand of this work has been a turn towards walking as means of 'theorising the world through consideration of the everyday pedestrian practices of others' (Bates & Rhys-Taylor, 2017). Walking as a method of inquiry enables observation, experience and sense-making of various phenomena. Urban environments are amenable to such methods as they afford an opportunity to experience physical change through walking, and other associated practices like riding bikes, travelling in the back of a taxi, or modes of public transport. In the social sciences, and other fields, the turn to mobilities has also reinforced the value of walking and other mobile methods. Encountering the contemporary city through walking 'induces a mobile, grounded perspective and foregrounds corporeal, sensual and affective matters', enabling the researcher to 'observe issues unfolding at street level, if only for a short time' (Bates & Rhys-Taylor, 2017, p. 4). There are always limitations to research methods relating to the length of time spent immersed in the subject context and the vagaries of urban life that change so frequently and with varying effects on those inhabiting – or visiting – the city. Yet, when aligned with other methods of 'documentation', the experience of walking can be a powerful embodied method – complementary with other data sets like those presented and triangulated across the *#RioZones-Approach*.

Our *#RioZones* walking method represents a hybrid research mobilities approach, balancing the act of walking-as-observation with photographic, auditory and video capture techniques. This approach adheres to Springgay and Truman's (2017) contention that walking can be 'merged with various other methods like photography and video, drawing, sensory methods, different mapping techniques, including GPS (Global Positioning System), and performance' [online] (also see Springgay & Truman, 2018). We sought to follow the prescribed routes and paths mapped out by MSE organisers before and during Rio 2016, experiencing the flows and circulations of visitors and residents alike. The walking method also enabled us to experience waymarking, barriers, security checks, and the use of sounds and senses to attract visitors to fixed spaces (whether entry was restricted by barriers or not). Including the sensory dimension is important as the city is increasingly viewed as a place where heightened collective sensory experiences are managed and curated (Gandy, 2017). Our approach sought to include soundscapes (music), smells (street traders and official vendors) and collective gatherings (atmosphere) because 'urban space is a stage that can be modified at will through light, sound, or other stimuli' (p. 365). We wanted to explore how the Olympic city was modified and atmospheres curated to produce the desired outcomes for organisers – and to what extent these might be resisted or negotiated. In order to ensure we were able to experience the 'right' spaces we undertook significant pre-event planning. Prior to arriving in Rio, the authors' examined official event publications (e.g. maps, visitor brochures) to familiarise themselves with the host city context, and more importantly: the locality, character and pre-event uses of the public spaces temporarily (re)engineered and used as a platform for Olympic sports, cultural and commercial activities.

For *#RioZones* we utilised a number of techniques to document our walking observations. We used photography, video, and audio descriptions to document the spatial transformations taking place in Rio before and during the early part of the Olympic Games. We imposed some structure on these observations by capturing or recording information related to criteria developed in response to the study's principal research questions. These themes were split down into:

i. What kind of public spaces were sequestrated and territorialised and why;
ii. How were these urban spaces spatially organised and zoned, and what human (e.g. security personnel, Games volunteers) and non-human actors (e.g. signage, barriers) actors were responsible for organising and affording such conditions;
iii. What were the associated spatial impacts on visitor flow, circulations and the creation of festival atmospheres across the Olympic city;
iv. To what extent did event spaces emerge as a platform for protestation, contestation and/or resistance efforts.

Across the 10-day observational period of *#RioZones* we walked to, from and within as many of the main host event zones and official venues as possible. These included: the (i) Olympic Stadium, (ii) Maracanã Stadium, (iii) Copacabana's temporary Beach Volleyball stadium and (iv) the Live Site at Porto Maravilha officially referred to as the 'Olympic Boulevard' (OB). The OB played host to live public screenings of ceremonies and sports activity, inclusive of the majority of Rio 2016 cultural programming and official sponsors, suppliers and supporters, and a series of local-regional retail and hospitality small enterprises. Attempting to mimic the visitor 'journey' and 'gaze' – and to facilitate immersion in the urban context – one of the authors' stayed within the main event zone of Copacabana and pre-booked tickets to experience Olympic sports, culture and commercial activity. Immersive in nature, our methodology was designed to bring the research as close as possible to the phenomena – and context – under investigation and expose it to a multi-dimensional and multi-sensory experience of event environments (Adams & Guy, 2007). Identifying, first-hand, the spatial dynamics of the MSE urban context was central to the effectiveness of our methodology, helping to unpack 'the systematic description of events, behaviours, and artefacts in the social setting chosen for study' (Marshall & Rossman, 1989, p. 79).

The walking methodology provided a way into understanding more about the dynamics of temporary Olympic enclaves as they interacted and replaced existing uses of public space. Experiencing Last Mile spaces, Live Sites and the spaces in-between further enabled the authors to observe how people subverted these spatial arrangements, exhibiting localised micro resistances that contested official doctrine. These included micro entrepreneurs selling products close to, or outside, official Games venues, infringing on the trading regulations in place for the Games (see Figures 1–6); Or, unofficial sponsors participating in ambush marketing activity within regulated advertising-free zones. The interactive nature of the walking approach, aligned with the documentation and capture of visual evidence of spatial change, produced powerful imagery that disrupts official narratives produced by MSE organisers. Over the duration of the fieldwork, over 400 videos and 2000 photos were generated. Below, we present a short slideshow of the kind of visual evidence collected, related to the criteria and key themes for analysis.

From our walking methodology, we were able to document and capture the relationship between the event, urban space and visitor interactions. Figure 7 illustrates how, within an 'existing territory' the Olympics temporarily takes over via the imposition of a 'project territory' afforded by the strategic deployment of temporary – often contingent – actors (Dansero, Mela, & Rossignolo, 2015). The aim of the (temporary overlaid) project territory is to reconfigure public space in a way to afford easy access for visitors to engage with – and consume within – official Olympic event zones. Existing travel and transport arrangements are reconfigured to suit the needs of external actors (e.g., the IOC VIPs and corporate sponsors). New transitory zones are created to funnel Olympic audiences from existing transport hubs to official Olympic enclaves. One good example is the 'Last Mile' transit spaces to and from venues required as part of the HCC. These Last Mile spaces represent the final stretch of the visitor's journey before going through airport style security, and ticketed entrance points to access the official venues. The Last Mile is an under-researched but vitally important and interesting spatial phenomenon in and of itself.

Figure 1. Illustrating the way in which physical barriers, security and official Games Makers play a role in zoning public space.

Figure 2. Illustrating how, once zoned, wayfinding signage is installed to influence economic and socio-spatial relations, visitor flows and circulations.

However, the recording and documenting of spatial changes is unsatisfactory unless a wider audience – particularly those who can influence and effect change – is engaged. If not, whilst walking observations can provide a window into localised effects, it may not – by itself – produce the necessary, disruptive change required to achieve more inclusive economic and socio-spatial outcomes. We now discuss the

Figure 3. The installation of airport style security on Copacabana beach, illustrating the line between public-civic space and newly privatised-commercial enclaves.

Figure 4. Illustrating the types of extraordinary militarised ways – and actors used – to create secure, 'safe' spaces and control visitor flows and circulations.

next central aspect of this article: the value of using participatory digital techniques to extend the reach and engagement with *#RioZones* and bring in influential stakeholders of whom may help support change during live staging and beyond the event itself.

Figure 5. Illustrating the kinds of micro-level attempts by locals to engage in entrepreneurial leveraging of (usually inaccessible) Olympic visitor economies, an example of resisting determined, private-commercial enclaves installed by Rio 2016.

Figure 6. One of many examples illustrating the way newly zoned public spaces become commodified and subject to corporatisation effects – usually leading to the exclusion of non-official product offerings (Authors own Figures 1–6.).

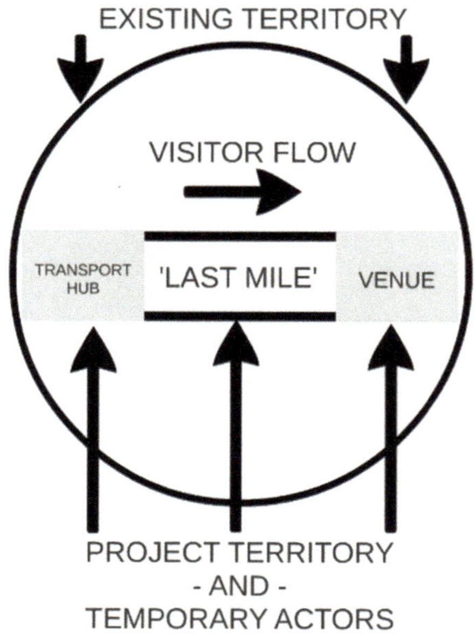

Figure 7. Conceptual overview of spatial territorialisation, economic and socio-spatial dynamics (Authors own.).

Participatory digital methods: extending reach and effects

Over recent years we have witnessed a (digital) ethnographic turn (Lupton, 2015, Pink et al., 2015) whereby the ethnographic ethos has been adapted to fit the affordances of digital culture. When studying MSEs, or any other touristic context, it is no longer satisfactory to deal in the binaries of offline and online, virtual and real. Instead, as Pink et al. (2015) suggest, 'we live and act in a context that is, today, almost always co-constituted and entangled with digital technologies, content, presence and communication' (p. 2). MSEs are open to (digital) forms of research because the changes they bring about to urban locations, over time, are amenable to observation and documentation. MSEs are also, increasingly, digitally mediated events which generate significant media content that can also be the source of valuable data. Instead of a reliance on text, digital (ethnography) differs from its analogue variant in its use of 'video, photography or blogging' (Pink et al., 2015, p. 3).

In utilising digital methods, knowledge can be captured and documented via online field notes, the production of microblogs and vlogs (public or private), and the use of social media as a data generation tool and dissemination technique. Lupton (2015) states that researchers can use digital tools to network and build conversations, help develop professional practice and, enable their research to reach a wider public. Linked to this idea of professional practice is the idea of openness, where 'digital sharing and collaboration become ways of being and relating to others' (Pink et al., 2015, p. 11) – this includes other academics, but also beyond to key industry and policy stakeholders. Digitally mediated research is undertaken with others, including those posting online, sharing videos, writing blogs and filtering photographs. Here, capturing and sharing photographs can be considered a knowledge making practice – though open to

Figure 8. Front page of *#RioZones* vlog.

adaptation through editing techniques and creative reimagining (filtering) – they also evoke feelings, emotions and perceptions of everyday publics on their urban locale.

How research is disseminated is also an important consideration. Websites, blogs, social media and press releases can be viewed as tools to broadcast research 'findings' in a manner that can often reinforce power relations between the researcher and the 'researched'. However, adopting a more open and collaborative process, the same 'platforms' can be used to invite participation in the research process. Yet, the devil (or the beauty) is in the detail: *How can this be afforded? And, does the researcher have the confidence and the ability to invite key stakeholders into knowledge production, dissemination and application?*

Despite the temptation to foreground the 'digital' in research, Pink et al. (2015) suggest that it is important not to be overly digital-centric. In the context of studying MSEs, there is value in understanding the changing dynamics of urban space, observing changes in the physical environment by being located in the city for an extended period of time, as described in our walking methodology. These methods need not be solely digital. However, there is also value in capturing those changes using the new tools that the digital environment provides. Geo-tagging the photographs taken by others in the Olympic city and using the meta-data to produce a chronology of changes can add to physical observations. Crowd-sourcing video footage using the medium of social media platforms can help the researcher extend their data gathering process with the help of non-academic actors. In sum, it is important for researchers to be reflective about how central digital methods might be to their work and identify when digital forms of data collection and/or dissemination can be most complementary or supplementary for achieving research aims.

In the *#RioZones* project, we integrated digital (blogging and social media) methods as knowledge creating and sharing practices. Complementing walking observation and sensory experiences, our digital methods extended the research to include stories generated by others and alternative voices (academic and resident). Prior to arriving in Rio, we set-up and disseminated information to academic and industry networks (like the Tourism Research Information Network (Tri-net), and JISC lists to promote the *#RioZones* project and encouraging – when live tweeting and vlogging started – debate, discussion and wider dissemination. The *#RioZones* Twitter and WordPress accounts served two key purposes. First, Twitter served as a medium for real-time dissemination of on-the-ground insights and enabled others living or visiting Rio to share their own representations using the hashtag. Second, the *#RioZones* vlog (see Figure 8) served as a platform for consolidating emerging findings and analysis generated every day from the walking methods and digital documentation. At the end of each day, the authors reflected on their observations of the Olympic city, recording a vlog, and sharing that with a wider audience of international academics and industry stakeholders. Conversations via Twitter were archived, representing another data set for analysis.

If using digital methods, it is important to assess reach and potential effects. In total, by June 2018, the *#RioZones* vlog had received over 800 unique visitors, and over 3500 views with a range of prominent authors in the MSE field contributing to the spatial foci of the *#RioZones* project. One of the significant (and lasting) benefits of creating a digital space like *#RioZones* for ongoing research enquiries into MSEs is that it can be used to maintain a level of scrutiny of legacy outcomes (i.e. spatial transformations post-Games) beyond singular event cases. This is particularly critical due to the inherent deficiencies in 'event case study' approaches (see Yin, 2013) that predominantly serve to capture a 'moment in time' and space across what are very complex project lifecycles (e.g. Duignan, 2018).

Beyond microblogging and vlogging activities, in a bid to increase research exposure, the authors co-wrote a series of short, but widely disseminated special-interest articles for 'The Conversation' (www.theconversation.com/uk). These were written for an official dedicated section on the Rio 2016 Olympic Games [references withheld for anonymity]. These articles were read approximately 24,000 times, cited and/or reposted verbatim by a number of other international platforms like 'Yahoo! News'. Furthermore (and partially as a result of these articles) the authors aimed to include a public engagement component to the *#RioZones-Approach* by leveraging physical and digital data findings and channels to generate international, national and regional media attention. As a result, the authors conducted media interviews via the BBC during the Games live from Copacabana [references withheld for anonymity] and after the Games for both regional and international media outlets like Australian national station: ABC [references withheld for anonymity]. These served as an ideal platform to share critical insights beyond traditional communicative routes. Media appearances included regional, national and international level radio and TV broadcasts – and a mixture of both pre-recorded and live segments. Yet, whilst such lines of dissemination do not constitute specific change and research impact, our efforts afforded a public window (which may help to increase public knowledge and shape public opinion) into detailed analytical insights and phenomena, alongside situational, contextual challenges that would rarely be reported across mainstream media channels. It is important to note that all of the findings discussed in this section were disseminated in English – a limitation we acknowledge and will return to in the conclusion.

Limitations, reflections and future considerations

It is important to not only view walking methodologies as simply a 'new' method for extracting already-existing data or to capture and record an environment like a neighbourhood or a defined urban space. Rather, as Springgay and Truman (2017) suggest, walking approaches to research methodology need to be about 'activating problems and concepts in the midst of the event' [online]. In this sense, reflecting on the limitations of our own practice, it is important to go beyond using walking as just another observational technique within which the researcher is armed with technology like smartphones or video cameras to record events. Instead, to be meaningfully disruptive, it is more important that walking approaches to research methodology generate new dialogue and bring about critical responses and actions. For example, in the context of MSE spatial dynamics, walking methodologies could be utilised to work with neighbourhoods affected by change, inviting residents to participate in reclaiming their environment in line with their identified needs. Additionally, if researchers are to prolong the life of the research and actively produce change, a strategic approach into industry and policy network formation is crucial. Although examining and describing the spatial organisation and ordering practices provides a unique insight and window into the phenomenon, to be truly disruptive, researchers need to identify agents for change within the wider international, national, regional and host city situation and context. As part of the ongoing project lifecycle of *#RioZones*, we explicitly targeted influential gatekeepers and key informants to comment on the key emergent themes of the project. We suggest that although MSEs are delivered in different social, economic political environments and cultural contexts, incorporating such stakeholders into MSE analyses can be achieved fairly easily due to the relative stability of individual and organisations who play a primary and secondary role in planning, developing and delivering the project.

Although we appreciate the logistical and access challenges of incorporating sometime conflictual perspectives into a case study analysis, the stable occurrence of all these units represents an opportunity for researchers to interact, research and share findings with those outside of the academy. By triangulating data and bolstering evidence bases – we argue that creative-disruptive methods, as outlined by the *#RioZones-Approach*, may help support greater internal and external validity of results to other host city contexts. Yet, adding to these internal and external complexities, we too need to be aware of our individual positionality across planning and collecting data, and throughout the analysis stages outlined above. Indeed, we could devote an entire research article to this endeavour and the multifaceted ways the prism of our (for example, and not exhaustive list of characteristics) Western, white, middle-class, non-local, non-Portuguese speaking language proficiency directly and indirectly, consciously and unconsciously determined the decisions made across the lifecycle of *#RioZones*. Yet, for now, we reflect on and acknowledge the influence of three key factors.

First, we identify 'limited temporal engagement' as a factor limiting the longitudinal perspective achievable from our study. We attempted to overcome this limitation by remaining highly engaged with the case study context after we had left the research field, using digital methods and through planning the next stages of primary data collection working with key stakeholders located in the city. Second, we identify 'cultural awareness' as a limiting factor because we are acutely aware that our involvement in Rio 2016 provided a relatively superficial understanding of the deeper cultural and institutional political, social and cultural context of the city. We attempted to address this limitation by utilising a series of secondary resources to familiarise ourselves with the environment and, as discussed above, included 17 in-depth interviews with key stakeholders located in Rio in the second phase of the research to provide greater local insights which complemented our own physical and digital approaches. Third, 'non-local perspectives' was also a limitation of the study, linked to deficits in (local) cultural understanding, which affected how we planned and gathered data across the case study setting. Research decisions were largely informed by official Olympic documentation outlining how, where, and when event spaces and transit zones across residential and commercial districts were to be transformed. Such influences impacted, for example, on the date of our arrival, departure, and the spaces we walked, observed and analysed at different time intervals throughout the project. We suggest that researchers, in order to potentially overcome these biases and cultural limitations, may wish to partner with either fellow researchers and/or residents within the case study context to help navigate the host city from the perspective of different population groups. Adopting the technique of supported 'localised walk' may help generate greater, deeper understanding of local settings, culture and the nuances therein that can be often smoothed out and/or ignored in MSE analyses. This practice may well have a key utility value – enabling researchers to access specific places and spaces they may not venture alone without safe and de-risked, and/or without eased passage. Indeed, this factor raises an important issue about the inclusivity and accessibility of our analysis and communication more broadly. Namely, we emphasise the need to work collaboratively with those of native tongue to translate research findings so as to ensure that both domestic and international audiences can engage with research insights generated.

Conclusion

Using the Rio 2016 Olympic Games as a vehicle for empirical analysis, the *#RioZones* project – and specific *#RioZones-Approach* – illustrates the multifaceted, creative ways researchers can interrogate the organisational and spatial arrangements and effects of MSEs. The approaches detailed in this article can be leveraged to disrupt the linear, sometimes entrenched epistemological approaches to generating knowledge about MSEs. We sought not to simply subvert conventional methodological approaches but seek to creatively synthesise new and existing approaches by integrating an amalgam of physical and digital methods together under the umbrella of an on-going case study.

Primarily, the *#RioZones-Approach* involved brought together embodied walking and participatory digital methods to document, capture and build a picture of the spatial organisation, arrangements, dynamics and ordering across the Rio 2016 Olympic city. Specifically, our methods generated insights into how civic public spaces are (temporarily) altered by capturing and documenting 'real time' changes as they occurred in an MSE host city. Real-time observational, photographic, video, and narrated analysis combined to provide the authors with an opportunity to share online microblogging analysis (i.e. via Twitter hashtag: *#RioZones*) with academic and non-academic stakeholders. Digital platforms like Twitter (microblogging) enabled the project, in a timely manner, to share on-the-ground live insights to disseminate initial findings and raw data (i.e. descriptive analysis, visual footage like photos and videos). If used effectively, these digital methods can democratise and extend access to knowledge beyond academia.

Leveraging alternative digital platforms for research enquiry, analysis and dissemination is an important contribution of this research. As a scholar, being receptive and open to opportunities to engage with non-traditional outlets is an additional learning point. By doing so, going outside the 'normal' lines of inquiry and debate, researchers can be disruptive and involve a whole set of stakeholders that they may not always invite for comment and/or for collaboration. Extending the research scope in this way produces an additional primary data set, the opportunity to critique and validate the researchers own descriptions and interpretations of the research setting, and a potentially powerful tool for dissemination and means of influencing both policy and practice within an Olympic host city, or other event and tourism setting. If implemented effectively researchers can bridge the void between physical and digital research activity to extend the reach of their work, invite wider critical commentary and reflection, and strategically align with influential stakeholders who may be able to initiate, and bring about, change.

Scholars of tourism and critical event studies may wish to apply the *#RioZones-Approach* to disrupt – or at least attempt to influence – how and why MSEs are bid for, planned and executed. This is particularly so as the approach goes part-way toward foregrounding the less-visible, often overlooked impacts of MSEs on the local spaces, dwellers and communities affected. Yet, thinking back to our earlier reflections and limitations of the study, there are immediate steps we can take to reconfigure the framework to either strengthen future analysis and/or mould to new case study settings both MSE and non-MSE related. We believe that opening up such insights – both the opportunities and challenges bestowed on host cities and communities – may have the potential to lead to greater democratic accountability and governance arrangements that promote fairer, more egalitarian MSE projects. For example, at the bid stage, utilising the networked publics enabled by digital platforms can lead to initial plans coming into the public domain more quickly, enabling affected groups and organisations to mobilise themselves at a faster rate. Additionally, competing claims to 'truth' (about why the bid will be beneficial) can be aired and debated (and analysed by researchers), and researchers can work with activists and social movements to fact check, informing the debates as they develop. For example, in recent years, bid books have now become public documents, partly because of increasing lobby groups highlighted earlier and leaks on social media.

Beyond the live, spectacular elements of event production, the *#RioZones* project illustrates the very specific ways researchers may maintain a level of examination and scrutiny past the Closing Ceremony. It is here, that the article touches on a central-critical point: the need to examine access to opportunity across the entirety of the Olympic (project) lifecycle – to examine, amplify and tackle inevitable urban struggles and exclusion so prevalent in the construction, lead up, live staging and legacy periods of MSEs. Leveraging digital platforms, projects like *#RioZones* can outlive the brief snapshot in time and duration of the 'event case study' itself. We suggest that whilst our research provides a much-needed window into such matters, academic and non-academic debate, and a method for engaging with influential event, industry stakeholders and policy makers we urge scholars to consider embedding themselves in host city contexts for more significant time-scales to access deeper interrogations. We believe this is of importance to critical studies more broadly (i.e. critical organisational and management studies), and more specifically in terms of sector studies like critical events and critical tourism studies too – wishing to breakdown structural barriers leading to the marginalisation or subjugation of side-lined social groups at the heart of Olympic cities.

Disclosure statement

No potential conflict of interest was reported by the authors.

ORCID

Michael B. Duignan http://orcid.org/0000-0003-1539-2310

References

Adams, M., & Guy, S. (2007). Editorial: Senses and the city. *Senses & Society*, *2*(2), 133–136. doi: 10.2752/174589307X203047

Bates, C., & Rhys-Taylor, A. (2017). Finding our Feet. In C. Bates & A. Rhys-Taylor (Eds), *Walking through social research*. London: Routledge.

Boykoff, J. (2012). *Celebration capitalism and the Olympic games*. London: Routledge.

Boykoff, J. (2015). Sochi 2014. In R. Gruneau & J. Horne (Eds), *Mega-events and globalization: Capital and spectacle in a changing world order*. London: Routledge.

Broudehoux, A.-M., & Sanchez, F. (2015). The politics of mega event planning in Rio de Janeiro: Contesting the Olympic city of exception. In V. Viehoff & G. Poynter (Eds), *Mega event cities: Urban legacies of global sport events* (pp. 109–123). London: Routledge.

Cade, N., Everett, S., & Duignan, M. B. (2019). Leveraging digital and physical spaces to 'de-risk' and access Rio's favela communities. *Tourism Geographies*, 1.). doi:10.1080/14616688.2019.1571099

Chevalier, J. M., & Buckles, D. J. (2013). *Participatory action research: Theory and methods for engaged inquiry*. Abingdon: Routledge.

Coaffee, J. (2015). The uneven geographies of the Olympic carceral: From exceptionalism to normalisation. *The Geographical Journal*, *181*(3), 199–211. doi:10.1111/geoj.12081

Dansero, E., Mela, A., & Rossignolo, C. (2015). Legacies of Turin 2006 eight years on: Theories on territorialization in the aftermath of the Olympic Games. In V. Viehoff & G. Poynter (Eds), *Mega event cities: Urban legacies of global sport events* (pp. 99–107). London: Routledge.

Duignan, M. B. (2018). London's local Olympic legacy: Small business displacement, 'Clone Town' effect and the Production of 'Urban Blandscapes'. *Journal of Place Management and Development*, doi:10.1108/JPMD-05-2018-0033

Duignan, M. B., Everett, S., Walsh, L., & Cade, N. (2017). Leveraging physical and digital liminoidal spaces: The case of the #EATCambridge Festival. Tourism Geographies, *1*, 1–22.

Duignan, M. B., & Pappalepore, I. (2019). Visitor (im)mobility, leisure consumption and mega-event impact: The territorialisation of Greenwich and small business exclusion at the London 2012 Olympics. *Leisure Studies*, *38* (2), 160–174. doi:10.1080/02614367.2019.1572212

Duignan, M. B., Pappalepore, I., & Everett, S. (2019). The 'summer of discontent': Exclusion and communal resistance at the London 2012 Olympics. *Tourism Management*, *70*, 355–367. doi: 10.1016/j.tourman.2018.08.029

Edensor, T. (1998). The culture of the Indian Street. In N. Fyfe (Ed), *Images of the street: Planning, identity and control in public space*. London: Routledge

Fletcher, T., & Platt, L. (2018). (Just) a walk with the dog? Animal geographies and negotiating walking spaces. Social and *Cultural Geography*, *19* (2), 211–219.

Frew, M., & McGillivray, D. (2008). Exploring hyper-experiences: Performing the fan at Germany 2006. *Journal of Sport & Tourism*, *13*(3), 181–198.

Gaffney, C. (2016). Gentrifications in pre-Olympic Rio de Janeiro. *Urban Geography*, *37*(8), 1132–1153. doi:10.1080/02723638.2015.1096115

Gandy, M. (2017). Urban atmospheres. *Cultural Geographies*, *24* (3), 353–374. doi:10.1177/1474474017712995

Getz, D. (2018). *Event evaluation: Theory and methods for event management and tourism*. Oxford: Goodfellow.

Giulianotti, R., Armstrong, G., Hales, G., & Hobbs, D. (2015). Global sport mega-events and the politics of mobility: The case of the London 2012 Olympics. *The British Journal of Sociology*, *66*(1), 118–140. doi:10.1111/1468-4446.12103

Gogishvili, D. (2018). Baku formula 1 city circuit: Exploring temporary spaces of exception. *Cities*, *74*, 169–178. doi:10.1016/j.cities.2017.11.018

Hagemann, A. (2010). From the stadium to the fan zone: Host cities in a state of emergency. *Soccer & Society*, *11*, 723–736. doi:10.1080/14660970.2010.510729

House of Lords (2013). Keeping the Flame Alive: the Olympic and Paralympic Legacy. Retrieved from: http://www.publications.parliament.uk/pa/ld201314/ldselect/ldolympic/78/78.pdf

Ingold, T., & Vergunst, J. L. (2008). *Ways of walking: Ethnography and practice on foot*. Burlington, VT: Ashgate.

Kirby, S. I., Duignan, M. B., & McGillivray, D. (2018). Mega-sport events, micro and small business leveraging: Introducing the "MSE-MSE leverage model". *Event Management*, *22*(6), 917–931. doi:10.3727/152599518X15346132863184

Klauser, F. (2011). The exemplification of 'Fan Zones': Mediating mechanisms in the reproduction of best practices for security and branding at Euro. *Urban Studies*, *48*(15), 3203–3219. doi: 10.1177/0042098011422390

Lupton, D. (2015). *Digital sociology*. London: Routledge

MacRury, I., & Poynter, G. (2009). *Olympic cities: 2012 and the remaking of London*. Farnham: Ashgate .

McGillivray, D. (2018). Sport events, space and the 'Live City'. Cities, https://doi.org/10.1016/j.cities.2018.09.007

McGillivray, D., Duignan, M. B., & Mielke, E. (2019). Mega sport events and spatial management: Zoning space across Rio's 2016 Olympic city. *Annals of Leisure Research*. doi:10.1080/11745398.2019.1607509

McGillivray, D., & Frew, M. (2015). From Fan Parks to Live Sites: Mega events and the territorialisation of urban space. *Urban Studies*, *52*(14), 2649–2663. doi:10.1177/0042098014550456

Marshall, C., & Rossman, G. B. (1989). *Designing qualitative research*. Newbury Park, CA: Sage.

Miller, P. (1987). *Domination and power*. London: Routledge and Kegan Paul.

Müller, M. (2015). The mega-event syndrome: Why so much goes wrong in mega-event planning and what to do about it. *Journal of the American Planning Association*, *81*(1), 6–17. doi:10.1080/01944363.2015.1038292

Müller, M. (2017). Approaching paradox: Loving and hating mega-events. *Tourism Management*, *62*, 234–241. doi:10.1016/j.tourman.2017.06.003

OECD. (2008). Local Development Benefits from Staging Global Events: Achieving the Local Development Legacy from 2012. Retrieved from http://www.oecd.org/regional/leed/46207013.pdf

Osborn, G., & Smith, A. (2016). Olympic Brandscapes: London 2012 and the seeping commercialisation of public space. In G. Poynter, V. Viehoff, & Y. Li (Eds), *The London Olympics and urban development: The mega-event city* (pp. 139–253). Abingdon, New York: Routledge.

Pappalepore, I., & Duignan, M. B. (2016). The London 2012 Cultural Programme: A consideration of Olympic impacts and legacies for small creative organisations in East London. *Tourism Management*, *54*, 344–355. doi:10.1016/j.tourman.2015.11.015

Pavoni, A. (2017). *Controlling urban events*, London: Routledge

Petersen, A. (1997). Risk, governance and the new public health. In A. Petersen & R. Bunton (Eds), *Foucault, health and medicine*. London: Routledge.

Pfeffer, J., & Salancik, G. R. (1978). *The external control of organizations: A resource dependence perspective*. New York, NY: Harper and Row.

Pink, S., Horst, H., Postill, J., Hjorth, L., Lewis, T., & Tacchi, J. (2015). *Ethnography in a digital world, digital ethnography: Principles and practice*. London: Sage.

Raco, M., & Tunney, E. (2010). Visibilities and invisibilities in urban development: Small business communities and the London Olympics 2012. *Urban Studies*, *47*(10), 2069–2091. doi:10.1177/0042098009357351

Reason, P., & Bradbury, H. (2013). *The Sage handbook of action research: Participative inquiry and practice* (2nd ed). London: Sage.

Rose, N. (1999). *Powers of freedom*. Cambridge: Cambridge University Press.

Smith, A. (2013). Borrowing public space to stage mega-events. *Urban Studies*, *21*, 247–263. doi:10.1177/0042098013489746

Smith, A. (2016). *Events in the city: Using public spaces as event venues*. London: Routledge.

Springgay, S., & Truman, S. E. (2017). On the need for methods beyond proceduralism: Speculative middles, (In) tensions, and response-ability in research. Qualitative Inquiry, *24*, 203–214.

Springgay, S., & Truman, S. E. (2018). *Walking methodologies in a more-than-human world: WalkingLab*. London, UK: Routledge.

Steinbrink, M. (2013). Festifavelisation: Mega events, slums and strategic city-staging – the example of Rio de Janeiro. *Die Erde*, *144* (2), 129–145.

Talbot, A., & Carter, T. (2018). Human rights abuses at the Rio 2016 Olympics: Activism and the media. *Leisure Studies*, *37*(1), 77–88. doi:10.1080/02614367.2017.1318162

Weber-Neuth, F. (2016). Landscapes of London 2012: 'adiZones' and the production of (corporate) Olympic space. In A. Tomlinson (Eds), *The Olympic legacy: Social scientific explorations*. London: Routledge.

Work Foundation. (2010). A Lasting Legacy: How can London Fully Exploit the Opportunities Created by the Olympic Games in the Recovery and to 2030?. Retrieved from http://www.theworkfoundation.com/Assets/Docs/olympics011210.pdf.

Yin, R. (2013). *Case study research: Design and methods* (5th ed.). London: Sage Publication.

'Que será, será!': creative analytical practice within the critical sports event tourism discourse

Richard Keith Wright

ABSTRACT

Sometimes it is easier to not think too long or hard about the future, and to just accept that 'what will be, will be' (that the future is pre-determined or out of our control). My paper invites you to consider the application of socially-constructed creative analytical practice (CAP) within the production and consumption of critical sports event tourism discourse. In January 2017, following months of media speculation about the future of several major tourism-generating sports events, 20 experienced event professionals located across multiple time zones were asked to respond to a piece of thought-provoking narrative foresight. The scenario offered several reasons as to why Auckland, New Zealand, could submit a bid to host the 2032 Summer Olympic Games. The consensus amongst the respondents was that the dream would never become a reality, but a number of alternative options were suggested offering some meaningful insight into the minds and mindset of those employed to deliver major sports event experiences. Having scripted a piece of creative non-fiction, I seek to temporarily transport you into the middle of an ongoing conversation occurring all over the globe around the long-term future of the world's biggest multi-sports events. My goal is to mix hindsight, insight and foresight. I hope to offer something evocative, provocative and, most of all, informative. The story concludes with a new piece of narrative foresight for you all to consider.

摘要

有时候, 不要太久或太深入地思考未来的事情, 对将来要发生的事情（有时候未来的事情是预先决定好了的, 或者不在我们的控制之下）接受就是了, 这样会更容易一些。我的论文邀请您考虑社会建构的创造性分析实践(CAP)在批判性体育赛事旅游话语生产和消费中的应用。2017年1月, 在媒体对几个主要的旅游体育赛事的未来进行了数月的推测之后, 我们要求来自多个时区的20名经验丰富的赛事专业人士对一篇发人深省的未来体育赛事的叙述做出回应。这一设想为新西兰奥克兰申办2032年夏季奥运会提供了几个理由。受访者的共识是, 这个梦想永远不会成为现实, 但他们提出了一些可供选择的方案, 以提供一些有意义的见解, 了解那些受雇提供重大体育赛事经验的人士的想法和心态。在编写了一篇创造性的非小说类作品之后, 我想暂时把您带入一场正在全球各地进行的对话之中, 这场对话围绕着世界上最大的多项体育赛事的长期未来展开。我的目标是把后见之明、目前的见解和远见结合起来。我希望能提供一些能引起共鸣、具有挑衅性的东西，最重要的是，能提供一些有益的信息。这个故事以一种全新的未来叙事视角结尾，值得大家深思。

1. The prologue: 'the show *never* stops'

> '… the popular wisdom among event and destination marketers is that an event must have or create a distinct brand if it is going to be a useful tool for destination branding… Destination marketers who seek to use events to build their brand must construct a portfolio of events. A single event – even one with a high profile – has only a passing effect on the destination brand'. (Chalip & Costa, 2005, p. 228)

The fate of cities and countries looking to host major or mega tourism-generating, sports events is typically decided by a small number of experts, empowered to assess the authenticity and accuracy of the promises provided during the bidding process (Gupta, 2009; Masterman, 2008; McGillivray & Turner, 2018; Oliver, 2014; Salisbury, 2017). The International Olympic Committee's (IOC) request that bidding nations to prove that their applications have the backing of the general public as well as cross-party political leaders became a formal requirement in 2014, following the creation of their Agenda 2020 roadmap (MacAloon, 2016; McGillivray & Turner, 2018; Muller, 2015). While the costs and benefits of hosting large-scale sporting fixtures have been widely documented, debated and occasionally disputed over the past two decades, the is little doubt that these events generate opportunities for the hosts to change perceptions (Chiam & Cheng, 2013; Preuss & Alfs, 2011) and profit, if only temporarily, from an increase in sports event tourism activity (Chalip & Costa, 2005; Tien, Huai-Chun, & Hsiou-Wei, 2011).

Sport event tourism is, according to Getz (2003, p. 49), 'internationally recognised as a substantial and highly desirable niche market' and destination marketers have long seen hosting one-off and recurring sports events as an effective means of enhancing the brand image of the host destination and enticing visitors (both domestic and international) into a region (Chalip & Costa, 2005; Getz, 2003; Green, Costa & Fitzgerald, 2003; Morgan & Wright, 2008). In 2004, three highly respected management scholars argued that sport tourism and event tourism were essentially one and the same, noting the significant amount of focus being placed on the movement and motivations of those who travel specifically to experience major sports events (Deery, Jago, & Fredline, 2004).

I consider 'sports event tourism' to incorporate all travel and tourism-related activity that occurs as a direct consequence of the tangible and intangible legacies established through hosting a sports event. This includes domestic and international travel to actively participate in a sports event (elite or amateur) and/or to experience a sports event (of any form) as a spectator. The inclusion of 'tangible and intangible legacies' allows me to include the thousands, if not millions, of people who travel to visit sites of sporting significance before and after they have hosted a major tourism-generating sports event (i.e. a pilgrimage to Indianapolis or Monaco, Augusta or St Andrews, Wembley or Wimbledon). My deliberately broad definition also acknowledges that the leveraging of inherent legacies can begin the moment that a city is confirmed as the host of a major sports event, and that tourists who are inspired to visit a place that has hosted one or numerous sports events should also be considered as sports event tourists. An inherent legacy is an outcome that can be considered an unavoidable consequence of hosting a major event (e.g. a new or upgraded sports venue, increased accommodation options, improved public transport etc.).

Despite the extensive library of sports events and event tourism research that exists within the academy (Preuss, 2015; Thomson, Cuskelly, Toohey, Kennelly, Burton, & Fredline, 2018), promises of lasting legacies and an influx of high yielding international visitors continue to be promoted during the event bidding phase only to be broken, or conveniently forgotten, once the rights to the event has been won, the event has been delivered and the eyes of the world shift from the present to the future (i.e. the next event). In 2012, Gibson, Kaplanidou, and Kang (2012) acknowledged the growing concerns regarding the long-term sustainability of positive legacies from hosting mega sports events, and the increasing amount of evidence that disputes the notion that hosting sports events helps destinations to attract tourists during the post-event years. Muller (2015) speaks about the emergence of, and the negative publicity attached to, a Mega Event Syndrome (MES), whilst McGillivray and Turner (2018) acknowledge an ever-expanding community of well-informed individuals, connected largely through social media, are publicly challenging those who dare adopt the mantra of 'if you build it, they will come'.

The core premise of MES is that the world has too many white elephants to believe the narrative foresight found in many of the mega event bidding documents produced during the past 30 years (Muller, 2015). It supports Chalip and Costa (2005) conclusion's that successfully staging international sport events can help tourist destinations secure the rights to host more in the future but raises questions around what is a 'successful' event and illustrates all the inherent legacies that have, over time, failed to meet the needs and expectations of the host community (McGillivray & Turner, 2018; Muller, 2015). A number of international sports organisations, for example, have recently found themselves having to adopt new strategies in order to find local communities willing to turn their fields into elite sports facilities (McGillivray & Turner, 2018). Furthermore, within the closely connected sports, event and tourism industries, there would appear to be a systematic failure when it comes to documenting and delivering the learnings gathered from the past (Werner, Dickson & Hyde, 2015; Wright, 2007).

Although there are plenty of case studies that focus on the positive and negative impacts of mega sports events, there remains a distinct lack of literature on the strategic planning that occurs prior to and during the sports event bidding process (Hartman & Zandberg, 2015; McGillivray & Turner, 2018). According to Milojevic and Inayatullah (2015, p. 161), narrative foresight 'seeks to investigate current modes of knowing the world, challenge detrimental and non-functioning schemas and open up alternatives'. It is 'an evolving pedagogy, a process that gives primacy to story and uses narratives to make specific strategies for change more viable' and 'a tool that enables those who utilise it to become more proficient creators of their own narratives' (p. 161).

To date, much of the existing sports tourism literature has targeted the thoughts of the sport event tourist before, during or not long after the consumption of a lived experience. Although some scholars have looked at the internal stakeholders, both paid and unpaid, few have sought the opinions of the event producers, without whom there would arguably be no event (Cashman & Harris, 2012; Parent, 2016; Parent, Rouillard, & Naraine, 2017). The aim of this paper is to articulate the vast amount of knowledge and expertise that exists within the world of sports event production, much of which is ignored when politicians decide to submit and support a bid for a major sports event. According to Matthews (2008, p. 9), the term 'producer' is often used interchangeably with 'manager' and 'planner', referring 'to the person responsible for coordinating and executing – and occasionally assisting with creating – the event, particularly the technical side that involves design, scheduling, staging, sound, lights, A-V, entertainment, and décor. In most event situations, the producer is not the event manager but is contracted by the event manager as defined above, who thereby becomes the 'client' of the producer'. The producers, and event production team, are employed once the bidding process has concluded and the a new host has been officially announced (Cashman & Harris, 2012; Matthews, 2008). Their voices are rarely heard, and they are very rarely seen during an event, preferring to work backstage and behind the scenes (Matthews, 2008). The best sport event producer's typically have their next employment position lined up prior to their current contract coming to an end, and it is rarely in the same country, let alone region (Cashman & Harris, 2012). In other words, they are not around to oversee the delivery of post-event legacies within the host community.

The future of New Zealand's tourism industry has proven a popular topic of academic discussion and public debate (Gillovic & McIntosh, 2015; Wallace & Riley, 2015; Yeoman et al., 2015; Yeoman, Palomino-Schalscha, & McMahon-Beattie, 2015). Since 2012, Auckland, home to approximately a third of the New Zealand's 4.5 million population, has been ranked within the world's top three sports event tourism destinations on three separate occasions, including a couple of first places in the small/medium sized categories of the ultimate sport event city awards (Auckland Council, 2016; Stuff, 2016). The gateway city is known locally and internationally as 'the City of Sails', having hosted two America's Cup regattas in 2000 and 2003. It is marketed domestically as a place where 'the show never stops', and established a major events portfolio that has proven attractive to sports fans from all over the world.

In 2016, I challenged a cohort of sports event management students to tell me why Auckland would never become an Olympic City, encouraging them to document their initial socially-constructed reaction to the statement, before critically examining the subject matter with an open mind and an objective, evidence-based, approach. They were asked to consider the various different scenarios that would have to eventuate for Auckland Council, the New Zealand government and the New Zealand public to support a bid for the rights to host the 2032 Summer Olympic Games. Their arguments, both for and against, were fiercely debated for over an hour in the classroom before a vote was taken (the outcome was closer than you may imagine). This debate inspired the creation of a scenario, more specifically a piece of thought-provoking narrative foresight.

In January 2017, the scenario created in a classroom on a university campus was emailed to 20 professional event producers located all over the world, all of whom had at least a decade of relevant major sports event experience. A personalised invitation was also attached, explaining the aims of the researcher and stressing the importance of them providing an honest and candid personal response based entirely on the benefit of hindsight. The experts were assured that their identities would remain hidden and that their subjective, socially-constructed, insight would be (re)presented in a manner that ensured they could not be identified. Fifteen agreed to participate, two politely declined and three never responded.

By turning the transcripts constructed from their responses into a piece of staged dialogue, I hope to transport you into the middle of an ongoing highly complex, equally political, global debate. I also seek to showcase the potential of employing creative analytical practice (CAP), and more specifically creative non-fiction (CNF), to share the thoughts of industry experts 'verbatim' (Saldaña, 2012). The ethnodrama created offers insight into the mind and mindset of those responsible for delivering inherent international sports events legacies (both tangible and intangible). I have not put words into their mouth, but I have employed some CAP when it comes to the construction of the narratives presented in this paper. In keeping with the principles and practice of producing creative non-fiction, for example, I have chosen to remove the ever-present temptation to re-interpret their responses or add my own bias and subjectivity into the conversation (Hackley, 2007).

The paper is split into five parts, including this introductory scene-setting prologue. In part 2, I discuss the theory and philosophies behind creative analytical practice (CAP), including the use of Creative Non-Fiction (CNF) and ethnodrama. In part 3, I present a piece of staged dialogue packaged as a private round-table conversation between the 15 experienced industry experts. Part 4 provides you with an overview of why and how the 2016 scenario was created, allowing you to join the dots and draw your own conclusions having already consumed those of the respondents (if you find yourself getting lost in part 3, please jump ahead to this part and then return to the staged dialogue at the end). I conclude part 4 with a discussion deliberately designed to leave you wanting the conversations to continue. Finally, part 5 arrives in the form of an epilogue, which covers the new realities (the known knowns) that have emerged over the last 12 months (i.e. since the creation of this ethnodrama). My paper concludes with a second piece of thought-provoking narrative foresight for you and my 15 respondents to take into the future.

2. Creative analytical practice; creative non-fiction & ethnodrama

As the 20th century drew to a close, Laurel Richardson published a trilogy of evocative essays that showcased several alternative ethnographical practices, all packaged within a genre that she referred to a creative analytical practice (CAP) (Richardson, 1995, 1997, 1999). In the decade that followed, Richardson supported her vision of the future through a couple more publications that focussed on the process, the power and the potential of both producing and consuming creative non-fiction, conversation, drama, fiction, hypertext, layered text, new journalism, poetry, reader's theatre, satire, songs, true fiction and writing stories (Richardson 2000, Richardson & St Pierre, 2005). This section provides an overview of the thinking and theories that provide the platform for this paper, and my non-traditional approach to engaging in critical sports event tourism research.

CAP is, according to Parry and Johnson (2007), founded on a politics of hope that encourages the reader to consider and create their own conclusions and allows a new generation of scholars to share their social explorations in a more consumer friendly, easily understood, manner. CAP has proven particularly popular with poststructuralists who wish to establish a personal connection with both the participants of their research activity and those who access/assess the outputs generated as a consequence (Denzin & Lincoln, 2005; Wright, 2018). It is perfect for authors and artists who want to draw their audience into a storied narrative (Wright, 2018).

Lincoln and Guba (2005, p. 211) define CAP narratives as being 'texts that seek to break the binary between science and literature, to portray the contradiction and truth of human experience'. Berbary (2015) suggests that the employment of CAP represents more than just a choice or a whim, but a distinctive mindset 'grounded in a thoughtful resistance and careful intention to be forthright about our abilities and inabilities to capture and re-present reality (p. 50). She notes how the storyteller can adopt dialogic and/or juxtaposed narratives to illustrate the complexity of the subject matter, sharing of life history, life story, testimonials and personal narratives to elicit deeper understanding of the subject area (Wright, 2018).

Creative non-fiction (CNF) can be produced in many different ways, including evocative stories that utilise creative writing techniques and other types of conversational, multi-vocal and critical representations (Caulley, 2008; Gerard, 2004; Hackley, 2007; Lott, 2000; Root, 2008; Smith, 2013; Wright, 2017). What separates CNF from other forms of narrative-based scholarship, however, is the degree to which the academic storyteller embraces the creative techniques used on a daily basis by journalists and the producers of literary fiction in order to transfer their ideas and opinions across to their audience (Gerard, 2004; Hackley, 2007; Smith, 2013; Wright, 2018). The goal is simply to evoke a reaction through the production of a narrative that is deemed truly unique, original, authentic and believable (Berbary, 2015; Richardson, 1995; Wright, 2018).

Lincoln and Denzin (2003, p. 377) define ethnodrama as a method capable of opening up 'institutions and their practices for critical inspection and evaluation'. The epistemology is based on the theoretical concept of nascency (Saldaña, 2005). In essence, the judgement of an ethnodrama's value and validity is not for the author to dictate, but for each individual audience member to determine in their own time (Saldaña, 2005). In 2010, Saldaña concluded that the majority staged dialogues were framed as works of CNF, based on truthful (re)constructions of stories extracted from interviews and fieldnotes. It is more than the reporting of an informal interview conversation or a structured question and answer session.

The producers of CNF and ethnodrama purposefully seek to enhance an individual's deeper understanding in a manner that changes attitudes and social behaviours, if not immediately then at some point in the future (Saldaña, 2012). Staged dialogue can be scripted verbatim, turning a participant's personal account of a lived experience into a narrative that requires the reader to immerse themselves into the story, filling in the gaps and joining up the dots when/where-ever necessary (Mienczakowski, 2001; Saldaña, 2012). Saldaña (2012) differentiates 'ethnodrama' from 'ethnotheater', defining the former as 'the written transformation and adaptation of ethnographic research data (e.g. interview transcripts, participant observation fieldnotes, journals, documents, statistics) into a dramatic playscript' (p. 284).

Although ethnodrama is frequently referred to as a new or emergent form of arts-based performance ethnography (Mienczakowski, 2001; Mienczakowski & Morgan, 2001; Morgan, Jones, Gilbourne, & Llewellyn 2013; Saldaña 2010), anthropologist Victor Turner experimented with something strikingly similar over thirty years ago (Saldaña, 2010). In 1982, Turner proposed that 'embodied re-enactment' offered students the opportunity to gain a deeper understanding, suggesting it would be 'more fun' to 'turn the more interesting portions of ethnographies into play scripts, then to act them out in class' (Turner, 1982, p. 89). According to Morgan et al, (2013, p. 522) the authors of ethnodramatic narratives exploit the well-established principles of participatory and interactive theatre, inviting their audience 'to negotiate and construct understanding and meaning from ethnographic performances in order to effect change'. They can be packaged and presented within a number of different forms. A staged dialogue, for example, is different to a scripted monologue, although both need to move a storyline from start to finish (Saldaña, 2010).

Morgan et al (2013, p. 523) argue that the principal purpose of an ethnodrama is to deliver eye-opening and easy to consume personal narratives that 'inform, disturb, confront and hold the attention of an audience'. Mienczakowski and Morgan (2001) also acknowledge the importance of producing evocative 'informant-led' scenarios that inspire personal reflection, internal confrontation, a period of critical self-examination, plus the desire for further deconstruction and future discussion. Barone (2002) notes the narrator's desire to provoke, as opposed to persuade, and to generate new understanding without providing all of the answers. In sum, it places the emphasis on learning through the construction and consumption of your own critical conclusions (Wright, 2018).

Gray et al. (2000) provide a compelling argument for the adoption of research-based theatre, noting how it can be used to address the difficulty of disseminating valuable, life-changing, academic research into society. They note 'the potential to present research material in a way that helps to clarify and transform social understandings', arguing that the most valid and valuable insight always occurs as a result of 'audience engagement' (p. 138). They also propose that scripted ethnodrama has advantages over traditional text-based approaches as it can present a more accurate/authentic account of a remains 'lived reality' (p. 138). It is not without its critics, however, and Stuttaford et al. (2006) outline the ethical challenges connected to producing an ethnodrama, raising concerns about transcript selection and participant confidentiality.

As noted above, the benefits of employing creative non-fiction, and more specifically the artful genre of ethnodrama, can be found in the hours, days and weeks after one has consumed it (i.e. the legacy created from its production). Reading this type of work can leave you inspired or frustrated. It should leave you better informed at about a certain situation or scenario. What it should not do is leave you feeling confused, or questioning the influence or bias of the author. In sum, it provides a place for the voices of professionals to be presented unfiltered and unedited by anyone other than themselves. The audience is therefore not left with an expert's socially-constructed interpretation of what was said, but with an expert's socially-constructed interpretation of what they were asked to consider and comment upon. The reader therefore gets an authentic account of what the respondent was thinking at that moment in time (verbatim).

One could argue that there are a host of likeminded phenomenological methodologies that can/could achieve the same goals, but what distinguishes creative analytical practices (CAP) ethnographies, an ethnodrama, from other personal narratives is the fact that they rarely conform with the rules and structures established within the academy (Richardson, 1999). Storytelling within the social scientists is not new, but Saldaña (2005, p.4) reminds his readers that 'Theatres in the western world has been telling stories for over 2500 years and, more often than not, representing social life on stage – interpreted artistically by playwrights and actors with perceptive insight into human condition'. He adds that 'Playwrights are and always have been ethnodramatists, for what other source for a drama *is* there besides social life?' (p. 4).

Academics wanting to adopt a narrative approach are typically directed towards the seminal work of Donald Polkinghorne, Jerome Bruner, Catherine Kohler Riessman or Thomas Schwandt. Those wanting to embrace CAP and ethnodrama can take inspiration from a much wider field, including the timeless work of William Shakespeare, William Wordsworth, Ernest Hemmingway, J. R. R. Tolkien and J. K. Rowling. My personal inspirations are fictional authors Ian Fleming, W. P. Kinsella, Nick Hornby and Ben Elton and sporting biographer Christopher Hilton, but there are literally thousands of places that an aspiring playwright can look for inspiration. The same applies to those who we want to read our work. History would suggest that industry practitioners do not read academic journal articles, unless they are completing a postgraduate qualification or an academic (past, present or in waiting). Academics looking for promotion are judged by the number of other academics who cite their work, but also the 'impact' that their work has had on society. The production of an evocative Ethnodrama therefore offers scholars a chance to produce something for an audience of millions, as opposed to one of likeminded peers (Gray et al, 2000; Saldaña, 2010).

The scripted dialogue presented in the following section is preceded by the 300 word scenario that my respondents were asked to assess on their own, in their own time, without forwarding it or discussing it with their peers. The response has been interwoven into a playscript for others, yourself included, to act out. It is presented as if all 15 of my respondents were invited to enter the same meeting room, sit around the same table, at the same time, and argue the merits and miscalculations of the same 300 word scenario. In reality, even with the advent of video-conferencing technology, a single focus group featuring all of these experts would have been impossible to orchestrate. Different work commitments on four different continents, within seven different countries and nine different time zones meant that I was never going to get these people together at the same time. Thankfully, however, that was never the intention.

As noted in the prologue, I seek to illustrate the power of employing and exploiting CAP to capture personal viewpoints often overlooked within traditional academic research. Ethical approval was obtained prior to the initial invitation being sent and the sports event producers were each offered the option to withdraw their response at any time during the analysis phase. Each respondent was encouraged to take their time, and edit their own response prior to returning it to me. When printed out, the length of the emails received between January and the start of March 2017 varied from a single sheet of A4 paper to seven and a half pages. Several of the respondents sent me more than one email, whilst some included attachments and links to news stories (some continue to do so to this day).

3. 'Stranger things have happened!'

Despite travel and tourism recently replacing dairy as being New Zealand's top export (Bradley, 2017), there has been little exploration into the importance of sustaining the geographically isolated nation's reputation as an award-winning host of international sports events. The following scenario was constructed as a direct consequence of that debate. Once you've consumed it, you may want to take a minute to consider its contents before you continue.

3.1. The 2016 scenario

On 13 September 2017, the city of Los Angeles, California, will be awarded the rights to host the 2024 Summer Olympic Games, following Paris' decision to negotiate a deal to stage the 34th edition four years later (without the need to submit another bid). Having delivered the best World Masters Games, the major events team of Auckland's award winning Tourism, Events and Economic Development Agency (ATEED) will be encouraged to attend the Gold Coast's Commonwealth Games (April 2018) to explore the feasibility of a New Zealand bid for the 35th Summer Olympic Games (2032). Knowing that New Zealand's conservative population is unlikely to support it, this initial work will be done in secret. The low-cost 'back to basics' bid will be built around the use of existing and temporary venues, many of which will have already been used to host international sports events.

To gain public support, the government will promise to use the bid to help secure the money needed to fast track a number of proposed infrastructure improvements across the Auckland region. The public will also be reminded that the IOC's Agenda 2020 document promises a sizable reduction in the financial risks associated with staging the world's largest sports event and promotes a genuine desire to make the games more accessible to smaller nations who are yet to host their showpiece competition. An Australian city is expected to be the strongest candidates to host the Games of the XXXV Olympiad in September 2032, with Brisbane or Perth already being touted as Australia's next Olympic City (following Melbourne and Sydney). The IOC, however, have already suggested that they would love to see New Zealand host an Olympic Games, and the games of the 35th Olympiad (2032) could potentially be Auckland's to lose!'.

3.2. The 2017 response

SARAH: '2032 won't be hosted by anyone in Europe or North America, and Asia's got it in 2020… so, assuming they want it, I'm sure that the Aussies will get their 3rd Olympic Games… I don't see it coming to Auckland, not in my lifetime!'.

JIM: '2032 could possibly go to a South African city… like Durban, who have the Commonwealth Games on the horizon, or Cape Town, or even somewhere else in South America… Like Buenos Aires, who are hosting next year's Youth Olympic Games. Personally, I think it will end up in Dubai, or somewhere that won't need to ask the IOC to help them pay for it. It will never reach New Zealand's shores, and that is not a bad thing either. We're too isolated. If it's going to go back to Australia then I can't see it going to anyone other than Melbourne… or, perhaps, even Sydney once again'.

FRANK: 'You could add Doha to the list of potential hosts for 2032, if they can deliver a decent FIFA World Cup in 2022… and Dubai will surely be interested, assuming they do a good job with the World Expo in 2020. If not 2032, then one of them will get it in 2036. It's just a matter of time. It's too soon to return to South America. Rio was a PR disaster for the IOC. The Aussies got it after it was in Atlanta and the smart money suggests it will be in LA in 2028. The Americans will concede 2024 before the French do, and they'll probably work it so 2028 is the most profitable games ever!'

WAYNE: 'I think it will be in LA in 2028, not Paris, and it'll no doubt be the biggest games ever seen. The hosts of 2032 will have a pretty tough act to follow. I think New Zealand is better off staying out of that game, and focusing on the returns it can get from other smaller sports events. Let the Aussies have it in 2032, in Melbourne or Sydney, not Brisbane or Perth. People could visit NZ on the way!'.

STEVEN: 'I'd be open to New Zealand investing a few million dollars in bidding for the event, but wouldn't want Auckland to win'.

FRANK: 'We've all seen the negative press. Mega events cost the local community millions, leaving debts that impact upon entire generations. I can't say that it will never be hosted in New Zealand, but why would anyone pay millions just to put their hand up for that kind of damage?'.

MARY: '… a bid would cost millions. It's just not worth the investment, and never will be. That's why they're struggling to find people wanting to host it… it wasn't that long ago that you would see five or six candidate cities all fighting for the rights, but not now. They can't give them away!'.

AMANDA: 'Unlikely, but I guess that stranger things have happened… New Zealand could ask the IOC to help them pay for it'.

FRANK: 'ATEED cannot afford to sit back and watch others take their place. There will be a bunch of them at the Gold Coast Commonwealth Games, and that they'll be looking to bring that back to Auckland'.

JIM: 'I can easily imagine a conversation occurring… between the Mayor of Auckland and the Chairman of ATEED… I can see a bid for that [the Commonwealth Games] in the not too distant future, but not an Olympics. Personally, I think the Aussies would be more interested in getting the FIFA World Cup in 2030 or 2034. Why would they want the Olympics back after all the issues that followed the Sydney Games?'.

SARAH: 'I'm guessing that the conversations have already started about bidding for the 2026 Commonwealth Games. The Olympics comes with a price tag that is well beyond the reach of our little country. The 2032 Games could end up in either Brisbane or Perth, assuming it goes to Australia… Melbourne doesn't need it, and Sydney definitely won't want it again'.

BOB: 'If done right, an unsuccessful Olympic bid would promote the country to billions of people and attract new visitors these shore [New Zealand]. We definitely don't need the hassle of having to host it. We just have to show the world that we would like to host it, and remind them of all the reasons why New Zealand would be a great place to visit. By 2025 we will have hosted the America's Cup at least once more, possibly twice, plus a couple of other smaller world cups or world championships. The 2026 or 2030 Commonwealth Games are the obvious next choice, or maybe a joint Trans-Tasman bid for the 2034 FIFA World Cup?'.

JIM: 'I just can't see the public buying into your dream of Auckland becoming an Olympic city. New Zealanders are way too risk averse, and too conservative. Even this conversation would scare a lot of them. Politicians are only interested in keeping their voters happy, and the public won't want to see their hard earned money being spent on an event that the majority of the country wouldn't be able to see. Trust me, it's just not going to happen!'.

MIKE: 'Stranger things have happened, but it won't be coming here [New Zealand] in my lifetime. Maybe sometime in the 50's or 60's when New Zealand has a population of over ten million and they've run out of other options? But 2032 is way too soon'.

AMANDA: 'What about Christchurch? The South Island is what most international visitors picture when they think of New Zealand, and the people down there have spent the past couple of decades helping to pay for all the events that Auckland has hosted'.

MARY: 'Maybe Christchurch should enter a bid for the 2026 Commonwealth Games. They need something to help them speed up the rebuild, and they have always loved their sport down there. They had it years ago, before Auckland'.

JIM: 'Sydney never showed any interest in it [the Olympics], until Melbourne started to assess the possibilities. If little old Christchurch did the same, then I am sure that Auckland's councillors would soon follow, pointing out all the reasons why it would just "have" to be up here. It's all politics!'

MIKE: 'The 2030 Winter Olympic Games could be something for New Zealand's South Island to seriously consider. If the IOC can award 2022 to Beijing, then Christchurch could be in with a shout for 2030, assuming 2026 was held somewhere in Europe. The NZ winter games is a great event, and regularly attracts the best athletes in the world. But New Zealand couldn't host the Commonwealth Games during the same year, so the people of Auckland and Christchurch would need to prioritise one over the other'.

MARY: 'I'd love to see a South Island bid for the 2030 Winter Olympics, but 2026 is too close. From memory, they've not got the infrastructure to host anything of that size, but it could easily be built up over the next decade in time for 2030. I think it would be good for Auckland, and the country as a whole. They've not sorted out who is going to host it after Beijing, so there is plenty of time to get all our ducks in a row'.

MARY: 'I have another idea for you… ATEED should look into the costs associated with hosting the Youth Olympics?'.

BOB: 'If Christchurch hosted the Commonwealth Games, then I'm sure we could get the local politicians excited about the thought of one day hosting an Olympics, even if it's only the Youth one. Stranger things have happened'.

SARAH: 'The 2018 YOG is in Buenos Aires and the four "potential" bidders for the 2023 summer games are Gaborone, Hong Kong, Rotterdam and Munich. They're going to make the decision the same day they announce whether LA or Paris get to go first in 2024. Auckland should look at a bid for 2027, before the Aussies decide they want it'.

CLIVE: 'Sometimes coming second is better than winning. An unsuccessful bid would get people thinking about why we lost, and could result in a change in thinking. We could get everything we need as a result of our bid being unsuccessful. That has happened elsewhere in the world'.

MARY: 'Any formal expression of interest would show the IOC that Auckland is at least open to the idea of becoming an Olympic city'.

ANDY: 'The IOC will have to make their events more affordable for cities to host it, assuming they want them to continue… New Zealand could host the Summer and Winter versions of the Youth Games. The summer one in Auckland in 2027, and the winter one down in Christchurch in 2029. Why not?'.

CLIVE: 'ATEED need something that captures the imagination, and generates some international headlines. A Netball or Hockey World Cup won't do that, and neither will the Women's Cricket World Cup or another America's Cup Regatta. I'd like to see them aim for something bigger… and nothing compares to an Olympics!'.

MIKE: 'It feels like Auckland has bounced from one major sports event to another over the past decade, without ever stopping to assess the legacies left by them all, like how we can leverage more repeat tourism activity. I think we've done pretty well, but there is only one way you can go from the top'.

JIM: 'Has Auckland peaked? Because New Zealand's not going to get the Rugby or Cricket World Cups back again for at least another 20 years. They've had a good innings, but it was never going to be sustainable over the long-term. There are only so many events to go around… Que será, será'.

JANE: 'These decisions are made by politicians and not the public or the people working in the events industry. I do worry that New Zealand's hospitality sector is going to struggle over the next couple of years. The hotels have also had it pretty good for the past decade. Perhaps an unsuccessful bid would force people to start asking some awkward questions about what's possible and what is not, and whether Auckland has aspirations to ever be seen as a global city?'.

ANDY: 'They could just focus on the events portfolio that ATEED has already. …try to get more repeat visitation. The public aren't stupid. They will want to see some tangible evidence of the benefits'.

JIM: 'New Zealand could use all the facilities built for the 2026 Commonwealth Games, and if they do give 2032 to the Aussies, then this conversation needs to be repeated in fifteen years' time. Auckland should aim for the 2044 Olympics, not that I'd be around to see it happen. But, then again, stranger things have happened!'.

4. 'Que será, será'

> 'Que Será, Será,
>
> whatever will be, will be,
>
> the future's not ours to see,
>
> Que Será, Será'
>
> (Livingston & Evan's, 1956)

Sometimes it is easier to not think too long and hard about the future, and to just accept that 'what will be, will be' (that the future is pre-determined or out of our control). When asked to consider the long term future of events, Donald Getz struggled to picture an increasingly mobile and global population without them, noting their importance when it comes to facilitating social and economic exchanges, promising highly desired experiences, embodying cultural differences, communicating symbolic meanings, and nourishing both individual and group identity (Getz, 2015). Their role in society may not have changed, but the practice of deciding where major sports events are hosted, how much they cost and what legacies they're expected to leave behind for the local community appears to be in a constant state of change. As noted in part one, to date, most of the sports event tourism literature has looked at the legacies as a post-event impact attached to something that has already occurred. Little has been done in terms of looking further into the future, trying to predict or plan for where these events may end up being hosted.

CAP is all about facilitating deeper learning through being critical, and not just consuming or copying what others have done (Wright, 2018). It's about asking the audience to fill in the blanks, to think critically about the narrative or script presented and, ideally, to add their socially constructed viewpoint to the discourse (Richardson, 2000). To date, traditional sports tourism research has tended to focus on sharing descriptions of past event experiences, making suggestions about the future in the form of equally descriptive recommendations. This paper asked a small group of experts to imagine what is going to happen in the future, and then to describe the potential impact of events that are unlikely to ever occur.

The lyrics created by Jay Livingston and Ray Evans, made famous by Doris Day in the 1956 Alfred Hitchcock movie 'The Man who knew too much', have been sung by many an English football fan in the stands of many an English football stadium, myself included. For those struggling to picture the scene, the third line is creatively changed to 'we're going to Wembley', signifying the hopes and dreams of all those who sing it (that this could be the year!). The song came to mind when thinking about a title to capture the essence of this script. Having looked back over the transcripts, I found it hidden in one of the statements made by one of the respondents (Jim). With the benefit of hindsight, I feel it adequately articulates the perceived lack of foresight evident within the planning of future sports event tourism scenarios.

Scenario planners attempt to remove the risks attached to employing a socially constructed 'Que será, será' approach to attracting sports event tourists, providing the tools required to test multiple scenarios created from professional insight gained through personal hindsight (Bussey, 2014; Lindgren & Bandhold, 2009; Milojevic & Inayatullah 2015; Postma & Liebl, 2005). In sum, scenario planners seek to ensure that the future is ours to see, to manage and to manipulate in order to meet the needs of those in the future. Paul Schoemaker, one of the leading advocates for the employment of scenario planning methodology, concluded that many of the mistakes of yesterday and today could have been avoided had those involved been encouraged to critique all the potential unknowns (known and unknown) (Schoemaker, 1995). He identified that;

> ... it is useful to consider three classes of knowledge: the things we know we know, the things we know we don't know and the things we don't know we don't know. Various bias – overconfidence, under- and over-prediction, the tendency to look for confirming evidence – plague all three, but the greatest havoc is caused by the third... (Schoemaker, 1995, p. 38)

The narrative foresight located in part two was constructed in a classroom in September 2016, but its origins lie in an online news report published on the 21 October 2011. The headline to that evocative narrative was 'NZ Could host Olympics – IOC Head', and the opening statement read 'New Zealand is not too small to host an Olympic Games, the head of the International Olympic Committee says' (NZ Herald, 2011). The story was published whilst Jacque Rogue was in New Zealand for the 2011 Rugby World Cup, and quoted the former IOC chief as saying;

> 'No it [New Zealand] is not too small, you can organise the Olympics with the population of only 4.4 million. Whether it would be useful for the country, only you can determine, it is not to the IOC to tell that. But it is a possibility. Finland organised the Olympic Games with a population of 5 million, Norway has hosted a Winter Olympic Games ... with the same population as New Zealand'. (NZ Herald, 2011)

Towards the end of 2015, following the online publication of a similar article entitled 'IOC Boss: NZ Could host Olympics', I began to formulate a plan to explore this subject area in more detail. The second article quoted Thomas Bach, Jacque Rogue's successor at the IOC, as saying "You cannot restrict the right to host the Olympics to just 20 countries. The Olympic Games are universal and we should open doors and windows' (Leggat, 2015). The article acknowledges Bach's involvement in the creation and launch of Agenda 2020, a 2013 document that promised to make the Olympic Games more attractive and affordable to new host cities.

The consensus amongst the 15 experts supported the outcome of my on-campus debate, although the end result was by no means as close as it was in the classroom. Those with over a decade's worth of experience in the event industry were unable to overlook the emergence of Muller's mega event syndrome, citing the high costs, the myriad uncertainties and the multitude of proven risks attached to being an Olympic candidate city (successful or not). The majority could clearly envision conversations occurring between the local politicians and the destination marketers around a possible bid for the 2030 Commonwealth Games, but not the 35th Olympiad. All 15 expressed their doubts that the New Zealand public would ever support a proposal to invest in bidding for something that has received so much negative media coverage over the past decade. Interestingly, however, the idea that politicians within the South Island's two biggest regions could come together to propose a bid for the 2030 Winter Olympic Games was tabled on several occasions, with several suggesting that such a bold statement of intent could also lead to those living in Auckland experiencing a sudden change of heart (i.e. choosing to back a bid to host a Commonwealth Games or a Youth Olympic Games). The expression 'stranger things have happened' was used on multiple occasions, whilst the option of not bidding for anything and letting the rest of the world worry about the positive and negative legacies attached to hosting tourism-generating sports events was also widely dismissed by the majority.

Looking to the future, it has been argued that creative non-fiction, including ethnodramas and other forms of creative analytical practice (CAP) can resonate with academic and non-academic audiences in a way that more traditional textual documents cannot (Richardson & St Pierre, 2005). As a result, various academic fields and discipline areas have begun to embrace and accept CAP approaches to ethnographical research, including education, anthropology, sociology, sport and health care (Saldaña, 2012). The aim of my paper was not to replicate the work of Laurel Richardson, but to personally invite you to assess its potential value and validity in terms of the production of critical sports event tourism research. Whilst I seek to inspire, to inform and to generate interest, I am not claiming to know what the future holds, nor am I trying to persuade you to follow in my footsteps. I merely seek to evoke a meaningful reaction amongst my peers, to provoke a memorable response, and to inspire a more critical assessment of the current academic and industry discourse. It is my belief that applying CAP provides sport, event and tourism scholars with a valid, equally valuable, way of documenting a disseminating the knowledge and discourse that already exists within these overlapping industries.

Morgan et al. (2013, p. 522) note the critical importance of being able to transition the reader's mindset 'from the cognitive to the emotional realm, where the attitudes and feelings of the audience can be tapped into in order to stimulate engagement'. They conclude that provoking audience participation can not only generate new understanding, but also stimulate critical self-reflection and a re-evaluation of the value/validity of one's established personal beliefs and past lived experiences (Morgan et al., 2013). My goal was to provide you with an alternative approach to presenting critical sports event tourism research. I also wanted to produce something that would provoke a reaction, positive or negative. Ultimately, the value of my contribution to this special edition is for you, my audience, to debate and determine in your own time. As identified in part three, the power and potential of CAP lies in its proven ability to engage its target audience's imagination, generating new knowledge through the new conversations and conclusions created in the future (Richardson, 2000). The staged dialogue presented in the previous section offers only a snapshot of the data captured. It was created to kickstart future conversations amongst those interested in this niche, special interest, subject.

5. The epilogue: 'back to the future'

In March 2017, one the eve of Auckland hosting arguably the best ever World Masters Games, the Commonwealth Games Federation (CGF) confirmed what the global media had been speculating for several months, that Durban would no longer be hosting their 2022 showpiece event and that they were now looking for an alternative location. Although Auckland's name was mentioned in a number of media stories, the city's mayor was quick to deny any possibility of entering the race and no bid was forthcoming (McMillan, 2017; Radio Sport, 2017). In December 2017, it was announced that the only city to have already confirmed their interest in hosting the 2026 event, Birmingham, had successfully secured the hosting rights for the event four years earlier than planned, having overcome a late challenge from Liverpool, a city located less than 160 kms away. Mackay (2018a) reports that Birmingham paid 28 million euros ($46.5 million New Zealand Dollars) for the rights.

On 11 July 2017, just as predicted, the International Olympic Committee (IOC) 'voted unanimously to move forward with a dual award for the 2024 and '28 Summer Games' (Maase, 2017). What I had failed to foresee in 2016, however, was that the IOC would change the rules midway through the contest, making an offer that the American's couldn't refuse. Maase (2017) opened his report of the unprecedented action, which essentially ensured that both bidders walked away happy, by stating that 'LA has decided its Olympic aspirations can wait four years'. He later revealed that the IOC had agreed to not only 'waive more than $50 million in fees and organizing costs' and 'contribute any of the IOC's profits of the 2028 Games back to the city of Los Angeles', but also 'advance the organizing committee $180 million over next five years to begin 2028 preparations'. Maase's (2017) cites several extracts from the official news conference in which the Los Angeles Mayor, Eric Garcetti, claimed that 'the deal' was 'too good to pass up' and that 'the Games in 2028 will give us the seed of a new Olympic legacy here in Los Angeles and around the world … because we can build that legacy right now instead of waiting until the Olympics are over' (Garcetti, 2017, cited in Maase, 2017).

On the 24 April 2018, having failed to attract any confirmed bidders and missed several deadlines related to the 2023 Youth Olympic Games, the IOC announced in yet another unprecedented move that they wanted to 'take the Youth Olympic Games 2022 to Africa' (IOC, 2018). In October 2018, the IOC announced that the proposed 2023 YOG would be hosted in 2022 in Dakar, Senegal. At the same session, they announced that the three confirmed bidders for the 2026 Winter Olympic Games were Calgary, Milan-Cortina' d'Amprezzo (a joint bid) and Stockholm, (Morgan, 2018). The following month, the Calgary bid was withdrawn following a plebiscite in which 56.4% of local residents voted against the bid (Palmer, 2018) and, on the 11 January 2019, the Swedish bid became a joint bid with Stockholm teaming up with an alpine ski resort located 620 kilometres away called Are (Morgan, 2019). Morgan (2019) reveals how the IOC had been forced to change the Olympic Charter, allowing both bids to be branded with more than one city.

To conclude, I hope that my exploration into these ongoing conversations has added something insightful and memorable to the existing sports event tourism discourse. The purpose of this epilogue was to showcase the extent to which the rules of engagement appear open to constant negotiation and unpredictable outcomes. Why is this important? Well, it suggests that destinations looking to become Olympic or Commonwealth Games cities over the next two decades need to make themselves known to the relevant people as soon as possible. The fact that the IOC has agreed to contribute a significant amount of money to the hosts of the 2022 YOG, the 2026 Winter Games and the 2028 Summer Games, plus allow joint bids from cities located in different regions, would suggest that it is now a buyer's market. Those considering the costs and benefits attached to bidding for the rights of the 2030 Winter and the 2032 Summer Olympic Games will now be able to argue for similar concessions.

With all the above in mind, as promised, I have constructed another piece of thought provoking narrative foresight for you and my 15 experts to contemplate. It was written on the Tuesday 29 January 2019 (Auckland's Anniversary Day). I would be very happy to receive your responses via email, should you feel the need to join the conversation and contribute to the critical sports event tourism discourse.

5.1. The 2019 scenario

The winning bids for both the 2026 and 2030 Commonwealth Games are currently scheduled to be decided and announced in September 2019, with the deadline for expressions of interest having been recorded as the end of December 2018 (Mackay, 2018b). At the time of writing (February 2019), Adelaide are the only city to have publicly announced an interest in hosting the 2026 Games, despite media speculation about Kingston and Kuala Lumpar (Mackay, 2018b). Despite the cities of Victoria and Edmonton both failing to gain the local support needed to bid for the 2022 or 2026 Games, the Canadian Commonwealth Games Committee have asked the CGF to award their nation the 2030 Commonwealth Games without them having to go through the bidding process (Mackay, 2018c). Interestingly, neither Auckland nor Christchurch's name has been included as a possible bidder for either the 2026 or 2030 games. Whether the conversations are occurring in private, however, remains to be seen. I would not be shocked to see Auckland enter the race before the September deadline, or after the deadline is extended. The announcement could possibly be made to coincide with the Black Caps winning the Cricket World Cup on the 14 July, or the All Blacks winning their third straight Rugby World Cup on the 2 November. Stranger things have happened.

In 2021, whilst Acklander bask in the glory of hosting its third America's Cup regatta and the Women's Rugby World Cup, Snowsports New Zealand, working alongside the three relevant regional development agencies in Canterbury and Otago will in 2021 submit an ambitious proposal to bid for the 2030 Winter Olympic Games to New Zealand's South Island. The bid will divide the nation, but be based on the promise of significant investment from the IOC who, at the time, will be under pressure to explain how the 2022 Winter Olympics was awarded to Beijing following their failure to deliver on many of the legacies promised prior to the 2008 Summer Games. In 2021, the IOC will also be keen to show that their event can be hosted in the Southern Hemisphere, by a small country with a population of less than five million (on an Island with a population of just over one million). The New Zealand cities of Christchurch and Dunedin are between 250 and 500 and kilometres from the popular Ski Resorts in Queenstown and Wanaka, which is considerable closer than the Swedish cities who will be awarded the right to the 2026 Winter Olympic Games.

The overwhelming favourite to win the right for the 2030 Winter Olympics will be the 1972 host city of Sapporo, Japan, who withdrew their bid for the 2026 game in September 2018 in order to focus on the 2030 event. Based on the assumption that the bid will prove unsuccessful, New Zealanders will reluctantly agree to support the 2030 South Island Winter Olympic bid, with the local politicians in Auckland ensuring that they receive similar support when it comes to them bidding for the 2034 Commonwealth Games. They will also demand that they get to host both semi-finals and the final of the 2023 FIFA Women's World Cup should they enter the race and be announced the winner on 5 June 2019.

The Australian government will agree to support the New Zealand bid, even if New Zealand challenges them for the rights to the 2023 FIFA Women's World Cup and, at some stage in the bidding process, the mayors of both Melbourne and Sydney will offer to host the opening and closing ceremonies (should Christchurch or Dunedin struggle to find the funds needed). The Australians will also support this bid in return for New Zealand's support when it comes to winning their bid for a third Olympic Games (2032) and their first FIFA World Cup (2034).

Ultimately, the offer to help will not be needed as the IOC will opt for the safer option of hosting it in Sapporo, offering New Zealand the chance to host the 2024 or 2028 Youth Winter Olympic Games before they bid to host the 2034 or 2038 Winter Olympics. Such an offer will capture the imagination of the politicians in Auckland, who will argue that they should also have aspirations of becoming an Olympic City, and lead to them submitting a formal expression of interest in bidding to host the Summer Olympics of 2032 (a bid they will not expect to win, especially once Melbourne enters the race). The sole intention of challenging for the rights to host the 2032 Olympics would be to get the IOC to award them the Youth Summer Olympic Games in 2031. Liverpool will turn its failed 2022 Commonwealth Games bid into a bid for the 2027 Youth Olympic Games, taking it into Europe for the first time, and ensuring it is no longer hosted during the same year as the Winter Olympics, the FIFA World Cup and the Commonwealth Games.

'Que sera, sera. What will be, will be!'

Acknowledgements

The author would like to acknowledge and thank the contribution of the 15 industry experts who embraced the opportunity to both think and talk about the future.

Funding

The author received no funding for this study.

Disclosure statement

No potential conflict of interest was reported by the authors.

References

Auckland Council. (2016). Auckland wins at Ultimate Sports Cities Awards. Retrieved from http://ourauckland.aucklandcouncil.govt.nz/articles/news/2016/04/auckland-wins-at-ultimate-sports-cities-awards/

Barone, T. (2002). From genre blurring to audience blending: Reflections on the field emanating from an ethnodrama. *Anthropology & Education Quarterly*, *33*(2), 255–267. doi:10.1525/aeq.2002.33.2.255

Berbary, L. (2015). Creative analytic practices: Onto-epistemological and theoretical attachments, uses, and constructions within humanist qualitative leisure research. *International Leisure Review*, *2*, 27–55.

Bradley, G. (2017). Tourism roars past diary as NZ's biggest export earner. Retrieved from https://www.nzherald.co.nz/business/news/article.cfm?c_id=3&objectid=11847120

Bussey, M. (2014). Extending the scenario horizon: Putting narratives to work. *Journal of Futures Studies*, *18*(3), 95–100.

Cashman, R., & Harris, R. (2012). *The Australian Olympic caravan 2000 to 2012: A unique Olympic events industry*. Sydney: Walla Walla Press.

Caulley, D. (2008). Making qualitative research reports less boring: The techniques of writing creative nonfiction. *Qualitative Inquiry*, *14*(3), 424–449. doi:10.1177/1077800407311961

Chalip, L., & Costa, C. (2005). Sport event tourism and the destination brand: Towards a general theory. *Sport in Society*, *8*(2), 218–237. doi:10.1080/17430430500108579

Chiam, M., & Cheng, E. (2013). Residents' perceptions of the inaugural Youth Olympic Games 2010: A cluster analysis. *Event Management*, *17*(4), 377–389. doi:10.3727/152599513X13769392444620

Deery, M., Jago, L., & Fredline, L. (2004). Sport tourism or event tourism: Are they one and the same? *Journal of Sport & Tourism*, *9*(3), 235–245. doi:10.1080/1477508042000320250

Denzin, N. K., & Lincoln, Y. S. (2005). Introduction: The discipline and practice of qualitative research. In N. K. Denzin, & Y. S. Lincoln (Eds.), *Handbook of qualitative research* (pp. 1–33). Thousand Oaks, CA: Sage Publications.

Gerard, P. (2004). *Creative nonfiction: Researching and crafting stories of real life*. Long Grove, IL: Waveland Press.

Getz, D. (2003). Sport event tourism: Planning, development, and marketing. In S. Hudson (Ed.), *Sport and adventure tourism* (pp. 49–85). New York, NY: Haworth Hospitality Press.

Getz, D. (2015). The forms and functions of planned events: Past and future. In I. Yeoman, M. Robertson, U. McMahon-Beattie, E. Backer, & K. A. Smith (Eds.), *The future of events and festivals* (pp. 20–35). Oxon: Routledge.

Gibson, H., Kaplanidou, K., & Kang, S. J. (2012). Small-scale event sport tourism: A case study in sustainable tourism. *Sport Management Review*, *15*(2), 160–170. doi:10.1016/j.smr.2011.08.013

Gillovic, B., & McIntosh, A. (2015). Stakeholder perspectives of the future of accessible tourism in New Zealand. *Journal of Tourism Futures*, *1*(3), 223–239. doi:10.1108/JTF-04-2015-0013

Gray, R., Sinding, C., Ivonoffski, V., Fitch, M., Hampson, A., & Greenberg, M. (2000). The use of research-based theatre in a project related to metastatic breast cancer. *Health Expectations*, *3*(2), 137–144. doi:10.1046/j.1369-6513.2000.00071.x

Green, B. C., Costa, C., & Fitzgerald, M. (2003). Marketing the Host City: Analyzing exposure generated by a sport event. *International Journal of Sports Marketing and Sponsorship*, *4*(4), 48–352. doi:10.1108/IJSMS-04-04-2003-B005

Gupta, A. (2009). The globalization of sports, the rise of non-Western nations, and the impact on international sporting events. *The International Journal of the History of Sport*, *26*(12), 1779–1790. doi:10.1080/09523360903172390

Hackley, C. (2007). Special report: Auto-ethnographic consumer research and creative non-fiction: Exploring connections and contrasts from a literary perspective. *Qualitative Market Research: An International Journal*, *10*(1), 98–108. doi:10.1108/13522750710720422

Hartman, S., & Zandberg, T. (2015). The future of mega sport events: Examining the "Dutch Approach" to legacy planning. *Journal of Tourism Futures*, *1*(2), 108–116. doi:10.1108/JTF-12-2014-0002

IOC. (2018). IOC announces Youth Olympic Games 2022 evaluation commission. Retrieved from https://www.olympic.org/news/ioc-announces-youth-olympic-games-2022-evaluation-commission

Leggat, P. (2015). IOC boss: NZ could host Olympics. Retrieved from https://www.nzherald.co.nz/sport/news/article.cfm?c_id=4&objectid=11443564

Lincoln, Y. S., & Guba, E. G. (2005). Paradigmatic controversies, contradictions, and emerging confluences. In N. K. Denzin, & Y. S. Lincoln (Eds.), *Handbook of qualitative research* (pp.191–216). Thousand Oaks, CA: Sage Publications.

Lincoln, Y. S. & Denzin, N. K. (2003). The Revolution in Presentation, In Y. S. Lincoln & N. K. Denzin (Eds.), *Turning points in qualitative research: Tying knots in a handkerchief* (pp. 375–378). Walnut Creek, CA: Altamira Press.

Lindgren, M., & Bandhold, H. (2009). *Scenario planning-revised and updated: The link between future and strategy*. London: Palgrave-MacMillan.

Livingston, J., & Evans, R. (1956). *Que será, será (whatever will be, will be) [recorded by D. Day]*. New York, NY: Columbia.

Lott, B. (2000). Toward a definition of creative nonfiction. *Fourth Genre: Explorations in Nonfiction*, *2*(1), 192–200.

Maase, R. (2017). L.A. Olympic bid committee agrees to wait, giving Paris the 2024 Games'. Retrieved from https://www.washingtonpost.com/sports/olympics/la-olympic-bid-committee-agrees-to-wait-giving-paris-the-2024-games/2017/07/31/ec34eae4-7614-11e7-8839-ec48ec4-cae25_story.html?utm_term=.1f5ef4670a87

MacAloon, J. (2016). Agenda 2020 and the Olympic movement. *Sport in Society*, *19*(6), 767–785. doi:10.1080/17430437.2015.1119960

Mackay, D. (2018a). Birmingham paid £25 million to host 2022 Commonwealth Games. Retrieved from https://www.insidethegames.biz/articles/1063313/birmingham-paid-25-million-to-host-2022-commonwealth-games

Mackay, D. (2018b). Jamaica investigating bid for Commonwealth Games. Retrieved from https://www.insidethegames.biz/articles/1064122/jamaica-investigating-bid-for-commonwealth-games

Mackay, D. (2018c). Canada planning to ask to host centenary Commonwealth Games in 2030. Retrieved from https://www.insidethegames.biz/articles/1063231/canada-planning-to-ask-to-host-centenary-commonwealth-games-in-2030

Masterman, G. (2008). *Losing bids, winning legacies: An examination of the need to plan for Olympic legacies prior to the bidding*. K. Barney, M. Heine, K. Wamsley & G. MacDonald (Eds.), *Pathways: Critiques and discourses in Olympic Research* (pp.171-178). London: University of Western Ontario, International Center for Olympic Research.

Matthews, D. (2008). *Special event production: The process*. Oxon: Butterworth-Heinemann.

McGillivray, D., & Turner, D. (2018). *Event bidding, politics, persuasion and resistance*. Oxon: Routledge.

McMillan, C. I. (2017). Why Auckland could host the Commonwealth Games. Retrieved from https://www.nzherald.co.nz/sport/news/article.cfm?c_id=4&objectid=11809576

Mienczakowski, J. (2001). Ethnodrama: Performed research—limitations and potential. In P. Atkinson, A. Coffey, S. Delamont, J. Lofland, & L. Lofland (Eds.), *Handbook of ethnography* (pp. 468–476). Thousand Oaks, CA: Sage.

Mienczakowski, J., & Morgan, S. (2001). Ethnodrama: Constructing participatory experiential and compelling action research: Participative inquiry and practice. In P. Reason, & H. Bradbury (Eds.), *Handbook of action research: Participatory inquiry and practice* (pp. 219–227). London: Sage.

Milojevic, C. I., & Inayatullah, A. (2015). Narrative foresight. *Futures*, *73*, 151–162. doi:10.1016/j.futures.2015.08.007

Morgan, K., Jones, R. L., Gilbourne, D., & Llewellyn, D. (2013). Changing the face of coach education: Using ethno-drama to depict lived realities. *Physical Education & Sport Pedagogy*, *18*(5), 520–533. doi:10.1080/17408989.2012.690863

Morgan, L. (2018). Senegal officially awarded 2022 Summer Youth Olympic Games at IOC session. https://www.insidethegames.biz/articles/1070831/senegal-officially-awarded-2022-summer-youth-olympic-games-at-ioc-session

Morgan, L. (2019). Stockholm-Åre and Milan-Cortina d'Ampezzo submit bid books for 2026 Winter Olympics and Paralympics to IOC. Retrieved from https://www.insidethegames.biz/articles/1074098/stockholm-are-and-milan-cortina-dampezzo-submit-bid-books-for-2026-winter-olympics-and-paralympics-to-ioc

Morgan, M., & Wright, R. K. (2008). Elite sports tours: Special events with special challenges. In J. Ali-Knight, M. Robertson, A. Fyall, & A. Ladkin (Eds.), *International perspectives of festivals and events*. Oxon: Butterworth-Heinemann; Elsevier.

Muller, M. (2015). The mega-event syndrome: Why so much goes wrong in mega-event planning and what to do about it. *Journal of the American Planning Association*, *81*(1), 6–17.

NZ Herald. (2011). NZ could host Olympics - IOC head. Retrieved from https://www.nzherald.co.nz/sport/news/article.cfm?c_id=4&objectid=10760701

Oliver, R. (2014). The legacies of losing: Rethinking the 'failure' of Toronto's Olympic Games bids. *Sport in Society*, *17*(2), 204–217.

Palmer, D. (2018). Calgary City Council unanimously votes to end 2026 Winter Olympic bid. Retrieved from https://www.insidethegames.biz/articles/1072442/calgary-city-council-unanimously-votes-to-end-2026-winter-olympic-bid

Parent, M. M. (2016). Stakeholder perceptions on the democratic governance of major sports events. *Sport Management Review*, *19*(4), 402–416. doi:10.1016/j.smr.2015.11.003

Parent, M. M., Rouillard, C., & Naraine, M. L. (2017). Network governance of a multi-level, multi-sectoral sport event: Differences in coordinating ties and actors. *Sport Management Review*, *20*(5), 497–509. doi:10.1016/j.smr.2017.02.001

Parry, D. C., & Johnson, C. W. (2007). Contextualizing leisure research to encompass complexity in lived leisure experience: The need for creative analytic practice. *Leisure Sciences*, *29*(2), 119–130. doi:10.1080/01490400601160721

Postma, T. J., & Liebl, F. (2005). How to improve scenario analysis as a strategic management tool? *Technological Forecasting and Social Change*, *72*(2), 161–173. doi:10.1016/j.techfore.2003.11.005

Preuss, H. (2015). A framework for identifying the legacies of a mega sport event. *Leisure Studies*, *34*(6), 643–664. doi:10.1080/02614367.2014.994552

Preuss, H., & Alfs, C. (2011). Signaling through the 2008 Beijing Olympics using mega sport events to change the perception and image of the host. *European Sport Management Quarterly*, *11*(1), 55–71. doi:10.1080/16184742.2010.537362

Radio Sport. (2017). Goff rules out Auckland Commonwealth Games. Retrieved from https://www.nzherald.co.nz/sport/news/article.cfm?c_id=4&objectid=11809532

Richardson, L. (1995). Writing-stories: Co-authoring "the sea monster": A writing-story. *Qualitative Inquiry*, *1*(2), 189–203. doi:10.1177/107780049500100203

Richardson, L. (1997). *Fields of play: Constructing an academic life*. New Brunswick, NJ: Rutgers University Press.

Richardson, L. (1999). Feathers in our CAP. *Journal of Contemporary Ethnography*, *28*(6), 660–668. doi:10.1177/089124199129023767

Richardson, L. (2000). Evaluating ethnography. *Qualitative Inquiry*, *6*(2), 253–255. doi:10.1177/107780040000600207

Richardson, L., & St Pierre, E. A. (2005). Writing: A method of inquiry. In N. K. Denzin, & Y. S. Lincoln (Eds.), *The SAGE handbook of qualitative research* (3rd ed., pp. 959–978). Thousand Oaks, CA: Sage.

Root, R. (2008). *The nonfictionist's guide: On reading and writing creative nonfiction*. Lanham, MD: Rowman & Littlefield Publishers.

Saldaña, J. (Ed.). (2005). *Ethnodrama: An anthology of reality theatre*. Walnut Creek, CA: AltaMira.

Saldaña, J. (2010). Writing ethnodrama: A sampler from educational research. In M. Savin-Baden & C. Howell Major (Eds.) *New approaches to qualitative research: Wisdom and uncertainty* (pp. 61–69). London: Routledge.

Saldaña, J. (2012). Ethnodrama. In L. M. Given (Ed.), *The SAGE encyclopedia of qualitative research methods* (pp. 284–285). Thousand Oaks, CA: Sage.

Salisbury, P. (2017). An analysis of Glasgow s decision to bid for the 2014 Commonwealth Games. *Sport in Society*, *20*(12), 1870–1887. doi:10.1080/17430437.2017.1232365

Schoemaker, P. J. (1995). Scenario planning: A tool for strategic thinking. *Sloan Management Review*, *36*(2), 25–40.

Smith, B. (2013). Sporting spinal cord injuries, social relations, and rehabilitation narratives: An ethnographic creative non-fiction of becoming disabled through sport. *Sociology of Sport Journal*, *30*(2), 132–152. doi:10.1123/ssj.30.2.132

Stuff. (2016). Sport: Auckland named one of world's top sports cities for second time. Retrieved from http://www.stuff.co.nz/auckland/local-news/manukau-courier/79206753/sport-auckland-named-one-of-worlds-top-sports-cities-for-second-time

Stuttaford, M., Bryanston, C., Lewando Hundt, G., Connor, M., Thorogood, M., & Tollman, S. (2006). Use of applied theatre in health research dissemination and data validation: A pilot study from South Africa. *Health: An Interdisciplinary Journal for the Social Study of Health, Illness and Medicine*, *10*(1), 31–45. doi:10.1177/1363459306058985

Tien, C., Huai-Chun, L., & Hsiou-Wei, L. (2011). The economic benefits of mega events: A myth or a reality? A longitudinal study on the Olympic Games. *Journal of Sport Management*, *25*(1), 11–23. doi:10.1123/jsm.25.1.11

Thomson, A., Cuskelly, G., Toohey, K., Kennelly, M., Burton, P., & Fredline, L. (2018). Sport event legacy: A systematic quantitative review of literature. *Sport Management Review*, *22*, 295–321. doi:10.1016/j.smr.2018.06.011

Turner, V. (1982). *From ritual to theatre: The human seriousness of play*. Baltimore, MD: PAJ Publications.

Wallace, S., & Riley, S. (2015). Tourism 2025: An industry perspective. *Journal of Tourism Futures*, *1*(1), 53–57. doi:10.1108/JTF-12-2014-0021

Werner, K., Dickson, G., & Hyde, K. (2015). Coopetition and knowledge transfer dynamics: New Zealand's Regional Tourism Organizations and the 2011 Rugby World Cup. *Event Management*, *19*(3), 365–380. doi:10.3727/152599515X14386220874841

Wright, R. K. (2007). Planning for the great unknown: The challenge of promoting spectator-driven sports event tourism. *International Journal of Tourism Research*, *9*(5), 345–359. doi:10.1002/jtr.640

Wright, R. K. (2017). Looking back (and forth): Acknowledging the people who make personal narratives plausible. *Sport Management Review*, *20*(1), 81–91. doi:10.1016/j.smr.2016.11.004

Wright, R. K. (2018). 'Doing it for Dot': exploring active ageing sport tourism experiences through the medium of creative analytical practice. *Journal of Sport & Tourism*, *22*(2), 93–108. doi.org/10.1080/14775085.2018.1466349 doi:10.1080/14775085.2018.1466349

Yeoman, I., Andrade, A., Leguma, E., Wolf, N., Ezra, P., Tan, R., & McMahon-Beattie, U. (2015). 2050: New Zealand's sustainable future. *Journal of Tourism Futures*, *1*(2), 117–130. doi:10.1108/JTF-12-2014-0003

Yeoman, I., Palomino-Schalscha, M., & McMahon-Beattie, U. (2015). Keeping it pure: Could New Zealand be an eco paradise? *Journal of Tourism Futures*, *1*(1), 19–35. doi:10.1108/JTF-12-2014-0017

OPEN ACCESS

Why is research–practice collaboration so challenging to achieve? A creative tourism experiment

Nancy Duxbury, Fiona Eva Bakas and Cláudia Pato de Carvalho

ABSTRACT

Within tourism research, there has been little attention to research–practice knowledge exchange during the research process nor to practice-based research. This article examines a research-and-application project on creative tourism in which research–practice collaboration is explicitly foregrounded and made central. Through a reflexive process, the challenges this hybrid approach embodies and the pragmatic dilemmas that accompany the complexities of building closer research–practice relations and capturing practice-based knowledge are examined in three strategic areas: developing spaces for ongoing knowledge exchange, enabling practitioners to take on the role of co-researcher, and fostering researchers' close attention to the application side of the project. In the context of the CREATOUR project, hybrid roles question who can do research, reinforce consideration of the added value of research processes for practitioners, and lead researchers to go beyond traditional research activities, with this 'disruptive' context causing tensions, uncertainties, and dynamic co-learning situations. Ongoing interactions over time are necessary to build relations, understanding, and trust, while flexibility and responsiveness are vital to address emerging issues. Training on research–practice collaboration, knowledge transfer, and mentorship techniques for both researchers and practitioners is advised. Challenges in integrating practice-based knowledge directly into research articles suggest a customized communication platform may be a useful 'bridging' mechanism between practice-based and academic knowledge systems.

摘要

摘要:在旅游研究领域, 很少关注研究过程中的研究与实践的知识交流, 也很少关注基于实践的研究。本文研究了一个创意旅游方面的研究应用项目, 该项目明确以研究与实践的合作为中心展开研究。本文通过一个反思的过程从三个方面（发展持续进行知识交流的空间,推动实业界认识承担合作研究人员的视角以及促进研究者密切关注项目的应用面）思考了这种混合方法表现出来的挑战以及伴随建立更紧密的研究实践关系, 获取实践性知识的复杂所提出的实际困难。在创意旅游项目的背景下, 多元混搭角色, 用这种产生紧张、不确定性和动态共同学习状况的"颠覆性"背景, 质问了谁可以做研究，强化研究过程给从业者带来附加价值的想法，以及带领研究人员超越传统的研究活动等方面的问题。为了建立关系、理解和信任，需要不断地进行互动，同时灵活性和反应能力对于解决新出现的问题至关重要。建议为研究人员和实践者提供研究-实践协作、知识转移和辅导技巧方面的培训。将基于实践的知识直接集成到研究文章中所面临的挑战表明，定制的交流平台可能是基于实践的知识系统和学术知识系统之间有用的“桥梁”机制。

Introduction

The connections (or lack thereof) between research and practice have been debated and explored in the educational, business, health, and social work fields for some time (e.g. Palinkas, 2010; Walsh, Tushman, Kimberly, Starbuck, & Ashford, 2007), and practice-based and practice-led research methodologies have been extensively discussed in artistic fields (e.g. Freeman, 2010; Leavy, 2015). Within the tourism field, however, there is very limited research on the methodological intricacies of practice-based research and knowledge development. Developing closer links between research and practice in tourism has been discussed and, indeed, there is a call for academics to simplify and share the findings of research projects that may inform strategic decision-making by developing an understanding of the stylistic requirements for translating their knowledge to industry (Walters, Burns, & Stettler, 2015). To this end, some research does produce industry-sensitive materials; however, most methodological proposals focus on knowledge transfer and mobilization of academic findings in the post-research period, rather than the transfer occurring *during* the research period (Anderson & Sanga, 2018).

Collaborative research focuses primarily on the perspectives, interests, and needs of participants, requiring that these participants influence the entire research process, from project design, data gathering, and data analysis, to the presentation of final results (Lassiter, 2005). This entails close consideration of the complementarities of academic research and practice-based knowledges (Jeannotte & Duxbury, 2015), and the development of methodologies to weave them together. Practice-based research means that questions and methodology emerge through making, doing, and testing things out (Hope, 2016). Practitioner-researchers do not just think their way out of a problem, they 'practice to a resolution' (Nelson, 2013, p. 10), with practice viewed as more than just doing – as Wenger (1998) remarked, it is 'doing in a historical and social context that gives structure and meaning to what we do' (p. 47). Within the tourism field, however, the potential of practice-based research approaches has also not been explored.

The research-and-application project that informs this article aims to explicitly foreground research–practice relations, with the imperative to bridge research and practice made central to the project. In this context, knowledge exchange and mobilization are seen as integral dimensions to catalyze and manage during the research project, not only in the post-research period, and practice-based experiences and insights are a valued dimension of the research. This experimental approach promises to provide insights for researchers seeking to develop closer research–practice relations and to capture practice-based knowledge within collaborative tourism research projects.

This article reports on a reflexive process that has examined the complexities of building closer research–practice relations and capturing practice-based knowledge within a research-and-application project. Guided by the long-standing question, 'Why has the task of closer research–practice collaboration been so challenging to achieve?', it focusses on the difficulties this hybrid approach embodies and the pragmatic dilemmas that accompany such work. In particular, the article examines and reflects on the 'grubby business' of research–practice collaboration in three areas that are central to this project:

1. *Developing spaces for ongoing knowledge exchange*, fostering informal discussion, learning, and knowledge-building, in which researchers and participating organizations develop relationships and opportunities to interweave complementary types of knowledge;
2. *Enabling practitioners to take on the role of co-researcher*, involving participating organizations in research tasks and knowledge co-creation, changing the norms of researcher–participant relations and expanding upon the concept of reciprocity; and
3. *Fostering researchers' close attention to the application side of the project*, requiring researchers to attend carefully to 'application' and 'implementation' as an integral part of the overall project and, potentially, to act 'beyond' their usual research work roles.

Altogether, these aspects produce implications for the research processes and the roles and identities of researchers in this context. Within the research process, the question of who can do research is challenged, a reinforced questioning of the added value of the research process takes place, and researchers must reflect upon the complexities of making knowledge useful for practitioners and their wider communities. Consequently, the processes and strategies implemented in a research-and-practice project, and the issues realized in the design and implementation of these approaches, instigate both systemic and individual/professional dilemmas, discomforts, and messiness, with the act of stepping outside one's comfort zone becoming a part of the process.

This article was developed through a reflexive process involving three of the researchers in the project.[1] Reflexivity is an interpersonal process through which a person considers the relational and inter-subjective processes they are involved in, gives new meaning to the processes, and recognizes the active role they assumed in guiding events. Using a reflexive process as a method, true learning is only perceived to occur after one has been through a learning experience and taken the time to make sense of the experience. Understanding reflexivity as a process (Finlay, 2002), the authors of this article explore the mutual meanings shared within relationships, focusing on the situated and negotiated nature of these meanings. Drawing on lived experiences of participating within the research-and-application project, the researchers reflect on the uncomfortable aspects of how actions unfolded in praxis, and why things happened in various aspects of the overall process. In using the method of the reflexive process (Cheek, Lipschitz, Abrams, Vago, & Nakamura, 2015), this article aims to avoid sanitizing the reporting of the research process, creating qualitative research that shines a spotlight on how researchers and practitioners navigate tensions in designing and carrying out research together.

To contextualize the research-and-application approach of this project, the article begins by presenting a synthesis of research areas that address the issue of research--practice collaboration, from macro/systemic perspectives to micro/researcher perspectives. This is followed by a brief overview of creative tourism, the CREATOUR research-and-application project, and the participating organizations within it. Then, the article examines the three dimensions outlined above and discusses issues encountered in aiming to design strategies to encourage research–practice collaboration. In closing, it brings together the insights from these three examinations to comment on the ways in which a research-and-practice project disrupts methodologies and research practice more generally.

Bridging research and practice: a literature review

Discussions of knowledge exchange during research projects promoting research–practice collaboration are addressed in research from a macro/system perspective in the fields of knowledge transfer and mobilization, and stakeholder theory. Regarding the promotion of more egalitarian roles between researcher and practitioner, research relating to engaged scholarship, para-ethnography, researcher reflexivity, and reciprocity provide insights on approaches to disrupting 'traditional' research approaches.

Knowledge transfer and mobilization

Scanning broadly, there is a seemingly exponential growth of both research and practical activity in the field of knowledge transfer and mobilization, most prominently within the medical research field (Ward, 2017) and education (Vanderlinde & van Braak, 2010). The growing literature includes both a research-related approach and a more practical/application approach (Ferlie, Crilly, Jashapara, & Peckham, 2012; Greenhalgh & Wieringa, 2011; Oborn, Barrett, & Racko, 2013). Within this literature, only limited efforts have explored knowledge transfer and mobilization strategies in the tourism field (e.g. Tribe & Liburd, 2016) and there are only limited empirical studies on knowledge management in the hospitality industry (Hallin & Marnburg, 2008). Knowledge transfer and mobilization research in the tourism field has tended to focus on the commercial added-value that academics can bring to the public and private sector as consultants (Ryan, 2001), the transfer of academic knowledge to the industry via tourism graduates (Anderson & Sanga, 2018), and calls for universities to reconsider their current incentive systems that focus predominantly on journal publications that achieve minimal reach outside the academy (Walters et al., 2015).

The dominant knowledge transfer paradigm, reflected in these examples, has been criticized for reinforcing a hierarchical and linear model of knowledge creation. In this paradigm, 'knowledge is transferred from knowledge 'producers' to knowledge 'users' such as practitioners, government, and industry actors and less often the lay public, who are all perceived as deficient in scientific understanding' (Anderson & McLachlan, 2016, p. 297; Estabrooks et al., 2008). Attempting to address this issue, knowledge mobilization is put forward to provide a less hierarchical framework, based on 'the reciprocal and complementary flow and uptake of research knowledge between researchers, knowledge brokers and knowledge users – both within and beyond academia – in such a way that may benefit users and create positive impacts...' (SSHRC, 2018). Anderson and McLachlan (2016) take up the concept of knowledge mobilization to frame collaborative research processes in which academic researchers 'work to valorize multiple ways of knowing in the co-production of knowledge that is mobilized in intentional processes of social change, ... engag[ing] with community and social movement actors as co-enquirers through horizontal processes of research, learning, and action' (Kemmis, McTaggart, & Nixon, 2014, pp. 297–298).

Nonetheless, knowledge mobilization can sometimes reproduce an emphasis on only research-to-practice knowledge flows. For example, Ward (2017) refers to knowledge mobilization as the process of moving knowledge to where it can be most useful, relating across different sectors of society and with different fields of power relations, and problematizing knowledge power hierarchies (Hart et al., 2013). This work focuses on the need to bridge knowledge production between scientific, practitioner, and policy-related contexts and shines a spotlight on two important aspects: the roles of researchers and practitioners in the process of knowledge creation, and how knowledge users can access and use this knowledge. Stakeholder theory perceives stakeholders, defined as people with the right and capacity to participate in the process, as part of a network of mutual dependencies where their individual interests can be jeopardized in diverse ways, and the satisfaction of all of them is necessary for maintaining the balance of the ensemble (Nilsson, 2007). However, a stakeholder approach is problematic in terms of initially identifying stakeholders, resistance to cooperation by stakeholders, and the fact that building consensus is a time-consuming and costly process (Ooi, 2013). Thus, more collaborative approaches to research that use reflexive understandings of long-term partnerships with diverse stakeholders are currently being developed to promote sustainable tourism development (Cockburn-Wootten, McIntosh, Smith, & Jefferies, 2018).

Greenhalgh (2010) argues for an extension beyond the 'impoverished' notions of knowledge exchange and transfer towards 'engaged scholarship', during which knowledge emerges dialectically when academics and practitioners (or policymakers) must converge to address a problem (Van de Ven, 2007; Van de Ven & Johnson, 2006). Engaged scholarship is a movement that has been growing since 1995, which offers a new way of bridging gaps between the university and civil society, with the values of citizenship and social justice at the core of this process (Beaulieu, Breton, & Brousselle, 2018). The scholarship of engagement means creating a special climate in which academic and civic cultures communicate more continuously and more creatively with each other, enriching the quality of life for all, with the aim to democratize knowledge. Accordingly, researchers' roles are not limited to knowledge production but expand to become 'actors of change' who participate actively in creative intellectual activities with various stakeholders. Experiential learning is core to engaged scholarship and researchers actively seek to contribute to the common good and consider their role of citizen as intrinsic.

Para-ethnography, researcher reflexivity, and reciprocity

Viewing 'researcher-as-expert' approaches as intrusive and detracting from creating knowledge, para-ethnography explores what it means for researchers to take seriously the efforts of their informants in producing academically relevant knowledge. Para-ethnographers are particularly concerned about 'the new predicaments of expertise', and tend to strongly 'reject suppressive idolatries of control' (Wilson & Hollinshead, 2015). Participants are not ethnographers per se, however, and their forms of reflexivity may be distinct from those of researchers hence the term 'para' ethnography (Beech, MacIntosh, & MacLean, 2010). The 'para' element makes itself felt in the multiple ways in which knowledge is produced in a 'side-by-side' way. Para-ethnography promotes forms of researcher–informer collaboration to promote reflexivity among both parties, by treating informants as both partners and observers in theory building. Building on community-engaged principles, the concept of 'knowledge hybridity' is important, whereby the knowledge produced is somewhere in the middle of the interaction between researcher and researched (Diver & Higgins, 2014).

Researcher reflexivity acknowledges the agency of the researchers, researched, academic audiences, and others in producing knowledge (Tribe & Liburd, 2016), and stresses the importance of establishing a non-exploitative and effective working relationship with significant others (Olsen, 2011). Reflexivity promotes the idea that researchers should expose the politics of representation implicated in research in order to represent their participants better (Pillow, 2003). It is also important to help validate qualitative research (Bakas, 2017) and generate 'valid' knowledge (Ateljevic, Harris, Wilson, & Collins, 2005). Research on researcher positionality and reciprocity, investigating questions of how to authentically reciprocate participants' efforts throughout the research process, is a relatively new study area (Trainor & Bouchard, 2013). The most common form of reciprocal act is 'giving-back' to participants by conducting research that benefits these individuals. Extreme reciprocity is 'the help given to research participants that goes beyond simply empowering them through knowledge production and which could perhaps be of use to them in the long-run' (Bakas, 2017, p. 130), while dynamic reciprocity is an *ongoing* practice of exchange of mutual benefit between academic and community research partners in the context of community-engaged research (Diver & Higgins, 2014). In practice, researchers' ability to engage with the researched on a reciprocal basis is often compromised since academic reality is characterized by an increasing pressure for the individual to 'publish or perish' (Bakas, 2017).

Creative tourism

Creative tourism is a relatively young and dynamic field of tourism research and practice, generally defined as a type of tourism that offers travelers the opportunity to develop their creative skills and potential through active participation in creative experiences which are characteristic of the place where they are offered (Duxbury & Richards, 2019a; Richards & Raymond, 2000). It is a niche tourism area that emerged both as a development of cultural tourism and in opposition to the emergence of mass cultural tourism; its activities commonly incorporate four dimensions: active participation, creative self-expression, learning, and community engagement (Bakas, Duxbury, & Castro, 2018). Creative tourism responds to the contemporary motivations of travelers seeking meaningful and transformative experiences, and to be actively involved in the 'everyday culture' of the place that they visit through interactive experiences that are connected to that place (Richards, 2018).

Creative tourism does not fit well within traditional tourism research paradigms because it is not a mass market with significant economic impacts in urban and traditional holiday locales. However, in more and more places (and particularly in peripheral, rural places), creative tourism is fostering significant 'soft' impacts (Duxbury & Richards, 2019a). Consequently, a growing range of theoretical and disciplinary perspectives are being brought to studies of creative tourism, including by many researchers from outside the tourism field, producing an interdisciplinary nexus informed by the disciplines of tourism, cultural heritage, cultural development, creativity, cultural geography, and local and regional development, among other areas.

Research on creative tourism has included contributions from both academic researchers and practitioners offering observations, insights, and reflections on developments and issues in a wide variety of international contexts. Academic research accompanying the development of creative tourism has situated itself quite closely to the practices unfolding on the ground. Despite this situation, however, there continues to be a gap between theory and practice that has been evident in much scholarship on creative tourism, and there has been a lack of integrated research approaches that use a research-and-application framework to address this gap. Consequently, perhaps, in the creative tourism field we still know relatively little about 'the profile, motivations, and experiences of those who provide or co-create the supply of creative experiences' or 'the business models adopted by creative tourism entrepreneurs' (Duxbury & Richards, 2019b, p. 184). Related to this, one of the most challenging aspects of attending to practice-based issues has related to a lack of previously documented 'successful case studies' and guides (Bakas, Duxbury, & Castro, 2018; Duxbury, forthcoming).

Creative tourism and CREATOUR in Portugal

Tourism is currently one of the main drivers of the Portuguese economy and has been growing rapidly in recent years. In line with the trends of diversification of the tourism sector worldwide, creative tourism initiatives are now emerging in Portugal. Creative tourism is a new idea for tourism in Portugal and therefore not yet well understood. As a diversified area of practice, creative tourism encourages innovative practices in tourism (i.e. inspiring new types of creative and artistic activities and new perspectives on cultural resources and locale). Creative tourism initiatives highlight and articulate the local, the vernacular, the specificities of particular places. The small-scale and interactive nature of activities encourages experimentation, flexibility, changing offers, co-learning between locals and visitors, and networks of artisans and small entrepreneurs. Local design, implementation, and control allows for diverse sites of experimentation with content, models, and approaches. In this context, the CREATOUR project is piloting experimental practices and pilot trials to more fully understand its issues and potential in small cities and rural contexts in Portugal.

CREATOUR: Creative Tourism Destination Development in Small Cities and Rural Areas is a national three-year (2016–2019) interdisciplinary research-and-application project in Portugal. The project aims to connect the cultural/creative and tourism sectors through the development of an integrated research-and-application approach to catalyzing creative tourism in small cities and rural areas throughout the country. It involves five research centers working with 40 participating organizations located in small cities and rural areas across Portugal in the Norte, Centro, Alentejo, and Algarve regions (see Figure 1). Figure 2 presents an overview of the project's main elements, actions, and outputs.

The overall structure of the project is guided by OECD advice that creative sector development can be enhanced by policy measures and programs designed to build knowledge and capacity, support content development, link creativity to place, and strengthen network and cluster formation (OECD, 2014). These dimensions form the framework for CREATOUR's approach.

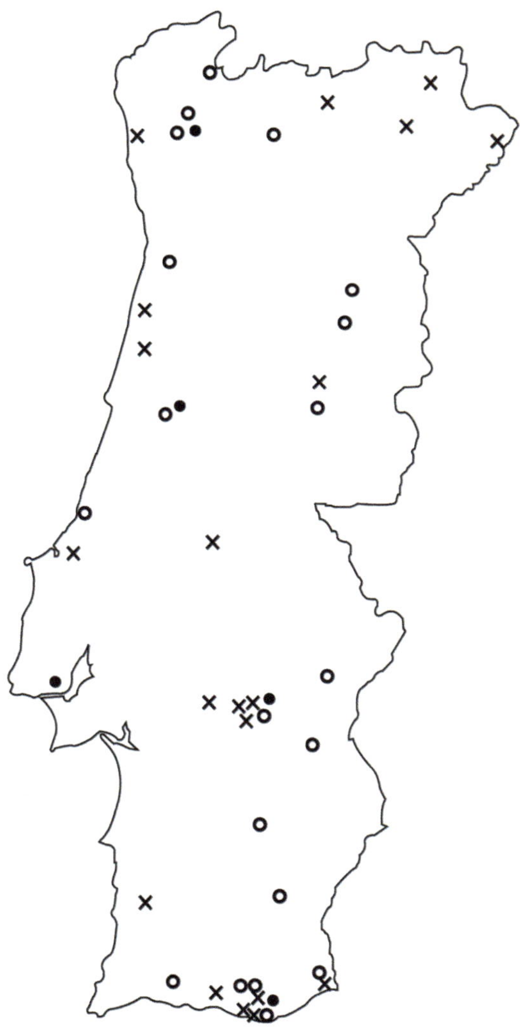

Figure 1. Map of the 40 pilots and 5 research centers involved in the CREATOUR project, Portugal. *Source*: Authors. Legend to map: o: 1st call CREATOUR pilots; x: 2nd call CREATOUR pilots; ●: Research centers.

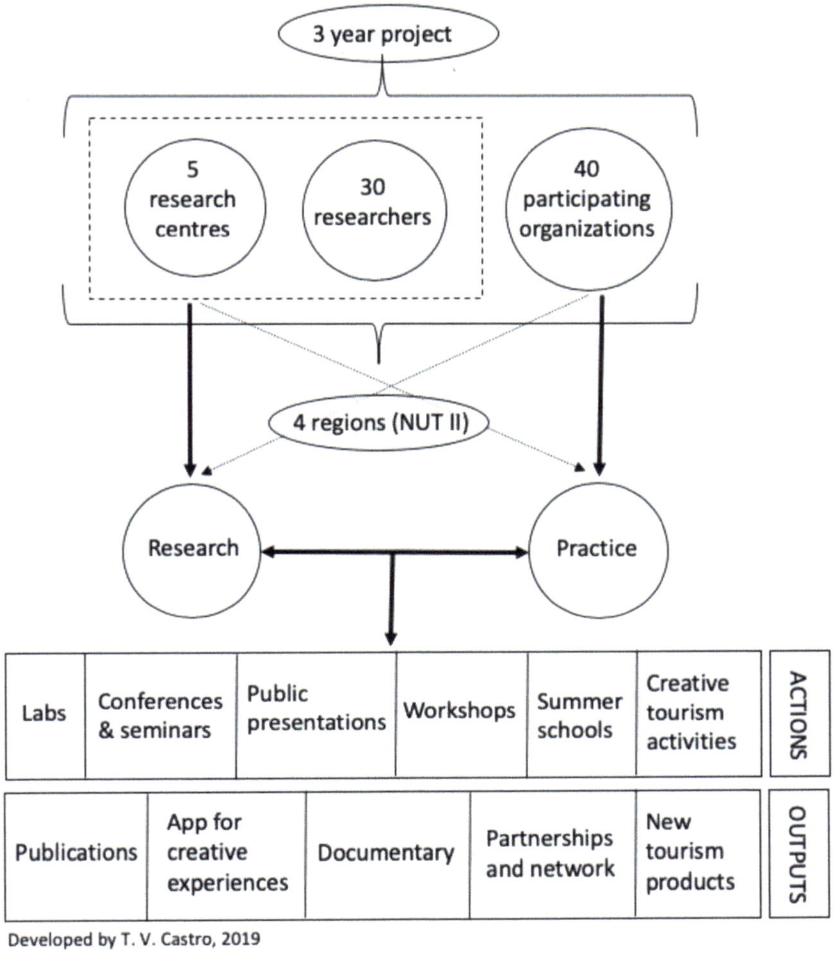

Figure 2. CREATOUR's main elements, actions, and outputs. *Source*: Tiago Vinagre de Castro (used with permission).

On the research side, the project aims to examine, monitor, and reflect on the creative tourism activities, including development patterns, reception experiences, and community impacts. Figure 3 lists the types of methodologies included in the project. The overall design of the project is rooted in the principles that there are multiple ways of knowing (Wilson & Hollinshead, 2015), that practice-based knowledge should interweave with academic research and be mutually informing (Jeannotte & Duxbury, 2015), and that research methodologies should be flexible and welcome epistemological ruptures (Kara, 2015). This approach acknowledges that an 'ecology of knowledges' exists (Santos, 2007) and that knowing is thus fundamentally based on inter-knowledge exchange.

On the practice side, the CREATOUR project aims to catalyze creative tourism offers in small cities and rural areas in Portugal, inform and learn through these initiatives, and link them with each other through the development of a national network. The project design incorporates network-building activities, among the pilots themselves (on a regional and national basis) and to connect with relevant public agencies.

- Multidisciplinary literature reviews
- Content analysis of grey literature/reports
- Review of tourism and culture statistics available for extra-metropolitan areas (Portugal)
- Content analysis of creative tourism websites (international)
- Cultural mapping through selected objects evocative of place
- Design-thinking and collaborative design
- Facilitated ideation processes
- Guided articulation and reflection on plans through canvas models
- Semi-structured one-on-one interviews (face-to-face, and via skype)
- Facilitated focus group discussions
- Informal discussions
- Actor network analysis / stakeholder analysis
- Development of self-assessment local impact criteria
- Questionnaires (for creative tourists)
- GIS mapping of territory of creative tourism activities and collection of participants' photos
- Development of baseline data to monitor activities (statistics and general information)
- Site visit observations
- Participation in creative tourism pilot activities (participant observations)
- Co-thinking and co-writing of texts
- Case studies
- Content analysis of reflective texts
- Photo elicitation and analysis
- Practice-based learning and reflection
- Participatory-action research
- Idealab observations
- Reflexive analysis of memories of lived experiences (e.g., idealabs and site-visits)
- Longitudinal para-ethnography

Figure 3. List of methodologies and approaches implemented within CREATOUR. *Source*: Authors.

CREATOUR functions as more than simply a network to market creative tourism offers, but also to help its pilot projects by providing opportunities to build and share knowledge, network with others, improve their tourism offers, and create strategies to enhance community benefit. The project aims to offer visibility through critical mass and connectedness as well as support through research, co-learning, and capacity building. The project includes research on and articulation of post-project sustainability options, with researchers working with the participating organizations to design and create a network that continues after the end of the project.

Participating organizations and pilot activities

The participating organizations in the CREATOUR project were selected through two national open calls for pilot projects (with deadlines in January and November of 2017), with five organizations selected in each region (Norte, Centro, Alentejo, and Algarve) during each call. Applications were reviewed according to an array of criteria: the cultural value and creativity of the activities proposed; capacity of tourism attractiveness; community development potential; human resource capacity; and commitment to work with research team during the project. Shortlisted proposals were also considered as a group to ensure a diversity of focus among the projects.

The open call and selection of the participating organizations occurred within the project, which meant that the actual participating practitioners could not be known in advance. The selected organizations include not-for-profit art and cultural associations, small entrepreneurial businesses, municipalities, regional development associations, and a few multi-organizational partnerships developed for this opportunity. Given the diversity of organizations – from the tourism and culture sectors as well as more general development associations and municipalities – the project brings together different skills, capacities, and strengths, but also differences of perspective and operating contexts.

The creative tourism projects designed and implemented within CREATOUR are varied, informed by the places that inspire and shape them and the relationships between people and place in these varied settings. For example, the pilot activities include workshops involving learning traditional arts and crafts as well as techniques of contemporary artistic expression inspired and informed by unique landscapes, local natural resources (e.g. marble), cultural heritage assets (e.g. Roman mosaics), and traditional activities (e.g. artisanal fishing, gastronomy, festivities, architecture) (for further details, see Duxbury, Silva, & Castro, 2019; CREATOUR, 2019).

For most participating organizations, creative tourism is a new addition to a portfolio of other tourism, craft, or artistic activities. They are looking to create a new creative tourism business or public initiative, often in conjunction with other work and activities they currently organize and conduct in a small community or rural area. Municipalities and regional development associations (and some independent entrepreneurs) are organizing local networks of creative tourism offers in collaboration with a range of independent individuals or organizations. Altogether, developing such enterprises and networks is much more complex than the delivery of simple workshops, taking time and ongoing attention, demanding more planning efforts, and needing reference models and possibly financial and other sources of support for start-up enterprises. This situation also brings higher expectations of the project's researchers and the assistance that the project can bring to these entrepreneurs.

Re-positioning research–practice exchange and mobilization – in practice

By shifting real-time application and 'knowledge mobilization' from the borders of a research project and positioning them to be more central and integrated into the project design and management, researchers gain closer relations with practitioners and a greater awareness of 'on-the-ground' realities and challenges. This article focuses on three central strategies:

1. *Developing spaces for ongoing knowledge exchange*, fostering informal discussion, learning, and knowledge-building, in which researchers and participating organizations develop relationships and opportunities to interweave complementary types of knowledge;

2. *Enabling practitioners to take on the role of co-researcher*, involving participating organizations in research tasks and knowledge co-creation, changing the norms of researcher–participant relations and expanding upon the concept of reciprocity; and
3. *Fostering researchers' close attention to the application side of the project*, requiring researchers to attend carefully to 'application' and 'implementation' as an integral part of the overall project and, potentially, to act 'beyond' their usual research work roles.

Developing spaces for ongoing knowledge exchange

The creation of communication spaces is especially important within a rural context where practitioners are geographically dispersed with a need for ongoing connectivity and access to knowledge (Ortiz, 2017). In the context of the project's challenges of distributed geography – across four regions and involving many small cities and rural contexts – regular and multimodal forms of communication, in face-to-face fora and online, are imperative. For the ongoing exchange, mobilization, and building of knowledge to occur, a multiplicity of activities and communication channels were developed within CREATOUR to stimulate communication and dialog, with both formal and informal dimensions. These spaces include regularly scheduled IdeaLabs (regional and national face-to-face meetings, described further below), regional workshops, annual conferences, on-site visits, regular in-person and Skype-enabled conversations, a private Facebook group, a listserv, and other email correspondence. The intent is that through longitudinal and regular contact, collegial connections and trust are built up among participants.

The project's activities are organized and presented as elements of a shared co-learning journey. From a research perspective, the longitudinal (over a period of three years) and close involvement of participants and researchers in the research-and-application processes integrated within CREATOUR sees the evolution of a type of para-ethnography (Islam, 2015; Vangkilde & Rod, 2015). The nature of the research–participant relationships emerging from this para-ethnographic methodology of investigating creative tourism in small city and rural contexts are varied, as would be expected given the large number of researchers (30) and their multi-disciplinary backgrounds as well as the diversity of practice settings (geographically and institutionally) and pilot initiatives. This intentional diversity enriches the CREATOUR project with a plurality of perspectives and insights, while simultaneously challenging attempts at synthesis. Researchers are confronted with both complementary and conflicting perspectives, observations, and interpretations to be discussed and negotiated within the project – a rich though challenging base from which to operate.

A total of 24 regional IdeaLabs and 3 national IdeaLabs are organized within the CREATOUR project lifespan, bringing together project researchers and participating organizations for 1–2 day working meetings.[2] The IdeaLabs provide regular points of face-to-face contact, providing interactive exercises to support content development, sessions to discuss issues and challenges – and positive surprises, and activities to articulate linkages between the pilot projects and their place and to encourage stronger relations between culture, tourism, and local/regional development. Within the IdeaLabs, participant organizations present and actively discuss their projects (including their aspirations and doubts) and share developments in implementation, while researchers bring in complementary research findings, contextual information, ideas, and reflections to inform these plans and actions.

The activities organized within the IdeaLabs are adapted to the time period they are held in the course of the project and to the needs of the pilots. For example, initial IdeaLabs included a design-inspired cultural mapping exercise to contextualize and help 'make visible' the cultural and natural resources of the places where the pilots are based and from which they are inspired (Figure 4). This activity served as background for thinking about how the pilot projects are embedded in their locales and in developing strategies for linking creativity to place. Idea generation/refinement sessions complemented this work, probing and elaborating business, communication, and community impact plans from the initial pilot project ideas proposed. Subsequent IdeaLabs have focused on collaboratively reviewing implementation tests and experiences, reflecting on surprises and lessons learned, presenting and sharing collectively gathered information from the participating organizations as well as from outside sources (to consider individual experiences within wider contexts), and planning changes for the future.

Figure 4. IdeaLab cultural mapping presentation session in Coimbra, Centro region, Portugal, Spring 2018, in which participating organizations presented 12 objects evocative of their place inspired by a framework of adjectives (participants pictured). *Source*: K. S. Alves (used with permission).

The meetings are structured to allow for multiple informal discussions, as well as more 'formal' presentation and group feedback sessions. For example, during the process of working through activity-based exercises such as developing a canvas model business plan, creating a map of strengths and weaknesses, or a project's alignment with creative tourism principles, pilot-participants and researchers informally discuss the information being developed, with researchers dropping by the organizations' posters being developed and discussing ideas as well as responding to questions and doubts (Figure 5). Later on, each pilot-participant presents their poster to all IdeaLab participants, which is then commented on by both researchers and pilots. These open and lively discussions frequently provide additional ideas and suggestions for the participating organizations to consider when elaborating their initiative, as well as additional themes for research.

Although it takes time for researchers and practitioners to understand each other and their various operating contexts – a challenge keenly felt when dealing with 40 diverse participating organizations – close and ongoing contact reveals the differing working goals and imperatives of project participants as well as inherent contradictions between research and practice, such as researchers' focus on presenting and writing research articles while operational sustainability is top of mind for the practitioners; differing time scales for developing and reporting results; and different perceptions of importance for different tasks (such as distributing questionnaires or conference presentations), among other dimensions. Furthermore, differences among practitioners' contexts and needs are also revealed, such as the operating issues and priorities of a municipality versus that of an independent entrepreneur.

Figure 5. IdeaLab canvas model development and mentoring session in Coimbra, Centro region, Portugal, Spring 2018 (researchers and organizational participants pictured). *Source*: K. S. Alves (used with permission).

In the process of designing and implementing the IdeaLabs, challenges have arisen related to the maximization of the face-to-face encounters in the face of multiple and diverse priorities, and more generally addressing an emergent need for a temporal 'balancing' between 'time together' and 'time alone' spent planning and implementing the pilot activities:

Maximization of face-to-face encounters

Time limitations of face-to-face interaction, and management of the schedule to maximize both research and practitioners' needs for information gathering and exchange, as well as to incorporate informal time for networking, continues to be challenging. With multiple competing priorities, there are no simple prioritization systems and each meeting is carefully prepared to balance the needs of research and practitioners. Compromises in time allocations have occurred, for example, to enable formal, regularly held one-on-one interviews with each participating organization rather than to allocate additional time to foster networking and discussions among the organizations to discuss inter-organizational cooperation possibilities. It is also important to note that the dynamics of each regional meeting differ, with impromptu discussions addressing emerging issues and concerns, which can alter the pre-planned agenda. Enabling this flexibility and responsiveness is imperative in order to address questions and concerns, to progress relations among researchers and practitioners, and to continue to advance the overall project.

Balancing 'time together' with 'time alone'

The IdeaLabs were conceived and designed as a central element within multiple channels and spaces for communication. The organization of three intensive project meetings (IdeaLabs) annually in addition to annual site-visits[3] and interviews, and regular electronic communications (via email, a listserv, and a private FaceBook group) was originally felt to be sufficient to maintain connections with participating organizations. However, in practice, a desire for additional regular communications was expressed. Some participating organizations felt there was insufficient individualized follow-up with organizations following each intense IdeaLab, with the consequence that the motivation and energies stirred up in these meetings dissipates afterwards, when the organizations feel 'alone' again. This perceived need for additional, individualized follow-up was not anticipated in the original project design. Some research centers responded by organizing additional seminars and meetings in their regions, then in response to ongoing pilot calls for even more regular contact and support from researchers in the project, regular in-person and Skype conversations (depending on geographic proximity) were implemented to support further connectivity and exchanges among pilot-participants and researchers. However, organizing monthly meetings between researchers and pilots to discuss developments and plans and to provide support and constructive feedback has become a challenging endeavor in many cases. Both researchers and practitioners are busy with multiple obligations and coordinating schedules, cancelations, and subsequent rescheduling has been difficult. Consequently, the timeliness and regularity of connections has suffered.

Beyond these operational challenges, an emerging concern is researchers' limited use (to date) of the practice-based knowledge shared within the activities and discussions in research publications. The exchanges among researchers and practitioners have been central to the project's operations and to building the knowledge base of all involved in the project. The knowledge, observations, and insights shared within the IdeaLabs are documented through posters, researchers' notes, photographs, and video recordings of presentations. As well, an 'instant report' of each IdeaLab is developed during each event that contains key points and images from all activities and discussions, which is presented at the end of the event and made available to all project participants for reference following. This knowledge-sharing and documentation has been invaluable to inspire and inform research topics, to articulate issues to address in practice, and to share good practices. Practitioners' insights have also been used to inform the development of book chapters reflecting on the development of the project, to highlight the needs and issues of the participating organizations, and to inform the general management of the project. However, to date, researchers have made only limited use of the practice-based knowledge and insights gained in these meetings in their academic journal articles.

We acknowledge that the researcher–practitioner exchanges are, by their nature, messy and largely open-ended, but the IdeaLab sessions can be compared to interactive focus groups. However, researchers have tended to incorporate methods such as (additional) interviews within the journal articles they have developed, to better ensure a clear correspondence between academic conceptual frameworks and research findings. In part, this is due to the characteristics of academic publishing which prefers, for example, clearly defined and more traditional research methodologies, such as standardized interviews, as the basis of data collection. In part, this situation also stems from a greater need to know how to effectively bridge practice-based insights and knowledge with academic interests, and the greater time required to sort through multiple sources of documentation and wide-ranging discussions, and relate it to recognized conceptual frameworks and theories. This seems to indicate that while practice-based insights can provide important advancements in understanding development processes and 'front-line' observations and insights not otherwise available, the nature of this knowledge may be different from what researchers aim to write for research journals and that some 'misalignment' is present in this process. The situation also suggests that research publications may not be the most suitable outlets for articulating and bringing this knowledge forward, that a different communication platform is needed as a 'bridging' mechanism, and that integrating practice-based knowledge into research journals may be a subsequent activity.

Encouraging hybrid roles: co-researching and 'going beyond' research

With a project that is intentionally trying to foster different relations between research and practice, the 'traditional' roles, perspectives, and practices of research are also intentionally disturbed: hybrid roles have been encouraged as practitioners become 'co-researchers' and researchers are asked 'go beyond' typical research activities to support the participating organizations. The positioning of practitioners as co-researchers aims to equalize power differentials within the project, contributes to the recognition and valuation of practice-based research, and foregrounds the desirability to weave together different types of knowledge. At the same time, researchers are called upon to explicitly consider and respond to the needs of the participating organizations, challenging the traditional researcher-as-observer role. Altogether, this hybridity of roles means that the research design practically and conceptually responds to the need to better link tourism research and practice, on an ongoing basis, and strives to represent and address practitioners' challenges. The implications of this 'disruptive' context and 'new' relationship is that the researcher and participant-practitioner roles are problematized and hybrid researcher–practitioner roles emerge, the liquidity of which results in dynamic knowledge production but can often be confusing for the parties involved. The practice of these roles is an emergent process.

Enabling practitioners to take on the role of co-researcher

Within CREATOUR, practitioners (from the participating organizations) are called upon to be *co-researchers* (see Figure 6). On an operational level, they share in research tasks, such as distributing and managing the completion of questionnaires by their visitor-participants and maintaining statistics on the number of workshops/activities organized, number of participants, brief observational notes, and so forth. On a research design level, they provided input into the design of a visitor-participant questionnaire and are co-developing impact assessment criteria, working with a research team to select organization-specific criteria that align with the objectives of each organization's initiatives. The participating organizations' insights and experiences

Practice-based knowledge sharing ↑

- Co-writing texts documenting and assessing the development and implementation of each pilot project
- Practice-based knowledge sharing through IdeaLab discussions
- Developing reflexive texts on experiences
- Collaboration in customizing criteria for measuring impacts of pilot activities in each of their communities
- Recording 'front lines' data on activities implemented (e.g., title, brief description of activity, date, length of activity, number of participants, fee charged, language of activity, etc.) in Excel form, and making brief observations/comments on each activity
- Providing feedback on the design of the questionnaire for tourist-participants
- Copying and distributing questionnaires to tourist-participants

↓ **Operational tasks**

Figure 6. Types of activities of the practitioner co-researchers. *Source*: Authors.

conceiving and steering developments both 'on the ground' and 'in motion' during the duration of the project are akin to practice-based research providing knowledge through practice, and are making important contributions to informing the field of creative tourism in terms of 'supply' development. In this vein, discussions among researchers and practitioners are moments of co-learning through sharing different perspectives and knowledges. For example, in the autumn IdeaLabs, the participating organizations' observations and reflections on summer-season activities are presented, discussed, and documented as complementary insights into the summer's activity implementation experiences, alongside the visiting researchers' observations and field notes.

While the overall structure of this approach has been productive, some issues have arisen in the adoption of the co-researcher role by some practitioners. For example, some participating organizations had problems distributing questionnaires to creative tourists largely due to time constraints and logistics issues at the event they organized but also due, in part, to viewing this task as something 'extra' rather than integrated into the activity plans. These organizations saw this as the researcher's task, with themselves only responsible for organizing the event. In other situations of co-creating research tools, communication difficulties between researchers and practitioners led to varied engagement in following up after a group session. In such cases of more advanced input into research instruments (for example, customized impact criteria assessment grids), more personalized approaches to involving each practitioner may be needed, such as jointly working on these materials on-site, rather than leaving this to be completed as 'follow-up' work. These issues suggest that further training and discussion relating to research–practice relations and the project's hybrid roles, for both researchers and practitioners, would have clarified expectations, offered more operational knowledge and support, and helped decrease the uncertainties of acting in these new situations.

Fostering researchers' close attention to application (or researchers' actions beyond research)

Collaborative research brings researchers into a more reflexive position where intensive, ongoing reflection is needed on the consequences faced by all parties involved, but a research-and-application project may go beyond this. It can also entail a series of actions beyond those that can be understood as 'research' activities and methodologies, especially when pilots need more in-depth mentorship and training. With the continuance and sustainability of the pilot projects very much top of mind, the hybrid project structure demands the incorporation of activities and methods that specifically address the practice-related issues and concerns that arise in the processes of implementation. In this 'practice-foregrounded' context, research-and-application projects require researchers (who adopt an engaged scholarship approach to knowledge creation) to attend to more-than-research aspects of the global project, to deeply listen and respond to practitioner needs and expectations in regards to assisting the 'practice' side of the project, and to add to or adjust operational actions to attempt to address these needs.

As researchers' roles move beyond that of a 'traditional' researcher whose main aim is to gather and analyze data, they actively practice a form of 'extreme' reciprocity in terms of making concerted efforts to develop, exchange, and mobilize knowledge within a practice- and application-based research setting, in the process extending, disrupting, and challenging 'traditional' research–practice interactions. Some successful practices in this vein include researchers adopting a mentoring role for participant organizations, helping them develop their business ideas during IdeaLabs, regularly posting tourism marketing news on a shared social media page, and organizing presentations (involving researchers and participating organizations) at the national tourism trade fair. Furthermore, researchers began to learn to alter the way in which they presented their knowledge, interpreting data collected during the project (e.g. obtained through questionnaires and on-site interviews) in a format that would be useful to the practitioners, aiming to provide 'practice-actionable' information that may be valuable to them in the long run. Such 'extreme' reciprocity acts provide indications of ways in which research power dynamics can be equalized and point towards the future emergence of hybrid researcher–-practitioner roles. Such roles appear to be essential and integral to striving towards more egalitarian research relationships that result in better representation and investigation of contemporary tourism phenomena.

However, altering traditional boundaries and 'comfort zones' to develop closer research–practice relationships is an ongoing and challenging practice. Within IdeaLab sessions and other meetings, spaces have been organized for expressing and discussing participants' expectations of the project, whether they are being addressed, and how researchers might subsequently try to fulfill these expectations to their best ability. While the very nature of research-and-application projects means that practitioners enter the project with business-development related expectations, challenges can emerge in meeting these. In the project discussed here, since the level of development is so varied among the participating organizations' pilot initiatives, expectations also vary, ranging from start-up mentorship needs, to 'bringing more clientele to them', to more nuanced expectations of 'constructing a network of partners and complementary offers', among an array of other topics.

We have found that participating organizations' expectations of a range of pragmatic development assistance and support has been difficult for the researchers to address, with many business and marketing consultancy expectations beyond the expertise of the researchers involved in the project, and highlighting an important gap in forging this relationship. This challenging situation is a function of a combination of practitioners' high and varied expectations of the project, a lack of researchers' expertise and capacities for the types of training required, limited experience in mentorship roles involving practitioners, and project design constraints (e.g. time limitations and resource inflexibility to directly address the greater-than-anticipated start-up support needs).[4] This suggests heightened consideration of addressing the emergent pragmatic needs of participating practitioners, carefully managing expectations while also remaining flexible to address emerging issues and opportunities, providing greater training of researchers on knowledge transfer and mentorship techniques, and including dedicated partners or funding to provide pragmatic training sessions within such a project.

Conclusions

We believe significant advances in understanding creative tourism development in small cities and rural areas are emerging through this research-and-application project, through embracing the practice of engaged scholarship, blurring the traditional boundaries between researchers and practitioners, and foregrounding the importance of practice-based knowledge. Furthermore, within the complex context of this project, questions have arisen of how reciprocal relationships can be maintained within a research-and-application project, how the practice side of the project can be attended to, and how the knowledge co-created in this process can benefit both researchers and practitioners. The reflexive process method employed on the lived experience of being a researcher within this research-and-application project has highlighted the complexity of this experiment and the need for researchers to know more about how to effectively bridge practice-based insights and knowledge with academic interests and contexts. While there is no simple formula for fostering closer research–practice relations, in this concluding section, we synthesize key findings from the three central strategies implemented within the CREATOUR project. We close with a few insights from others involved in collaborative research that reassure us that the challenges we have faced and have aimed to address are an integral part of collaborative processes more broadly.

Key findings

Developing spaces for ongoing knowledge exchange

A multiplicity of activities and communication channels were developed within CREATOUR to stimulate communication and dialog, such as IdeaLabs, regional workshops, annual conferences, on-site visits, regular in-person and Skype-enabled conversations, a private Facebook group, a listserv, and email correspondence. The application of these multiple approaches highlighted that it takes time for researchers and practitioners to understand each other and their various operating contexts, and to build trust. This explicit recognition of time and process in relationship-building is essential in order for participant organizations to feel their needs are being taken seriously and attended to, for researchers to feel less pressured by time restrictions, and for the overall research-and-application process to be more effective. Within a multichannel communications environment, communication preferences and complementarities must be strategically used to address a temporal 'balancing' between participants' 'time together' and 'time alone' working independently on different activities. The careful design of face-to-face meetings (of individuals and as groups) is imperative to illuminate and advance this relationship-building and knowledge-sharing process. In the face of multiple and diverse priorities, contents and approaches/formats, different perspectives and values, and networking possibilities inherent in any activity must be considered jointly. Overall, an ongoing spirit of care, flexibility, and responsiveness is vital in order to address emerging questions, issues, and concerns; to foster relations among researchers and practitioners over time; and to continue advancing the overall project.

Enabling practitioners to take on the role of co-researcher

The positioning of practitioners as co-researchers aimed to equalize power differentials within the project, contribute to the recognition and valuation of practice-based research, and foreground the desirability to weave together different types of knowledge. Overall, this approach emphasized that the observations, experiences, and practice-based knowledge of the practitioners 'on the front lines' (planning and implementing creative tourism pilot projects, interacting with tourists, and evaluating and reflecting on these experiences) were essential and central aspects of researching creative tourism development in small cities and rural areas. While the overall structure of this approach has been productive, some issues have arisen in the operational adoption of some tasks by some practitioners, and the design of some collaborative exercises by researchers. The issues encountered suggest that further training and discussion about the intentionally designed 'practitioner as co-researcher' role, for both researchers and practitioners, would have clarified expectations, offered further operational knowledge and support, and helped decrease the uncertainties of this new situation.

Fostering researchers' close attention to the application side of the project

Many researchers adopted an engaged scholarship approach to knowledge creation, attending to aspects beyond the project's research dimension by deeply listening and responding to practitioner needs and expectations as well as conducting operational actions to attempt to address practitioner needs. However, this situation also caused tension because many business and marketing consultancy expectations arose which were often beyond the competencies and areas of expertise of the researchers involved in the project. While researchers aimed to address pilots' expectations and needs, they struggled to adequately address all desirable actions and consultancy/mentorship interventions. As researchers were challenged to stretch beyond the 'usual' researcher role, they were confronted with issues of self-assessing their competency and capacity to take on some of these 'expected' extended actions. For researchers, this led to tensions as well acts of personal growth and transformation as they moved towards adopting a hybrid 'researcher–consultant' role.

The reflexive process undertaken for this article has highlighted that juxtaposing and resolving researchers' focus on presenting and writing research articles and practitioners' priority for operational sustainability is a central issue in the productive design and management of research–practice relationships. Challenges in sharing research data quickly with practitioners as well as integrating shared practice-based knowledge into research publications suggest that a customized knowledge/communication platform may be useful as a 'bridging' mechanism between practice-based and academic knowledge systems, one that values, articulates, and shares various knowledges and perspectives in a manner that aims to respond to the knowledge needs of both researchers and practitioners. Dedicated resources and expertise must be devoted to this process for its effective realization.

In closing, revisiting the question, 'Why has the task of closer research–practice collaboration been so challenging to achieve?' the following remarks on collaborative practice resonate strongly with us regarding our experiences and reflections:

- It challenges us to trust.
- It is often surprising.
- It is often difficult.
- Sometimes there is tension.
- It takes time.
- It demands personal growth.
- It requires acknowledgment of others.
- It asks us to question our own points of view.
- It thrives in the in-between spaces.
- There is no one way.
- It is an act of transformation. (Maddox, 2019)

Notes

1. We acknowledge that this article is written from the researchers' perspective, and that a complementary examination of the practitioners' perspectives is warranted. This, however, is beyond the scope of the current article and will be the focus of subsequent investigation within the research-and-application project.
2. Each year, two regional IdeaLabs take place in each project region (in Winter/Spring and Autumn), with an annual national IdeaLab organized with the annual conference (May 2017, June 2018, and October 2019).
3. The on-site visits, held during the time of selected creative tourism implementations, offer moments for participant observation, documentation, and *in situ* interviews with the pilot-organizers, and occur (at least) annually with all pilots. These events provide an opportunity for the researcher to 'live' the creative tourism project and have an immersive experience which allows for a better understanding of the participant organization, its product, and its reception by visitors. Simultaneously, on-site visits provide the opportunity for participating organizations and others to ask any questions they may have about the CREATOUR project.
4. Sometimes even 'best' intentions to address these conditions do not work out in the end. For example, after a series of IdeaLab meetings where participating organizations strongly asked for more help with marketing their products, researchers attempted to address these needs by organizing a digital marketing bootcamp weekend to be presented by a well-regarded communications consultant/trainer working in the cultural and tourism sectors. However, in the end, even though the program was customized to their needs and the price arranged was very competitive, practitioners did not sign-up for the bootcamp, with some expressing that it should have been offered 'within the project' rather than 'in addition' to the regular meetings – a reminder of the messiness and uncertainty of these efforts.

Acknowledgements

The authors would like to thank CREATOUR researchers and participants for their precious contributions, and the anonymous reviewers for their inspirational and useful critiques and suggestions.

Disclosure statement

No potential conflict of interest was reported by the authors.

Funding

This article was developed within the framework of a research project on creative tourism in small cities and rural areas, entitled 'CREATOUR: Creative Tourism Destination Development in Small Cities and Rural Areas' (Project 016437), which is supported by the Portuguese Foundation for Science and Technology (FCT/MEC) through national funds and cofunded by FEDER through the Joint Activities Programme of COMPETE 2020 and the Regional Operational Programmes of Lisbon and Algarve.

References

Anderson, C. R., & McLachlan, S. M. (2016). Transformative research as knowledge mobilization: Transmedia, bridges, and layers. *Action Research*, *14*(3), 295–317. doi:10.1177/1476750315616684

Anderson, W., & Sanga, J. J. (2018). Academia–industry partnerships for hospitality and tourism education in Tanzania. *Journal of Hospitality & Tourism Education*, *31*(1), 34–48. doi:10.1080/10963758.2018.1480959

Ateljevic, I., Harris, C., Wilson, E., & Collins, F. L. (2005). Getting 'entangled': Reflexivity and the 'critical turn' in tourism studies. *Tourism Recreation Research*, *30*(2), 9–21. doi:10.1080/02508281.2005.11081469

Bakas, F. E. (2017). 'A beautiful mess': Reciprocity and positionality in gender and tourism research. *Journal of Hospitality and Tourism Management*, *33*, 126–133. doi:10.1016/j.jhtm.2017.09.009

Bakas, F. E., Duxbury, N., & de Castro, T. V. (2018). Creative tourism: Catalysing artisan entrepreneur networks in rural Portugal. *International Journal of Entrepreneurial Behavior & Research*, *25*(4), 731–752. doi:10.1108/IJEBR-03-2018-0177

Beech, N., MacIntosh, R., & MacLean, D. (2010). Dialogues between academics and practitioners: The role of generative dialogic encounters. *Organization Studies*, *31*(9-10), 1341–1367. doi:10.1177/0170840610374396

Beaulieu, M., Breton, M., & Brousselle, A. (2018). Conceptualizing 20 years of engaged scholarship: A scoping review. *PLoS One*, *13*(2), e0193201. doi:10.1371/journal.pone.0193201

Cheek, J., Lipschitz, D. L., Abrams, E. M., Vago, D. R., & Nakamura, Y. (2015). Dynamic reflexivity in action: An armchair walkthrough of a qualitatively driven mixed-method and multiple methods study of mindfulness training in schoolchildren. *Qualitative Health Research*, *25*(6), 751–762. doi:10.1177/1049732315582022

Cockburn-Wootten, C., McIntosh, A. J., Smith, K., & Jefferies, S. (2018). Communicating across tourism silos for inclusive sustainable partnerships. *Journal of Sustainable Tourism*, *26*(9), 1483–1498. doi:10.1080/09669582.2018.1476519

CREATOUR. (2019). *CREATOUR pilots and projects*. Coimbra: Centre for Social Studies, University of Coimbra.

Diver, S. W., & Higgins, M. N. (2014). Giving back through collaborative research: Towards a practice of dynamic reciprocity. *Journal of Research Practice*, *10*(2), M9.

Duxbury, N. (Forthcoming). Catalyzing creative tourism in small cities and rural areas in Portugal: The CREATOUR approach. In K. Scherf (Ed.), *Creative tourism and sustainable development in smaller communities*. Calgary: University of Calgary Press.

Duxbury, N., & Richards, G. (2019a). Towards a research agenda for creative tourism: Developments, diversity, and dynamics. In N. Duxbury & G. Richards (Eds.), *A research agenda for creative tourism* (pp. 1–14). Cheltenham: Edward Elgar Publishing.

Duxbury, N., & Richards, G. (2019b). Towards a research agenda in creative tourism: A synthesis of suggested future research trajectories. In N. Duxbury & G. Richards (Eds.), *A research agenda for creative tourism* (pp. 182–192). Cheltenham: Edward Elgar Publishing.

Duxbury, N., Silva, S., & Castro, T. V. (2019). Creative tourism development in small cities and rural areas in Portugal: Insights from start-up activities. In D. A. Jelinčić & Y. Mansfeld (Eds.), *Creating and managing experiences in cultural tourism* (pp. 291–304). Singapore: World Scientific Publishing.

Estabrooks, C. A., Norton, P., Birdsell, J. M., Newton, M. S., Adewale, A. J., & Thornley, R. (2008). Knowledge translation and research careers: Mode I and mode II activity among health researchers. *Research Policy*, *37*(6-7), 1066–1078.

Ferlie, E., Crilly, T., Jashapara, A., & Peckham, A. (2012). Knowledge mobilisation in healthcare: A critical review of health sector and generic management literature. *Social Science and Medicine*, *74*(8), 1297–1304. doi:10.1016/j.socscimed.2011.11.042

Finlay, L. (2002). Negotiating the swamp: The opportunity and challenge of reflexivity in research practice. *Qualitative Research*, *2*(2), 209–230. doi:10.1177/146879410200200205

Freeman, J. (2010). *Blood, sweat & theory: Research through practice in performance*. Faringdon: Libri Publishing.

Greenhalgh, T. (2010). What is this knowledge that we seek to "exchange"? *The Milbank Quarterly*, *88*(4), 492–499. doi:10.1111/j.1468-0009.2010.00610.x

Greenhalgh, T., & Wieringa, S. (2011). Is it time to drop the 'knowledge translation' metaphor? A critical literature review. *Journal of the Royal Society of Medicine*, *104*(12), 501–509. doi:10.1258/jrsm.2011.110285

Hallin, C. A., & Marnburg, E. (2008). Knowledge management in the hospitality industry: A review of empirical research. *Tourism Management*, *29*(2), 366–381. doi:10.1016/j.tourman.2007.02.019

Hart, A., Davies, C., Aumann, K., Wenger, E., Aranda, K., Heaver, B., & Wolff, D. (2013). Mobilising knowledge in community – university partnerships: What does a community of practice approach contribute? *Contemporary Social Science*, *8*(3), 278–291. doi:10.1080/21582041.2013.767470

Hope, S. (2016). Bursting paradigms: A colour wheel of practice-research. *Cultural Trends*, *25*(2), 74–86. doi:10.1080/09548963.2016.1171511

Islam, G. (2015). Practitioners as theorists: Para-ethnography and the collaborative study of contemporary organizations. *Organizational Research Methods*, *18*(2), 231–251. doi:10.1177/1094428114555992

Jeannotte, M. S., & Duxbury, N. (2015). Advancing knowledge through grassroots experiments: Connecting culture and sustainability. *Journal of Arts Management, Law and Society*, *45*(2), 84–99. doi:10.1080/10632921.2015.1039739

Kara, H. (2015). *Creative research methods in the social sciences: A practical guide*. Bristol: Policy Press.

Kemmis, S., McTaggart, R., & Nixon, R. (2014). *The action research planner: Doing critical participatory action research*. Singapore: Springer

Lassiter, L. E. (2005). *The Chicago guide to collaborative ethnography*. Chicago, IL. University of Chicago Press.

Leavy, P. (2015). *Method meets art: Arts-based research practice* (2nd ed.). New York, NY: The Guilford Press.

Maddox, D. (2019). Introduction. In D. Maddox (Ed.), *What I mean when I talk about collaboration. What is a specific experience collaborating on a project with someone from a different discipline or "way of knowing"?* New York: The Nature of Cities. Retrieved from https://www.thenatureofcities.com/2019/02/28/mean-talk-collaboration-specific-experience-collaborating-project-someone-different-discipline-way-knowing/?utm_source=The+Nature+of+Cities+Newsletter&utm_campaign=361d013524-EMAIL_CAMPAIGN_2019_03_10_09_08&utm_medium=email&utm_term=0_65cc822135-361d013524-1402919

Nelson, R. (2013). *Practice as research in the arts*. Basingstoke: Palgrave Macmillan.

Nilsson, P. Å. (2007). Stakeholder theory: The need for a convenor. The case of Billund. *Scandinavian Journal of Hospitality and Tourism*, *7*(2), 171–184. doi:10.1080/15022250701372099

Oborn, E., Barrett, M., & Racko, G. (2013). Knowledge translation in healthcare: Incorporating theories of learning and knowledge from the management literature. *Health Organization and Management*, *27*(4), 412–431.

OECD. (2014). *Tourism and the creative economy*. Paris. OECD.

Olsen, V. (2011). Feminist qualitative research in the millenium's first decade. In N. Denzin & Y. S. Lincoln (Eds.), *The SAGE handbook of tourism studies* (pp. 129–146). Los Angeles, CA: SAGE.

Ooi, C.-S. (2013). Tourism policy challenges: Balancing acts, co-operative stakeholders and maintaining authenticity. In M. Smith & G. Richards (Eds.), *Routledge handbook of cultural tourism* (pp. 67–74). New York, NY: Routledge

Ortiz, J. (2017). Culture, creativity and the arts: Building resilience in northern Ontario (Doctoral thesis). University of the West of England, UK.

Palinkas, L. A. (2010). Commentary: Cultural adaptation, collaboration, and exchange. *Research on Social Work Practice*, *20*(5), 544–546. doi:10.1177/1049731510366145

Pillow, W. (2003). Confession, catharsis, or cure? Rethinking the uses of reflexivity as methodological power in qualitative research. *International Journal of Qualitative Studies in Education*, *16*(2), 175–196. doi:10.1080/0951839032000060635

Richards, G. (2018). Cultural tourism: A review of recent research and trends. *Journal of Hospitality and Tourism Management*, *36*, 12–21. doi:10.1016/j.jhtm.2018.03.005

Richards, G., & Raymond, C. (2000). Creative tourism. ATLAS News, no. (23), 16–20.

Ryan, C. (2001). Academia–industry tourism research links: States of confusion. The Martin Oppermann Memorial Lecture 2000. *Pacific Tourism Review*, *5*(3–4), 83–95.

Santos, B. D S. (2007). Beyond abyssal thinking: From global lines to ecologies of knowledges. *Review*, *30*(1), 45–89.

Social Sciences and Humanities Research Council of Canada (SSHRC). (2018). Definition of terms – Knowledge mobilization. Retrieved from http://www.sshrc-crsh.gc.ca/funding-financement/programs-programmes/definitions-eng.aspx#km-mc

Trainor, A., & Bouchard, K. A. (2013). Exploring and developing reciprocity in research design. *International Journal of Qualitative Studies in Education*, *26*(8), 986–1003. doi:10.1080/09518398.2012.724467

Tribe, J., & Liburd, J. J. (2016). The tourism knowledge system. *Annals of Tourism Research*, *57*, 44–61. doi:10.1016/j.annals.2015.11.011

Van de Ven, A. (2007). *Engaged scholarship: A guide for organizational and social research*. New York, NY. Oxford University Press.

Van de Ven, A., & Johnson, P. E. (2006). Knowledge for theory and practice. *Academy of Management Review*, *31*(4), 802–821. doi:10.5465/amr.2006.22527385

Vanderlinde, R., & van Braak, J. (2010). The gap between educational research and practice: Views of teachers, school leaders, intermediaries and researchers. *British Educational Research Journal*, *36*(2), 299–316. doi:10.1080/01411920902919257

Vangkilde, K. T., & Rod, M. H. (2015). *Para-ethnography 2.0: An experiment with the distribution of perspective in collaborative fieldwork*. Paper presented at Research Network for Design Anthropology, seminar 3: Collaborative Formation of Issues, January 22–23, 2015. Retrieved from https://kadk.dk/sites/default/files/14_paper_kasper_tang_vangkilde_morten_hulvej_rod.pdf

Walsh, J. P., Tushman, M. L., Kimberly, J. R., Starbuck, B., & Ashford, S. (2007). On the relationship between research and practice: Debate and reflections. *Journal of Management Inquiry*, *16*(2), 128–154. doi:10.1177/1056492607302654

Walters, G., Burns, P., & Stettler, J. (2015). Fostering collaboration between academia and the tourism sector. *Tourism Planning & Development*, *12*(4), 489–494. doi:10.1080/21568316.2015.1076596

Ward, V. (2017). Why, whose, what and how? A framework for knowledge mobilisers. *Evidence & Policy: A Journal of Research, Debate and Practice*, *13*(3), 477–497. doi:10.1332/174426416X14634763278725

Wenger, E. (1998). *Communities of practice: Learning, meaning, and identity*. Cambridge: Cambridge University Press.

Wilson, E., & Hollinshead, K. (2015). Qualitative tourism research: Opportunities in the emergent soft sciences. *Annals of Tourism Research*, *54*, 30–47.

The case for linguistic narrative analysis, illustrated studying small firms in tourism

Lucia Tomassini, Xavier Font and Rhodri Thomas

ABSTRACT

We advocate the adoption of more expansive and creative methodological approaches the study of tourism. More specifically, we argue that by examining how individuals narrate their experiences and social practices, researchers can gain an insight into the meanings actors attach to their actions. Considered from this perspective, narratives become performative; they prompt actors to take actions that (they feel) actualise the story they are seeking to tell. To illustrate its value, we use linguistic narrative analysis to explore how the owner-managers of small values-based tourism firms narrate the operation of their business. A dataset of first person accounts made of both narratives in storified form featuring a chronological order with beginning, middle, and end, and narratives without a storified form largely recounting opinions, feelings, and points of view, are interpreted to offer new perspectives on the behaviours of small firms in tourism. We argue that narrative approaches should complement methods used routinely by tourism scholars to examine this constituency of actors (and others).

摘要

我们提倡采用更广泛和创造性的方法来研究旅游。更具体地说，我们认为，通过研究个体如何叙述他们的经历和社会实践，研究人员可以洞察个体行动者赋予他们行为的意义。从这个角度来看，叙事变成了一种表演，叙事促使行动者采取行动，（感觉）实现了他们企图讲述的故事。为了阐明这种方法的价值，我们使用语言叙事分析来探索基于价值的小型旅游公司的业主也就是管理人员如何叙述他们业务运作过程。本研究以第一人称叙述的资料集为基础，既有按照时间顺序展开的故事式叙述，又有主要表达观点、感想和看法的非故事性叙述。这种故事叙述方法为阐释小型旅游企业的行为提供了新的视角。我们认为，这些叙事方法应该可以补充旅游学者经常使用的其它方法，以检查这一群体的行动者(和其他群体行动者)看法的一致性。

Introduction

Narrative research methods are under-utilised in tourism studies despite their acknowledged strengths and growing application in other fields of enquiry (Riessman, 1993,

Smith & Sparkes, 2008). In this article, we seek to show how a narrative approach can generate new knowledge by revealing the meanings actors ascribe to their actions. We discuss research undertaken with owner-managers of small firms in tourism that define themselves in terms of their ethics to illustrate our argument. Following others, 'smallness' refers to the personalised nature of organisational management, which is reflected by the employment of a maximum of 50 people (Walmsley, 2019). We show how narrative enquiry enhances knowledge generated about such firms using more 'standard' or 'conventional' methodologies and methods.

Narrative research lends itself to a dynamic, holistic, qualitative and interpretive understanding of complex systems (Tsoukas & Hatch, 2001). Its strength lies in allowing multiple connections across time and linking disconnected elements as related parts of a whole. The narrative approach is not a search for truth but a search for meaning (Rhodes & Brown, 2005). Narrative research embraces a postmodern perspective that challenges singular, and often essentialist, theories in favour of an interpretive approach that acknowledges the presence of fragmentation and tensions in the processes of sense making (Buchanan, 2013). Individuals and groups construct narratives in order to organise their experiences and 'map their reality' (Wilkins & Thompson, 1991). Its starting point is an acceptance that all social actors position themselves – and others – through stories about themselves and others (Davies & Harré, 1990).

Narratives are a mode of association enabling an alternative way of knowing. They are a distinct form of discourse that focuses on the biographical, social, cultural and historical situations that have conditioned life experiences (Denzin & Lincoln, 2005). Narrative enquiry, therefore, revolves around life experiences as narrated by those who live them. They are a way in which individuals create coherence and meaning, reflecting their understanding of events and behaviours over time (Chase, 2005; Denzin & Lincoln, 2005). While conceptually appealing, the methodological challenges of data capture are significant. The section that follows discusses one way of collecting narrative data (for a discussion of others see Andrews, Squire, & Tomboukou, 2013; Reissner & Pagan, 2013) and an approach to interpretation in tourism studies.

Narrative data collection and transcription

Informal and unstructured interviews allow for close observation of a few individuals so as to scrutinise their narratives carefully, and thus interpret the meaning they attach to them (Riessman, 1990; Bredvold & Skålén, 2016). The researcher (in this case the first author) is the listener and the founder/owner-manager the storyteller. For each participant, the researcher organised an interview via Voice over Internet Protocol (VoIP) - Skype audio call – since participants were located in different parts of Italy. Each interview lasted between 45 and 80 minutes. Through questions like '*Would you tell me in your own words the story of your firm? How have things changed over the time? How does it work now?*', the researcher elicits the conversation with a set of open and flexible questions as guidelines for the conversation (Riessman, 1993). The data collection resulted in a set of first person accounts in Italian – the mother tongue of both researcher and participants – which were translated into English only after being carefully transcribed and analysed. After the first rough transcription, the researcher proceeds with a careful reiterated reading of the text together with further listening to the audio file.

This second phase of the transcription process focuses on working on a more detailed rendering of the conversation, removing all the researcher's questions and interactions. This also means transcribing the extra linguistic features of the conversation in order to render the nuances and the complexity of a spoken language as closely as possible in the written text (Gee, 1991; Riessman, 1993). Making sense through narratives requires the analysis and interpretation of multiple potential meanings, tensions and contradictions, including examining the linguistic and extra-linguistic features of first person accounts. Hence, it requires re-transcribing the interviews, adding utterances, verbal emphasis, vocalisations, short pauses, long pauses, word repetitions, dysfluencies, deep breathes, laughs, sighs, and sniffs. The additional non-narrative linguistic details help to initiate the narrative analysis of each interview, since the analysis cannot be easily separated from the transcription (Riessman, 1993). This allows the identification of different parts of the text and potential boundaries among them.

The next step is to start 'unpacking' the narratives by identifying the beginnings and endings of the different narrative segments. The transcript of each participant's interview allows the identification of segments contributing to chronological ordering, as well as those mainly concerned with personal opinions and comments. This allows the differentiation of stories – stretches of narratives with a storified form and a plot with beginning, middle, and end – and stretches of narrative largely recounting opinions, feelings, and points of view without a temporal order (Cunliffe, Luhman, & Boje, 2004; Boje & Durant, 2013). With the exception of few segments with a chronological order and storified form, the data result in a set of first person accounts organised thematically and episodically.

Narrative data analysis

Narrative analysis is an umbrella phrase that describes several possible methods designed to frame the analysis of stories and narratives (Riessman, 1993; Smith & Sparkes, 2008). All commentators accept that narratives do not speak for themselves and, therefore, require interpretation when used as data in social research (Riessman, 1993; Riessman & Quinney, 2005; Smith, 2000). To illustrate the potential value of narrative research, we adopt Gee's (1986; 1991) linguistic approach to studying narratives episodically; sectioning them into idea units, stanzas and strophes on the basis of how the narratives are spoken. Hence, in this study we operate as a bricoleur, scrutinising, sectioning, linking, and interpreting a complex set of data through by building up holistic view.

A linguistic approach

In the linguistic approach to narrative, text can be 'unpacked' according to its different levels of structure and meaning (Gee, 1991; Riessman, 1993). The researcher organises the text, identifying pitch glides – movements in the tone of the speaker – and transfer them into the written text by marking the point of each variation of tone with a slash. Each pitch glide marks what Gee defines as an 'idea unit'; a movement in the tone of voice identifying a change of focus in the speaker's discourse. Each pitch glide corresponds to a different focus and to a new piece of information that forms a different idea unit. The researcher organises the text in numbered lines, each one consisting of one or more idea units separated by slashes (Gee, 1991). This means that – in general – every line features a central idea, articulated syntactically and intonationally.

Once the text is organised in terms of idea units and lines, the researcher groups lines into larger units, defined as *stanzas* (Gee, 1986; 1989; 1991; Riessman, 1993). A stanza consists of a group of lines about a single topic; it captures a single scene, a notional image on a specific person, event, claim, or piece of information. Hence, stanzas are the basic elements of long pieces of discursive narrative such as descriptions, arguments and personal opinions. Following the example of Gee (1991), stanzas on the same or similar topics are linked to make a larger unit called a *strophe*. Strophes, in turn, fall into still larger units called *parts*. It is these that assemble the narrative as a whole, representing the different chapters of the story.

This first level of analysis acts as a foundation on which the other levels of analysis of structure and meaning are built. Once the text is structured into idea units-lines-stanzas-strophes-parts, the researcher moves to a second level of analysis separating the main line of the plot from the remaining material. The researcher identifies the main line through the use of present tense verbs or historical present verbs; non-subordinate and non-embedded clauses.

This level of analysis is the basis for the third level, which focuses on the psychological subjects of main sentences. The psychological subject does not always coincide with the grammatical one; the researcher needs to examine them because 'such subjects are "psychological launching off" points in a stanza; they represent points of view from which the material of a stanza is viewed; they represent what the narrator is "empathising" with' (Gee, 1991, p. 23). The researcher needs to interpret their variety, their repetition, and the switch of viewpoints throughout the text.

The fourth level of structure and meaning examines the focusing system. Following Gee (1991) the researcher can identify the focused material across the stanzas of the whole narrative. The focused material consists of key images or themes recounted by the participants and contributing to establishing an overall interpretation of the whole narrative. The researcher can highlight the focused material – words, verbs, adjectives and rhetorical figures – of each stanza with the idea of performing a thematic interpretation supported by the previous levels of analysis. In this level, the role of the researcher as interpreter is crucial to connect and give sense to the whole structure of the text in terms of key images and themes. The identification of the focusing system and its thematic interpretation is an interpretative reading where the researcher is aware that there could be a number of other possible thematic readings and interpretations but also a number of impossible ones, due to the previous levels of structure (Gee, 1991). Focused material coming from different lines needs to be separated, in this case by a cross hatch ('#'), and material coming from different idea units but in the same lines, can be separated by a comma. An example can be seen in Table 1.

Table 1. Sample of a linguistic approach to data analysis - Participant 1.

Transcription

Then talking about Firm 1 EH we started these trips / that featured some key ingredients and that are, in fact, meeting with the locals EH / then a human component, a component of cooperation, of meeting and this is the first step. The second point was the classical one of responsible tourism, / that is accommodation, meals and local transportation and local property guides in that way the money spent stays in the local economy / much more than when you go on a trip with multinationals this is the second point. The third point has always been to have (p) enough time available, / it is an on-going travel flexibility that was never included, as far as I know, at least in Italy, in the founding points of - - of responsible tourism but in my opinion, if this dimension of flexibility and time availability is missing, / you go nowhere because things are muddled, yes, you know, somehow something always happens when travelling / so you have to figure things out and anyway you have to take some time. To us this means that, / it means that if you go to India for 15 days, / you can't go to twelve different places, it really doesn't make any sense (p) EH in order to get some quality / that is one of the fundamental elements of responsible tourism, you must at least, how can I say (p) achieve some deeper level - - you go physically deeper, for example, when you sleep at least two nights in the same place that is a small thing, but it is a fact, it is solid.

First level - Lines / Stanzas / Strophes / Parts

STROPHE 7 – Key ingredients
Stanza 21
91 then talking about Firm 1 EH we started these trips / that featured some key ingredients
92 and that are, in fact, meeting with the locals EH / then a human component, a component of cooperation, of meeting
93 and this is the first step.
Stanza 22
94 The second point was the classical one of responsible tourism, / that is accommodation, meals and local transportation and local property guides
95 in that way the money spent stays in the local economy / much more than when you go on a trip with multinationals
96 this is the second point.
Stanza 23
97 the third point has always been to have (p) enough time available, / it is an on-going travel flexibility
98 that was never included, as far as I know, at least in Italy, in the founding points of - - of responsible tourism
99 but in my opinion, if this dimension of flexibility and time availability is missing, / you go nowhere
100 because things are muddled
Stanza 24
101 yes, you know, somehow something always happens when travelling / so you have to figure things out
102 and anyway you have to take some time.
103 to us this means that , / it means that if you go to India for 15 days, / you can't go to twelve different places,
104 it really doesn't make any sense (p) EH
105 in order to get some quality / that is one of the fundamental elements of responsible tourism,
106 you must at least, how can I say (p) achieve some deeper level
- - you go physically deeper, for example, when you sleep at least two nights in the same place
107 that is a small thing, but it is a fact, it is solid

Second level - Main line / Off main line

Stanza 21 (91)
Then talking about Firm 1 EH we started these trips / that featured some key ingredients

Third level - Psychological subjects

Stanza 21 we (91)
Stanza 24 you (101), you (102), you (103), you (106)

Fourth level - Focusing system

STROPHE 7 – Key ingredients
Stanza 21 trips, key ingredients # meeting, human, cooperation # first step
Stanza 22 second point, accommodation, meals, local transportation, local property, guides # local economy
Stanza 23 third point, time available, flexibility # founding points
Stanza 24 to figure things out # take time # quality, fundamental # deeper level # small, fact, solid

The narratives of small tourism firms

A research project that examined the decision-making of socially and environmentally responsible tour operators is now used to exemplify the use of narrative research. The seven participating owner-managers studied were members of the Italian Association for Responsible Tourism (AITR). The association was established in the late 1980s as a non-profit consortium and now consists of 92 members. Making an explicit ethical commitment to addressing global inequalities is one of the conditions of membership. These firms state – in their websites, brochures or leaflets – that their products contribute to ethical goals such as poverty reduction, local development, environmental conservation and cultural heritage preservation. Examples of the strophes and stanzas used are shown as part of a process of identifying three narrative types: (i) the intellectual and educational, (ii) the professional and entrepreneurial and (iii) the empathic 'free speech'. As these are being used to illustrate the value of the approach rather than to report the findings of the study, we limit the discussion of the key characteristics of each narrative type. As is often the case, there is a degree of overlap between the three (Riessman, 1993; Bredvold & Skålén, 2016).

The intellectual and educational narrative type

The research findings reveal how in this narrative type, participants understand themselves and their firms as subjects critically reflecting on tourism, contemporary society and development. Participants describe the shift from consumerism to citizenship by characterising themselves neither as managers nor as entrepreneurs but rather as individuals and citizens concerned with global challenges (Elkington, 1998; Klein, 2000). This resonates with the articulation of a firm that – at first – was just an idea and an indistinct amalgam of Fair Trade movement and ethical consumerism theorisations (Moore, 2004; Nicholls & Opal, 2005). Participants understand the firm's identity as a private subject called to contribute to sustainable development and to educate people and travellers towards an authentic and responsible relationship with other cultures.

Participants' identity construction might be understood in in the context of a postmodern perspective or, perhaps, Bauman's (2000) liquid modernity; society is individualised and characterised by uncertainty. Participants understand their individuality and self-affirmation in opposition to a global and indistinct 'alterity'. Firms are recounted as subjects imbued with a sense of the value of artisan production and consumption that emphasises time, human relationships, cultural exchanges and local development.

Participant 1 (Table 2), for example, presents Firm 1 not as a tour operator but as a multifaceted association that includes Fair Trade, responsible tourism and cultural activities (lines 5–6). It is an idea about other worlds and their questions and themes (line 7). 'Stuff' means an indistinct amalgam. The word *idea* and its use play a key role, since Firm 1 originates from the *idea* of the 'yours truly' narrator (line 2). The *idea* in relation to Firm 1 is: 'stuff from other worlds', and themes and questions from a different world (line 7). The *idea* was to create a bridge connecting different cultures (line 10). Strophe 1 ends with the statement that dealing with responsible tourism is 'quite another idea'. For Firm 1, this means a critical discussion of tourism. In Stanza 02, the narrator makes clear that Firm 1 is, and always has been, a non-profit association (lines 04–06).

From its beginning, Firm 1's scope was ambitiously wide; Fair Trade, responsible tourism, cultural activities, and a critical debate about tourism. The choice of the firm's name is meaningful as it implies an indistinct amalgamation of things that are, nevertheless, impossible to divide (line 06). A bridge connecting different cultures and

Table 2. Participant 1.

STROPHE 1 –Stuff from other worlds

Stanza 1

02 because EH it was born as a group it actually it began as an idea of yours truly

03 that is called 'Name 1' EH

Stanza 2

04 and then in nineteen ninety the foundation of an association is formalised / association Firm 1, still existing

05 and dealing with fair trade, sustainable tourism, and cultural activities

06 these three things as EH as it happens in small groups are strictly connected to each other, / they are inextricably connected because we're dealing with Asia.

Stanza 3

07 The idea is Firm 1: / Firm 1 "Name 1" means, matters from the other world,

10 and the idea was to create a bridge between these different cultures

11 and the travel was one of the ways to escape from the commercial dimension through realities that had a meaning / and so we begin

Stanza 5

17 but for us dealing with responsible tourism means quite another idea: / first of all it means criticising tourism / it means then discussing tourism, first of all EH

different worlds is a key image and a powerful metaphor (line 10), as travelling is a way of escaping commercialism through the use of meaningful choices (line 11). Critics in the debate surrounding tourism are a constant feature of the entire narrative and also link Strophe 1 with Strophe 2. The 'quite another idea' of responsible tourism (line 17) is immediately recalled at the beginning of Strophe 2 (line 18) through the use of the pronoun 'this' to explain the origin of the Italian Association for Responsible Tourism.

Participant 2 (Table 3) contextualises Firm 2's vision within a larger perspective. The narrator presents the 'imprinting' that they received from the Fair Trade movement. She recalls being 'outsiders' in the tourism sector. The subject is mainly 'we', referring to the work team and the original founders. The first person singular appears three times; when describing their first encounter with the Director of the Italian Association for Responsible Tourism, when stressing the narrator's role in the Association, and when giving an example of their transparent and sincere relationship with travellers. Stanza 13 narrates the imprinting of Fair Trade values on the firm's vision for tourism (lines 86–87). The word 'imprinting' recalls the study of ethology, the first knowledge gained in life. It is a meaningful choice of word to explain how Fair Trade has inspired their vision of tourism. It was originally an ethical selling point (line 88) for them, informed by the values of the Fair Trade movement. They commercialised Fair Trade products and other goods created by cooperatives and social enterprises (lines 92–93).

The narratives from these two participants have some similarities with those of Participant 7 (Table 4). Strophe 13 details the reasoning behind participant 7's decisions by explaining his firm's vision: 'our work, which is very artisanal, really home-made' (line 148). The participant here links the firm's artisanal dimension with its capacity to provide unique products. He stresses that a 'unique piece' is the product of special, artisanal, and traditional knowledge. Even its imperfections make it extraordinary. Words such as 'unique-piece' (lines 152 and 153), 'artisanal' (lines 148, 150, 152 and 157), 'flaw' (lines 152 and 155), and 'virtue' (lines 155 and 156) are recurrent. This concept is expanded upon in Strophe 14, which explains why imperfections make artisanal products unique and valuable. The narrator stresses the importance of not

Table 3. Participant 2.

STROPHE 5 – Our values in tourism
Stanza 13
86 ours stems out of this kind of imprinting,
87 usually, our trips are EHH more like this, / in the sense that it is the development on tourism, of the principles of Fair Trade,
88 basically it began as an ethic selling point / because, it's was near the sustainable, in other words, classic, / well, there is *Altro Mercato* isn't it?
90 EH so, we couldn't handle it, as we had to keep our people's,
92 and with all the situations of social cooperatives producing work,
93 thus: prisoners who produced EHM MH leather goods, / disabled people EH making EH also there we had leather goods / and then another one producing and bookbinding instead, / a cooperative here making raku pottery,

becoming 'just a part of a big production chain' (lines 158). 'I wanted our strong and artisanal impact to stay, a very personalised relationship, also with our customers' (line 160); 'I still struggle to call them like this' (line 161), 'I call them participants' (line 162), 'I like it more' (line 163).

The professional and entrepreneurial narrative type

Some participants' narratives reflect an identity largely constructed in terms of entrepreneurship, efficiency and professionalism. First of all, these participants understand their entrepreneurial identity not only as the result of a long and peculiar path, but also as an innovative approach to the tourism and development agenda that goes beyond the constraints of profit/non-profit attitudes. The entrepreneurship model of innovation and disruption helps to locate participants' understanding and identity construction. Their self-introductions largely revolve around being 'wild spirits', striving to do new things, or things that have already been done but in a new way (Bredvold & Skålén, 2016; Ateljevic & Li, 2009). Their narratives disclose an identity constructed in terms of innovation and disruption in opposition to the literature presenting small tourism firms as more interested in perpetuating lifestyle choices (Shaw & Williams, 2004). Their success is narrated as the capacity to evolve towards new markets and opportunities and to put in place adaptive changes. This model of entrepreneurship reflects a society where different subjects operate in a creative, proactive and adaptive way following their core values, beliefs and goals.

Some aspects of Participant 3's narrative (Table 5) revolves around the firm's capability to legally and efficiently deliver high quality services compared to the weaknesses of other responsible tourism firms. The owner-manager speaks predominantly in the first person plural as he focuses on the firm's team. Nevertheless, the recurrent use of impersonal and generic subjects, such as people, travellers, travel agencies, and tour operators, alongside the use of 'you' make the narrative more persuasive in terms of its ascribing meaning. On occasions, this section assumes the tone of a theoretical enunciation. He highlights the strong relationship between the high quality services his firm provides and his commitment to social projects and the prices he charges customers. Respecting the law, being efficient, getting involved in social projects and providing expensive holidays for travellers are interconnected features of the firm (lines 54–55); 'respecting those [laws], you fit into a context, where, it would be totally illogical not to give a high-quality service' (line 45), 'because I have to charge it properly' (line 46).

Table 4. Participant 7.

STROPHE 13 – Artisanal work
Stanza 34
148 so it means: our work, / which is very artisanal, / really homemade,
149 even though, you make lots of mistakes, a lot of things, that could be done in a better way, for sure, and so on,
150 but, it's still an artisanal work,
151 well you know very well that, if you buy an Ikea piece of furniture, / EH it might be beautiful, / it's cheap
152 but, of course, the same piece of furniture made by an artisan / maybe has some flaws, but that's because it's a unique piece, right?
153 but it remains a unique piece,
154 and it's really different EHM
155 and - - and this, probably, well, can be judged both as a flaw, / and as a virtue,
STROPHE 14 – A virtue
Stanza 35
156 as far as I'm concerned, / I've always seen it as a virtue,
157 EH and this is why, however, EHM if an artisan, sells to Ikea/ just to keep our example / we don't have an artisan anymore, understood?
158 he becomes just a part of a big productive chain.
Stanza 36
159 Here, I didn't want neither Firm 7, / nor what Firm 7 had done, so the product, / to be a part of a big productive chain
160 I wanted a strong, artisanal, impact to stay, /a very personalised relationship also with our costumers,
161 that I still struggle to call them like this,
162 I call them participants,
163 I like it more,
164 and this, let's say, is somewhat then, you know, the idea right or wrong, EHM / this was the reason, see?

Participant 3 further argues that high quality services means high costs and the responsible tourism sector should accept this; 'we are our own worst enemy' (line 59), and 'the world of responsible tourism plays downward' (line 60). The narrator believes that being able to compete in the market system with big tour operators is crucial; 'we are forced to fight with teeth and crawls' (line 64), 'that ultimately lead us to be, ultimately, from an external point of view, exactly, like a traditional tour operator' (line 65). There are several references to prices and costs and frequent comparisons with big firms (lines 52, 53, 58 and 65), which frames the narrative with economic and pragmatic reasoning; 'tourism is an economic activity' (line 72). Competing with traditional tour operators and winning contracts to provide services for the big tour operators is the key to success; 'in my opinion, this is the key to success in the world of responsible tourism' (line 68). This is why the firm appears to be particularly efficient at overcoming the limits of responsible tourism.

Participant 4 (Table 6) describes himself as a self-made entrepreneur with narcissistic features (line 148). This opening represents a general introduction and what might be described as 'literary incipit', as he uses the third person singular. The narrator describes at length the wider contexts of each theme that he introduces. Stanza 41 outlines his personal satisfaction and fulfilment whilst at the same time, he claims: 'I am somehow, well, a slave to my own successes' (line 149). The words 'slave' and 'narcissistic' are used to describe the narrator's inability to recognise when a project – his 'creature' – can no longer survive (line 151). Nonetheless, the firm guarantees both his personal and economic fulfilment (line 153).

The focus of Participant 4's narrative moves from the narrator's astonishment at their economic achievements, to the money they have invested in developing

Table 5. Participant 3.

STROPHE 5 – High quality services

Stanza 15

45 and so we, first of all we respect those, / respecting those you fit into a context, where, it would be totally illogical not to give a high-quality service

46 because when I must: / have means of transportation paid / because I have to charge it properly,

47 I mean you find it always clean, / with the employees always trained, always clean, and always there

48 it is normal that if then if someone tells me: / "I want to spend five, ten thousand euros per person for a travel"

49 I don't say "no, because I am responsible",

50 but I answer, "but well bring three more people with you

51 because then, later, with the money you owe me, / I know how to employ it".

Stanza 16

52 So we have developed an attitude, extremely EHH efficient under this point of view, / EH, / to try, however, to bridge the gap that divides us from the big tour operators,

53 because, paradoxically, we cost more than the big ones.

54 I mean, Firm 3, because of the way it works,

55 namely, quality, compliance, and allocating part of the incoming money to social projects as well,

56 in the end, on some specific products, it costs more than a giant tour operator.

57 So eventually people can't understand,

58 many travellers can't understand this, they tell me: / "but how comes, I came to you who work in responsible tourism, / and a tour with you costs more than a tour with"

STROPHE 6 – I give high quality services

Stanza 17

59 here we enter an even more EH complicated matter, / where we are our own worst enemies

60 because our, the world of responsible tourism plays downward

61 I mean, it doesn't tend to sell quality/ but it just tends to sell ok?

Stanza 19

64 but MH as we have, let's say, substantial prices / we are forced to fight with teeth and crawls

65 and to promote ourselves, using techniques and strategies / that, ultimately lead us to be, exactly, from an external point of view, exactly, like EH a traditional tour operator

66 and this is why, agencies and other big tour operators contact us, / and willingly accept our particular attention for social issues,

67 but they don't get the difference between us and a different DMS

Stanza 20

68 and, in my opinion, this is the key to success in the world of responsible tourism

69 that is, to evolve the destination management company, on site,

70 and to invest in them as well trying to get them to adapt and conform to rules / in order to help them competing with mainstream tourism,

71 that is, you must give up somewhat the preconception that responsible tourism is not possible in good level facilities, / rather than, EHH I well, with appropriate services,

Stanza 21

72 because in the end, I mean, tourism is an economic activity, / is an economic activity, right?

73 It is defined that way even in the dictionary

74 so either you do it, / EHH trying to stay inside its own rules / and trying to offer a better service, / and emphasising also, your own peculiarities, / competing with everyone

75 or you are defeated before you start

countries, generating local economic development (Stanza 42–44, in Table 7). The fact that he only becomes aware of this when preparing data for public communications is significant because it implies that he does not often think about this aspect. Moreover, he claims that this amount is different from the 'development fund' generated through the shares project (line 155). His personal astonishment: 'F**k! We have made more than three million euros' (line 160) – reported as direct speech – is compelling and emphatic. The narrator highlights that this sum is the result of a long and enduring process; 'drops in the sea' (lines 161–162). In the end, it brings 'great satisfaction' (line 163).

Table 6. Participant 4.

STROPHE 12 – I'm a slave of my own success
Stanza 41
148 well, gratifying, because in someone who has always been, as I told you, somewhat his own boss MH / there is of course a narcissistic part, EH
149 and so, succeeding in / I am somehow well a slave of my own successes,
150 in the past, it often, happened to me, / I hope I won't come to this with Firm 4
151 that, sometimes, when you feel the creature is yours, then / even when it is time to let it go, / because it's falling apart, / just because you feel it's yours, / and you don't want it to finish it, / you stubbornly keep on trying
152 This happened to me in the radio world in the independent radios, / when I worked in the editorial staff of the radios EHH
153 so, there is an important personal reward aspect, of financial security / because well, being able to earn a pay check at the end of the month nowadays is not that easy.

Table 7. Participant 4.

Stanza 42
154 EHM, and well, sometimes I'm shocked too / when we calculate these statistics for public presentations, / and it comes out that, in all these years, well, we gave to the South of the world more than, now I think they are three million euros calculating EHM,
155 because one thing is the development found/ there you support some local projects
156 but EH consider that a great part of the touristic services that we buy / we buy them from providers that are an expression of the local community
Stanza 43
157 If we spend the night with a host family, / we pay our staying, to whom? To a family / and, thus, to the local community.
158 if we take a guide, who is a person I mean not, belonging to a Lugano based organisation, but who lives there, is from an organization that is based in Lugano, but who lives there,
159 anyway we estimate that this is always, financial support, against the purchase of services to the local community.
Stanza 44
160 When you try and put all the numbers together, you say: "fuck! We've made more than three million euro" / I mean they are impressive numbers, right?
161 Also because, they built up, as drops in the sea, / here, every trip a little, / every trip a little
162 but in fifteen years, they became quite impressive numbers
163 I mean on a personal level / well, it's a great satisfaction

The empathic 'free speech' narrative type

The empathic 'free speech' narrative type gives meaning to the emotional attachment that participants show for their firms, customers and local partners through a frank, open and critical self-introduction. Here, participants largely understand their emotional and empathic attachment to customers and partners in terms of a search for trust, legality and transparency. Participants see themselves as truth-tellers and free speakers concerning the challenges and constraints they experienced due to global economic inequalities, weaknesses and lack of transparency. The empathic 'free speech' narrative type resonates with the classical Greek notion of *parrhesia*, which literally means 'all-telling' and – by extension – 'to speak freely'. This notion pictures the obligation to speak the truth for the sake of the common good, even at personal risk (Sluiter & Rosen, 2004; Lardinois, 2011); hence, the notion of *parrhesia* and the role of the 'truth-teller' – the *parrhesiast* – implies an asymmetry of political power, whereby the *parrhesiast* is usually inferior to his/her recipient. Participants' *parrhesia* consists of the freedom to criticise others' moral apathy and ambiguities openly, juxtaposing this

Table 8. Participant 5.

STROPHE 5 – Sharing the same vision
Stanza 15
63 There, I don't know whether I can speak freely now / or you have questions
64 I mean maybe, stop me, if you need explanations
65 no, I was saying, that another possible question for you, that is also asked: / "but what are these trips?
66 That is, why are you called responsible tourism?"
Stanza 16
67 then, in my opinion, then I came to my own, personal EH / let's say, classification, I mean, decision.
68 So, essential is the choice of the partner,
69 I mean, of the people, of the community, I don't know, it is a cooperative, / it is also a tour operator, a travel agency that, with them you share some things, in my opinion, fundamental
70 I mean, the stronger this commonality of ideals is, / the more EH legs the thing, the relationship has
71 and so, also, actually, the possibility of making trips becomes more real

to their own moral rectitude and frankness (Colclough, 1999; Novak, 2006; Sharpe, 2007). The *parrhesiastics* attribute to themselves higher moral qualities and a deeper understanding of reality than others. These self-perceived moral qualities apparently permit them not only to know and tell the truth, but also to disclose to others what is 'the right thing to do'.

Participant 5 (Table 8), for example, poses the question of what makes tourism responsible as a sort of personal outburst; 'so, I don't know whether I can speak freely now or you have questions' (line 63), 'I mean maybe, stop me, if you need explanations' (line 64), 'then I came to my own, personal, let's say, classification, I mean, decision' (line 67). Of central importance is choosing a partner that shares your vision and values (lines 69–70). The narrator uses a vivid image to describe this point: 'I mean, the stronger this commonality of ideals is, the more legs the thing, the relationship, has' (line 70). The image of a relationship with 'legs', implying it is alive and dynamic. Thus, having legs means being able to proceed, to move on, and to concretely reach objectives (line 71).

Participant 6 (Table 9) describes his firm as quite 'grown-up' (line 11), 'a big boy' (line 12), and 'almost an adult already' (line 13). Stanza 03 personifies the firm by drawing a meaningful parallel between the firm and an adolescent maturing into adulthood. The repetition of the verb 'was born' (lines 14 and 15) reinforces this parallel as it describes the juvenile idea conceived by two young boys who were enthusiastic about nature and willing to create something new (line 16); 'at that time, the two founders were really young' (line 20). She emphasises the firm's naïve, inspired, and unprofessional beginning; 'it was born maybe even somehow by chance' (line 15). Indeed, it all began with a rudimentary idea based on the boys' wishes and desires. The repetition of the word 'idea' (lines 14, 16, 19, 22) is important as it emphasises their feelings of desire (lines 14, 16 and 22) and happenstance (lines 15 and 18).

Conclusion

This article has argued that narrative research offers the prospect of a more creative approach to the study of tourism. It rests on the premise that, as Garud, Gehman, and Giuliani (2014, p. 1182) note, 'narratives are performative: they trigger action as … (actors) … try actualizing the very arrangements they have proposed in order to

Table 9. Participant 6.

STROPHE 2 – Firm 6 is a big boy
Stanza 03
10 Firm 6, was born in eighty-eight,
11 so it has, a pretty history / let's say, it's already quite grown-up well
12 it is a big boy,
13 it is almost an adult already EHM
Stanza 04
14 it was born let's say, because of the interest, the idea of, in fact, Luca and Mario EEH about the willingness of explaining somewhat some natural phenomena, well
15 it was born, maybe even somehow by chance, well
16 from the idea, the desire, to have their own activity / that dealt, something to do with the natural environment
Stanza 05
18 well, MH so, EHH, it was born that way, / not I wouldn't say accidentally,
19 but not, Let's say, it doesn't stem from a specific idea of a project here,
20 then, in fact, at the time the two founders were very young, see?
22 and so, the idea of starting their own thing, with / because of the desire to recount nature,

generate meaning around their ventures'. The resulting insights may challenge and potentially disrupt knowledge generated through more widely used methodological approaches.

A study of small socially and environmentally conscious tourism businesses was used to demonstrate the application of the linguistic narrative approach and, in turn, to show how novel findings might emerge as a result. In this case, the findings challenge ideas that small tourism firms that are not commercially oriented and not interested in growth (Ateljevic & Doorne, 2000; Gray, 2002), do not have an entrepreneurial attitude (Shaw & Williams, 2004), and routinely face managerial limitations and constraints (Thomas, Shaw, & Page, 2011). The article argues that the methodological approaches routinely used in this aspect of tourism research have disclosed a partial understanding of the complexity of values that inform the decision-making of small socially and environmentally oriented tourism firms. The potential of exploring narratives offers an additional approach to interpreting the meanings ascribed by actors to their actions.

Most previous attempts to understand the values that inform the choices made by owner-managers of small tourism firms have used conventional methods such as questionnaires and interviews of various types, comparative analysis of secondary data or mixed methods (e.g. Ateljevic & Doorne, 2004; Sampaio, Thomas, & Font, 2011; Font, Garay, & Jones, 2016). While valuable, by paying insufficient attention to the individual socio-political dimension of the firms' owner-founders/managers, together with a well-rounded understanding of their driving values, identities, and sense of self, lead to less nuanced understanding than is now possible.

The application of a narrative analysis to the study of small tourism firms allows us to look beyond polar dichotomies since one participant's narrative can share elements of another. A creative and rigorous narrative approach allows us to see how small values-based tourism firms' owners-managers construct their identity around proactive critiques of contemporary society, the global economy and mainstream tourism. Owner-managers' first person accounts predominantly focus on values and beliefs rooted both in alternative paradigms of development and growth and in postmodern conceptualisations, such as citizenship over consumerism (Elkington, 1998; Klein,

2000), de-growth (Latouche, 2004), ethical consumerism (Moore, 2004; Nicholls & Opal, 2005), not-conventional entrepreneurship (Bredvold & Skålén, 2016; Shaw & Williams, 2004, liquid self-affirmation (Bauman, 2000), conviviality (Illich, 1971), and parrhesiastic frankness (Lardinois, 2011; Novak, 2006; Sharpe, 2007; Sluiter & Rosen, 2004). By adopting a narrative approach, the apparently disconnected elements of each first person account are seen as part of a whole meaningful understanding.

The particular novelty of the research reported in this paper lies in its moving beyond what the participants' report through stories and narratives to analysing and interpreting the subtle and unconscious understanding expressed via linguistic choices, non-verbal utterances, and the temporal and causal sequences of their storytelling. This allows for the 'unpacking' and interpreting of narrative data though a tailor-made narrative analysis (Riessman, 1993). By so doing, the study illustrates a new way of 'knowing' that has almost been disregarded within tourism studies thus far.

The reported empirical insights are deliberately limited; our goal was to use data for illustrative purposes as part of making a contribution to methodological debates in tourism rather than to research on tourism entrepreneurship. Some critics may highlight the limitations of narrative research in terms, perhaps, of positivist notions generalisability. This is not a criticism we would accept, arguing instead that the data and analysis reveal hitherto hidden insights that are generalisable to the theoretic (Yin, 2018).

Disclosure statement

No potential conflict of interest was reported by the author(s).

References

Andrews, M, Squire, C. & Tomboukou, M. (Eds.) (2013). *Doing narrative research* (2nd ed.). London: Sage.

Ateljevic, I., & Doorne, S. (2000). 'Staying Within the Fence': Lifestyle entrepreneurship in tourism. *Journal of Sustainable Tourism, 8* (5), 378–392. doi:10.1080/09669580008667374

Ateljevic, J., & Doorne, S. (2004). Diseconomies of scale: A study of development constraints in small tourism firms in central New Zealand. *Tourism and Hospitality Research, 5* (1), 5. doi: 10.1057/palgrave.thr.6040002

Ateljevic, J., & Li, L. (2009). Tourism entrepreneurship – concepts and issues. In J. Ateljevic (Ed.), *Tourism and entrepreneurship: International perspectives* (p. 9). London: Routledge.

Bauman, Z. (2000). *Liquid modernity*. Cambridge, UK: Polity Press.

Boje, D. M., & Durant, R. (2013). Free Stories!. *Tamara Journal for Critical Organization Inquiry, 5* (3), 19–37.

Bredvold, R., & Skålén, P. (2016). Lifestyle entrepreneurs and their identity construction: A study of the tourism industry. *Tourism Management, 56*, 96–105. doi:10.1016/j.tourman.2016.03.023

Buchanan, D. (2013). Getting the story straight: Illusions and delusions in the organizational change process. *Tamara Journal for Critical Organization Inquiry, 2* (4), 7–21.

Chase, S. E. (2005). Narrative inquiry: Multiple lenses, approaches, voices. In: N. K. Denzin & Y. S. Lincoln (Eds.), *The Sage handbook of qualitative research* (3rd ed., pp. 651–679). Thousand Oaks, CA, Sage Publications Ltd.

Colclough, D. (1999). Parrhesia: The rhetoric of free speech in early modern England. *Rhetorica: A Journal of the History of Rhetoric, 17* (2), 177–212. doi:10.1525/rh.1999.17.2.177

Cunliffe, A. L., Luhman, J. T., & Boje, D. M. (2004). Narrative temporality: Implications for organizational research. *Organization Studies, 25* (2), 261–286. doi:10.1177/0170840604040038

Davies, B., & Harré, R. (1990). Positioning: The discursive production of selves. *Journal for the Theory of Social Behaviour, 20* (1), 43–63. doi:10.1111/j.1468-5914.1990.tb00174.x

Denzin, N. K., & Lincoln, Y. S. (2005). *The SAGE handbook of qualitative research* (3rd ed.). Thousand Oaks, CA: Sage Publications.

Elkington, J. (1998). *Cannibals with forks: The triple bottom line of 21st Century Business*. Gabriola Island, BC: New Society Publishers.

Font, X., Garay, L., & Jones, S. (2016). Sustainability motivations and practices in small tourism enterprises in European protected areas. *Journal of Cleaner Production, 137*, 1439–1448. doi: 10.1016/j.jclepro.2014.01.071

Garud, R., Gehman, J., & Giuliani, A. P. (2014). Contextualizing entrepreneurial innovation: A narrative perspective. *Research Policy, 43*(7), 1177–1188. doi:10.1016/j.respol.2014.04.015

Gee, J. P. (1991). A linguistic approach to narrative. *Journal of Narrative and Life History, 1* (1), 15–39. doi:10.1075/jnlh.1.1.03ali

Gee, J. P. (1989). Two styles of narrative construction and their linguistic and educational implications. *Discourse Processes, 12* (3), 287–307. doi:10.1080/01638538909544732

Gee, J. P. (1986). Units in the production of narrative discourse. *Discourse Processes, 9* (4), 391–422. doi:10.1080/01638538609544650

Gray, C. (2002). Entrepreneurship, resistance to change and growth in small firms. *Journal of Small Business and Enterprise Development, 9* (1), 61–72. doi:10.1108/14626000210419491

Illich, I. (1971). *Deschooling society*. New York: Harper & Row.

Klein, N. (2000). *No logo: Taking aim at the brand name bullies*. Toronto: Knopf Canada.

Lardinois, A. (2011). The Parrhesia of young female choruses in ancient Greece. In L. Athenassaki, & E. Bowie (Eds.) *Archaic and classical choral song: Performance politics and dissemination* (pp.161–172). Berlin: De Gruyter

Latouche, S. (2004). Degrowth economics. Le Monde Diplomatique, 11.

Moore, G. (2004). The Fair Trade movement: parameters, issues and future research. *Journal of Business Ethics, 53* (1/2), 73–86. doi:10.1023/B:BUSI.0000039400.57827.c3

Nicholls, A., & Opal, C. (2005). *Fair trade: Market-driven ethical consumption*. London: Sage Publications.

Novak, D. R. (2006). Engaging Parrhesia in a democracy: Malcolm X as a truth-teller. *Southern Communication Journal, 71* (1), 25–43. doi:10.1080/10417940500503480

Reissner, S., & Pagan, V. (2013). *Storytelling in management practice. Dynamics and implications*. London: Routledge.

Rhodes, C., & Brown, A. D. (2005). Narrative, organizations and research. *International Journal of Management Reviews*, *7* (3), 167–188. doi:10.1111/j.1468-2370.2005.00112.x

Riessman, C. K. (1990). Strategic uses of narrative in the presentation of self and illness: A research note. *Social Science & Medicine (1982)*, *30* (11), 1195–1200.

Riessman, C. K. (1993). *Narrative analysis*. London: Sage Publications.

Riessman, C. K., & Quinney, L. (2005). Narrative in social work a critical review. *Qualitative Social Work: Research and Practice*, *4* (4), 391–412. doi:10.1177/1473325005058643

Sampaio, A. R., Thomas, R., & Font, X. (2011). Small business management and environmental engagement. *Journal of Sustainable Tourism*, *20*, 179–193.

Sharpe, M. (2007). A question of two truths? Remarks on parrhesia and the 'political-philosophical 'difference. *Parrhesia*, *2*, 89–108.

Shaw, G., & Williams, A. M. (2004). From lifestyle consumption to lifestyle production: Changing patterns of tourism entrepreneurship. In R. Thomas (ed.) *Small firms in tourism: International perspectives (advances in tourism research)* (pp. 99–111). Amsterdam: Elsevier.

Sluiter, I., & Rosen, R. M. (2004). *Free speech in classical antiquity:[Penn-Leiden Colloquium on Ancient Values, June 2002 at the University of Pennsylvania]*. Boston: Brill.

Smith, B., & Sparkes, A. C. (2008). Narrative and its potential contribution to disability studies. *Disability & Society*, *23* (1), 17–28. doi:10.1080/09687590701725542

Smith, C. P. (2000). Content analysis and narrative analysis. In H. Reis (ed.) *Handbook of research methods in social and personality psychology* (pp. 313–335). Cambridge, MA: Cambridge University Press.

Thomas, R., Shaw, G., & Page, S. J. (2011). Understanding small firms in tourism: A perspective on research trends and challenges. *Tourism Management*, *32* (5), 96.3–976.

Tsoukas, H., & Hatch, M. J. (2001). Complex thinking, complex practice: The case for a narrative approach to organizational complexity. *Human Relations*, *54* (8), 979–1013. doi:10.1177/0018726701548001

Walmsley, A. (2019). *Entrepreneurship in tourism*. London: Routledge.

Wilkins, A. L., & Thompson, M. P. (1991). On getting the story crooked (and straight). *Journal of Organizational Change Management*, *4* (3), 18–26. doi:10.1108/EUM0000000001194

Yin, R. K. (2018). *Case study research and applications: Design and methods* (6th ed.). Thousand Oaks, CA: Sage Publications.

Afterword

Any inquiry into mobilities or im-mobilities warrants the focus not just on the destination and the origin but on the journey itself. Similarly, as qualitative researchers our focus is on these journeys, lived experiences and the dynamic meaning making exercises that we engage with our participants. Our search for an inter-subjective emic perspective drives us to be conscious that reality, contexts, and politics keep changing both for the participants and for the researchers. The pandemic has brought home the deep understanding of what a privilege it is to have the means and resources to be immobile. This immobility masks the precarious mobilities the less privileged have to undertake to ensure that many of us can work within the safe confines of our homes.

Reflexivity is a corner stone of providing grounded understanding of people with whom we study the complex realities. It is an essential skill that needs to be honed in constant interaction and debates with a diverse set of stakeholders and constant (re)questioning of our own cultural frames and prejudices. To prevent reflexivity turning into a 'navel gazing exercise' we have to disrupt our methodological habits. Various creative and experiential methods from this volume provide us the tools to engage with reflexivity through a co-creation approach. By emphasising creativity and creative interventions in the knowledge generation process, researchers in this volume are disrupting the often held positivist benchmarks of representation, generalizability, and reproducibility of research. By disrupting the "sanitising effects" of positivist research we are opening up the possibilities to understand the messiness and rough experiences that surround any inquiry where we try to explore the multiple life words of our participants.

Qualitative research is not just about data collection. Through sustained, engaging, and respectful relationships in the field we co-create knowledge. It then becomes our responsibility that we interpret and communicate the rich information to diverse audiences. Following Professor Linda Tuhiwai Smith, we need to critically reflect on: Whose story are we telling? Who has the right to tell this story? Whose version of the story is it? and How does this story disrupt the imperial gaze? Such questions are integral to our embodied writing of diverse truths. With creative, experiential, and disruptive methodologies let us also push the epistemological boundaries by writing about truths and experiences that are uncomfortable, unfamiliar, repulsive, sorrowful, and painful. To have an holistic understanding of any situation we need to see both sides of the coin. I am sure the learnings from this book will give us the right directions to (as the editors suggest) *disrupt, subvert, confront, and go against the stream to co-create knowledge.*

Professor Ajay Bailey
Utrecht University, The Netherlands

Index

Page numbers in **bold** refer to tables and those in *italic* refer to figures.

Aalbers, M. 62
academic capitalism 61
academic knowledge production 102
academic legitimacy 64
academic tribes 67
active-passive participation continuum 152
activist tourism 80
Actor-Network Theory (ANT) 127, 129–30
Adaptive Methodologies for Eco-social Sustainability and Health (AMESH) 85
adaptive methods 84–5
adaptivity 91
Adkins, L. 41
'adventure holidays' 236
adventure tourism 240
advocacy coalitions 86
'Against Method' (2010) 126
agency 109–12
Ahmad, R. 40
Airey, D. 64, 65
Alaskan Walleye Pollock fishing 134
Alves, M. H. M. 257
Ancient Siebold's Beech *140*
Andéhn, M. 106
Anderson, C. R. 329
Anglo-American domination 61–6
Anglo-American tradition *68*
animal-based tourism services 102–4
'Annals of Tourism Research' 62
'anthropocene' concept 129
anthropology 104
anti-globalisation movements 80
Arlie Hochschild's notion of emotional labour 41
Armitage, D. 85
Arnab, S. 198
artificial nest-boxes 135
Asian tourism scholars 65
Association for Responsible Tourism (AITR) 355
Association of Master Trainers 175
Ateljevic, I. 65, 66, 215, 216, 230, 231, 257, 260
Atkinson, S. A. 91
Auckland's award winning Tourism, Events and Economic Development Agency (ATEED) 311
augmented cinema 89–90
Ayikoru, M. 64

Bajerski, A. 61, 62
Banks, E. 104
Bański, J. 61
Barker, A. 18, 19
Barone, T. 309
Bastian, H. 86
Bauman, Z. 356
Bays, A. 41
Becher, T. 67
Belk, R. W. 105, 106
Benckendorff, P. 70
Benedict XVI (2009) 59
Benediktsson, K. 127
Berbary, L. 308
Berg, L. D. 61, 65
Bianchi, R. V. 81
Birkeland, I. 221
Björk, P. 150, 151
Blakiston's Fish Owl 135, *136*
'bodily hexis' 40
Boneham, M. 260
Boniche, A. 61
Bosco, F. 173
Botteril, D. 66, 124–5
Bouchon, F. 62, 66
Bourdieu, P. 40, 58
Boyer, Ernest 83
Boykoff, J. 286
Bramwell, B. 193
'b-reel' footage, ethical consumer videos 106
Browne, K. 266
Brown, K. 104
Bruner, J. 203
Buller, H. 108
Burns, P. 273
Burt, G. 204
Buzinde, C. 62, 66

Çakmak, E. 40
Caliandro, A. 152

Campbell, M. 130
Canadian stories 29
Caplan, P. 39
Carlson, E. 31
Carney, S. 203
Carter, T. 287
Cartwright, D. 219
Casey, E. 238
Cashman, S. B. 95
Castoriadis, Cornelius 81
Cataldo, F. 257
Cater, C. 240
Chakrabarty, D. 129
Chakraborty, A. 135–6
Chalip, L. 305
Chambers, D. 62, 66
Chang, H. 248
chaotic communities 262–3
'Charlie's' memory 28–30
Chervany, N. L. 260–1
Cherednichenko, O. 85
Chesters, G. 83
Clarke, S. 219
Clark, N. 129
Clark, T. 262, 274
Cockburn-Wootten, C. 172
Cohen, E. 65
Cohen, S. A. 65, 215, 216, 218
collaborative research 342–3
collective memory work (CMW) 3; consciousness-raising pedagogical methodology 27–9; cultural institutions and language 21; development of 22–4; facilitation guide **27**; feminism 23–4; identity markers 22; participatory action-oriented methodology 21; unsettling methodology 17
Collins, F. L. 65, 215, 216, 230, 231
colonial domination 18
colonialism 16
commercial enterprises 150
commodification theory 58
The Commoners Choir *91*
Commonwealth Games Federation (CGF) 317–18
Communication Challenges Models *182*
community based participatory research (CBPR) 95
Compost Pile Model *181*
concept of 'affect' 217
The concept of play 174
constructionism 59
constructionist approach 66–8
constructivism 173–4
Cooke, L. 42
Copacabana's temporary Beach Volleyball stadium 290
Copernican paradigm 126
co-researchers *341*, 341–2, 345
Costa, C. 305
Côté, I. 264
'countertransference' 218–19, 230
Crang, M. 242
creative analytical practice (CAP) 307–11
creative non-fiction (CNF) 307–11
creative tourism: and CREATOUR in Portugal 331–4; description 330; emergence of 330; international contexts 331; organizations and pilot activities 334–5; 'soft' impacts 331
CREATOUR in Portugal 331–4
CriaAtivo Film School 267, 269
Critical Conversations element 94
critical event studies (CES) 83
criticality 3–4
critical qualitative research 123
Crookall, D. 194–5
Crossley, É. 5
crowd-sourcing video footage 295
Csikszentmihalyi, M. 175
Culala, H. J. 65
'Current Issues in Tourism' 66
cyberethnography 5
cyber stealth 150

Dann, G. M. S. 62
data analysis 46
data collection methods 248–9
Davies, A. 62
Debord, Guy 88
decolonial theories 3
decolonization 30–1
Decroup, A. 105, 261
"deep reflexivity" 5
Deleuzian deterritorialization 86
Demeritt, D. 127, 128
Denzin, N. K. 33, 261, 308
Desbiens, C. 61
Desmond, J. 239
Deterding, S. 194
'Digi-cal' model 259, 275–7, *276*
digital and physical methods 263
digitalized dashboard 200
digital platforms 5–7
digital technology 6
discourse theory 58
disruption 10, 85–6, 102
disrUpt project 87–8
dissensus: critical assessments 123; defined 123; natural-or-eco-tourism destinations 123; non-human voices 130–2
Dixon, D. 194
Dixson, C. 80
Dorothy Smith's feminist activism (2007) 83
Dowling, R. 105
Draper, D. 50
Dredge, D. 65
drug trafficking 258

'ecotourism planning' 134
Edensor, T. 240, 247
Einstein's Theories of Relativity 126

Elliott, E. 260
embodied tourism experience 239–40
embodied utopias and nondiscursive pleasures 240
Emmel, N. 260, 262
empathic 'free speech' narrative type 361–2
engaged scholarship 83, 329
English-speaking countries **63**
'enterprise society' 60
epihenomenal discoveries 249
Erickson, B. 20
ethical consumerism theorisations 356
ethico-politics 112–14
ethics of lurking 151–3
'ethnic outsider' 242
ethnodrama 307–11
ethnography 4
Evanson, P. 257
Everett, S. 260, 261, 262, 275
experiential learning 329

face-to-face encounters 339
facilitator 91
Fair Trade movement 356
Farkic, Jelena 5
Favelas 264
Feigenbaum, A. 83, 93
Feighery, W. 215
Female Sexualization: A Collective Work of Memory (1987) 22–3
feminine 40–2
Fennell, D. A. 108
Ferenc, M. 61
Feyerband, P. K. 58, 126
film making workshop 90–2
Fine, M. 95
Finlay, L. 214, 216
Finnish Lapland context 102, 103
Fletcher, G. 171–2
Foreign Commonwealth Office (FCO) 258
'formal' presentation and group feedback sessions 338
Foster, C. 193
Foucault, M. 58, 60, 238
Fox, K. M. 240
Fox, M. 95
Frankfurt School 3
Freire, Paulo 83, 91
Frenzel, F. 83, 93
Frew, M. 283
Frosh, S. 218, 219, 230
Fuller, D. 61

Gaffney, C. 286, 287
Galilean paradigm 126
game board 200
Game Dashboard and Play Board with miniatures *202*
game facilitator 196, 199
gaming 8
Gandini, A. 152
Garcia-Ramon, M. D. 61
Garud, R. 362
gatekeepers 262
Gauntlett, D. 173
Gayle, R. 130
Gay, R. 257, 274
Gee, J. P. 354
Gehman, J. 362
gender and feminist theories 3
general data protection regulation (GDPR) 7
gentrification 283
geo-filters 160
geographers 151
geography 104
Getz, D. 304
Gibson, C. 61–2
Gibson, H. 305
Gilbert, K. R. 219
Giuliani, A. P. 362
globalization process 61
Global Positioning System 289
Godlevsky, M. 85
Goffman, E. 47
Gold Coast's Commonwealth Games (April 2018) 311
Goldring, J. E. 260
Gray, R. 309
Greenhalgh, T. 329
Greenhill, A. 171–2
Gregson, N. 61, 71–2
Griffiths, M. 171–2
Grimwood, B. S. R. 3, 139
Grobler, A. 59
Grosfoguel, R. 58, 60
Guba, E. G. 125, 308
Gulley, N. Y. 27
Gurney, J. N. 275
Guttiérez, J. 61

Haber, H. F. 238
Habermans, J. 58
Hacking, I. 128
Hägerstrand, T. 70–1
Hakim, C. 41, 42
Hall, B. 103
Hall, M. 62, 64, 66, 215, 216, 218
Hall, S. 67, 241
Halvorsen, S. 86, 87
hanging 152–3
Haraway, D. 107
Harper, Stephen 19
Harrington, B. 51
Harris, C. 65, 215, 216, 230, 231, 260
Harrison-McKnight, D. 260–1
Haug, Frigga 22–5
Hausser, D. 260
Heidegger, M. 237, 241–2, 247

Helmefalk, M. 204
Henderson, N. R. 263
Herman, T. 173
hidden population groups 260
Hietanen, J. 105–7, 109
Higgins-Desbiolles, F. 33
Hodgson, A. 85, 86
Hollinshead, K. 193
Hollway, W. 218, 219
Holmes, K. 171–2
Hooks, B. 72
Host City Contract (HCC) 285
Hughes, J. A. 170
Huijbens, E. H. 65, 127
human/nonhuman duality 129
Humberstone, B. 239
Huppatz, K. 40, 41
Husserl's phenomenological method 58

IdeaLab cultural mapping presentation session *337*, *338*
'Ideal' WWOOFing Experience Model *183*
'Identity and Landscape kit' 176–7, *177*
ideologies and legitimacies 215
Inayatullah, A. 305
Indian Residential School system 19
'indigenous problems' 20
indigenous scholarship 3
individuals and groups construct narratives 352
inductive reasoning 125
Ingold, T. 239, 288
Institute for Scientific Information (ISI) 61
intellectual and educational narrative type 356–8
intensive short-term ethnography 246–7
'International Journal of Hospitality Management' 66
'International Journal of Tourism Research' 66
International Olympic Committee (IOC) 285, 318, 304
'Internet-mediated access' approach 268
intersectional reflexivity 214, 215
Iron Curtain in 1989 1

Jackson, S. F. 95
Jamal, T. 65
Jefferson, T. 219
Jenkins-Smith, H. C. 86
John Paul II (1981) 59
Johnson, C. W. 3, 24, 25, 27, 216, 308
Johnson, R. B. 260
Johnston, L. 239
Jokinen, E. 239
Jones, M. D. 86
journaling 45–6

Kang, S. J. 305
Kanouse, Sarah 86, 87
Kaplanidou, K. 305
Kauppinen-Räisänen, H. 150
Kay, J. J. 84–5
Kennedy, H. W. 91
Kerasidou, A. 261
Kerr, Roslyn 4, 5
Khaled, R. 194
Kim, Y. 61
kinetic empathy 239
King, V. T. 65
The Kino-Cine bomber 88–9, *89*, 94
Kishigami, N. 130
Kitchin, R. 61, 62
Klapper, R. 203
knowledge exchange 344
'knowledge mobilization' 328–9, 335–7
Kopnina, H. 129
Koro-Ljungberg, M. 132
Kowzan, P. 86, 87
Kozinets, R. V. 106, 150, 151
Kuebler, D. 260
Kuhn, T. 58

Laird, N. 24
Lamborn, C. C. 130
Lamond, I. R. 83, 85
Larsen, H. G. 65
The Last Mile 290
Latour, Bruno 130
Lazzarato, M. 60
learning mechanism 8
Lee, Emma 17
LEGO® SERIOUS PLAY® methodology: application techniques 176; 'charged atmosphere' 173; customised starter kit **178**; facilitated workshop technique 170; hotel industry 172; 'Identity and Landscape kit' 176–7, *177*; imagination 174–5; kinesthetic forms 171; metaphorical depictions 171; metaphors 175; methodology 184–7; organisational performance 171; tensions 170; The theory of flow 175–6; traditional qualitative research methods 172; unethical and/or inhospitable environment 172; workshops and participants **177**; WWOOF programme 172; *see also* The concept of play; constructivism
leveraging digital methods *271*
Lew, A. 62
Liebman Parrinello, G. 62
Lie, R. 40
Lincoln, Y. S. 51, 125, 261, 308
linguistic approach 353–5, *356*
linguistic narrative analysis 9
Li, P. 61
'Live Sites' 283
Lloyd, K. 105
location-based geo-filters 155
Lo, M.-C. M. 38, 41, 42, 48
Longhurst, R. 47

López-Nieva, P. 61
Lorimer, J. 105, 111
Lowman, E. B. 18, 19
Low, T. 260
Löytönen, T. 132
Lucey, H. 219, 222, 229
Lugosi, P. 102, 149
Lund, K. A. 127
Lupton, D. 294
lurkers 150
Lury, C. 41

Maasai tribe *225*
Maase, R. 318
MacCannell, D. 221
Mackay, D. 204, 318
Macoun, A. 21
Mair, H. 25, 27
Maldonado-Torres, N. 58
Mannheim, K. 58
Maracana Stadium 290
Maramwidze-Merrison, E. 264
Marschke, M. 85
Mata, J. 203
Matthews, D. 306
Mayer, I. 203
McCabe, S. 40, 193
McCormack, C. 24
McCurdy, P. 83, 93
McGillivray, D. 283, 305
McIntosh, A. J. 172
McKarcher, B. 64, 67
McLachlan, S. M. 329
McLean, R. 171–2
media ethnography 5
mega-sport events (MSEs) 283, 284, 305
Melody, J. 219, 222, 229
memory and retrieval 154
Mercier, P. 70
Meriläinen, S. 62
Merriam-Webster Dictionary Online 123
Mertens, D. M. 39, 51
Merton, R. K. 58
'metrification' 65
microblogging 296
Midgley, G. 85, 86
Mienczakowski, J. 309
Mignolo, W. D. 58
Milojevic, C. I. 305
Minca, C. 62
Mkono, M. 151
moments of discomfort 242–6
Mooney, S. 25
Morgan, J. 241
Morgan, K. 309, 317
Morgan, N. 66, 257, 260
Morgan, S. 309
Mostafanezhad, M. 80
'motherhood capital' concept: Bourdieu's concept of capital 40; and feminine 40–2; fieldwork 42–3; inuit communities 38; power inequalities 38; researcher and acknowledgement 39
Motivation, Opportunity and Ability Model' (MOA) 156
MSE academic-activism 286–8
Muldoon, M. 25, 27
Muller, M. 305
Muller's mega event syndrome 316
Mura, P. 62, 65, 66

Nacke, L. E. 194
narrative research methods 351–2
'natural heritage' 132
nature-based tourism activities 43
Nebuka, M. 139
Negative WWOOFing Experience Model *179*
neo-liberal political rationality 59
neoliberal publication regime 65
neo-Marxist approach 59
netnographer *150*, 152, 153
netnography 149, 150
network ethnography 5
Newtonian paradigm 126
New Zealand's tourism industry 306
Nind, M. 95
non-human stakeholders 125
non-linear thinkers 2
Northwest Passage cruise travellers 43
Nyaupane, G. P. 258

Oboler, R. S. 42, 43
Occupy encampment 80
Ogden, L. A. 103
Olds, K. 61
'Olympic Boulevard' (OB) 290
Olympic (project) lifecycle 285
Olympic Stadium 290
Olympic visitor economies *293*
'open persons' 47
'orthodox methodologies' 241
outdoor and social spaces 247
Oxford English Dictionary Online 123

Paasi, A. 61
Page, S. J. 64
Palmer, C. 273
Papert's constructionism (1991) 173
para-ethnography 329–30
Parry, D. C. 308
Participant Enquiry Sheet (PES) 272
Participant Information Form (PIF) 272
participant profiles *156*
participatory digital methods 6
Patterson, T. 27
'peer interviewers' 260
Penn, A. 85
'penultimate frontier for tourism' 236
percentage of authors **64**
'performative definition' 127

'performative dimension (agency)' 130
Perlman, J. 258, 262, 263
Pernecky, T. 66
personal reflexivity 214
Petr, C. 105
Pezzullo, P. C. 80, 86
photo-elicitation interviews 227–30
photo-elicitation led group discussions 83
physical capital 40
Piaget, J. 174
Piaget's constructivism (1955) 173
Pickering, M. 249
Pillow, W. 216
pilot project 25
Pink, S. 241, 294, 295
playwrights 310
Plummer, R. 85
Police Pacification Unit 265
Pons, Obrador 240
Pontifical Council for Justice and Peace 59
Porter, B. A. 39, 42, 48
positional reflexivity 214, 215
position factor 67
Positive WWOOFing Experience Model *180*
post-modernist conceptions 58
post-positivism 215, 286–8
post-research period 326
poststructuralism 3
post-visit narratives 149
Potter, R. 61
Poudel, S. 258
'power-brokers' 67
practice-based experiences 326
Prazeres case: digital and physical spaces 266–70; Morro dos, location of 264, *265;* physical access 270–4; research participants 265; self-made community projects 265; south zone favelas 264–5
Presence, S. 91
Pritchard, A. 66, 257, 260
privacy 157–9
privacy, memory and ephemerality 153–4
Privileged Access Interviewers (PAI) 260
privilege human-centered interpretations 123
professional and entrepreneurial narrative type 358–61
Prusinowska, M. 86, 87
psychoanalytic theory 218
"psychological launching off" 354
psychosocial research 217
Puar, J. K. 80
Punch, M. 274
Punch, S. 249

qualitative research 7
qualitative tourism research 124–5
'quantum turn' 126
'Que será, será' 315–17
Quinton, S. 149

Ramírez, R. 203
reciprocity 330
the *Red Command (Commando Vermelho)* 258
'reflexive disclosure statements' 215
reflexivity 5; definition 214; formulaic and predictable format 216; interpersonal process 327; leisure activities 223; Maasai village 226–7; methodology 221–2; psychosocial approaches 217–21; qualitative research 214; researcher and research participants 262; in tourism studies 215–17; 'uncomfortable' process 216; volunteer tourist photography 223–6
Regan, P. 20, 30
Ren, C. 130
representation theory 58
reproduction theory 58
researcher-participant relationships 155
researcher reflexivity 330
research locations, northern Canada *44*
research-practice collaborations 7–9, 327
Rich, A. 41, 42
Richards, G. 40
Richardson, Laurel 307
'RioZones-Approach' 284, 287
Ritchie, B. W. 273
robust anthropocentric bias 129
Rodríguez-Pose, A. 62
Rokka, J. 105–7, 109
Room, G. 126
Ross-Smith, A. 40, 41
Rouzrokh, M. 25, 27

Said, E. 58
Saldaña, J. 309, 310
Santa Teresa neighbourhood 257
Saunders, M. 263
Saville Young, L. 218–20, 230
scenario planners 316
Schnackenberg, A. K. 261, 272, 275
Schoemaker, Paul 316
Schoorman, F. D. 261
scripted dialogue 310
'self-retrieval' 154
Selin, C. 203
Seregina, A. 105–7
'serious gaming' method: applicability in other domains 196–7; challenges 199–200; conceptual implications 201–3; context 200–1; definition 194–6; future avenues 205, **205**; gameplay sessions 196; participants' learning outcomes **197**, 197–8; pitfalls 200; practical implications 204–5; research frames **195**; theoretical implications 203–4
settler colonial invasion 18
settler colonialism 18, 19, 24
settler identity 19, 29
settler stories 29–30
Sharples, M. 193
Sharrock, W. W. 170

Shefner, J. 257, 274
Sheppard, S. R. J. 199
Shirakami Mountains World Heritage 137–141, *138*
Shiretoko Peninsula 133–7, *134*
Short, R. 61
Simonsen, K. 61, 62, 71–2
'single dialectical process' 217
Sinha, S. 275
Sirmilik National Park 43
Sixsmith, J. 260
Skeggs, B. 41
Slater, D. 61
Small, J. 24
Smith, D. E. 22, 84
Smith, J. W. 130
Smith, L. T. 16, 39, 58
Snapchat 152, 154–5, 159–60
Snapchatters 149
snowball effect, WhatsApp *270*
social artifacts 125
social constructionist approach 127–9
Socialist Women's Association 22
socially-constructed creative analytical practice 4
social messaging platforms (SMPs) 6, 259, 263–4
spatial affordances 239
sport event tourism 304
Spracklen, K. 83
Springgay, S. 289
staged dialogue 308–9
stakeholder theory 329
stanzas 354, 356
Stebbins, R. 242
Steller's Sea Eagle (Haliaeetus pelagicus) 134, 135
Stewart, E. J. 4, 5, 44, 46, 48, 50
Stirling, A. 126
story and stimulation (SAS) approach 196
Stoudt, B. G. 95
Strakosch, E. 21
strophes 354, 356
Stuttaford, M. 309
subtraction process 88
Suchet-Pearson, S. 105
Suttie, N. 198

Takenaka, T. 135–6
Talbot, A. 287
'tangible and intangible legacies' 304
Tanita, K 103
Taylor, S. 217
Teghtsoonian, K. 83, 84
Tesar, M. 132
The theory of flow 175–6
'The Truth about Tourism' 67
the *Third Command* 258
Thomas, R. 62
Thor Johannesson, G. 130
Thrift, N. 239
Tienari, J. 62
'time together' with 'time alone' balancing 339–40
Tomlinson, E. C. 275
top journals 66
Torabian, P. 25, 27
Torre, M. E. 95
Tourism Activity Game (AGT) 201
Tourism and/or Hospitality and/or Travel 63
tourism destinations 201–2
tourism ethnographers 4
'Tourism Management' 66
tourism research. *see also* 'serious gaming' method
Tourism Research Information Network (Tri-net) 295
tourist created content (TCC) 149
transcription process 353
transference 218, 230
Tribe, J. 64, 65, 67, 68, 70, 260
Truman, S. E. 289
trustworthiness 261–2
Tseng, W.-C. 171
Tuck, E. 17, 31
Tucker, H. 33, 65
Turner, D. 305
Turner, Victor 309
The 2016 scenario 311
The 2017 scenario 312–15
The 2019 scenario 319–20

UNESCO's World Natural Heritage program 132
Upham, P. 203
Urbanik, J. 109
Urry, J. 81, 153

Vaiou, D. 61, 71–2
'value-free' research 124
van der Duim, R. 130
van der Heijden, K. 204
Veijola, S. 239
Veracini, L. 18
Vergunst, J. L. 288
Verheijdt, C. 204
videography 104–7
'Vienna 1918' 69
Viken, A. 62, 71
Vincent, C. 43
virtual ethnography 5, 150
Voice over Internet Protocol (VoIP) 352
volunteer tourist photography 223–6

Wachowich, N. 38
Walkerdine, V. 219, 222, 229
Walker, M. 86
walking method 288–9
Waltner-Toews, D. 84–5
Ward, V. 329
Warren, S. 43

The Washington Post 80
Web 2.0 digital ethnography 6
Weiss, G. 238
Wengel, Y. 172
Wenger, E. 326
"western" academic discourses 3
'Western World' model 60
Wijesinghe, S. N. 62, 65, 66
Wilkins, E. 130
Williams, A. 62
Williamson, Theresa *267*
Wilson, E. 65, 193, 215, 216, 230, 231, 260
Wittgenstein, L. 69
Wolf, E. 238
Women's Liberation Movement 22
'The World is Coming to Me' Model *183*
World Natural Heritage Sites (WNHS) 124
World Trade Organization (WTO) 80
World Wide Opportunities on Organic Farms (WWOOF) programme 176
Wright, Richard Keith 4
WWOOFers 176, 182, 184
WWOOFing 179–80
Wylie, J. 239

Xiao, H. 70

Yang, W. K. 17, 31
Yudina, O. 139
Yusoff, K. 129

Zhang, J. 65
Zielińska, M. 86, 87
Zybertowicz, A. 58, 71